Toxicity in Fishes

The Author

Dr Krishan Lal Jain (PhD), a renowned Professor of Zoology and is retired from the department of Zoology and Aquaculture of 'The CCS Haryana Agricultural University, Hisar' in Haryana (India). He has been grossly associated with various aspects of water pollution and fish toxicology during his valuable last about 10 years of active service.

He did his BSc (1969) from The Punjab University Chandigarh, MSc (1971) from Punjabi University Patiala, and PhD (1986) from CCS Haryana Agricultural University, Hisar, as a Senior Research Fellowship of ICAR, New Delhi. Dr. Jain has a total teaching and research experience of about 37 yrs for teaching UG and PG classes and has guided many M Sc and Ph D students. Dr. Jain had completed many research projects on various aspects of 'Environmental and Applied Biology' and had more than 100 research publications in his credit on Bee Biology, Bee Ecology and Toxicology, Acarology and Fish Toxicology. His research capabilities are substantiated by his excellent work on Bee Biology, Bee Ecology and Behavior and Bee toxicology. In 1986, in recognition of their unit work, The ICAR had honored the whole unit for their work on 'The study of Ecological and Biological Aspects of Some Crop Specific Pollinators' with the most prestigious Agricultural Sciences Award 'The Rafi Ahmed Kidwai Memorial Prize'. He had been further honored many a times with appreciation letters from various hon'ble Vice Chancellors of CCS HAU and The USDA Department of USA for his pioneer research work on bees. His major work on fish toxicology included symptomatic, behavioral, biochemical, hematological, enzymological and some histo-pathological studies in fishes chronically exposed to pesticide and heavy metal toxicants in water.

Dr Jain also worked as The Head of the Zoology and Aquaculture Department in The CCS HAU, Hisar from Oct 2003 to Oct 2007. Keeping in view the sincerity and the achievements of Dr Jain as Head of The Department, the then Vice Chancellor had also honored him with a strong worded appreciation letter in 2006.

Toxicity in Fishes

– Author –

Dr K L Jain

Ex. Professor and Head (Retired)
Department of Zoology,
CCS Haryana Agricultural University,
Hisar – 125 004, Haryana

2019

Daya Publishing House®

A Division of

Astral International Pvt. Ltd.

New Delhi – 110 002

ISBN: 9789389569063 (Int. Edition)

Published by : **Daya Publishing House®**
A Division of
Astral International Pvt. Ltd.
– ISO 9001:2015 Certified Company –
4736/23, Ansari Road, Darya Ganj
New Delhi-110 002
Ph. 011-43549197, 23278134
E-mail: info@astralint.com
Website: www.astralint.com

Dedicated to

The Ever-Cherished Memories of My Wife
Late Smt. Sheela Jain
(1950-1994)

Preface

This book provides glimpses into the recent researches on 'Mechanisms of Toxicity in Fishes' especially the heavy metals, made in the field in past few decades especially in India and many other countries. An exclusive literature survey on water quality is discussed. The pollution of groundwater and other freshwater systems and the problems arising from various wastes are highlighted. Particular attention was drawn to the harmful effects of heavy metals on water quality and the eventual ill effects on fishes. Each chapter concludes with a comprehensive review of literature including latest references particularly in context with the studies undertaken in India. Heavy metals today constitute the most hazardous category of toxic chemicals in water.

This book is the compilation of about 2000 quality research and review articles contributed by leading experts. This book begins with Part-1 providing a brief description of water as the precious natural resource, factors affecting the quality of ground and river water and other aspects like 'Scope of water pollution and the important definitions', and chapters on 'Major water contaminants, Metal epidemiology' and The Quantitative aspects of metal contaminations in water; followed by a chapter explaining metal availability and biomagnifications *etc.* Part-2 contains chapters on various qualitative aspects, such as 'descriptions of metal induced behavioural anomalies, biochemical changes in fish tissues, such as the tissue carbohydrates, proteins, lipids, major hematological and immunological parameters and metabolic enzymes in fishes, role of stress proteins and the oxidative stress on their exposures to metal contaminated waters'.

Hope to be very popular among the fish scientists, ecologists and Zoologists, working in various State/Central Agricultural and Traditional Universities, Pollution Deptts, State Fisheries Deptts., IARI Delhi and other ICAR Institutes.

Dr K L Jain

Contents

Chapter 1

Water Pollution: Scopes and Definitions

The Earth is so far the only one such planet in the galaxy on which life has originated because of the availability of two most basic components, vital to life *i.e.* water and oxygen. Water, however, being an essential commodity for the survival and growth of all kind of animals and plants including fisheries; and the humans has been the most pressing resource due to its further contamination with many natural or manmade pollutants as a result of increasing natural soil erosion, leaching and weathering processes and mass industrialization. Today, whole world is facing crisis due to fast increasing pace of water pollution and its subsequent impact on total living world including humans, agriculture and fisheries. Natural disasters although too pollute the climate to some extent, but being manageable naturally as the nature has its ways to maintain an ecological balance, these do not pose major threat to animal life. Industrialization, however, has proved as one of the man's worst endeavors of the last century it has especially come in for a great deal of concern since World War-II. Chemicals from industrial effluents along with flow of various domestic and agricultural waste waters have been continuously affecting and deteriorating the quality of water in both the lentic and the lotic systems. In spite of the fact that these water bodies possess self purifying capacity due to the natural process of recycling, these bodies have become heavily contaminated thus, affecting life.

Humans though have been interfering in nature since centuries through mining *i.e.* extracting natural resources and discarding residues into the environment, thus, deteriorating continuously the water. An industry means an increase in output for more gains but with irrational input of sources like variety of synthetic chemicals, simultaneously increasing the volume of effluents with more toxic chemical wastes into the natural water bodies; may be lakes, rivers, oceans, as well as into the groundwater. Anthropogenic manipulations and environmental exploitations during the past few decades have reached the limits that it has become impossible to recover back to a healthy environment. Human impact since has significantly changed (or halted) the growth of biodiversity on one hand and had accelerated on the other hand the rate of species extinction on account large scale destruction of natural habitats and the breeding grounds of various terrestrial and aquatic animals such as birds and the freshwater and marine fishes. It had resulted in a heavy decline in their natural populations. Presently, there are no such ecosystem termed free of such human activity; even the areas which are still unknown to human inhabitation are found contaminated by the pollutants being carried by air or the water currents. Technological advancements and profit oriented capitalism throughout the world have further increased mining as well as production and application of large number of new industrial chemicals at such a high alarming rate that most of these chemicals find their way only into the water. In this new era of development of technologies, there has been further large scale of transformations on earth, such as the development of heavy industries, mega-infrastructures; along with simultaneous increase in deforestation and burning of fossil fuels, as well as heavy exploitation of various other natural resources. Scientists have named the current era as "**Anthropocene *i.e.* the age of man**" (Zalasiewicz *et al.*, 2008). 'Anthropocene' is an informal geologically chronological term that serves to mark the evidence, the extent of human activities that have caused a significant global impact on the Earth's ecosystems. This term was coined by an ecologist 'Eugene Stoermer'; it was however, popularized widely by the Nobel Prize-winning atmospheric chemist 'Paul Crutzen'. Some environmentalists see the 'Anthropocene' as a disaster by definition, since they find all human interferences as signs of 'degradation of a pristine Eden'.

Industrialization though on one hand has helped a nation in strengthening its economy; on the other hand it has been responsible for the continuous discharge of various toxic wastes and chemical rich effluents in water and the air. Industrial developments in fact being further associated with the human greed to maximize profits in the name of economical development and also an illusion for luxury and comforts, have caused heavy cost to degradation of natural water resources. Continuously increasing demand for the brighter colors in textiles, toys, hoe paintings and automobiles, and in a variety of food products has further been of serious concern, as more the incorporation of these bright colors in the environment means more the use of toxic chemicals. As per the archeological evidences, water pollution with fiber colors started around 4-5 thousand years ago, when the early man especially in India started coloring fibers first with dyes, prepared basically from plants. Use of heavy metals started with the development of special dyes in the mid of 20th century when new "high tech" fibers, such as nylon and polyester were arrived in the market. To color these fibers, dyes containing heavy metals and other toxic compounds were used. Continuous increase in the demand of various textile and leather products has caused an simultaneous increase in textile printing with colorful and fluorescent dyes and leather tanning especially in economically undeveloped countries like Bangladesh. The developing countries has comfortably shifted such heavily polluting industries to these countries in the name of economic help thus, leading to an alarming increase in release of more toxic chemicals in water bodies. This makes the modern world though much beautiful but simultaneously it encourages a chemical rich environment affecting worst the human life as well as all other living organisms existing on land and in water. If the humans just resort to the use of the natural colors as their only choice, as was being done in the past, they can make the earth free from many such toxic chemicals. The world's most famous and centuries old "Ajanta and Ellora Caves" in Maharashtra (India) are the best example of man's past endeavor, who had used natural colors in the wall paintings inside these caves. These rock-cut caves illustrate excellent artistry of the ancient Indian craftsmen who had demonstrated it several hundred years ago. Ajanta Caves dates from 100 B.C. while Ellora Caves are younger by some 600 years.

Occurrences of some other noticeable chemical mis-happenings due to human errors both in India and other countries are the world's most disastrous 'Nuclear bomb blasts in Japan' known to contaminate the environment and cause havoc to the millions of people and the wild life. Such events had not only killed thousands in a few days, these also had left millions to suffer for years to come next. In India, escape of highly toxic gas methyl-isocyanate (MIC) from the premises of 'The Union Carbide India Limited' plant in Bhopal (Madhya Pradesh) in Dec. 1984, led to about 7000 deaths and left another about 10000 people in its immediate aftermath, and a further about 1.5 lac expected to suffer over the next 20-40 years from various critical chemical injuries and diseases. This massive gas leak has infact made more than 2 lac people to suffer continuously from serious health problems as a consequence of their exposures to various hazardous chemicals an water and air. Likewise, excessive applications of pesticides in Agriculture has been the examples of occurrence of certain disasters like the one that occurred in the middle of last century in the coastal waters of Louisiana State of USA, famously known as The Pelican State'. It was reported to have worst effects of DDT poisoning in water, causing a significant decline in pelican's population. DDT poisoning had been known to cause thinning of their egg shells thus, lowering chick production. There had been further occurrence of a number of incidences of oil spills. Today, approximately 10 per cent of the already existing species of plants and animals are said to have become already extinct due to such calamities.

Around the world the rivers, lakes, and other aquifers are further fast dwindling, than at the rate at which the 'Mother Nature' can replenish them. Natural and cultivated human commodities are further at the state of collapse as a result of global warming. As per recently published 'The World Fish Center' report in Feb. 2011 (Macfadyen *et al.*, 2011), the climatic changes are going to threaten and ruin ocean reefs, push salt water into freshwater habitats and produce more coastal storms; thus, may affect millions of struggling people especially in fisheries dependent nations as in Africa, Asia and South America. Unprecedented rains and floods in many countries in the last years of twentieth century as well as the beginning of twenty first century, as it happened in many areas of Australia, Brazil, Pakistan, Philippines, Russia and Sri Lanka, as a result of global warming; resulting a severe fall in agricultural production, are evidently the indications of collapse of civilizations.

As per recent estimates, five top countries contributing to global warming are-China, USA, India, Russia and Canada; out of these China alone contributes to 24 per cent and USA to 18 per cent of total carbon pollution. In eastern and southern Africa, rising temperatures in freshwater lakes over the last century have already reduced fish stocks. Recently, there had been reports of severe decline in shrimp production in most of the south-east countries including Bangladesh. Shrimp production in these countries is a most valuable export commodity and in Bangladesh alone which had about 20000 shrimp enclosures spread over about 58000 ha of its wet land in only one sub district and producing shrimps valued at approximately BDT 1000 crore, has been said to be under a severe threat due to industrial heat stress (Satkhina, 2010). Laser Brown, an environmentalist and founder of the 'World Watch' in his just released book "World on the Edge" (Brown, 2011) has documented that overloading the atmosphere with carbon dioxide, the mankind has pushed the civilization to the brink of collapse by bleeding aquifers dry and over harvesting the land to feed an ever growing population. Bill Clinton, the earlier USA President while quoting 'Laser Brown' once said that "He tells us how to build a world and save the planet in a practically straight forward way". It thus, deemed necessary to visualize and monitor the degree of occurrence of toxic chemicals in natural waters and hazards of Global warming to humans, fish and other living organisms. Further climate changes are expected to worsen this trend, also leading to lower water levels due to decreased rain and increased evaporation on one hand and melting glaciers and rising sea level on the other hand.

1. Facts about Public Awareness

The environmental facts were although identified about two centuries ago, but public awareness about the toxic effects of the environmental pollution and suggesting ways monitoring of gross environmental pollutants or testing their toxicity on various aquatic and other animals *etc.* is only recent. Since the whole world has been concerned about the environmental health, there had been various 'International Summits' as well as 'Formulation of new Policies' to regulate environmental pollution. There has been, however, little progress as the industrial world is in no mood to compromise at the cost of their productivity and the economy. The environmentalists and the policy makers throughout the world are still focusing on the subject 'How to save the planet', and how to keep our rivers and other freshwater bodies clean and safe for humans and the wild life, and further how to keep our food free from the toxics? The most important event since 1973 in this direction has been 'Observing June 5' as the 'World Environment Day'; and since 1993, 'March 22 as the World Water Day'. These 'two days' are to generate interest and increase public awareness about the pollution related problems as well as to promote all such activities to save the environment and the water resources. Further important has been identification of certain terms related to environmental pollution, water toxicity, eco-toxicology *etc.*

Most important term in use has been **"Pollution'** itself, which means introduction of any contaminant or contaminants into the environment that causes instability, disorder, or discomfort to the life of a living organism in their ecosystem. Hynes (1966) described pollution as a biological process as it affects life. Water pollution occurs when chemicals are discharged directly or indirectly into a water body without adequate treatment for the removal of all the harmful constituents.

Pollution can result in due to any chemical substance or any other physical entity like energy in the form of noise, heat, or light or even any such naturally occurring substance that has exceeded its normal level in the atmosphere. Such substances, that already exists in nature as well as even the trace elements when present in excess inside a living organism, are termed as 'First category pollutants', whereas the elements/chemicals that do not occur in nature but are made to enter into the environment deliberately or due to carelessness of the human beings are put under 'Second category pollutants'. Later includes industrial effluents and all the manmade chemicals that are being introduced regularly into the environment. Since, water contaminants had an equally high potential risk for the health of humans and the other animal populations, protection from the toxic effects of such water pollutants particularly at their low toxicity levels and the likely subtle biological effects (chronic effects) in the living aquatic organisms inhabiting in such polluted waters, are of major concern to the environmentalists and the aquatic biologists. **Water pollution** thus, has a wider definition, and is described as: any physical, chemical or a biological change in water quality, affecting adversely a living organism, or making water unsuitable for its desired use by any living organism.

Literally, the word **'toxicology'** (derived from the Greek words: toxico- and logos) refers to the study of nature and the adverse effects of a poisonous chemical on any living organism. Toxicology includes the study of poisons, their identification, chemistry, degree of toxicity and the physiological actions in a recipient animal or a plant. It involves basically identification of various symptoms and the mechanisms of their action as well as treatments of the people or the animals poisoned. The term '**Environmental toxicology**' classically refers to the study of the direct effects of environmental chemicals especially on the human beings. The newly defined term '**Aquatic toxicology'** has emerged further as a potential issue in today's reference to the water pollutants. The aquatic toxicology, though a younger discipline and until 1970's it had been aimed majorly at determining only the maximum amount of a toxicant that can be permitted in water, without causing any significant harm to the resident biota. Thereafter, aquatic toxicology has been increasingly in use as an important tool to the animal biologists and the physiologists, with a sole purpose to understand that why a fish or an invertebrate is debilitated inside toxic water. History of aquatic toxicology has been reviewed by Macek (1980). Aquatic toxicology today is defined as the study of the effects of environmental contaminants on aquatic organisms, such as the effects of chemicals like pesticides on the health of a fish or any other aquatic organisms.

Another recently introduced term '**Ecotoxicology'** aims to assess the impact of chemicals on a ecological system. It refers to the scientific study of the potential effect of a anthropogenic compounds upon any organism, released into the natural environment and it concerns more with their toxic effects on a species and its ecosystem. Other subjects such as: Chemical toxicology, Forensic toxicology, and Medical toxicology are further newly designed terms depending upon their specific application. Some more specialized disciplines within the field of toxicology such as 'Toxic genomics' have also been introduced by the toxicologists. It involves applying molecular profile approach to the study of toxicology.

2. Water Pollution and the Fisheries

Fish being a potential source of protein rich food and a livelihood for large number of human beings constitute one of the major components of the global food production technologies and advances. Fisheries, one of the most rewarding activities now a day in most of the developing countries, including India is of greater concern for sustaining aquaculture production.

Freshwater systems with tropical rivers and inland fisheries are said to possess a large economic potential, and are estimated to generate about 250 billion rupees annually worldwide, with an additional yearly wet land derived resource benefits of the value of other millions of rupees. The aquaculture sector alone thus, is said to make a significant contribution to income, employment creation and food security in a country. As per the 2011 survey report of the World Fish Center (Macfadyen *et al.*, 2011), the aquaculture has been identified as the fastest growing sector in the world, and it continues to outplace population growth with per capita supply from aquaculture; increasing from 0.7 kg in 1970 to 7.8 kg in 2008, on an average annual growth rate of 6.6 per cent. Egypt's aquaculture production (705,490 ton in 2009) was by far the largest of any African country and it has placed it 111^{th,} in terms of global production. Likewise, in Calcutta (India) alone about 25000 ha of ponds were estimated to produce approx. 70000 tones of fish every year, catering the needs of about 20 per cent of the cities. In Asia small-scale fisheries is said to contribute about 25 per cent of the total fisheries productions in Malaysia, Philippines, Thailand, and Taiwan for the decade ending in 1997. The importance of small scale fishing is still greater in West Africa, contributing three fourth of the region's total fish catch (Kura *et al.*, 2004). As per the estimates of the Food and Agriculture organization, over 90 per cent of the 15 million people working in the word's coastal waters are small-scale fishers (FAO, 2002).

The development of fisheries in inland water bodies, however, is becoming increasingly difficult due to a variety of environmental constraints, posed by various anthropogenic stresses. Although, water is a naturally renewable resource, yet these days the increasing contamination due to a continuous inflow of all the poisonous industrial effluents and the city sewage systems are creating a serious problem in the development of fisheries as well as the health of humans. Dumping of solid wastes exceed the limit of self purification of water bodies and alter its physico-chemical and biological characteristics. Global fisheries, however, is reported to exhibit a constant decline in fish stocks, both in coastal and inland water resources on account of continuously increasing water pollution. All the world famous rivers are today heavily polluted and thus, disturbing natural ecological harmony and destroying the biodiversity, as well as posing a serious threat to even the existence of humans itself. Increasingly use of all the rivers around the whole world simply as the natural dustbins for discharging all sort of domestic and industrial wastes has turned these rivers only into noxious sewers, which appear more as the death traps for all the inhabiting aquatic life. The inland natural water resources in almost all the Asian countries are highly threatened further due to indiscriminate dumping of the industrial wastes.

Such realizations about pollution hazards in aquatic systems, have provided an impetus to the prodigious amount of researches on aquatic fauna both in lentic and lotic habitats and the intensity and composition of the water pollutants. The fish today are seriously endangered by water borne pollutants transferred along the food chain (Hoo *et al.*, 2004; Ayas *et al.*, 2007; Kumar and Achyuthan, 2007; Shukla *et al.*, 2007; Srivastava and Srivastava, 2008). The instances of destruction of fish breeding grounds and impediments in migration are well documented (Jhingran, 1991). Stresses exerted by the industrial effluents culminate in the mortality of a fish and the fish food organisms. The results are witnessed in loss of aquatic biota and a steady decline in the fish yield. Large-scale episodal mortality among the freshwater fishes of certain lakes (the Sankey Lake) of Bangalore City, Karnataka State, India as reported to occur in June 1995, was the result of a sudden and considerable fall in dissolved oxygen (DO) levels in some locations, caused by sewage let into the lake which had resulting in asphyxiation (Benjamin *et al.*, 1996). Reports on trout catches since 1980 in Switzerland alone have indicated their fall by 60 per cent. Findings viewed by the Federal Institute for Environmental Science and Technology and the Agency for the Environment, Forests and Landscape (Eawag and Saefl Report 1999) were sufficient grounds for launching the fish net project in 1998. At the same time, health problems were identified for fish in different streams. The project leadership has concluded that the principal reasons for the decline in stocks were poor quality of habitat like deficiencies in shelters in constructed stream courses, deteriorating water quality and the proliferative kidney disease (PKD). Poor habitat quality including environmental constraints, coupled with residual wastes of inorganic fertilizers and pesticides in public health and agricultural sectors (Ghosh *et al.*, 2000) have been further identified as the factors continuously impeding development of fisheries in inland water bodies. To the humans who regularly consume fish and other sea food items, water pollution is most concerned aspects for their health. Without healthy water for drinking, cooking, fishing and farming, the human race thus, may perish. In Bangladesh, as per the report made public by the Department of Environment about 20000 tons of waste comes only from tanneries and about 60000 tons from textiles, dyeing, printing, washing and other industries. About 80 per cent per cent of the land here is wet and around 80 million people are exposed in and around Dhaka alone to high levels of water toxicity, contaminated with arsenic and other such toxic chemicals discharged from around 7000 local industries. These industries are discharging about 1.5 million cubic meters of waste every day.

Currently, water pollution alone is suggested as the leading worldwide cause of human deaths and diseases, and is said to account for the deaths of more than 14,000 people daily. As per the World water quality facts and statistics report, published to commemorate the 'World Water Day 2010', every day 2 million tons of sewage and industrial and agricultural waste are discharged into the world's water (UN WWAP, 2003), the equivalent of the weight of the entire human population of 6.8 billion people. It is thus, a very high time to make the people aware of their right to maintain ecological balance in the environment and about the safety of the water bodies. One most important way is to acquaint with all such practical approaches and the methods useful in analyzing the water quality or the fish health in order to cultivate only the healthy fish in healthy waters and also to know how to remediate the unhygienic conditions, if any in any of the water bodies. There is a Chinese saying that 'Give a fish to a man and you feed him for a day; but you teach him how to breed and catch a healthy fish, you have fed him for a life time.' Aquaculture, a major industry in many parts of the world and an important source of revenue, is a science, art and business of farming or cultivating fish under healthy conditions.

REFERENCES

Ayas Z., Ekmerkci G., Yerli S.V. and Ozmen M., 2007. Heavy metal accumulation in water, sediments and fishes of Nallihan Bird Paradise, Turkey. J. Environ. Biol., 28, 545-549.

Benjamin Ranjeev, Chakrapani B.K., Devashish Kar, Nagarathna A.V., and Ramachandra T.V., 1996. Fish mortality in Bangalore lakes, India. E-Green J., Dec., 1996 Issue (6).

Brown L.R., 2011. World on the Edge: How to Prevent Environmental and Economic Collapse. Earth Policy Institute, Washington DC: 174pp.

Eawag and Saefl Report, 1999: Federal Institute for Environmental Science and Technology (EAWAG) and the Agency for the Environment, Forests and Landscape (SAEFL) Report- Project Network Declining Fish Yields Switzerland: http://www.fischnetz.ch/content_d/publ.

FAO., 2002. The State of World Fisheries and aquaculture 2002.Rome; FAO Fisheries Department.

Ghosh P.B., Saha T. and Bandyopadhya T.S., 2000. Distribution of lead, cadmium and chromium in the waste water of Calcutta canals. Res. J. Chem. Environ., 4: 33-36.

Hoo L.S., Samat A. and Othman M.R., 2004. The level of selected heavy metals (Cd, Cu, Fe, Pb, Mn and Zn) at residential area nearby Labs river system riverbank, Malaysia. Res. J. Chem. Environ., 8: 24-29.

Hynes H.B.N., 1966. Biology of polluted waters. Liverpool Univ. Press. 202 pp.

Jhingran V.G., 1991. Impact of environmental stress on fresh water fisheries resources. J. Inland Fish. Soc. India, 23(2): 20-32.

Kumar A.K. and Achyuthan H., 2007. Heavy metal accumulation in certain marine animals along the East Cost of Chennai, Tamil Nadu, India. J. Environ. Biol., 28: 637-643.

Kura Y., Revenga C., Hoshine E. and Mock G., 2004. Fishing for answers: making sense of the global fish crisis. Washington-DC: world Resources Institute.

Macek K.J., 1980. Aquatic Toxicology, facts or fiction? Environ. Hlth. Prospect., 34: 159.

Macfadyen G., Mohamed Nasr Allah A. and 7 others., 2011. Value chain analysis of Egyptian Aquaculture. Project report 2011-54. The World Fish Center, Penang, Malashia. 84pp.

Satkhina, 2010. The Daily Star, Dhaka, News, May 30:12.

Shukla V., Dhankhar M., Prakash J. and Sastry K.V., 2007. Bioaccumulation of Zn, Cu and Cd in Channa punctatus. J. Environ. Biol., 28: 395-397.

Srivastava Rama and Neera Srivastava, 2008. Changes in nutritive value of fish, Channa punctatus after chronic exposure to zinc. J. Environ. Biol., 29: 299-302.

UN WWAP, 2003. United Nations World Water Assessment Programme. The World Water Development Report 1: Water for People, Water for Life. UNESCO: Paris, France.

Zalasiewicz J. and others, 2008. "Are we now living in the Anthropocene?". GSA Today, 18 (2): 4–8. P2

Chapter 2

Major Industrial and Non-Industrial Water Contaminants

1. Industrial Wastes and the Chemicals

'Industrial development' has been a major factor contributing to fast degeneration of species. Today, approximately 10 per cent of the already existing species of plants and animals are said to have become extinct due to industrial pollution. It thus, deemed necessary to visualize the degree of occurrence of these toxic chemicals in nature, their sources and impact assessment, as well as finding ways and means to identify toxicants and their amelioration. An increase in human population is contemporary to increase in industrialization and subsequent increase in water pollution. Hundreds of millions people are known to severally suffer from polluted water throughout the world. In India, there are about 4 thousand chemical industries, releasing large quantity of chemicals in the form of gas, liquid and solid wastes into the water. Major loci of these industries in India are Bombay, Kolkata, Kanpur, Delhi, Jamshedpur, Bokaro (Jharkhand state), Hyderabad, Vishakhapatnam, Chennai, Baroda, Cochin and Ganjam (Orissa State). Konar, 1981 stated that aquatic systems are fast losing their productivity and causing large scale fish mortality due to water pollution.

The confluence of industrial effluents and the sewage wastes in the nearby water bodies conglomerates the entire aquatic biota due to the physico-chemical and biological changes during its course depending upon the intensity of the polluting contents. Waste generation due to the expansion of industrial activity in India has thus, a serious impact on the water resources. Commonly known water polluting industries are the textiles and dyeing industries, tanning industries, agro-chemical industries, insecticide industries, paints and dye industries, chlor-alkali metal industries as well as industries involving electrical and electronic items. Among these industries, the textile industries are one of the most chemical intensive and are the next high water polluter on earth after agriculture. These industries discharge heavy load of effluents in drainages and rivers. It utilizes about 500 gallons of water to produce enough fabric to cover only one sofa. In addition to high water temperature and pH, the mill effluents contain chemicals such as formaldehyde (HCHO), chlorine, heavy metals and many other chemicals, which causes significant impact on environmental degradation and human illnesses (EHTI, 2006). Natural waters also appear polluted by either excess of erosion of surrounding upland (silting), deforestation or volcanic eruptions of gases and the flow of 'Lava' (Hot mineral rich flow from the volcanoes) into the water.

Organic pollutants like PAHs (mostly associated with petroleum), polychlorinated dibenzo-p-dioxins, dibenzofuranes, polyhalogenated biphenyls (PCBs) and a variety of other insecticides and herbicides are also of major concern for fish physiology. Gossett *et al.* (1983) analyzed 27 organic compounds in livers of fish collected in the vicinity of Los Angeles (USA) wastewater treatment plant. PCB's are frequent in use worldwide as agricultural pesticides. Fantastic arrays of new chemicals are being produced every year and eventually almost all of which find their way into the environment. Rains dissolve the toxins therein and contaminated water percolates down to the aquifers to reach ultimately the sea. This problem is persisting and cannot be ignored given that it can have a serious long-term impact on the national economy and human health. The National Water Quality Inventory 1996 Report to Congress (EPA, 1998), refers to urban runoff as the leading source of pollutants causing water quality impairment related to human activities in ocean shoreline waters and the second leading cause in estuaries across the nation. Rivers, lakes, reservoirs close to the industries continue to be dirty and contaminated resulting in the decline of the quality of the water. Urban runoff was also a significant source of impairment in rivers and lakes. The per cent of total impairment attributed to urban runoff is substantial. This impairment constitutes approximately 5,000 square miles of estuaries, 1.4 million acres of lakes, and 30,000 miles of rivers. Major assessed area of streams, lakes, and bays and estuaries were classified as heavily polluted with chemicals and almost all substances were toxic beyond a certain level of an animal's tolerance to that chemical.

In case a chemical hazard is manifested as an acute lethal episode, the end result is immediately apparent as death of the animal in general and investigations for their prevention become prompt. Whereas, in case of exposures to sub lethal concentrations, the "signs and symptoms" become more subtle; any toxic damage usually remains undetected for long periods and at times a remedial action may become more difficult or even impossible. The pollutants like pesticides, detergents and heavy metals even when present in trace amounts can cause abnormalities in fish physiology, morphology as well as behavior. The state of Karnataka has about 2000 perennial and about 30,000 seasonal tanks with a total water spread area of 30 lakh hectares. The average annual fish yield from these tanks is estimated to be about 350 kg fish per hectare per year. The total water spread in India (Benjamin *et al.*, 1996) is about 4.5 million hectares and the inland aquaculture resources cover about 3 million hectares. These include about 0.72 million hectares of natural lakes and 2.0

million hectares of constructed reservoirs. Benjamin *et al.*, 1996 reported large-scale episodic mortality among the freshwater fishes of certain lakes of Bangalore City in Karnataka State, India in June 1995. Studies made in Sankey Lake (Banglore) have revealed that the fish-kill was due to a sudden and considerable fall in dissolved oxygen (DO) levels in some locations, caused by sewage let into the lake resulting in asphyxiation. Measuring dissolved oxygen (DO) levels in water in fact is a direct measure of availability of sufficient amount of oxygen needed to support the aquatic life; any reduction in its amount in water beyond certain limits is not only hazardous to life in water, and it also threatens the aquatic ecosystem in general.

According to the survey of the US Department of Health, Education and Welfare (HEW), 8 different categories of water pollutants have been differentiated, *viz.*, Sewage and Waste products, Plant nutrients, Particulates, Radioactive substances, Minerals and chemical substances, Heat, Organic chemical exotics and the infectious agents. The studies carried out by National Institute of Oceanography (NIO) have shown that the coastal pollution has a global impact. While speaking at Tata Energy Research Institute, New Delhi, about marine pollution, Elrich de Sa, Director, (NIO) described major pollutants and their annual quantum in our seas within the coastal zone as:

- ☆ Sediments – 1600 million tonnes
- ☆ Industrial effluents – 50 x 106 m^3
- ☆ Sewage (largely untreated) – 0.41 x 109 m^3
- ☆ Garbage and other solids – 34 x 106 tonnes
- ☆ Fertilizer residues – 5 x 106 tonnes
- ☆ Synthetic detergents residues –1, 30,000 tonnes
- ☆ Pesticides residue – 65,000 tonnes
- ☆ Petroleum Hydrocarbons (Tar balls residue) 3,500 tonnes
- ☆ Mining Rejects, Dredged Spoils and Sand Extractions 0.2 x 106 tonnes.

Many instances of fish mortality are due to adverse effects of various water pollutants in the aquatic ecosystem and their inclination to get magnified in the food chain components (James *et al.*, 1998). It is thus, also important to monitor their bioaccumulation and bio-magnification in a aqua-biological system, in order to assess their ultimate possible impact on human health since they heavily consume the contaminated fish (Kotze *et al.*, 1999). Our rivers appear converted as refuge carriers that carry their load to dump it in the sea and in addition, we have specialized in polluting the atmosphere near the coasts via our smoke belching industries. It resolves to the need for finding both their sources and discharge nationwide (Biney *et al.*, 1994). The Yamuna River, which is the lifeline of Delhi, is one of the most-polluted rivers in the country. About 85 per cent of the pollution is caused by domestic and industrial sources. A wide range of contaminants are continuously introduced into the river and their toxicity is a problem of increasing significance for ecological, evolutionary and environmental reasons. Recently, studies have revealed heavy metals as the important contaminants and these due to their toxicity, accumulation and non-degradable nature, constitute one of the most dangerous groups (Malik *et al.*, 2014).

These industrial wastes are known to contain a wide range of hazardous elements, such as petroleum hydrocarbons, chlorinated hydrocarbons and heavy metals, which are toxic to the bio–organisms. There are innumerable reports on the biological hazards of the heavy metals, drugs, pesticides and the fertilizers affecting metabolic functions, breeding, survival and diversity of the living organisms. Moore and Ramamoorthy, 1984 have stated that discharge of heavy metal rich industrial wastes as pollutants into water has tremendously increased during recent years. Taking a note of the high level of pollution the government did start the system of monitoring the level of pollution in water, but they cannot gauge holistically the impact of pollution on aquatic life. An organism faces the torture of pollution and show drastic chemical changes including the genetic changes as well. The genetic imprints of a species are further carried generation to generation.

Textile and Dying industries are further reported to contribute to 10-15 per cent of about 1000 different chemicals in water, leading to an increase in both the Biological oxygen demand (BOD) and Chemical oxygen demand (COD) levels, needed to support biological entities like bacteria and algae, and to oxidize chemicals; thus, adversely reducing DO contents in water, and creating anaerobic conditions for the aquatic fauna (Kumar, 2002; Jain and Sharma, 2003). Today, as many as 2,000 different chemicals are used in the textile industry. The World Bank estimates that 17-20 per cent of industrial pollution comes from textile coloring and treatments only. They've also identified 72 toxic chemicals in our water solely from textile dyeing, 30 of which are permanent. About 40 per cent of globally used colorants contain organically bound chlorine, a known carcinogen. Traditionally produced fabrics contain residuals of chemicals that evaporate into the air we breathe or are absorbed through our skin. Some of the chemicals are carcinogenic or may cause harm to children even before birth, while others may trigger allergic reactions in some people. According to a June 5, 2005 article in Business Week, the population that is allergic to chemicals will grow to 60 per cent by the year 2020. What's more, traditional dyeing uses an astonishing amount of water. Estimates vary, but to color just one piece of fabric can take anywhere from seven to 75 gallons of water per pound of fabric (26 to 284 liters per 0.45 kilo). About 7-8 hundred dyeing and bleaching units are located only around Tiruppur and Coimbatore districts in Tamil Nadu State. These industries consume substantial amounts of about 10000 different dyes and pigments, and chemicals like, bleaching liquid, soda ash, caustic soda, sulphuric acid, and sodium peroxide along with large amounts of water (Zollinger, 1987).

Likewise, the traditional process of tanning of leather is causing havoc in poisoning the water. Daily discharge of wastes from these tanneries is about 18000 liters of liquid waste, 115 tones solid wastes during peak time; about 60,000 ton of raw hides and skins are processed in these tanneries every year, a process which release nearly 9,500 liter of untreated effluents into the open environment daily. Chrome tanning since is the most popular technique in leather industry; these effluents contain chromium along with cadmium, lead, ammonia, chlorides, sulphides *etc.*

and a large quantity of other organic compounds. Tanning industries also uses a variety of other chemicals, such as lime, sodium chloride, sodium carbonate, sodium bicarbonate, chrome sulphate, dyes and many oils. It has been observed that every 10 kg of raw skin require at least 350 liters of water for tanning and about 30-40 liters are taken out as the waste water rich in chemicals. Reev and Mariappan (1972) stated that high tannin and chlorides had affected the growth of the crops. Chlorides in waste water of tanneries caused chlorosis of plants and high sodium content increased the sodium absorption ratio (SAR), thereby impairing soil structure and plant growth (Mahedeswara, 1983; Narayanmurthy, 1987). The untreated effluents from these tanneries have considerably affected the quality of ground water in this area, thus, causing damage to soil fertility and irrigation to crops (Singaram and Pothiraj, 1989), thus, declining the availability of good quality irrigation water in the area (Venkattakumar, 2000).

Electroplating an important industrial phenomenon is known for the release of chromium (Cr); detergents and fungicides for Ni (Nickle), fossil fuel and gasoline for lead (Pb), fertilizers, fungicides and Ni-Cd batteries for cadmium (Cd), natural soil erosion and insecticides for arsemic (As), galvanizing, paint and insecticides for zinc (Zn) and many insecticides and fungicides for Pb, As, Ni, Hg and Zn *etc.* are the major sources of metal contamination. Extensive use of chromium in tanning industries have resulted in occurrence of chromium contaminated soil and ground water at their discharge sites, thus, posing a serious threat to human health, fish and other aquatic organisms (Turick *et al.*, 1996).

Mukhopadhyay *et al.* (1994) have studied the overall impact of pollution stress on aquatic organisms' *viz.*, fish, plankton and zoo benthos. They concluded that heavy metals form an important group of pollutants which play an adverse role on the production potentiality of an ecosystem, particularly in an industrial zone. Even the human wastes are known to contribute to iron (Fe), zinc (Zn) and copper (Cu) contamination in the soil and water. Heavy metals become toxic when they are not metabolized by the body and accumulate in the soft tissues. Clarkson (1995) has made it evident that now the increasing human activities have modified the global cycle of heavy metals and metalloids, including the toxic non-essential elements like As, Hg, Cd and Pb.

Industries though are the compelling need of an increasing world population to fulfill the basic needs in terms of cloth, shelter and food; their indiscriminate growth and dumping of the solid and liquid wastes in soils and water, however, have been a source of great annoyance and harm to the aquatic ecosystems. In an industrial area, there is an acute possibility for percolation and migration of the chemical pollutants via soil in to drinking wells (Zyka, 1988 and Larn *et al.*, 1994). It is estimated that at least 20 million hectares in 50 countries are irrigated with raw or partially treated wastewater. Natural water resource contamination with untreated industrial effluents has been a rich source of release of heavy metals in the water. About 20 per cent of metal toxicity in soil and water ecosystems is because of industries. Now-a-days the question, whether to eat a fish or not has even made fish consumption a real dilemma. On one hand, more and more people think about the positive effects of eating fish instead of meat as a healthy fish contains high amounts of fatty acids such as omega-3 and omega-6. On the other hand, studies have shown the potential adverse health effects of fish consumption contaminated with heavy metals.

2. Metals and Metal Compounds as the Xenobiotics

Life in general is much inorganic and some metals play an important role as the biological ions in living organisms, as there is no life that can survive without the participation of such metal ions. Pollution of natural environment by metals is a worldwide problem because these metals are indestructible and many of them have toxic effects on living organism, especially when they exceed certain threshold. Metals and metal compounds though are the natural constituents of all ecosystems, moving in between atmosphere, hydrosphere, lithosphere, and biosphere (Bargagli, 2000), these are being further added frequently and at such an alarming rate especially since World War-II, that today some of the heavily toxic heavy metals are of every body's concern as the 'Devils in disguise'. Hydrosphere pollution by various heavy metals such as cadmium, lead, zinc, nickel, copper and iron *etc.* is accelerated dramatically further during the last few decades due to mining, smelting manufacturing, use of fertilizers and pesticides in agriculture, municipal waste and traffic emissions. India has been rapidly expanding its industrial potential and this has simultaneously led to a fast increase in environmental pollution by various toxic metals, since there has been excessive use of such poisonous metals in industries (Lokhande and Kelker, 1999). They cause greatest threat to the health of Indian aquatic ecosystem (Joshi *et al.*, 2002; Ohe *et al.*, 2004; Jaishankar *et al.*, 2014). Dumping of industrial wastes and discharge of effluents containing major heavy metals *i.e.* lead (Pb), cadmium (Cd), mercury (Hg), and inorganic arsenic (As) alter the physicochemical and biological characteristics of water. As per an estimate about 50 per cent of the existing species of flora and fauna may disappear in the next few decades if the metal pollution in the environment continued increasing at the same rate.

Animal's exposure to these heavy metals is potentially harmful especially to those metal-compounds, which do not have any physiological role in the metabolism, later apparently became a health issue and concern for millions of people starting c. 200 years ago. A **xenobiotic** is a foreign chemical substance found within an organism that normally is not naturally produced by or expected to be present within that organism. There are 35 metals that concern us because of occupational or residential exposures; 23 of these *i.e.* antimony, arsenic, bismuth, cadmium, cerium, chromium, cobalt, copper, gallium, gold, iron, lead, manganese, mercury, nickel, platinum, silver, tellurium, thallium, tin, uranium, vanadium, and zinc are termed as the heavy elements or "heavy metals" (Glanze, 1996). "Heavy metals" are metallic elements which have a high atomic weight and a density much greater (at least 5 times) than water. Some well-known toxic metallic elements with a specific gravity of 5 or more are: arsenic, 5.7; cadmium, 8.65; iron, 7.9; lead, 11.34; and mercury, 13.55.

Heavy metals *viz.*, lead (Pb), copper (Cu), cadmium (Cd), chromium (Cr), zinc (Zn), nickel (Ni) and arsenic (As) have adverse effects on human metabolism and health. Sharma *et al.* (2011) have compared metal composition in 88 medicinal plants, and was reported that a maximum permissible level (MPL) of Pb was exceeded in 21 plant medicine species, Cd in 44 species, and Hg in 10 species. They have contended that metal contamination in these medicinal plants may be due to contamination of the area from where the samples were collected. Permissible exposure limit (PEL) for heavy metals is 30 µg/l for lead, 0.1 mg/l for mercury, 10 µg/l for arsenic, and 2.5 µg/l for cadmium. There have been reports of contamination of children blood with heavy metals living near the urban waste incinerators (Fernandez *et al.*, 1992). Asdeo and Loonker (2011) assessed the impact of heavy metal contaminated vegetables irrigated with polluted untreated effluent water from the sewage in Vinayakiya Nallah region of Jodhpur district and found that concentration of each metal *viz.*, lead, cadmium, copper, nickel, zinc and iron, exceeded the natural limits in all vegetable samples. Leafy vegetables accumulate higher metal contents than others (Al Jassir *et al.*, 2005).

3. Trace Metals

The trace metals essentially play an important role in various physiological functions in animals, such as maintaining the body metabolism and growth; and these elements or any of their form, are commonly found in natural foodstuffs, fruits and vegetables, and in commercially available food supplements and the multivitamin products. Metals like sodium (Na), potassium (K), magnesium (Mg), and calcium (Ca) are the essential elements to sustain life, as these are useful in many biological functions; whereas, metals like Cu, Fe, Co, Mo, Sn, Se and Zn being essential as trace metals in biological systems becomes toxic when appear in excessive amounts in cells and tissues; and most of these even prove to be potentially teratogenic in humans, causing embryo toxicity and the birth defects. Nema *et al.*, 2012 reported presence of certain trace metals in medicinal plants; such as copper (Cu), chromium (Cr), manganese (Mn), iron (Fe) and nickel (Ni) as well as heavy metals including arsenic (As), lead (Pb) and mercury (Hg), though these were present within the permissible limit. They had determined these metals in eight different plant species including *Aloe vera, Centella asiatica, Calendula officinalis, Cucumis sativus, Camellia sinensis, Clitoria ternatea,* Piper betel and *Tagetes erecta* cultivated in West Bengal, India and are frequently used in herb medicines in India.

Toxic trace metals concentrations in soil exert a decisive impact on soil quality and its use in food production particularly in an industrial area. An attempt was made to study toxic metals such as Cr, Cu, Ni, Pb, Zn, including Ba, Co and V in representative soil samples from Patancheru industrial area near Hyderabad, Andhra Pradesh (Dasaram *et al.*, 2011), It is a known polluted area and is one of the most contaminated regions where about 260 small and large-scale manufacturers of pharmaceuticals, paints, pesticides, chemicals, steel and metallic products have been functioning for over several decades. The various indices showed that residential soils were contaminated with Cr, Ni and Pb (Cu to some extent). The agricultural area, although were invariably enriched in these toxic metals, showed comparatively less contamination possibly due to uptake by plants. The trace metals can even produce cumulative toxicity in small doses over a long period of time and acute toxicity in higher doses.

No matter how many good health supplements or procedures one takes, heavy metal overload will be a detriment to the natural healing functions of the body. They tend to accumulate in the food chain and in the body and can be stored in both soft (kidneys) and hard tissues (bones). Being metals, they often exist in positively-charged form and can bind onto a negatively-charged organic molecule to form complexes. The body in fact, needs approximately 70 friendly trace metals, some of which are naturally found in the body and are essential to human health. Iron, for example, prevents anemia, and zinc is a cofactor in over 100 enzyme reactions. The ingestion of such metals via food or water could modify the metabolism of other essential elements like: Zn, Cu, Fe and Se (Abdulla and Chmielnicka, 1990). Furthermore, most metals are capable of forming covalent bonds with carbon called chelation, means binding of an organic molecule to the metal ions at two or more points (and hence quite a strong binding), forming a special type of metal-organic complex called a *Chelate.* The term *chelation* comes from the Greek *ch-l-* meaning a crab or a lobster claw, *i.e.* a way in which the metal is gripped in at least at two places by the organic groups. Such metal-organic chelate compound transformations by methylation/alkylation, influence metal mobility and accumulation in tissues, as well as their toxicity (Fergusson and Kim, 1991; Cook *et al.*, 2005; Florea, 2005). Chelating agents are used to produce stable compounds with relatively low toxicity and also to enhance excretion of the metals.

4. Non-Industrial City Wastes

Discharge of untreated city sewage wastes into the water bodies is one major non-industrial factor contributing to water pollution in India (Saikia *et al.*, 1988; Kaviraj, 1989 and Ghosh *et al*, 1990). Large amount of organic and inorganic wastes are discharged into water bodies by agricultural, industrial and the domestic activities. Around 10000 cubic million m of waste water is being generated in India alone, of which about 60 per cent comes from cities, is treated with various primary treatment systems (Juwarker *et al.*, 2002). These mainly contain acids, pesticides, detergents and the heavy metals. These wastes are being indiscriminately thrown into the natural water bodies and the level of wastes exceeds the self purifying capacity of the water bodies. Dumping of these wastes exceed their limit of natural degradation and alter the physiochemical and biological characteristics of water, affecting ultimately growth and survival of the existing millions species of plants and animals including fishes and the humans. Use of waste water for irrigation also affects ground water quality (Sial *et al.*, 2005; Naz Nasrullah *et al.*, 2006). There are reports of water pollutants affecting aquatic biotic components useful to humans in a variety of ways. For example, *Bellamya bengalensis*, an aquatic gastropod had drawn the attention due to its nutritional and economical importance as a source of food protein for a large section of human population of India. It is widely consumed by the rural folk of West Bengal, Bihar and Orissa. According to Ray and Chattopadhyay, 2002, its natural habitat is at great risk due to seasonal contamination of waters by the agricultural runoff containing pesticides, as these affect seriously their immunological responses to the xenobiotics and the pathogens, thus, affecting their survival in contaminated waters. The molluscan haemocytes are said to perform diverse functions for internal defense, such as phagocytosis, encapsulation and killing of parasites during animal's exposure to the polluted waters and other bio-unsafe environments.

Ship breaking is another major curse amounting to environmental pollution and water contamination withhundreds type of toxic chemicals. In addition to discharge of industrial effluents or other poisonous wastes in the water bodies, water pollution also includes the impact of discharge of large quantities of water wash from the ships. As per recent survey, every year about 500-700 small and big ships are sent to beaches of south Asian countries like China, India, Bangladesh and Pakistan. Ship breaking means dismantling of obsolete vessel's structure for scrapping, but without proper storage or disposal of toxic wastes, most of which goes to soil or directly into water causing acute and long term pollution. These scraps and the water run offs contain a variety of toxic chemicals, such as asbestos, polychlorinated biphenyls (PCBs), polyvinyl chloride (PVCs), polychlorinated hydrocarbons (PAHs) and heavy metals like mercury, chromium, lead and cadmium. Ship paints, coatings and electrical equipments also contain heavy metals. Most of these chemicals are carcinogenic and affect reproductive and neurological functions in living beings. Asbestos a highly toxic material is either banned or strictly restricted for use by a number of states in USA. PCBs being highly persistent organic pollutants are gradually accumulated in fatty tissues in living organisms (Permissible exposure limit (PEL) - 1.0 mg/l). PVC products are hazardous as when buried in soil these release highly toxic dioxin, and also carbon monoxide if burnt. Burning of certain painted materials also release another highly toxic water contaminant, tributylin (TBT). It is an organotin compound and is widely used in anti-fouling paints in ships. Organotin compounds contain Tin forming a bond with carbon, Organotin stabilizers are used to prevent changes in polyvinyl chloride upon exposure to light and heat. A number of organotin compounds are major ingredients in biocides and fungicides. Its PEL value is 0.1 mg/l.

Putreciable organic materials are other major constituents of untreated or inadequately treated domestic and industrial wastes, which are being dumped in continuously everyday in all major rivers, lakes or the sea. Industrial and municipal sewage containing heavy metals, hydrocarbons, and some halogenated hydrocarbons accumulate in aquatic food webs causing chronic or acute effects in fish populations. Many times even the land is not spared and landfills are created on porous depressions and filled with solid, urban and industrial waste. In Delhi alone about 200 million liter of waste water flows in Yamuna river and 160 million liter from other sources like the open drains (Mukherjee, 1996). The studies have revealed that when the Yamuna river enters the Delhi city its waters is already polluted and contain about 7500 *Coliform* organisms per dl, and when it leaves the city its water appear contaminated with around 24 million of these organisms per dl. Sharma and Kaur (1994) reported that in 1955-56 about 40 thousand people of Delhi were found to be the victims of infective hepatitis, while several died due to the use of contaminated Yamuna water. These rivers today are thus, implicating millions of people and the domestic animals throughout their course of flow due to their contaminated waters and making them to suffer from several contagious and non-contagious ailments and digestive disorders.

5. Water Acidification

Water acidification is another one of the most important environmental factors affecting fish (Kossakowski Korwin, 1988). Acid rains, affecting the living organisms directly as a pollutant and also by changing water pH resulting in acid pollution, are major source of metal pollution in water. These have become widespread in recent times due to heavy air contamination with fuel and sulpher gases as well as from mine drainages in many parts of the developing and developed countries (Morris *et al.*, 1989). Acid rains are further said to mobilize certain metals in the rocks. The hydrogen-ion concentration is an important quality parameter of both natural and waste waters. It is used to describe the acid or base properties of wastewater. A pH less than 7 in wastewater effluent is an indication of septic conditions, while values less than 5 and greater than 10 indicate the presence of industrial wastes and its non-compatibility with biological operations. Water pH as the medium has wider implications as it influences the activity of almost all enzymes associated with osmo-regulatory, metabolic growth and reproductive functions or even the embryonic development in fish. An indication of extreme pH is known to damage biological processes in biological treatment units (EPA, 1996; Gray, 2002). The pH concentration range for the existence of biological life is quite narrow (typically 6-9). Dheer *et al.* (1987) reported mortality, loss of weight, increase in blood glucose and liver glycogen levels at low pH in *C. punctatus*. Rask *et al.* (1990) reported delayed spawning in perch, *Perca fluvicetilis* in acidified lakes due to decreased gonadal maturation. Titanium dioxide rich industrial effluents in Finland are reported to contain Zn, Cd, Pb, and Cu, and the higher water acidity results in their subsequent precipitation, proving thus, deleterious to the bottom feeding fish (Larsson *et al.*, 1980). A brown colored precipitate of iron-titanium complex was reported to appear on the gills of these fish, even if the fish were cultured in diluted effluents.

Reddy *et al.* (2008) evidenced a relationship in between the fish energetics and the pH of the medium in *C. carpio*. They found carp fry surviving better at alkaline Ph in water as it enhanced oxidative metabolism in fry. Fish fry was found unable to survive at Ph 6.5. Similarly the effects of surface water acidification on fish populations have been shown to be mediated through the presence of metals, as acidification increases bioavailability of metals to the aquatic organisms. Laboratory exposures over 1.3 generations showed that trace metal mixtures played a dominant role in the reproductive failure of American flag-fish (*Jordanella floridae*) in acidified (pH 5.8) water (Hutchinson and Sprague, 1986). Detergents like alkylbenzene sulfonates (ABS) and linear alkylate sulfonates (LAS) also brings significant change in the water pH, thus, increasing metal (Baker 1982) and ammonia toxicity in water (Lloyd, 1992). The major acids of concern are the sulfuric acid and the nitric acid, which are the result of frequent discharge of industrial H_2S and NO gases in the atmosphere. The water bodies surrounded by base mineral (Ca/Mg carbonates/bicarbonates) rich soils however, remain immune to changes in pH due to acid precipitation. Acid pollution is also to cause severe impact on fish immunological functions. Lymphocyte proliferation and antibody production were reported suppressed in channel catfish exposed to acidified waters (pH 4) for 2 weeks, possibly due to failure of B-cells to process the antigen.

6. Oil Spills and Oil Residues

Oil spills *i.e.* leakage of huge quantities of crude oil from tankers/vessels onto the sea, oil residues, sludge and bilge, and ballast water being rich in oil and cargo residues when released in water further poses serious hazards to the living beings in a marine water body and are considered as the worst water affecting pollutants. Frequently occurring oil spill all over the world are the major disastrous events, resulting in severe damage to coastal waters and exerting a number of direct and indirect adverse effects on coastal fisheries. In the past about 50,000 tons of crude oil spilled into the surrounding waters. The oil tanker Amoco Cadiz ran aground on Portsall Rocks, 5 km (3.1 mi) from the coast of Brittany, France, on 16 March 1978, and ultimately split in three and sank, all together resulting in the largest oil spill of its kind in history to that date. Amoco Cadiz contained 1,604,500 barrels (219,797 tons) of light crude oil. The Sea Empress oil spill occurred at the entrance to the Milford Haven Waterway in Pembrokeshire, Wales on 15 February 1996. Over the course of a week, she spilt 72,000 tons of crude oil into the sea. The Exxon Valdez oil spill occurred in Prince William Sound, Alaska, on March 24, 1989, when Exxon Valdez, an oil tanker bound for Long Beach, California, struck Prince William Sound's Bligh Reef at 12:04 a.m. local time and spilled 11 to 38 million US gallons (260,000 to 900,000 bbl; 42,000 to 144,000 m^3) of crude oil. Likewise, there has been Deep water Horizon oil rig explosion in USA in 2010 and the China's largest reported oil spell in July, 2010 due to bursting of an oil pipe at the busy northeastern port. Later was the China's latest environmental crisis spread off to a large surface area, as emptied beaches along the Yellow Sea as oil spread to about 430 sq. km area. It hurled oil shipments from part of China's strategic oil reserves to the rest of the world emerging as a severe threat to affect water quality and the marine life as well as the sea birds. Recently, there has been a huge oil spill in the Gulf of Mexico in May, 2010 from the deep sea oil well has been few major examples of man's casual attitude towards nature. These have been some of the major tragedies of this century which have deteriorated the natural ecological harmony for the coming many decades to the extent that thousands of animal and plant species including the aquatic birds and fishes may fell prey to these disasters. In December 2010, there has been an oil spill in The Bay of Sitakunda in Chittagong (Bangladesh). It was over 3 kilometer long and 300-400 feet wide with a reddish black layer.

As per the available estimates more than three million metric tons of oil contaminates the sea every year. Since 1984, 12 major oil-spill-related seabird mortality events have occurred along the California coast. On Sep. 6, 1999 the dredge M/V Stuyvesant spilled at least 2100 gallons of intermediate fuel Oil 180 into the Pacific Ocean near the mouth of Humboldt Bay, near Eureka, California. The figures of effluents and sewage that is discharged in to the sea at Mumbai alone are a gargantuan 2200 million liters a day. Situation is no better at Kolkata, Chennai and Visakhapatnam ports. The industrial hubs in Gujarat, Pondicherry and Orissa add to the misery of the biota, where the coastal and estuarine waters remain constantly in degraded conditions. Petroleum hydrocarbon content in Arabian Sea are said to be ranged from 1.8 to 11.1μg/l in water, 1.84 to 5.81μg/l dry weight in sediments and 0.33 to 3.67 μg/gram wet weights in fish.

The composition of the crude oil is complex of many organic chemicals and the majority of these chemicals like aliphatic hydrocarbons, cyclic paraffin hydrocarbons, aromatic hydrocarbons, naphthenic-aromatic hydrocarbons, resins, and many other metallic compounds are highly toxic to the aquatic life. Quoting earlier studies (Joshi, 2007), worst sufferer is the marine biota. Oil remains in the water column and trapped around and within the sediment particles of the substrate. Storms and practices such as dredging may suspend oil again that has been buried under the sediments. Crude oil sinks to the bottom and cover shellfish. Bird feathers coated with oil inhibit flight and prevent the feathers from insulating the birds against the cold. These spills are reported to adversely affect breeding, roosting and migration of brown pelicans (Joshi, 2007Anderson *et al.,* 1996). The pelican is an icon of the Gulf coast, and is the state bird of Louisiana. Tens of thousands of the birds breed winter or live here year-round.

The oil also kills vast quantities of plankton, which form the base of the oceanic food chain. Oil also enters marine and aquatic environments through nonpoint sources such as sewer systems from homes and businesses in which oil products have been improperly disposed. Runoff from roads and bridges may also contain oil from motor vehicles. In time, oil dissipates from the environment as it is metabolized (broken down) by bacteria and dissipated by wave action and scientists have also identified some oil-eating bacteria that may assist in "clean up" of oil spills. As per the U.S. policy, as stated in the Clean Water Act 1972, there should be no discharges of oil or hazardous substances into or upon the navigable waters of the U.S., on adjoining shorelines or into or upon the waters of the contiguous zone, or which may affect natural resources belonging to or pertaining to or under the exclusive management or authority of the U.S. Scientists have identified some oil-eating bacteria that may assist in "clean up" of oil spills.

These crude oil spills had immediate and long lasting impact on the surrounding environment and its inhabitants. Since the density of oil is less than water it easily floats over water surface forming a thick impermeable layer, thus, proving highly hazardous to marine organisms including fishes and the seabirds and sea mammals which are regular visitor to sea surface to capture fish for feeding. Most important are the coral reefs, which are home to 25 per cent of the all marine life and are among the world's most fragile and endangered ecosystems. Oil coverings on fish gills and bodies cause them to suffocate. Fish exposed to crude oil experience two sets of challenges: one, the crude oil on the top of the water column reduces oxygen diffusion and shades the water column, thus, limiting photosynthesis; what further causes a reduction of dissolved oxygen, and secondly the affected fish experiences a significant respiratory distress, as well as changes at other biological levels or the detoxification mechanisms. Therefore, the animals exposed to crude oil activates all adaptive mechanisms directed towards increasing oxygen uptake, including a significant reduction of intra-erythrocyte levels of modulators of Hb-O_2 affinity. Mining and crude oil transport in the northern Latin America has potentially affected the fish of the Amazon due to physical environmental disturbances that primarily include extreme reduction in dissolved oxygen and ion-poor acidic waters (Val and de Almeida-Val, 2004). As many fish of the Amazon breathe air or depend on the water-air interface to uptake oxygen, an oil spillage is a deadly environmental situation. To copy with these conditions, tambaqui fish of the Amazon evolved a suite of behavioral, physiological and biochemical responses exhibited by healthy individuals to maintain their organic homeostasis, under the changed environmental conditions. These are so-called stress responses to environmental stressors. The tambaqui (*Colossoma macropomum*) is a freshwater fish of the family Characidae. It is also known by the names pacu, black pacu, black-finned pacu or giant pacu. The tambaqui is found in the Amazon and Orinoco basins in its wild form. However, its pisciculture form is widely distributed in South America. Tambaqui fish when face only a thin oil slick on the top of the water column, *i.e.*, an oil slick of 1mm, its high sensitivity in hypoxia relates to the specific behavioral adaptation, that is its ability to expand the lower lips to skim water surface when exposed to low oxygen, though it also increases the amount of crude oil taken in. In the case of the obligatory air-breathing fish like pirarucu, *Arapaima gigas*, the situation is even drastic, as these fish must surface at regular time intervals to breathe. This fish reduces breathing intervals what results in an increasing oxygen debt and in a reduced locomotion. Impairment of gonad development and failure in spawning, fertilization, hatching and survival in fish is further associated with aquatic hypoxia.

Naphthalene is one of the prominent diaromatic fractions of crude and refined oils and it is a polycyclic aromatic hydrocarbon (PAH). Generally PAHs are less sensitive to photo-oxidation and therefore are more persistent in water (Rand and Petrocelli, 1985). Fish exposure to the crude oil water accommodated fraction (WAF) and chemically dispersed crude oil WAF is reported to show changes in certain metabolic enzymes like cytochrome-C oxidase (CCO) and lactate dehydrogenase (Cohen *et al.,* 2001). There was a significant stimulation in CCO activity in the gills and livers of fish exposed to the WAF of Bass Strait crude oil and chemically dispersed crude oil, compared to the control treatment. In addition, LDH activity was significantly stimulated in the liver of fish exposed to the dispersed crude oil WAF, compared to the crude oil WAF. Fish exposed to the dispersed crude oil WAF treatment had significantly higher oxygen consumption, as measured by oxygen depletion in a sealed chamber, than fish exposed to the crude oil WAF and control treatments. Naphthalene (PAH), a major fraction of crude and refined oils is rapidly accumulated by aquatic organisms, reaching levels higher than those in the ambient medium and affecting normal function of the aquatic life. Studies have clearly indicated that naphthalene has diverse effects on the gill tissues of the fishes, such as the histopathological damage to gills in the form of typical epithelial-capillary separation, proliferation of epithelial cells and consequent fusion of lamella, aneurysm, hemorrhages and the presence of parasites (Santos *et al.,* 2011). The number of lesions caused by naphthalene is related to its concentration. The naphthalene also makes the fishes more susceptible to diseases and parasites. Even the EROD (ethoxyresorufin-O-deethylase) activity increased was found to go up to 60 pmol mg $^{-1}$ min^{-1} in tambaqui fish exposed to crude oil for 24 hours, compared to animals from unpolluted sites that had an activity for this enzyme varying between 1.76 and 22 pmol mg $^{-1}$ min^{-1} (Matsuo *et al.,* 2006). Peculiarly, in specimens of tambaqui acclimated to humic acid and further exposed to crude oil for 24 h, Ethoxyresorufin-O-deethylase (EROD) activity was found to double relative to crude oil exposure alone. EROD activity is recorded in fish as a biomarker of chemical exposure.

REFERENCES

Abdulla M. and Chmielnicka J., 1990. New aspects on the distribution and metabolism of essential trace elements after dietary exposure to toxi metals. Biol. Trace Elem. Res., 23: 25-53.

Al-Jassir M.S., Shaker A. and Khaliq M.A., 2005. Deposition of heavy metals on green leafy vegetables sold on roadsides of Riyadh city, Saudi-Arabia. Bull. Environ. Contam. Toxicol., 75: 1020-1027.

Anderson D.W., Franklin Gress and Michel Fry. 1996. Survival and dispersal of oiled brown Pelican after rehabilitation and release. Mar. Pol. Bull., 3(100): 711-718.

Asdeo A. and Loonker S., 2011. A Comparative Analysis of Trace Metals in Vegetable. Res. J. Environ. Toxicol., 5: 125-132.

Baker J., 1982. Effects on fish of metals associated with acidification. In Acid rain Fisheries, Johnson R.E. Ed., American Fisheris Soc. Bethesda MD: 165.

Bargagli R., 2000. Trace metals in Antartica related to limate hange and increasing human impact. Rev. Environ. Contam. Toxicol., 166: 129-173.

Benjamin Ranjeev, Chakrapani B.K., Devashish Kar, Nagarathna A.V. and Ramachandra T.V., 1996. Fish Mortality in Bangalore Lakes, India. Electronic Green J. Issue 6.

Biney C., Amazy A.T., Calamari D., Kaba N., Mbome I.L., Naeve H., Ochumba P.B.O., Osibanju O., Radegonde V. and Saad M.A.H. 1994. Review of heavy metals in the African aquatic environment. Ecotoxicol. Environ. Safety., 31: 134-159.

Clarkson T., 1995. Health Perspect., 103: 9-13.

Cohen A., Nugegoda D. and others, 2001. Metabolic responses of fish following exposure to two different oil spill remediation techniques. Ecotoxicol. Environ. Saf., 48(3): 306-10.

Cook A.G., Weinstein P. and Centeno J.A., 2005. Health effects of natural dust: role of trace elements and compounds. Biol. Trace Elem. Res., 103: 1-15.

Dasaram B., Satyanarayanan V., Sudarshan V. and Keshav Krishna A., 2011. Assessment of Soil Contamination in Patancheru Industrial Area, Hyderabad, Andhra Pradesh, India, 3 (3): 214-220.

Dheer J.M.S., Dheer T.R. and Mahajan C.L., 1987. Haematological and haematopoetic response to aid stress in an air breathing freshwater fish, *Channa punctatus* (Bloc). J. Fish Biol., 26: 475-481.

EHTI., 2006. "Environmental Hazards of the Textile Industry," Environmental Update #24, published by the Hazardous Substance Research Centers/South and Southwest Outreach Program, June 2006.

EPA., 1996. U.S. Environmental Protection Agency, American Society of Civil Engineers, and American Water Works Association. Technology Transfer Handbook: Management of Water Treatment Plan Residuals. EPA/625/R-95/008. Washington DC.

EPA, 1998. National Water Quality Inventory: 1996 Report to Congress. Urban Storm Water Preliminary Data Summary. *water.epa.gov/./stormwater.*

Fergusson J.E. and Kim N.D., 1991. Trace elements in street and house dusts: sources and speciation. Sci. Total Environ., 100: 125-150.

Fernandez Miguel A., Martinez Liuis, Segarra Merce, Garcia Jose C. and Espiell Ferran, 1992. Behavior of heavy metals in the combustion gases of urban waste incinerators. Environ. Sci. Technol., 26 (5): 1040–1047.

Florea A-M., 2005. Toxicity of alkylated derivatives of arsenic, antimony and tin: cellular uptake, cytotoxicity, genotoxicity effects, perturbation of Ca2+ homeostasis and cell death. Aahen: Shaker Verlag.

Ghosh B.B., Mukhopadhyay M.K. and Bagchi M.M., 1990. Status of organic pollution in the Mathabhanga Churni river system, West Bengal. J. Inland Fish. Soc. India, 22(1 and 2): 16-21.

Glanz J., 1996. Physicists advance into biology. Science, 272: 646-648.

Gossett R.W., Brown D.A. and Young D.R., 1983. Predicting the bioaccumulation of organic compounds in marine organisms using octanol/water partition coefficients. Mar. Pollut. Bull., 14: 387- 392.

Gray F.N., 2002. Water Technology: An Introduction for Environmental Scientists and Engineers. Butterworth-Heinemann Oxford,: 35-80.

Hutchinson N. J. and Sprague J. B., 1986. Toxicity of Trace Metal Mixtures to American Flagfish (*Jordanella floridae*) in Soft, Acidic Water and Implications for Cultural Acidification. Canadian Journal of Fisheries and Aquatic Sciences, 43(3): 647-655

Jain K.L., Sharma M., 2003. Toxic effects of mercury and cobalt on biochemical composition of freshwater fish C. mrigala. Proceedings of 3rd Interaction Workshop December 17-18, 2003., 203-207.

Jaishankar Monisha, Tseten Tenzin, Anbalagan Naresh, Mathew Blessy B., and Beeregowda Krishnamurthy N., 2014. Toxicity, mechanism and health effects of some heavy metals. Interdiscip Toxicol., 7(2): 60–72.

James R., Sampath K. and Selvamani P., 1998. Effect of EDTA on reduction of copper toxicity in Oreochromis mossambicus (Peters). Bull. Environ. Contam. Toxicol., 60: 487-493.

Joshi P.K., Bose M. and Harish D., 2002. Haematological changes in the blood of Clarias batrachus exposed to mercuric chloride. J. Ecotoxicol. Environ. Monit., 12: 119-122.

Joshi V.K., 2007. In: Environmental Biomarkers (www.boloji.com)

Juwarkar A.A., Juwarkar A.S., Thwale P.R. and Khanna P., 2002. Land treatment and disposal of wastewater- An ecofriendly management approach. In Kumar A. (Ed) Ecology of polluted waters Vol. 1. APH Publ. Corp Corp., New Delhi, 185-196.

Kaviraj A., 1989. Heavy metal concentrations of fish and prawn collected from Hoogly estuary. Sci. Cult., 55: 257.

Konar S.K., 1981: Pollution of water by pesticides and its influence on aquatic ecosystem. Ind. Rev. Life Sci., 1: 139-165.

Kossakowski Korwin M., 1988. Larval development of carp, *Cyprinus carpio* in acidic water. J. Fish. Biol., 32: 17-26.

Kotze P., Du Preez H.H. and Van Vuren J.H.J., 1999. Bioaccumulation of copper and zinc in Oreochromis *mossambicus* and *Clarias gariepinus,* from the Olifants River, South Africa. Water SA., 25(1): 99-110.

Kumar P., 2002. Studies on the effects of cobalt toxicity in some freshwater fishes. M.Sc Thesis. Department of Zoology and Aquaculture. CCS HAU, Hisar.

Larn R.F., Brown J.P., Fan A.M., and Miles A., 1994. Chemicals in California drinking water: Source contaminants, risk assessment, risk management and regulatory standards. J. Hazard Water, 39: 173-192.

Larsson Ake., Lehtinen K. John and Haux Carl, 1980. Biochemical and hematological effects of a Titanium Dioxide industrial effluent on Fish. Bull. Environm. Contam. Toxicol., 25: 427-435.

Lloyd R., 1992. Pollution and freshwater fish. Fishing News Books., Oxford.

Lokhande R.S., Kelker N., 1998. Studies on heavy metals in water of Vasai Creek (Maharastra). Ind J Environ Protect, 19: 664-668.

Mahedeswara S., 1983. Effect of heavy metals on Green gram. Leather Sci., 30: 177-181.

Malik Darshan, Singh Sunita, Thakur Jayita, Singh Raj Kishore, Kaur Amarjeet, and Nijhawan Shashi. 2014. Heavy Metal Pollution of the Yamuna River: An Introspection. Int.J.Curr.Microbiol.App.Sci., 3(10): 856-863.

Matsuo Aline Y. O., Woodin Bruce R., Reddy Christopher M., Val Adalberto L. and Stegeman John J., 2006. Humic Substances and Crude Oil Induce Cytochrome P450 1A Expression in the Amazonian Fish Species *Colossoma macropomum*(Tambaqui). *Environ. Sci. Technol.*, 2006, *40* (8): 2851–2858.

Moore J.W. and Ramamoorthy S., 1984. Heavy metals in natural waters In: Applied Monitoring and impact assessment, Springer Verlag, NY.

Morris R., Taylor E., Brown D. and Brown D.(Eds). 1989. Acid Toxicity and Aquatic Animals, Cambridge Univ. Press, NY.

Mukherjee Biswarup, 1996. Environmental Biology. Tata McGraw-Hill Publ. Co, Ltd., New Delhi. pp 502-525.

Mukhopadhyay M.K., Ghosh B.B. and Bagchi M.M., 1994. Toxicity of heavy metals to fish, prawn and fish food organisms of Hoogly estuarine system. Geobios., 21 : 13-17.

Narayanmurthy Asha, 1987. Murder by poisoning and death of Palar, flight for suevival. People's Acton for environment. Ed. Anil Agarwal, D. D'Monte and U. Samrath, Thomson (India) Press, New Delhi.

Naz Nasrullah R., Bibi H., Iqbal M., and Durrani M.I., 2006. Pollution load in industrial effluent and ground water of gadoon Amazai industrial estate (Gaie) Swaie, NWEP. J. Agricul. and Biol. Sci., 1(3): 18-23.

Nema N.K., Maity N., Sarkar B.K. and Mukherjee P.K., 2012. Determination of trace and heavy metals in some commonly used medicinal herbs in Ayurveda. Toxicol.Indust. Hlth. (Pubmed, in press)

Ohe T., Watanabe T. and Wakabayashi K., 2004. Mutagens in surface waters. A review. Mutat. Res., 567: 109-149.

Rand G.M. and Petrocelli S.R., 1985. Fundamentals of Aquatic Toxicology. 1st Edn., Hemisphere Publishing Coorporation, Washington, Philadelphia and London, 666pp.

Rask M., Vuorinen P.J. and Vuorinen M., 1990. Delayed spawning of perch, Perca fluviatilus (Linn.) in acidified lakes. J. Fish. Biol., 36, 317-325.

Ray S. and Chattopadhyay S., 2002. Recognition of hemocyte surface of freshwater snail from contaminated habitat by antiserum raised against murine lymphocytes. In Kumar A. (Ed) Ecology of polluted waters Vol.1. APH PUbl. Corp., New Delhi. Pp 61-65.

Reddy A.S., Reddy M.V. and Radhakrishnaiah K., 2008. Impact of copper on the oxidative metabolism of the fry of common carp, *Cyprinus carpio* (Linn.) at different pH. J. Environ. Biol., 29(5) 721-724.

Reev M. and Mariappan M.1972. Seminar on Tretment and disposal of tannery and slaughter house. Central Leather Research Inst., Madras. P 29.

Saikia, D.K., Mathur, R.P. and Srivastava, S.K., 1988. Heavy metals in water and sediments of upper Gange. Indian J. Environ. Hlth., 31 : 11-17.

Santos T.C.A., Gomes V., Josh A. Maria, Passos C.R., Rocha A.J.S., Salaroli R.B. and VanNgan P., 2011. Histological alterations in gills of juvenile Florida pompano *Trachinotus carolinus* (Perciformes, Carangidae) following sublethal acute and chronic exposure to naphthalene. Pan-American j. Aquatic Sci., 6(2): 109-120.

Sarma H., Deka S., Deka H. and Saikia R.R., 2011. Accumulation of heavy metals in selected medicinal plants. Rev. Environ. Contam. Toxicol., 214:63-86.

Sharma B.K. and Kaur H., 1994. Water pollution. Goel Publ. House., Meerut (India).

Sial J.K., Bibi S. and Qureshi A.S., 2005. Environmental impacts of sewage water irrigation on ground water quality. Pakistan J. Wat. Resourc., 9(1): 51-53.

Singaram P. and Pothiraj P., 1989. Soil health care in tannery polluted areas of North Arcot Dist. Nat. Sym. Environ. Cons. Abdul Hakim College, Melvishasam, Tamil Nadu.

Turick C.E., Apel W.A., and Carmiol N.S., 1996. Isolation of hexavalent chromium reducing anaerobes from hexavalent chromium contaminated and non-contaminated environments. Appl. Microbiol. Biotechnol., 44: 683- 688.

Val Adalberto Luís and de Almeida-Val V. M. F., 2004. Crude oil, copper and fish of the Amazon. Proc. 6th Intl. Cong. Biol. Fish, Manaus, Brazil (August, 2004): 1-6.

Venkattakumar R., 2000. Impact of tannery effluent pollution on agriculture and socio-economic status of the farming community. PhD Thesis, TNAU, Coimbatore-3.

Zollinger H., 1987. Colour chemistry- synthesis, properties and applications of organic dyes and pigments. VCH Publ. NY., pp 92-100.

Zyka V., 1988. Trace elements (Lead, copper zinc, cadmium, arsenic) in drinking waters of Czechoslovakia. Geology of Pruzkum, 30: 162-167.

Chapter 3

Epidemiological Significance of Heavy Metals

1. Metal Toxicity in Humans

There are more than 20 toxic metals known to be major water and land pollutants but 12 are very poisonous which are said to directly interfere to the body enzyme systems and the metabolism in humans. Out of these the deadly toxic four metals: lead, cadmium, mercury, and inorganic arsenic are of particular concern to human health. According to the U.S. Agency for Toxic Substances and Disease Registry, these four heavy metals are among the top six hazardous chemicals present at the toxic waste sites. Industrial exposures accounts for a common route of exposure for adults. In USA, determination of arsenic, lead, and cadmium absorption levels, in up to 5 years old 1774 children, living in 19 towns, having primary nonferrous metal smelters, showed an increased systemic absorption of arsenic (Baker *et al.*, 1977). They are known to cause damaging effects even at very low concentrations. In general, heavy metals are systemic toxins with specific neurotoxic, nephrotoxic, hepatotoxic, fetotoxic and teratogenic effects. Heavy metals can further directly influence human behavior by impairing mental and neurological functions, influencing production of neurotransmitters and their utilization, and altering numerous other metabolic body processes. The association of symptoms indicative of acute toxicity are not difficult to recognize because the symptoms are usually severe, rapid in onset, and associated with a known exposure (Ferner, 2001): cramping, nausea, and vomiting; pain; sweating; headaches; difficulty breathing; impaired cognitive, motor, and language skills; mania; and convulsions. In humans, high levels of lead, copper, manganese, or mercury have been found to be associated with attention deficit hyperactivity disorder (ADHD), memory deficits, impulsivity, anger, aggression, inability to inhibit inappropriate responding, juvenile delinquency, and criminality. Heavy metals act primarily as neurotoxins and there is a synergistic effect between all neurotoxins which is responsible for the illness producing effect. Mercury in particular has been found to be a factor in anger, aggressive behavior, depression, obsessive compulsive behavior (OCD), attention deficit disorder (ADD), autism, schizophrenia, suicidal behaviors, learning disabilities, anxiety, mood disorders, and memory problems (NRC, 2000). Chromium (Cr) is also a potential pollutant and is well known for its mutagenicity, and carcinogenicity effects in humans, animals and plants.

Heavy metal toxicity can also result in damaged or reduced mental and central nervous function, low energy levels and damage to blood composition, lungs, kidneys, liver, and other vital organs. Long-term exposures may result in slowly progressing physical, muscular, and neurological degenerative processes that mimic Alzheimer's disease, Parkinson's disease, muscular dystrophy, and multiple sclerosis. Further studies in humans (Goyer, 1995; USEPA, 1995) have shown that heavy metals such as mercury, cadmium, lead, and tin affect chemical synaptic transmission in the brain and the peripheral and central nervous system. They also have been found to disrupt brain and cellular calcium levels that significantly affect many body functions, such as calcium levels in the brain affecting cognitive development and degenerative CNS diseases, calcium-dependent neurotransmitter release, resulting in depressed levels of serotonin, nor-epinephrine, and acetylcholine. These chemicals are related to mood and motivation *etc.* Deficiency of certain essential nutrients has also been commonly reported depleted due to exposure to metals such as magnesium and zinc, exhibiting increased neurological damage (Govani and Memo, 1979; Nriagu, 1990; Windham, 1999). Depression, headache, trigeminal neuralgia, seizures, and increased pain levels indicate redistribution of metals into the central nervous system (CNS). A recent paper on Parkinson's disease (Powers *et al.*, 2003) have revealed that just by eating iron and manganese-containing foods such as spinach or taking supplements containing Mn or Fe, the risk of developing the disease increased almost 2 fold. This demonstrates that even dietary supplements or organically grown foods are amongst the possible culprits in metal toxicity.

Today, more than one million references on heavy metal toxicity are available worldwide. The metals ultimately affect humans through their food chain both in water and on land. Metals can exist in the body with different kinds of chemical bonds and as different molecules. Metals like mercury, lead and aluminium are extremely toxic, even in the minutest dose these have negative physiological effects. For some heavy metal toxicity levels can be just above the concentrations found naturally. Heavy metals in general may enter a human body either through food, water, air, or absorption via skin, when they come in direct contact or via inhalation and ingestion in agricultural, pharmaceutical, residential or other industrial settings. If these heavy metals are mostly not metabolized within tissues, and these accumulate in tissue faster than the body's detoxification pathways resulting in their gradual buildup inside a body making them potentially toxic. As a rule, acute poisoning is more likely to result from inhalation or skin contact of dust, fumes or vapors, or materials in the workplace. Other systems in which toxic metal elements can induce impairment and dysfunction include the blood and cardiovascular, eliminative pathways (colon, liver, kidneys, skin), endocrine (hormonal), energy production pathways, enzymatic, gastrointestinal, immune, central and peripheral nervous, reproductive and urinary functions (ATSDR, 1999). Zinc is known to yield an unpleasant taste, irritability, muscle stiffness, loss of appetite and nausea in humans, whereas lead causes kidney damage, encephalopathy, miscarriage and sperm abnormality. Other common symptoms of metal toxicity are diarrhea, vomiting, sporadic fever, anemia, hypertension, paralysis, ulceration, convulsions and vertigo. Arsenic is known to cause a variety of skin lesions, cancers, respiratory problems, cardiac diseases and nervous and reproductive defects in humans. Metals such as chromium (Cr), nickel (Ni), cadmium (Cd), and barium (Ba) are also referred as potentially carcinogenic; Be, Cr, and radionuclides like Sr^{90} and Pu^{239} are characterized as mutagenic, causing chromosomal defects. Thomas *et al.*, 2009 stated that exposure to heavy metals may cause kidney damage. They described the major three metal elements interacting with humans are the metallic mercury vapors which primarily escape from dental amalgam fillings; cadmium from car fumes, cigarette smoke and pigment in oil paint; lead in out gassing from paint, residues in earth and food chain, gasoline and the contaminated drinking water. Allergies are not uncommon and repeated long-term contact with some metals or their compounds may even cause cancers (International Occupational Safety and Health Information Center, 1999).

Burry *et al.* (2003) has termed essential metals as the transition metals and essential for the health of most organisms since these are associated in forming integral components of many tissue proteins. Their ubiquity is governed by their ability to form a wide range of coordination geometries and redox states, which allows these elements to interact with many cellular entities, performing pivotal roles in cellular respiration, oxygen transport, protein stability, free radical scavenging, and the many cellular enzyme actions and in DNA transcription. Metals in excess bind to inappropriate biologically sensitive molecules or form dangerous free radicals. Consequently, there is a fine balance between metal deficiency and surplus and it is vital for organisms to maintain metal homeostasis **via** balancing absorption and excretion. Biologically active metal ions are known to be present at the active sites of certain macromolecules like the enzymes: carboxypeptidase, alkaline phosphatase and carbonic anhydrase; the cytochrome-C, hemoglobin, ferredoxin *etc.* and can sometime be displaced from an active site of a compound by other metal ions. These active sites are the potential locus for interaction of toxic metals. Among such groups are the imidazole, sulfhydryl, hydroxyl, carboxyl, amino, guanidine and peptide group of proteins, and heterocyclic bases, ribose, hydroxyl and phosphate group of the nucleic acids. Metal toxicities at the cellular level cause disturbances in reproduction, differentiation and maturation as exemplified by teratogenesis. They mainly affect the permeability of the cell membrane and disturb the energy metabolism, and also decrease the stability of lysosome membrane which disrupts cell functions by releasing various hydrolases at the molecular level.

Stated below are the major sources of origin and the known toxic effects of important heavy metals in humans.

a. Aluminium (Al)

Aluminium is the third most abundant element (after oxygen and silicon), and the most abundant metal in the Earth's crust. It makes up about 8 per cent by weight of the Earth's solid surface (Verstraeten *et al.*, 2008). Aluminium, a trivalent cation element is found in its ionic form in most kinds of animal and plant tissues and in natural waters everywhere (Jiang *et al.*, 2008). Aluminium though is not a heavy metal (specific gravity of 2.55-2.80). Workers in the automobile manufacturing industry also have concern about long-term exposure to aluminium (contained in metal working fluids) in the workplace and the development of degenerative muscular conditions and cancers. A critical appraisal of existing water sources becomes significant for assessment of sustainable development of any resource. Aluminium metal is also said to be too reactive chemically to occur natively. Instead, it is found combined in over 270 different minerals. It is readily available for human ingestion through food additives, antacids, buffered aspirin, astringents, nasal sprays, and antiperspirants; from drinking water, automobile exhaust and tobacco smoke; and from using aluminium foil, aluminium cookware, cans, ceramics, fireworks *etc.*, leading to a concern about its accumulative effects. In the home, we are in constant contact with aluminium in foods and in water; from cookware and soft drink cans; from consuming items with high levels of aluminium (*e.g.*, antacids, buffered aspirin, or treated drinking water; or even by using nasal sprays, toothpaste, and antiperspirants).

Due to its reactivity, aluminium in nature is found only in combination with other elements. Ancient Greeks and Romans used aluminium salts as dyeing mordent and as astringent for dressing wounds; alum, an alloy is still used as a styptic pencil. A styptic pencil is a short piece of an anti-hemorrhagic compound, usually alum or silver nitrate based, that stops bleeding from small cuts and nicks. In very high doses, aluminium can cause neurotoxicity and is reported to be associated with altered function of the blood-brain barrier (Bolt and Hengstler, 2008; Verstraeten *et al.*, 2008), and a possible connection with Alzheimer's disease (Campbell, 2002; Andrasi *et al.*, 2005). However, the World Health Organization (WHO, 1998) concluded that, there were studies that demonstrate a positive relationship between aluminium in drinking water and Alzheimer's disease, but there is no conclusive evidence for or against aluminium as a primary cause for Alzheimer's disease.

Most researchers agree that it is an important factor in the dementia component and most certainly deserves continuing research efforts. Acute exposures are more likely in the workplace, *e.g.* unintentional breathing of aluminium-laden dust from manufacturing or metal finishing processes, low levels of airborne aluminium dust and handling aluminium parts during assembly processes over many years. Aluminium toxicity has been recognized in areas where exposure is heavy or prolonged and the renal function is limited, or where a previously accumulated bone burden is released in stress or illness. Target organs for aluminium are the central nervous system, kidney, and digestive system. General toxicity may include: encephalopathy, gait disturbance, myoclonic jerks, seizures, coma, abnormal EEG, osteomalacia or aplastic bone diseases associated with painful spontaneous fractures, hypercalcemia, and timorous calcinosis, proximal myopathy, increased risk of infection, increased left ventricular mass and decreased myocardial function, microcytic anemia with very high levels, sudden death (Ward *et al.*, 1978; Gilbert-Barness *et al.*, 1998; Yokel and McNamara, 2001; Becaria *et al.*, 2002). In **microcytic anemia,** the red blood cells (erythrocytes) are usually also hypochromic, meaning that the red blood cells are paler than usual.

b. Arsenic (As)

Arsenic has been known to humans since the ancient civilization, and is one among the most important toxic heavy metals. Arsenic is ubiquitous with metalloid properties in the earth's crust and is distributed in air, water and food in nature, to which humans are frequently exposed. It is widely distributed in the earth's crust and ordinarily is a steel gray metal like material that sometimes occurs naturally. The arsenic content of the earth's crust is 1.5-2 mg/kg; it ranks 20th in abundance in relation to other elements (NAS, 1977). The principal natural reservoirs of arsenic are the rocks. Arsenic in nature is widely distributed in a number of minerals, mainly as the arsenide of copper, lead, nickel, silver, or gold or as sulfide. The concentration of arsenic, as determined in natural iron oxy-hydroxides and oxide minerals extracted from soils at the Ashanti Gold mines in Ghana (Bowell *et al.*, 1994), was found to vary from 2 to 35,600 mg/kg. Highest arsenic concentrations occurred in these phases in neutral-pH oxidized clay-rich soils and in the oxidized surface portion of mine tailings. High metal concentrations in mine drainage result from the oxidation of metal-bearing sulphide minerals, most commonly the pyrite (FeS_2O), which host metallic and metalloid elements, such as arsenic, cadmium, copper, lead and zinc (Younger *et al.*, 2002). Nagorski *et al.* (2003) showed that aqueous concentrations of As, Cu, Fe, Mn, and Zn in a small and large mining-impacted stream in Montana, USA were highest following spring runoff and short term storm events. Weis and Weis, 1999 reported chromatic copper arsenate treated wood panels a good source of arsenic residue. Arsenic along with other metals like mercury, lead, and cadmium are reported to occur in a variety of plant and animal tissues (Backaret, 1978). Arsenic residues up to 0.1 ppm were found present in marine food (fish fillet) and somewhat higher levels in crustaceans and bivalves. It is usually found combined with other elements like oxygen, chlorine and sulpher, and is commonly referred as inorganic arsenic. Soils may contain arsenic levels between 0.1 and 40 ppm (Yan-Chu, 1994), if the soil is not disturbed by any natural calamity. Chromated copper arsenate, sodium arsenate, and zinc arsenate are used as wood preservatives (Lansche, 1965). Some phenylarsenic compounds such as arsanilic acid are used as feed additives for poultry and swine and to combat certain diseases in chickens. As per WHO (1981) report, arsenic compounds are mainly used in agriculture and forestry as pesticides, herbicides, and silvicides; smaller amounts are used in the glass and ceramics industries and as feed additives. In agriculture, compounds such as lead arsenate, copper aceto-arsenite, sodium arsenite, calcium arsenate, and organic arsenic compounds are used as pesticides. Arsenic is released into the environment by the smelting process of copper, zinc, and lead, as well as in the manufacturing of certain chemicals and glasses. Arsine gas is a common by-product produced in the manufacturing of pesticides that contain arsenic. Other sources are paints, rat poisoning, fungicides, and wood preservatives. Erosion of natural deposits, runoff from orchards or from glass and electronic wastes majorly releases arsenic residues in water. Arsenic is also an essential component in the production of semi-conductor chips. Arsenic combined with carbon and hydrogen is termed as organic arsenic, later, however, being less harmful to the animals. Currently about 90 per cent of the arsenic available is used as wood preservative (Chromate copper arsenate- CCA) or in lead-acid batteries, medicines and some pesticide formulations. In water, arsenic is usually found in the form of arsenate or arsenite. The most important commercial compound, arsenic (III) oxide, is produced as a by-product in the smelting of copper and lead ores.

A summary of documented ground water arsenic incidents are reported by Chakraborti *et al.* (2002). There has been no recent public health disaster comparable to what people living in West Bengal in India, Bangladesh, United States, Taiwan, China and Sweden are experiencing due to high concentration of arsenic in their drinking water (Smith *et al.*, 2000). In India, arsenic rich groundwater mostly occurs in the Bengal Delta Plain, covering the state of West Bengal, extending to the adjoining states of Jharkhand, Bihar, Uttar Pradesh, Assam and other North-eastern states, as well as the neighboring country of Nepal (Singh, 2006). In Assam, out of 24 districts arsenic was detected in 21 districts. Arsenic poisoning in Bangladesh, is said to be worst than in West Bengal, as about 50 million people are reported to be affected with arsenic in water. Dr. Dipankar Chokraborti, Director (Research) at School of Environmental studies in Jadavpur University Kolkata, was the first who had discovered arsenic contamination in India and had done extensive research on this poisonous chemical He has stated that many people in the Gange- Meghna Brhammaputra basin die due to arsenic induced cancers. From the hydro-geological point of view, a younger deltaic deposit brought by the river Gange and sediments on the riverside is the main source of water contamination with arsenic in West Bengal, as well as in the neighboring country of Bangladesh (Mallick and Rajagopal, 1996). In West Bengal, arsenic problem in groundwater was first recognized in the late 1980s, where people in general are frequently exposed to arsenic contaminated underground water particularly those who drink water from the tube wells. Majority of the tube well waters in Bangladesh (Mitra *et al.*, 2002; Rahman, 2002) are reported to be contaminated with arsenic and the dose levels were reported to exceed the maximum permissible limits of 10µg/L, as prescribed in the World Health Organization (WHO, 1998) guidelines. A few investigators (Dhar *et al.*, 1997; Mandal *et al.*, 1998) have reported the aquifers put less than 300m deep (mostly <100m) in Bangladesh and West Bengal supply more than 90 per cent of drinking water and these water contain more than 50 $\mu g l^{-1}$ of arsenic thus, adversely putting health of millions of people at risk. MacArthur *et al.* (2001) further analysed these waters and revealed presence of more than 50µl of arsenic in the delta plain of gange-Meghna-brahmaputra river, because FeO-OH (in peat, partially decayed vegetable matter and organic matter in human faecal matter) is microbially reduced and it releases its sorbed load of arsenic to groundwater. The Bangladesh permissible limit for arsenic is prescribed up to 50µg/L. Currently, about 100 million people alone in Bangladesh are said to be affected with arsenic contaminated water.

Chakraborti *et al.* (2003) reported ground water contamination with arsenic in many parts of India, such as Chandigarh and many adjoining villages of Punjab and Haryana, and dug wells and hand tube wells of Chhattisgarh and West Bengal. The arsenic contamination of the ground water, however, derives mainly from the geological strata underlying countries, like Taiwan, Chile, and Argentina which had been recognized since long time, and some other countries like Nepal and Vietnam, which have also been recognized more recently. In Kolkata, a factory manufacturing a pesticide, Paris green (K-acetoarsenite), was reported to contaminate the sub-surface and even the underground water in the surrounding area of the factory, resulting in prevalence of keratosis in the affected persons (Guha-Mazumder *et al.*, 1997). Similarly, a lead factory was found responsible to contaminate the area with several other toxic metals including arsenic (Chatterjee and Banerjee, 1999). Paikaray *et al.*, 2005 reported pyrite rich-shales with high carbon sorbing more arsenic; whereas the organic carbon rich black soil demonstrated significantly lower arsenic sorption, possibly due to lesser condensation of the organic matter as compared to the pyrite rich shales. Arsenic-contaminated groundwater is often used in Bangladesh and West Bengal (India) to irrigate crops used for food and animal consumption, which could potentially lead to arsenic entering the human food chain. Al Rmalli *et al.* (2005) analyzed food vegetables and fish *etc.* imported into United Kingdom from Bangladesh and found the mean and range of the total arsenic concentration in all the vegetables 54.5 and 5–540 μg/kg, respectively. The freshwater fish was reported to contain total arsenic levels between 97 and 1318 μg/kg. The arsenic content of the vegetables from the UK was approximately 2- to 3-fold lower than those observed for the vegetables imported from Bangladesh.

Environmental survey with epidemiological studies conducted in between 1987-1994 in the southern parts of Thailand, where mining was the primary source revealed prevalence of chronic arsenic poisoning (Choprapawan and Rodcline, 1997). The burning of coal and smelting of metals are major sources of arsenic in air. Suzuki *et al.* (1974) reported concentrations of arsenic of up to 2470 mg/kg in soil near a smelter in Japan. The arsenic contents of surface waters in unpolluted areas vary but typical values seem to be a few micrograms per liter or less. In a study of river waters in the USA, about 80 per cent of the samples were found to contain arsenic levels of less than 0.01 mg/liter (Durum *et al.*, 1971). Quentin and Winkler (1974) found an average value of 0.003 mg/liter in river water and 0.004 mg/liter in lake water in the Federal Republic of Germany. A mean arsenic concentration of 0.0025 mg/liter was reported in some Norwegian rivers (Lenvik *et al.*, 1978). Much higher values have been reported from some areas including Antofagasta, Chile, where the average arsenic level in a river water supply of drinking water between 1958 and 1970 was 0.8 mg/liter (Borgono *et al.*, 1977). Lunde (1970) showed values of 10-100 mg/kg dry weight in marine algae from the Norwegian coast. Penrose *et al.* (1977) reported that sea-water ordinarily contains arsenic concentrations ranging from 0.001-0.008 mg/liter. Marine algae and seaweed usually contain considerable amounts of arsenic. The provisional tolerable weekly intake (PTWI) for inorganic arsenic is 15 μg/kg bw/week and the organic forms of arsenic present in sea foods need different consideration from the inorganic arsenic in water. There are, however, no reports of toxicity in man or animals from the consumption of organo-arsenicals in seafood.

In occupational settings, workers are often monitored for arsenic exposure levels by measuring arsenic contents in biological samples, such as hair, urine, or blood, to help reduce exposure before the onset of the overt clinical manifestations of arsenic poisoning (Peyster and Silvers, 1995). No data are available which indicate that long-term accumulation of arsenic exists in animals. However, some data have indicated that arsenic levels in the lungs of smelter workers, several years after exposure, were about six times those of controls. Many persons had been reported to develop malignancies such as arsenical keratosis due to long term use of such medicines (Saas *et al.*, 1993). Arsenic present in various metal ores or coal is released during the smelting process or in coal burning, which produces stack dust and flue gas to contaminate the soil and water with arsenic. Rencher *et al.* (1977) examined the lung cancer mortality at a smelter with high levels of airborne arsenic in the USA. People working in copper smelters and art glass industry are frequently exposed to inhalation of arsenic fumes, since it is found as sulfide in various ores as ubiquitous contaminant (Rahman *et al.*, 1996). Smelter workers exposed to inorganic arsenic compounds may have urine values of a few hundred micrograms per liter. The concentrations of arsenic in human tissues seem to be log-normally distributed and the highest levels are generally found in hair, skin and nails. The biological half-time of arsenic in rats is long (60 days), because of its accumulation in erythrocytes. In other animals and in man, most inorganic arsenic is eliminated at a much higher rate, mainly via the kidneys.

Arsenic in India is still in use as 'Ayurvedic system of medicines' to control blood counts in patients with hematological malignancies (Ireleaven *et al.*, 1993). Inorganic arsenic is an element considered as the most potential human carcinogen, and humans are exposed to it almost at places wherever it occur as a water pollutant. Until recently, different arsenic (As) compounds like As-iodide, and As-trichloride had been in use as medicines for a variety of ailments (WHO, 1981; Azcue and Nriagu, 1994) such as: syphilis and African sleeping-sickness. Several organic compounds or herbal/homeopathy products containing arsenic are still in use as human medicines. A routine survey and screening of herbal products in China by the National Fish and Wildlife Forensic Laboratory in Ashland (USA) has revealed many of the herbal medicine balls contained potentially dangerous amounts of As^{2+}, and As^{3+} occurring in herbicides, fungicides, and insecticides can attack –SH groups in enzymes and thus, inhibit their bioactivities (Ogwuegbu and Muhanga, 2005). Studies on animals and man have shown that both trivalent and pentavalent inorganic arsenic compounds in solution are readily absorbed after ingestion. Fish products like fishmeal and fish oil have been identified as major sources of feed contamination with arsenic. The maximum level of arsenic in fish feed is reported as 6 mg/kg and for inorganic arsenic is 2 mg/kg. Recently, import of many arsenic containing Indian herbal medicines have been banned by the USA.

A variety of skin lesions, cancers, respiratory problems, cardiac diseases and nervous and reproductive defects are reported in humans exposed to arsenic. The skin is quite sensitive to arsenic poisoning, and skin lesions are some of the most common and earliest non-malignant effects related to chronic arsenic exposure. Many workers have reported dose-response relationship between arsenic exposure and skin lesions, hypertension *etc.* in humans, such as in USA (Valentine *et al.*, 1992), Japan (Tsuda *et al.*, 1995), India (Guha-Mazumder *et al.*, 1998), Bangladesh (Tondel *et al.*, 1999), and in China (Guo *et al.*, 2001). Guo *et al.*, 2001 reported in areas of inner Mongolia, rates of arsenic induced skin lesions were 48.8 per cent, where 96.2 wells were detected with arsenic over 0.05 mg/L in water. Arsenic chiefly causes skin damage or problems with circulatory systems, and may have increased risk of getting cancer. The increase of prevalence in the skin lesions in humans is reported to occur even at a range of 0.005-0.01 mg/l of arsenic in drinking water (Yoshida *et al.*, 2004). Lung cancer is well known to result from arsenic exposure in occupational settings (Yoshida *et al.*, 2004). Inorganic arsenic in the trivalent state can give rise to skin lesions in man. Today throughout the world millions of people are at risk of cancer, heart disease and diabetes because of chronic exposure to arsenic compounds. It is further known as the most serious ground water pollutants in most of the south-east countries including India and Bangladesh. The characteristic effects that result from ingestion of arsenic-contaminated drinking-water are slowly manifested as skin lesions, *i.e.* diffused melanosis, followed by spotted melanosis, hyper-pigmentation, and keratosis. Hyper-pigmentation is much prominent on the trunk. Perspective studies in large area endemic arsenic poisoning like Bangladesh, China, and Taiwan, where the rate of malignancies is expected to increase within the next several decades, will essentially help in larger way to clarify the dose-response relationship between arsenic exposure levels and adverse health effects. Saha (1995), after a survey of about 61 villages in seven districts of West Bengal during 1983-1987, detected 1214 cases of chronic arsenical dermatosis (melanosis and keratosis). Three districts reported arsenic contaminated in water up to 0.5 ppm. Arsenic ingestion can cause severe toxicity through ingestion of contaminated food and water. Ingestion causes vomiting, diarrhea, and cardiac abnormalities. Lesions of the upper respiratory tract including perforation of the nasal septum, laryngitis, pharyngitis, and bronchitis have frequently been encountered in workers in the smelting industries on exposer to high levels of arsenic. Severe effects of exposure to arsenic have been demonstrated in Japan in the form of neurological effects and in Chile and China (Province of Taiwan) in the form of severe vascular disorders. Peripheral neurological damage has been observed in persons consuming arsenic-containing anti-asthmatic preparations on a long-term basis. The inherent complexity in dose-response pattern of arsenic toxicity, however, make this metal unpredictable in assessing its implications as a carcinogen. Epidemiological studies provide evidence that the dose response relationship for arsenic is nonlinear and that the current approach using linear extrapolation may overestimate arsenic cancer risk in humans exposed to arsenic.

Keratosis on the palms and soles of the feet are the most common signs of arsenic poisoning in humans. Mitra and colleagues (2002) revealed that 82 per cent of patients had moderate to severe skin lesions, and 72 per cent were young adults. The United States Environmental Protection Agency (U.S.EPA., 1995) has reported that ingestion of 50 µg/L of arsenic, results in skin cancer risk in 1, in 400. In 1993, WHO has recommended lowering of arsenic in drinking-water up to10µg/L.

Long-term ingesting inorganic arsenic in diet and artesian well water has also been demonstrated to induce black foot disease (BFD) in southwestern coastal area of Taiwan (Chen *et al.*, 2001). It is a unique peripheral vascular disease that ends with dry gangrene and spontaneous amputation of affected extremities. Arsenic concentration in these BFD area pond waters has been reported to show wide spatial variations; ranging from 8.1 to 251.7 µg L^{-1} (Singh 2001; Liao *et al.*, 2003). Cizick *et al.*, 1992 correlated the prevalence of malignancy with arsenic intake. Tseng (1977) established a positive dose-response relationship between the contents of arsenic in well water and the prevalence rate for skin cancer. A highly significant increase in mortality from cancer of the respiratory system (ICD 160-164)[a] was found for the whole group of former smelter workers under study (SMR = 304.8), which had a roughly linear relationship with the estimated time-weighted average total life-time exposure. Arsenic is also atherogenic and neuro-pathogenic (Tseng, 2002). The term **atherogenic** is used for substances or processes that cause atherosclerosis. In addition there are reports of arsenic induced disorders causing ischemic heart disease, stroke, hypertension, sub-clinical neuropathy, neurobehavioral dysfunctions and type-1 diabetes mellitus through mechanisms of induction of insulin resistance and beta-cell dysfunction (Tseng, 2004).

In persons exposed to water containing high levels of arsenic, whole blood levels exceeding 50µg/liter of arsenic have been reported. No data, however, are available on the influence of dietary habits on arsenic levels in blood. Palaniappan and Vijayasundaram (2008) determined the LC_{50} value for arsenic as 124.5 ppm. Disturbances of liver function and cirrhosis have been observed in both man and animals after chronic exposure to inorganic arsenic. Peripheral vascular disturbances have also been reported in some areas of the world where heavy exposure due to ingestion of inorganic arsenic has occurred, *e.g.*, in Chile, China (Province of Taiwan) and the Federal Republic of Germany. Inorganic arsenic can exert chronic effects on the peripheral nervous system in man. The carcinogenic potential of inorganic arsenic in smelter environments is evident from many epidemiological studies. There is substantial epidemiological evidence of respiratory carcinogenicity in association with exposure to mainly inorganic arsenic in the manufacture of arsenic-containing insecticides. Skin cancer has been associated with inorganic arsenic exposure in reports from many parts of the world. Several types of neoplastic changes of the skin, including Bowen's disease and basal and squamous cell carcinomas, have been associated with arsenic exposure. Methylated arsenic compounds, such as di- and tri-methylarsines, occur naturally in the environment as a result of biological activity. Data from studies on the experimental animals indicate that trivalent inorganic arsenic is more toxic than pentavalent. Following ingestion or inhalation of inorganic arsenic, the major forms of arsenic excreted in human urine are dimethylarsinic acid and methylarsonic acid, accounting for about 65 per cent and 20 per cent of excreted arsenic, respectively. Substantial amount of methylarsonic acid and dimethylarsinic acid are used, however, as selective herbicides. *In vivo* methylation of inorganic arsenic has been demonstrated in both animals and man. Trivalent (As(III)) and pentavalent (As(V)) arsenic entering in animals are further said to be rendered less toxic by methylation to monomethylarsonic and dimethylarsinic acids (Newman and Unger 2003; Sharma and Sohn, 2009).

c. Cadmium (Cd)

Cadmium is another major industrial toxic metal pollutant and a cumulative toxicant to higher life-forms, and is at number 7 on ATSDR's "Top 20 list" (ATSDR, 1999). It is although a rare element, but is widely spread in nature in low concentrations in rocks and soils. In the past Cd was recognized mainly as a health hazard to miners and the exposed workers. Today, this metal is considered as highly toxic and a serious environmental contaminant that may affect all forms of life because of the absence of homoeostatic control for this metal in human body. Cadmium enters the environment mainly from industrial processes or fertilizer application and it can be transferred to the food chain by plant uptake. Its major emission into the environment is from smelters and refining of ores, combustion of coal and other fossil fuels, improper disposal of batteries and other Cd containing products, and application of phosphate fertilizers. Because of many industrial applications Cd is present frequently in industrial effluents, phosphate sludge and even in cigarettes which are also known to contain cadmium. Cadmium may be found in reservoirs containing shellfish. Lesser-known sources of exposure are dental alloys, electroplating, motor oil, and exhaust. It is obtained usually from phosphate rocks as an inevitable by-product of the mining and smelting of lead and zinc, since these metals occur naturally within the raw ore. However, once collected the cadmium is relatively easy to recycle. It is used in nickel-cadmium batteries, PVC plastics, and paint pigments. It can be found in soils because insecticides, fungicides, sludge, and commercial fertilizers that use cadmium are used in agriculture. Cadmium in agriculture soil is mainly derived from fertilizers, fungicides and sewage sludge applied in the crop area. The roots of corn, oat, soybean, tomato and alfalfa as well as aerial parts of carrot, lettuce, tomato and potato are said to accumulate highest levels of Cd when grown in Cd contaminated soil. The most significant use of cadmium is in nickel/cadmium batteries, as rechargeable or secondary power sources exhibiting high output, long life, low maintenance and high tolerance to physical and electrical stress. Cadmium coatings provide good corrosion resistance, particularly in high stress environments such as marine and aerospace applications, where high safety or reliability is required; the coating is preferentially corroded if damaged. Other uses of cadmium are as pigments, stabilizers for PVC, in alloys and electronic compounds. Cadmium is also present as an impurity in several products, including phosphate fertilizers, detergents and refined petroleum products. David and Thomas (1989) has stated that cadmium^{113m} enter the environment as a consequence of testing of nuclear weapons in the atmosphere. Earlier it was demonstrated that Cd^{113m} could be determined accurately in water, soils, and sediments (Zhavoronkov *et al.*, 1987). In Lake Michigan, the present concentrations of Cd^{113m} were 350 ± 150 μBq l^{-1} ($n = 16$) in the water, and 10.0 ± 5.5 mBq g^{-1} ($n = 12$) in the sediments. In undisturbed soils near Lake Michigan, the mean Cd^{113m}/Cs^{137} activity ratio was $2.0 \pm 1.4 \times 10^{-2}$ (n = 4), implying a total Cd^{113m} input to the lake of 2.7 ± 1.3 TBq.

Cadmium is bio-persistent; once absorbed by an organism, remain resident for many years (over decades in humans) although it is eventually excreted. Cadmium has a high profile in human toxicology because cases of Cd poisoning exist wherever it has been transferred at harmful concentrations in humans through food chain. Cd poisoning was first evident in 1955, when it was found responsible for a mysterious disease called Itai-Itai in Japan. For more than a century, Cd had been known as an extremely toxic element among the inhabitants of Toyama, Japan due to the outbreak of this Ita-Ita disease (Ishizaki, 1965), due to consumption of Cd contaminated fish from Jintsu river. The disease was characterized by severe pain in back, joints and lower abdomen, accompanied with kidney disorders and loss of calcium leading to multiple bone fractures in humans. The average daily intake for humans is estimated as 0.15µg from air and 1µg from water. Smoking a packet of 20 cigarettes can lead to the inhalation of around 2-4µg of cadmium, but levels may vary widely. The estimated daily intake of Cd in different countries ranges from 25-60 µg/day, while its tolerable dose intake is about 57-72 µg/day. Skin neoplasm has been reported to occur in 2-3 years old salmons on a number of Russian fish farms during the past 2-3 years. Inhalation accounts for 15-50 per cent of absorption through the respiratory system; 2-7 per cent of ingested cadmium is absorbed in the gastrointestinal system. Target organs are the liver, placenta, kidneys, lungs, brain, and bones (Roberts, 1999; ATSDR, 1999).

Its long-term exposure in humans is associated with renal dysfunctions, pulmonary lesions, anemia, hypertension, liver cirrhosis and kidney damage, diarrhea, atrophy of gonads, retarded growth, disturbed carbohydrate metabolism and inhibition of drug-metabolizing liver enzymes (Larsson, 1975; Klassen, 1991a). Nausea, vomiting, salivation, diarrhea and abdominal cramps are commonly known symptoms of poisoning with this metal (Klasseen, 1991a). High exposures can lead to obstructive lung disease and has been linked to lung cancer, although data concerning the latter are difficult to interpret due to compounding factors. Cadmium may also produce bone defects (Osteomalacia and Osteoporosis) due to poor bone mineralization in humans and animals.

d. Chromium (Cr)

Chromium though a metal of biological interest as an essential nutrient, probably having a role in glucose and lipid metabolism (Langard *et al*, 1979; Witmer *et al.*, 1991), and also as one another most hazardous metals to human health. It is extensively used in metal alloys and pigments for paints, cement, paper, rubber, and other materials, as well as to tan hides in the leather industry. Presence of abundant chromium anions in the water are generally a result of the industrial wastes, which are discharged in large quantities into water bodies and also as gaseous wastes in to the environment, thus, yielding adverse effects on living organisms present in the vicinity. Tanneries alone contribute up to 30ppm of Cr salts in water. In India, more than 50 per cent of these tanneries are located in Tamil Nadu alone, and the Palar river basin of southern India in Kolar district of Karnataka has itself around 700 tanneries. It has been revealed that due to pollution of the Palar river by tannery effluents and the Cr contaminated ground water, there was an increase in the respiratory and skin diseases among the villagers. Same effects were reported by Venkataramani, 1987.

Chromium mainly affect sugar metabolism by potentiating insulin action via interaction with the insulin receptors on the cell surface, and participation in lipid metabolism by inhibition of hydrxymethyl-glutaryl CoA reductase with a hypo-lipidemic effect (Zima *et al.*, 1998). It causes deleterious effects on non-target aquatic organism resulting imbalance of an ecosystem (David and Thomas, 1989; Kumar Arun *et al*, 2004). Chromium exists in two major stable oxidation states as trivalent chromium- Cr (III) and hexavalent chromium- Cr (VI) (Cohen *et al.*, 1993). Intake of Cr (VI) up to certain limits, however, could be safe as it is quickly reduced in the human stomach to Cr (III) form under the gastric acidic conditions. There are also some reports suggesting pharmacological action of trivalent chromium in promoting the action of insulin on cells (Anderson, 1993; Merz, 1993).

Chronic exposures to Cr particularly the hexavalent form, are strongly associated with various adverse health effects, such as lung cancer and bronchial asthma, nephritis, ulceration of the gastro-intestinal tract and some other neurological disorders. Low-level exposure can irritate the skin and cause ulceration. Long-term exposure can further cause damage to kidney and liver and to circulatory and nerve tissues. Cr (VI) resembles sulphate and phosphate ions, thus, is said to be mutagenic and carcinogenic (Forstner and Wittman, 1979). Cr (VI) is actively absorbed into the cells. Once inside the cell, it is reduced to trivalent form, which prove highly toxic and it forms stable DNA adducts. The trivalent form, however, does not readily enter into a cell. It is further said to induce apoptosis. Apoptosis is the process of Programmed cell death (PCD) that may occur in multicellular organisms. Characteristic cell changes which occur in the process of PCD include blebbing, cell shrinkage, nuclear fragmentation, chromatin condensation, and chromosomal DNA fragmentation. In cell biology, a bleb is an irregular bulge in the plasma membrane of a cell caused by the localized decoupling of cytoskeleton from the plasma membrane. Treating human lymphoma U_{937} cells with 20 μM of Cr (VI) for 24h was reported to produce nuclear morphological changes and DNA fragmentation (Hayashi *et al.*, 2004). Production of hydroxyl radicals and intracellular increase in the concentration of calcium ions were also evident. It was suggested that intracellular reactive oxygen species (ROS) are important in this Cr (VI) induced apoptosis of the U_{937} human lymphoma cells. Cr (VI) and CrO_3 are reported to yield strong putrefaction effects that may cause ulcerations of skin and nasal mucosa, and perforation of nasal septum. The chronic adverse health effects are respiratory and dermatologic (Viessman and Hammer, 1985). Davidson *et al.*, 2004 also observed a dose dependent increase in number of UV-induced skin tumors, greater than 2 mm, in mice on simultaneous treatment with potassium dichromate (K_2CrO_4), suggesting that exposure to hexavalent chromium in drinking water or through inhalation could be an health hazard as it could elevate susceptibility to carcinogenesis in animals.

e. Copper (Cu)

Copper though an essential element in low concentrations, and is necessary for the normal biological activities of amine oxidase and tyrosinase enzymes; in high doses it can cause anemia, liver and kidney damage, and stomach and intestinal irritation. Copper is used in industries like organic chemicals, fertilizers, iron and steel works, electrical works, antifouling paints, pulp and paper industries, pesticides, fungicides and automobile industries, and is frequently discharged into freshwater environments in large concentrations as an effluent severely affecting the freshwater fauna, especially fishes (Venkataramana and Radhakrishnaiah, 2001). Corrosion of copper pipes of the household plumbing systems and erosion of natural deposits as well as from additives designed to control algal growth, are the main sources of Cu pollution in drinking water. Oil mining has been another potential source of copper pollution in the Amazon river (Brazil), as the production water from the Urucu petroleum mining plants was found to contain as much as 240 mg of copper per liter. Metals from mining sites and from industries may drain into nearby aquatic ecosystems; once there they can be transported considerable distances downstream.

Copper on short term exposures, may cause hypertension, uremia, coma, or sporadic fever and gastrointestinal distress, and its long term exposure results in liver or kidney damage. Excessive intake of Cu may cause neurological disorders, hemolysis, hepatotoxicity and nephrotoxic effects or accelerated ageing. People with Wilson's disease are at greater risk for health effects from over exposure to copper. Recommended daily intake is 2-3 mg per day for humans, but a deficiency of Cu can lead to high serum cholesterol and an increased risk of cardiovascular diseases (Reiser *et al.*, 1987).

f. Lead (Pb)

Lead, number 2 on the ATSDR's "Top 20 list" (ATSDR, 1999). is one of the most toxic heavy metals found in the environment (Needleman and others, 1990) and is a widely distributed toxic heavy metal in the urban environment, as a result of its extensive use as an antiknock additive in gasoline. Natural origin of lead in the environment occurs from weathering of rocks, mining and the anthropogenic sources. Lead in fact is known to be in use since the times of the ancient civilizations, such as of the Romans and Egyptians. They used lead extensively to line their utensils and main water channels, resulting in widespread residual lead toxicity. There is a saying that extensive applications of lead in those times had resulted in lead poisoning to the extent that it had been the cause of fall of those ancient civilizations. Most cases of lead poisoning occur following exposure to very elevated lead concentrations, as may be found in the vicinity of mines, waste dumps, and industrial plants. Every year, industry produces about 2.5 million tons of lead throughout the world. Lead being malleable and ductile and a very soft metal it is easily shaped, and is used in making batteries, in rolled and extruded products, alloys, pigments and compounds, cable sheathing, cable coverings, plumbing, shot and ammunition and as petrol additives (now no longer allowed in the EU), Even today almost all major industrial and electronic items are lead polished. Other uses are as paint pigments and in PVC plastics, x-ray shielding, crystal glass production, and pesticides. Lead is frequently used in the manufacturing, construction and chemical industries. Millions of homes built before 1940 still contain lead in the painted surfaces. Recently there are reports of heavy lead toxicity in children in China and also at many other countries due to lead in paints and toys. Poisoning from lead shots in water birds is further well documented globally and in some countries, legislation exists to combat lead toxicities at wetlands and/or in water birds. Fishera *et al.* (2006) identified fifty-nine terrestrial bird species that have so far been documented to have ingested lead or suffered lead poisoning from ammunition sources, including nine globally threatened or near threatened species.

Near lead and zinc smelters, elevated levels of lead and cadmium in hair provided evidence of external exposure to these elements. About 2 lac tones of lead are added from automobiles every year, as the lead-gasoline alone accounts for 20 per cent of road-side lead pollution. It is frequently used in pipes, drains, and soldering materials for many years. As per estimates, rivers run off alone contribute to about 2.5 lac tones of Pb per yr in the oceans. Sludge from sewage treatment facilities is often added to agricultural land and can be a source of lead pollution (Pain, 1995; Pattee and Pain, 2003), and lead in paint chips has been shown to poison even the captive birds, such as the sand hill crane, *Grus canadensis* (Kennedy *et al.*, 1977). Lead is among the most recycled non-ferrous metals and its secondary production has therefore grown steadily in spite of declining lead prices.

Human exposures can occur through lead contaminated drinking water, food, air, soil and dust from old paint containing lead. Weathering, flaking, chalking, and dust had resulted in large scale of lead poisoning in water. In the general non-smoking adult population, the major exposure pathway of lead is from food and water. Average daily lead intake for adults in the UK is estimated at 1.6 μg from air, 20 μg from drinking water and 28 μg from food. The natural lead contamination in water has been reported to be up to 10 μg/l, and daily dietary intake of lead ranges between 100-500 μg/day (U.S.EPA., 1989b). Food and beverages usually contain lead as these items absorb it from the improperly glazed containers. Leafy vegetable plants such as lettuce, spinach, potatoes, beans, are likely to absorb lead from the soil. Lead as air pollutant further contributes to its high levels in food through deposition of dust and the rain water containing this metal, on crops and the soil.

Lead produces negative effects on general health, reproduction, behavior, and potentially leading to death. Lead accounts for most of the cases of pediatric heavy metal poisoning (Roberts 1999). Klinghardt (2010) has stated that "Lead poisoning" has strong effects on the brain and in the bones; lead poisoning causes malfunction in the formation of blood and thus, leukemia, anemia, lymphomas, all the tumors involving the hematopoietic system play a part. Food, air, water and dust/soil are the major potential exposure pathways for infants and young children. For infants up to 4 or 5 months of age: air, milk formulae and water are the significant sources. Ingestion and inhalation are the two most common entry routes of lead into animals (Eisler, 1988). Because of physical and electrostatic (size and charge) similarities, lead can substitute for calcium and is incorporated in bones. That is the reason that children are especially susceptible to lead because developing skeletal systems require high calcium levels. Lead that is stored in bone is not harmful, but if high levels of calcium are ingested later, the lead in the bone may be replaced by calcium and mobilized. Once free in the system, lead may cause nephrotoxicity, neurotoxicity, and hypertension. Lead poisoning apparently does not produce any visible symptoms in humans; it can, however, result in a wide range of biological effects depending on the level and duration of exposure. Primarily, lead is known as a hemotoxic metal, it also target bones, brain, blood, kidneys, and thyroid gland (International Occupational Safety and Health Information Center, 1999. In case of chronic lead toxicity it affects synthesis of hemoglobin in blood and cause damage to CNS, kidneys and the bones (Klassen, 1991a,b). Major lead poisoning occurs due to inhibition of active transport mechanism including ATP molecules and suppressing cholinesterase activity in nerves and protein synthesis. Various effects occur over a broad range of doses, within a developing fetus and the infant are more sensitive than the adults. High levels of exposure may also result in toxic biochemical effects in humans, which in turn cause problems in the kidneys, gastrointestinal tract, joints and reproductive system. At intermediate concentrations, there is persuasive evidence that lead can have small, subtle, sub clinical effects, particularly on neuropsychological developments in children and anemia, vomiting, appetite loss, convulsions, liver, kidney and brain damage *etc.* in adults. Some studies suggest that there may be a loss of IQ points up to 2 for a rise in blood lead levels from 10-20 µg/dl in young children.

g. Mercury (Hg)

Mercury is today's one of the most important and most frequently available water pollutants. Mercury being liquid at room temperature is further a unique metal and is known for its complex and unusual chemical and physical properties. Mercury, number 3 on "Top 20 List" is said to occur as residue in the water since the beginning of increased mining, excessive burning of fossil fuel and wide spread applications of raw materials containing Hg in the industries. Approx 3000 to 6000 t of Hg are released in the environment each year due to human activities and 6000 to 10000 t is added further as a result of volatilization and sublimation, and another 84000 mt from the earth surface due to degassing of the earth's crust and from volcanic emissions. Weathering of mercury bearing rocks releases about 3500 t/yr of mercury, while 25-150 thousand tones is released in gaseous form during volcanic activities, and 2000-4500 t is added due to various anthropogenic activities and from electrical and medical equipments every year. The burning of fossil fuels, primarily coal, is other best characterized and most dominant anthropogenic source of mercury emissions (USEPA, 1997; Seigneur *et al.*, 2003), representing over 30 per cent of mercury emissions from domestic sources and 29 per cent of global anthropogenic emissions (Seigneur *et al.*, 2003). Burning of fossil fuels further add about 3000 t in a year, Approximately 40 per cent of the 68 metric tons of Hg from the coal burned domestically remains in the ash and scrubber residues, while 60 per cent is emitted to the atmosphere.

Mining operations, chlor-alkali plants, and paper industries are other significant producers of mercury (Goyer, 1997). For each ton of chlorine/NaOH produced in a chlor-alkali industry, of the about 250 to 500 g of Hg consumed, are discharged majorly in water as waste. Mercury compounds were also being added as paint used as a fungicide until 1990. These compounds are now banned; however, old paint supplies and surfaces painted with such old supplies still exist. Environment Expert panel has concluded that deposition, even at remote locations contains a significant fraction of Hg, and it is possible that the global cycling of Hg has been impacted not only by anthropogenic activities, but also by increase in the atmospheric ozone concentration since the industrial revolution. Global scale Hg cycling models (Mason *et al.*, 1994) provide useful information and perspectives on the behavior and fate of Hg in environment, especially presence and production of Reactive Gaseous Mercury (RGM). RGM being all Hg(II) species and water soluble, could significantly influence Hg uptake in ecosystems. RGM are reported at remote locations ranging from Polar Regions to the open ocean (Temme *at al.*, 2003; Laurier *et al.*, 2003). Elemental gaseous Hg is removed slowly from the atmosphere via wet and dry deposition because of its very low solubility, where as Hg species like RGM are removed rapidly due to their high solubility and reactivity with surfaces. (Ariya *et al.*, 2004) The importance of attributing the sources of Hg in deposition is now widely recognized and is critical for development in most of the countries. Deposition even at remote locations contains a significant fraction of Hg, there is an overall increase in Global Reactive Gaseous Mercury (RGM) cycling at the air-sea interface, appearing to be very significant with 5.7 of the 13 mmol yr^{-1} evading from the oceans and returning to the sea surface in deposition. It is further interesting to find that even reduction in anthropogenic activities will not produce any considerable change in Hg deposition due to continuous emission of Hg from large stores at Earth's surface. Particulate Hg is another species contributing to Hg deposition in the atmosphere. There are reports that Hg emissions in Asia, Africa and Australia are increasing in a dramatic fashion over the last few years (NADP, 2006). Among the Asian countries, China is now regarded as the largest anthropogenic Hg emission source. Recent studies have demonstrated the soils of the Amazonian region rich in natural mercury (Silva-Forsberg, 1999; Roulet *et al.*, 1998) serve as primary source of the contamination of the water and the organisms.

India is also one of the biggest consumers of mercury in the world and most of the Indian water bodies are severely polluted with mercury (Kalra, 2004). The chlor-alkali industries are the most significant source of Hg contamination in water (Saffi, 1981; Lodenius and Tulisalo, 1984). In India, about 200 ton of mercury and its compounds are introduced into the environment annually as effluents from industries (Saffi, 1981). According to Larsson (1970), Sweden alone contributed 15-19 metric tons of Hg from its alkali-chlor industries in 1967. Consequences of mercury pollution are well established in Japan and many of the other countries. Japanese were in fact the first to have the bitterest experience of mercury toxicity in their rivers in the mid of last century, when there was an outbreak of a neurological disorder, termed as the MinaMata disease (Fujiki, 1972). The fish in these waters was found to contain up to 40 ppm of Hg residue. It resulted in 121 cases of poisoning and 41 deaths. Release of Hg in these waters was from the alkali and plastic industries involved in the production of acetaldehyde. Takeuchi (1968) also reported human's exposure to mercury due to consumption of Hg contaminated fish, shellfish and sea mammals. Examples of data on total mercury and methyl-mercury exposures primarily from fish diets and many other sources in different parts of the world, including Sweden, Finland, the USA, the Arctic, Japan, China, Indonesia, Papua New Guinea, Thailand, Republic of Korea, Philippines, the Amazonas and French Guyana, are most common as described in UNEP (2002) report. In a study of a representative group of about 1700 women in the USA (aged 16-49 years) for years 1999-2000, about 8 per cent of the women had mercury concentrations in blood and hair exceeding the levels corresponding to the US EPA's reference dose. Data further indicate exposures are generally higher in Greenland, Japan and some other areas compared to the USA. The fish in Hg polluted waters was found to contain Hg up to 40 ppm. Consuming fish contaminated with mercury is the major cause of a variety of human ailments (Clarkson *et al.*, 2003).

Hg is said to exist in three forms: elemental mercury and organic and inorganic mercury. According to Mitra (1986) Hg appears in industrial discharges in five principal forms, divalent Hg ($Hg^{2}+$), metallic Hg (Hg^{o}), Phenyl-Hg (C_6H_5Hg), alkosyl alkyl-mercury (C_3H_7OHg) and methyl-Hg (CH_3Hg+) of which methyl mercury is the most toxic form. Exposure to each species results in both specific and general toxicological effects in children and adults. Recently, Counter *et al.*, 2004 provided a summarized toxicological account of the effects in children and adult humans of these three relevant mercury species. Mercury is mostly present in the atmosphere in a relatively underactive form as a gaseous element. The long atmospheric lifetime (of the order of 1 year) of its gaseous form means the emission, transport and deposition of mercury is a global issue. The main sources of mercury emissions are from the manufacture of chlorine in mercury cells, non-ferrous metal production, coal combustion and crematoria. In addition to these industries, the usages of mercury are widespread in various other products (*e.g.* batteries, lamps and thermometers), electrical apparatuses, thermostats, and dental amalgams. Many researchers suspect dental amalgam as being a possible source of mercury toxicity (Omura *et al.*, 1996). Medicines such as mercurochrome and merthiolate (a mercury-containing antiseptic), and algaecides and childhood vaccines are also potential source of mercury. Certain germ killing pharmaceutical products in use as against slime mold also contribute to Hg poisoning in water.

Mercury is a toxic substance and it has no known biological function in human physiology and also does not occur naturally in living organisms. Mercury contamination today is thus, a big concern in many countries due to its toxic effects on biotic and abiotic components of the ecosystem. It is one of the most toxic metals to the organisms at the top of a food chain (Guentzel *et al.*, 2007). Mercury is such a highly toxic non-essential persistent and non biodegradable and immutable heavy metal that it under goes bioconcentration and biomagnifications during its transfer through food chain. Whilst there has been a decline in the level of European emissions of mercury, emissions from outside of Europe have started to increase – increasing the level of ambient Hg concentrations in the continent. Atmospheric mercury is dispersed across the globe by winds and returns to the earth in rainfall, thus, accumulating in the aquatic food chains including the fish in lakes (Clarkson, 1990; Goyer, 1997). A random survey conducted to evaluate the fish consumption rate and health status among the people, revealed fish consumption rate of 34g/day, which was slightly higher than the national average of 30g/day. The hazard index calculated with the mean concentration of total mercury was 2.09, which indicates a high risk to human beings. The main pathway for mercury to humans is through the food chain. Inhalation is also a frequent cause of exposure to mercury. The organic form is readily absorbed in the gastrointestinal tract (90-100 per cent); lesser but still significant amounts of inorganic mercury are absorbed in the gastrointestinal tract (7-15 per cent). Target organs are the brain and kidneys (Roberts 1999; ATSDR (1999). Children are particularly vulnerable to Hg intoxicants, which may lead to impaired central nervous system, as well as pulmonary and nephritic damage. The elevated level of mercury are known to cause serious neurotoxic and genotoxic effects. The heavy metals in animals in general result in abdominal cramps, nausea, salivation, diarrhea, hypertension and sometimes may prove carcinogenic or bring about mutagenic or teratogenic effects also (Jain and Mittal, 2004). Neurological signs of Hg toxicity in mammals typically include lethargy, ataxia, limb paralysis, tremors, convulsions, and ultimately death. In freshwater aquatic ecosystems warm, shallow, organic rich lake systems are often important zones of net methylation. Inorganic mercury poisoning in adults is mainly associated with damage to the brain and the central nervous system, while fetal and postnatal exposures give rise to abortion, congenital malformation and development changes in young children, tremors, gingivitis and/or minor psychological changes, together with spontaneous abortion and congenital malformations. Abdominal pain, headache, diarrhea, haemolysis and chest pain are some of the other common symptoms of Hg poisoning in humans. Chronic poisoning may cause liver damage, neural damage, and teratogenesis (U.S.EPA., 1987).

Mercury in the form of methyl-mercury (MeHg) poses further great risk to humans. Methylation of mercury drastically alters its properties. It loses polarity and ceases to behave as a typical metallic ion and it becomes much more water fat soluble and less water soluble (Craig, 1986a). Natural biological processes can produce methylated forms of mercury which bioaccumulate over a million-fold and concentrate in living organisms, especially fish. These forms of mercury: mono-methyl mercury and di-methyl mercury are highly toxic, causing neuro-toxicological disorders. To protect human health, the U.S. Environmental Protection Agency has set a generalized, default fish tissue mercury residue criterion for freshwater and estuarine fish at 0.0175 mg per kg of fish per day (U.S.EPA. 2001). Various aerobic and anaerobic microbial populations are known to possess enzyme degradation systems that react with mercury species. These systems impart to the host some resistance to the toxic effects of MeHg, and result in the cleaving of MeHg, forming CH_4 and Hg-II. MeHg has been recognized as a significant risk to the viability of natural systems associated with Hg contamination in aquatic resources (Gnamus *et al.*, 2000; Hammerschmidt *et al.*, 2002). Fish and fish meal had been identified as the major sources of human exposure to methyl-mercury. In fish methylmercury may comprise 80 per cent or more of total mercury and it occurs primarily in muscles as a cystein complex (Craig, 1986b). The main source of exposure to mercury in human beings is the consumption of contaminated fish, shellfish and sea mammals (Bortoli *et al.*, 1995; Clarkson, 1998; Clarkson *et al.*, 2003). Methyl mercury through bio-magnification processes has commonly been found in prey fish at concentrations toxic to piscivorous birds and mammals; mono-methyl mercury has been most widespread source of Hg exposure and most commonly the result of consumption of contaminated foods, primarily the fish. Schober *et al.* (2003) have revealed that nearly 60,000 children each year are born at risk for neurological problems due to methyl mercury exposure in the womb. One in 12 U.S. women of childbearing age has potentially hazardous levels of mercury in their blood as a result of consuming fish. Mahaffey *et al.* (2004) calculated that blood total mercury levels > 3.5 µg/L in mothers could be associated with increased risk to the developing fetal nervous system. Fishes collected in India from The Bhaba Atomic Research Center area and from other beaches of Mumbai revealed presence of 0.1 to 0.15 ppm of Hg (USEPA., 1999), and nearly all fish contain some amount of methyl mercury. Effects on Wildlife MeHg concentrations generally increase with trophic level and with increased size and age of a given organism. Fish eventually sequester MeHg in skeletal muscle, reducing exposure to the central nervous system (Wiener *et al.*, 2002).

The intake of methyl mercury calculated in the study was 2.85 microgram MeHg/kg body weight/week which is much higher than the reference value suggested by FAO/WHO–JECFA, US, Canada and Japan. At East Fork Poplar Creek, Tannessee, USA, 250 tons of inorganic Hg was found released to water and soils from a USDA energy faculty in the 1950s and early 1960s (Gerlach *et al.*, 1995). In southern Florida a burgeoning population of invasive Burmese pythons measuring about 6 m in length, are said to be contaminated heavily with Hg. How these have become sinks for the toxic Hg, has been a point of serious concern for the scientists of the region. Long-term application of sulfur-rich agricultural fertilizers on the sugarcane fields of southern Florida has been reported to result in significant MeHg loadings to fish and birds downstream, within the Everglades wetland system (Bates *et al.*, 2002). Conversely, in their study of total mercury bioaccumulation in fish, the United States Geological Survey (U.S.G.S., 2001) found a significant negative correlation of mercury bioaccumulation with sulfate in water within 20 river basins nationwide. In addition to the abiotic processes of MeHg photo degradation in lakes (Seller *et al.*, 1996), these include production of HgS within soil pore water (Benoit *et al.*, 1999), and dissolution of HgS in the presence of humic and fulvic acids (Ravichandran *et al.*, 1998). Concern over potential human health and risk of consumption of mercury contaminated fish has led many nations to issue consumption advisories and limits for the consumption of fish and shell fish.

According to Jahed *et al.* (2005) seafood represents one of the major sources of non-occupational mercury exposure to human, as the MeHg readily crosses the placenta and the blood-brain barrier. Recently, US Food and Drug Administration (U.S.EPA., 2001) have recommended that pregnant women should not eat sea foods including shark, swordfish, king mackerel, and tile fish known to contain elevated levels of methyl mercury, an organic form of mercury. Shark, sword fish and king mackerel are reported to contain high levels of Hg. Mean Hg was found to be three times in sword fish and tuna of Canada than reported for US (Dabeka *et al.*, 2004). When mercury enters water it is often transformed by microorganisms into the toxic methyl mercury form. Symptoms of acute MeHg poisoning are pharyngitis, gasteroenteritis, vomiting, nephritis, hepatitis, and circulatory collapse. Generation, bioaccumulation, and bio-magnification of MeHg within aquatic systems has been studied for decades, following the identification of severe neurological and teratogenic impacts to humans associated with the consumption of fish contaminated with MeHg, such as in Minamata Bay in Japan in 1950s (Eisler 1987; Ninomiya *et al.*, 1995). As per USEPA., 2004 report MeHg is one of the most widespread waterborne contaminants. MeHg has been linked to potential reproductive and immune system effects in humans and wildlife (Wiener, 1996; Round *et al.*, 1998). MeHg contamination has more recently been portentously linked to long-range transport of emissions from fossil fuel combustion sources throughout the industrialized world (UNEP, 2002). Numerous laboratory studies have established a link between the ingestion of MeHg at sublethal levels and subtle visual, cognitive, and neurobehavioral deficits in small mammals (Wiener *et al.*, 2002). They also reported numerous reproductive and behavioral effects on wild avian and mammal populations associated with ingestion of realistic concentrations MeHg. In addition, controlled experiments with mink (*Mustela vison*) and otter (*Lutra canadensis*) established that dietary MeHg concentration of 1 mg/g leads to death in less than a year.

h. Nickel (Ni)

Nickel is a ubiquitous element and its compounds have long been classified as human carcinogen (IARC, 1990). Nickel is abundantly found in the environment and is primarily combined with oxygen (oxides) or sulfur (sulfides). It is found in all soils and is emitted from volcanoes. Nickel has properties that make it very desirable for combining with other metals such as iron, copper, chromium and zinc to form alloys. These alloys have important uses in the making of metal coins and jewelry and in industry for making items such as valves and heat exchangers. Most nickel is used to make stainless steel. Nickel compounds are used for nickel plating, to color ceramics, to make some batteries and as substances known as catalysts to increase the rate of chemical reactions. Nickel is also released into the atmosphere during nickel mining, by industries that convert scrap or new nickel into alloys or nickel compounds or by industries that use nickel and its compounds, as well as by oil-burning and coal-burning power plants and trash incinerators. The nickel that comes out of the stacks of the power plants is attached to small particles of dust that settle to the ground or are taken out of the air in rain. It usually takes many days for nickel to be removed from the air particularly if it is bound to very small particles. Acidic conditions render nickel more mobile in soil and may lead to seepage into groundwater. Exposure to nickel occurs through breathing nickel contaminated air or smoking tobacco, eating food (the major source of exposure for most people), drinking water, or even handling coins and other metals containing nickel. However, the major sources of nickel exposure are tobacco smoke, auto exhaust, fertilizers, super-phosphate, food processing, hydrogenated fats/oils, industrial wastes, stainless steel cookware, testing of nuclear devices, baking powder, dental work and bridges and combustion of fuel oil. Ni is also used in alloy metal plating, batteries and certain fungicides. It is usually present in food stuffs in low levels (1 mg/kg, Underwood, 1971), and up to 100 µg/l in the wine and bear (CEC, 1979). The burning of petroleum based fossil fuel in power plants results in the emission of a type of particulate matter termed as Residual oil fly ash (ROFA), containing relatively high quantities of transition metals, more frequently the Fe, Ni, and Vanadium metals, along with Ca and Mg, mostly in the form of soluble sulphates (Dreher *et al.*, 1997). In some instances the amounts of Ni reaches up to 35 per cent of the amount of all metals (Hamada *et al.*, 2002).

Nickel like other essential metals, is also required to maintain health in animals such as to produce red blood cells, its large doses however, have harmful effects. Short-term exposures to nickel further are known not to cause any health problems, but long-term exposure can cause decreased body weight, heart and liver damage, and skin irritation. Nickel can accumulate in aquatic life, but its presence is not magnified along food chains. Animal studies show that breathing high levels of nickel compounds may result in inflammation of the respiratory tract. The most common adverse health effect of nickel in humans is an allergic reaction. People exposed to Ni contamination at their work place may develop Ni-itching, nausea, dizziness, headache or chest pain, which are the usual common symptoms of Ni poisoning. Exposure to water soluble Ni rich ROFA has been found to be responsible for the majority of pulmonary diseases in humans (Dreher *et al.*, 1997; Kodavanti *et al.*, 1997). The EPA does not currently regulate nickel levels in drinking water. Its recommended dietary intake however, is about 300-400 µg/day (Dara, 1993). Ni is also known to affect gene regulation via several different mechanisms. It can induce gene silencing through epigenetic modulations of DNA and chromatin remodeling by affecting histone methylation and acetylation (Yan *et al.*, 2003) or modulation of various transcription factors, which result in the up-regulation and down regulation of specific genes (Salinikow *et al.*, 2000; Zhao *et al.*, 2004a). Ni is said to mimic hypoxia due to activation of hypoxia inducible factor-1 alpha (Zhao *et al.*, 2004b). Hypoxia signaling pathway is believed to be important in the initiation and progression of carcinogenesis. Ambient air polluting particulate matter is highly injurious to human health on the basis their elevated levels, epidemiologically linked to many respiratory and cardiovascular diseases (Pope *et al.*, 2002). Long term exposure to combustion related fine particles in air is considered a major environmental risk for even lung cancer mortality in humans. Salnikow *et al.*, 2004, were able to observe Ni induced hypoxia like stress and production of interleukins in human airway epithelial cells. In the rainbow trout, acute and chronic waterborne Ni exposure leads to marked renal Ni accumulation (Calamari *et al.*, 1982; Pane *et al.*, 2003, 2004 a,b). Acute Ni exposure causes a respiratory acidosis in the rainbow trout (Pane *et al.*, 2003), and given that the kidney is an important extra-branchial organ of acid-base regulation. Ishihara *et al.*, 2002 also reported acute bronchitis induced by 5-days inhalation of nickel chloride aerosols in Wister rats.

People can become sensitive to nickel when jewelry or other things containing this metal come into direct contact with the skin. Once a person is sensitized to nickel, further contact with it will produce a reaction. Symptoms include burning, itching, redness and bumps or other rashes. Eye or lung effects, including chronic bronchitis and reduced lung function, have been observed in workers who inhaled large amounts of nickel. High exposure can cause cough, shortness of breath and fluid in the lungs, which is sometimes delayed for 1 to 2 days after exposure. Exposure to Nickel can cause a sore or hole in the bone dividing the inner nose (septum). Nickel and certain nickel compounds may reasonably be anticipated to be carcinogens. Cancers of the lung, nasal sinus and throat have resulted when workers breathed dust containing high levels of nickel compounds while working in nickel refineries or nickel processing plants. Nickel may even damage the developing fetus. Nickel has been shown to cause lung cancer in animals. Additionally, epidemiologic studies have shown that nickel compounds cause nasal and lung cancers in exposed workers. Nickel-containing compounds induced tumors and enhanced tumor-genesis in various experimental animals and human cell culture via several different types of exposure (Kasprzak *et al.*, 1983; Kang *et al.*, 2003). DNA-protein cross-links and chromosomal aberrations were observed in mammalian cells in culture to which nickel compounds were added (Patierno *et al.*, 1985; Sen *et al.*, 1987). In addition, nickel causes oxidative damage to isolated DNA and chromatin, possibly due to the formation of reactive oxygen species (Kasprzak, 1991; Kang *et al.*, 2003; Kaur and Dani, 2003). Recently, several reports showed that nickel-induced various toxicological, physiological and histopathological alterations in different animal species (Pane *et al.*, 2003; Bersenyi *et al.*, 2004; Brix *et al.*, 2004; Doreswamy *et al.*, 2004; Gupta *et al.*, 2006).

i. Zinc (Zn)

Zinc is one of the most important trace elements in the body, and it participates in the biological function of several proteins and enzymes (Maity *et al.*, 2008). Excessive zinc enter the environment as a result of human activities such as mining, purification of zinc, lead and cadmium ores, burning of coal and burning of waste. Major industrial releases are from smelting, electroplating and metal manufacturing. Zinc is used in various forms which eventually find its way into the river or sea. It is also one of the most widely used metals in the world (DWAF, 1996). It occurs widely in nature as sulfides carbonates and hydrated silicate ores, and frequently accompanied by other metals like iron and cadmium (Alabaster and Lioyd, 1980). It is also frequently released in urban discharges, waste waters, fertilizers, wood treatment plants and incineration of municipal waste. Zn is used in numerous alloys and as a protective coating for other metals to prevent oxidation (rusting), and the Zn compounds are used in a variety of other consumer products and health aids, like dietary supplements, sun blocks, skin rash ointments, anti-dandruff shampoos, paints, batteries rat poisons *etc.* Vomiting, renal cell damage and cramps are common symptoms of Zn toxicity. High water temperature, pH and hardness enhances its toxicity effects, its combination with other metal like Cu also results in additive effect. Water with high amounts of Zn appears milky in color, and Zn deficiency may leads to rashes, diarrhea, and infections.

j. Antimony (Sb) and Barium (Be)

Antimony and Barium are other important metals which could also be active pollutants when present in excess in water. Antimony causes an increase in blood cholesterol and decrease in blood sugar levels. It is discharged from petroleum refineries, fire retardants, ceramics, electronics and iron solder. Barium causes excessive salivation, vomiting, diarrhea, paralysis, colic pain or increase in blood pressure. It is discharged from drilling wastes, from metal refineries and its erosion from natural deposits. The symptoms of toxicity resulting from chronic exposure (impaired cognitive, motor, and language skills; learning difficulties; nervousness and emotional instability; and insomnia, nausea, lethargy, and feeling ill are also easily recognized; however, they are much more difficult to associate with their cause.

REFERENCES

Al Rmalli S.W., Haris P.I., Harrington C.F. and Ayub M., 2005. A survey of arsenic in foodstuffs on sale in the United Kingdom and imported from Bangladesh. Sci. The Total Environ., 337(1–3):23–30

Alabaster J.S. and Lloyd R., 1982. Water quality: Criteria for Freshwater fish. London., Butterworths, 361pp.

Anderson, R.A., 1993. Recent advances in the clinical and biochemical effects of chromium deficiency. Prog. Clin. Biol. R., 380: 221-234.

Andrasi E., Pali N., Molnar Z. and Kosel S., 2005. Brain aluminium, magnesium and phosphorus contents of control and Alzheimer-diseased patients. J. Alzheimers Dis., 7(4):273-84.

Ariya P.A., Dastoor A.P., Amyot M., Schroeder W.H. and 7 others., 2004. The Arctic: a sink for mercury. Tellus series B- Chemical and physical meteorology, 56: 397-403.

ATSDR., 1999. Agency for Toxic Substances and Disease Registry, Top twenty Hazardous substances. Division of Toxicology, Mailstop E-29, Atlanta, GA.

Azcue J.M. and Nriagu J.O., 1994. In: Arsenic in Environment, Part 1: Cycling and characterization (ed. Nriagu J.O.), John Willey and Sons Inc. pp 1-15.

Backaret W.F., 1978. Mercury, lead, arsenic and cadmium in biological tissues. U.S.EPA., Office of Res. And Develop. Environmental monitoring and Suport La. Vegas, USA.

Baker E.L. (Jr.), Hayes C.G., Landrigan P.J., Handke J.L., Leger R.T., Housworth W.J. and J.M. Harrington., 1977. A Nationwide Survey of heavy metal absorption in children living near primary copper, lead, and zinc smelters. Am. J. Epidemiol., 106: 261–273.

Bates J.K., HarmonS.M., Fu T.T. and Gladden J.B., 2002. Tracing sources of sulfur in the Florida Everglades. J. Environ. Qual., 31: 287-299.

Becaria A, Campbell A, Bondy S.C. 2002. Aluminium as a toxicant. Toxicol and Health., 18(7):309-20.

Benoit J.M., Gilmour C.C., Mason R.P. and Heyes A., 1999. Sulfide controls on mercury speciation and bioavailability to methylating bacteria in sediment pore waters. Environ. Sci. and Technol., 33: 951-957.

Bersenyi A., Gy Fekete S., Szilagyi M., Berta E., Zoldag L. and Glavits R., 2004. Effects of nickel supply on the fattening performance and several biochemical parameters of broiler chickens and rabbits. Acta. Vet. Hungarica, 52: 185-197.

Bolt H.M., Hengstler J.G., 2008. Aluminium and lead toxicity revisited: mechanisms explaining the particular sensitivity of the brain to oxidative damage. Arch Toxicol., 82(11):787-8.

Borgono J.M., Vicent P., Venturino H.and Infante A., 1977. Arsenic in the drinking water of the city of Antofagasta: epidemiologica and clinical study before and after the installation of the treatment plant. Environ. Health Perspect., 19: 103-105.

Bortoli C. de, Chailley-Heu B. and Bourbon J.R., 1995. Production of transforming growth factor (TGF) beta by fetal lung cells. Biol. Cell, 84: 215–218.

Bowell R.J. 1994. Sorption of arsenic by iron oxides and oxyhydroxides in soils. Appl. Geochem., 9: 279-286.

Brix H., Gruber N. and Keeling C.D., 2004. Interannual variability of the upper carbon cycle at station ALOHA near Hawaii, Global Biogeochem. Cy., 18: GB-4019.

Bury N.R., Walker Paul A. and Glover Chris N., 2003. Nutritive metal uptake in teleost fish. J Exp Biol., 206: 11-23.

Calamari D., Gaggino F.F., and Pacchetti G., 1982. Toxico-kinetics of low levels of Cd, Ni and their mixure in long terms treatment on Salmo gaidneiri. Rich Chemosphere, 11: 59-70.

Campbell A., 2002. The Potential role of aluminium in Alzheimer's disease. Nephrology Dialysis Transplantation., 17 (suppl 2):17-20.

CEC., 1979. Commission of the European Communities Trace Metals: Exposures and health effects. Oxford, Pergamon Press.

Chakraborti D., Mohammed M.R., Paul K., Chowdhury V.K., Sengupta M.K., Lodh D., Chanda C.R., Saha K.C. and Mukerjee S.C., 2002. Arsenic calamity in the Indian subcontinent. What lesson have been learned ? Talanta., 58 : 3-22.

Chakraborti D., Sengupta M.K., Rahman M.M., Choudhary U.K., Lodh D. and 5 others., 2003. Ground water arsenic exposure in India. In: Arsenic exposures and health effects V. (Chappell W.R., Abernathy C.O., Calderon R.L.and Thomas D.J., Eds).1-24pp. Elsevier, The Nethyerlands.

Chatterjee A. and Banerjee R.N., 1999. Determination of lead and other metals in a residential area of greater Calcutta. Sci. Total Environ., 9: 173-185.

Chen C.J., Hsueh Y.M., Tseng M.P, Lin Y.C. and others., 2001. Individual susceptibility to arseniasis. In: Chappell WR, Abernathy CO, Calderon RL, editors. Arsenic exposure and health effects IV. Oxford, UK: Elsevier. p 135–143.

Choprapawan C. and Rodcline A., 1997. In: Arenic: Exposure and Health Effects (eds. Abernathy C.O. and others). Chapman and Hall UK: 69-77.

Clarkson T.W., 1990. Human health risks from methylmercury in fish. Environ. Toxicol. and Chem., 9 (7):957–961,

Clarkson T.W., 1998. Human toxicology of mercury. J. Trace Ele. Exp. Med., 11: 303-317.

Clarkson T.W. and Strain J.J., 2003., Nutritional Factors May Modify the Toxic Action of Methyl Mercury in Fish-Eating Populations. J. Nutr.,133:1539–1543.

Cohen M.D., Kargacin B., Klein C.B. and Costa M., 1993. Mechanisms of chromium carcinogenicity and toxicity. Crit. Rev. Toxicol., 23 : 255-281.

Counter S. A., and Buchanan L. H., 2004. Mercury exposure in children: a review. Toxicol. Appl. Pharmacol., 198: 209-230.

Craig P.J., 1986a. Ocurance and pathways of organometallic compounds in the environment, general considerations. In: Organometallic compounds in the environment, Principles and reactions, (Ed. P.J. Craig), Longman, Essex. UK, 368pp.

Craig P.J., 1986b. Organomercury compounds in the environment, Ch.-II. In: Organometallic compounds in the environment, Principles and reactions, (Ed. P.J. Craig), Longman, Essex. UK, 368pp.

Cizick J., Sasieni P., Evans S., 1992. Ingested arsenic, keratosis and and bladder cancer. Am. J. Epidemiol., 136: 416-417.

Dabeka R., McKenzie A.D., Forsyth D.S. and Conacher H.B.S., 2004 Survey of total mercury in some edible fish and shellfish species collected in Canada in 2002. Food Additives and Contaminants, 21(5): 434-440.

Dara S.S. 1993. A Textbook of "Environmental Chemistry and Pollution Control". S. Chand and Co. Ltd., Ram Nagar, New Delhi. 210 pp.

David Dunn and Thomas Tisue, 1989. Cadmium-113m in the water, sediments, and adjacent soils of Lake Michigan. Science of the Total Environ., 87-88: 305-314.

Davidson T., Thomas K., Fredric B. and 5 others., 2004. Exposure of chromium (VI) in the drinking water increases susceptibility to UV-induced skin tumors in hairless mice. Toxicol. Appl. Pharmacol., 196: 431-437.

Dhar R.K., Biswas B.K., Samanto G., Mandal B.K., Chakraborti D., and 8 others., 1997. Groundwater arsenic calamity in Bangladesh. Cirr. Sci. 73: 48-59.

Doreswamy K., Shrilatha B., Rajeshkumar T. and Muralidhara., 2004. Nickel-induced Oxidative Stress in Testis of Mice: Evidence of DNA Damage and Genotoxic Effects. J. Andrology, 25(6):996–1003.

Dreher K. L., Jaskot R. H., Lehman J. R., Richards J. H., and 3 others., 1997. Soluble transition metals mediate residual oil fly-ash induced acute lung injury. J. Toxicol. Environ. Health, 50: 283-305.

Durum W.H., Hem J.D. and Heidel S.G., 1971. Reconnaissance of selected minor elements in surface waters of the United States, October 1970, Washington, DC, US Department of Interior, Geological Survey Circular 643.

DWAF., 1996. Department of water affairs and forestry. South African Water Quality Guidelines – Second Edition. Volume 7: Aqua. Ecosys., 159 pp.

Eisler R., 1987. Mercury hazards to fish, wildlife, and invertebrates: a synoptic review. U.S. Fish and Wildlife Service Biological Report 85.

Eisler R., 1988. Lead hazards to fish, wildlife, and invertebrates: asynoptic review. US Fish Wildlife Serv. Biol. Rep., 85: 1–4.

Ferner D.J., 2001. Toxicity, heavy metals. eMed. J., 2(5):1.

Forstner U. and Wittaman G.T.U., 1979. Metal pollution in aquatic environment, Springer Verlag, New York, 486-532.

Fujiki M., 1972. The transitional conditions of Minimata Bay and the neighouring sea polluted by factory waste water containing mercury. 6th Intl. Water Pollut. Conf., Jerusalem: 902-917.

Fishera Ian J., Paina Deborah J. and Thomasb Vernon G., 2006. A review of lead poisoning from ammunition sources in terrestrial birds. Biol. Conser., 131: 421– 432.

Gerlach C.L., Dobb D., Miller E., and 5 others, 1995. Characterization of mercury contamination at the East Fork Poplar Creek, site Oak Ridge, Tannessee- a case study. National Service Center for Environmental Publications (NSCEP). U.S.EPA., North Carolina, EPA/600/R-95/110: 1-38 pp.

Gilbert-Barness E., Barness L.A., Wolff J. and Harding C., 1998. Aluminium toxicity. Arch Pediatr Adolesc Med., 152(5):511-2.

Gnamus A., Bryne A.R. and Horvat M., 2000. Mercury in the soil-plant-deer-predator food chain of a temperate forest in Slovenia. Environ. Sci. and Technol., 34: 3337-3345.

Govani S. and Memo M., 1979. "Chronic lead treatment differentially affects dopamine synthesis". Toxicology, 12: 343-49.

Goyer R.A., 1996. Results of lead research: prenatal exposure and neurological consequences. Environ Health Perspect, 104(10):1050–1054.

Goyer R.A., 1997. National Institute of Environmental Health Sciences. Toxic and essential metal interactions. Annu Rev Nutr.,17: 37-50.

Guentzel J.L., Portilla E., Keith M.K. and Keith E.O. 2007. Mercury transport and bioaccumulation in riverbank communities of the Alvarado Lagoon System,Veracruz State, Mexico. Sci. Total Environ., 338:(1–3): 316–324.

Guha-Mazumder D.N., Haque R., Ghosh N., De B. K., Santra A., Chakaraborty D. and Smith A. H., 1997. In: Arenic: Exposure and Health Effects (eds. Abernathy C.O. *et al.*) Chapman and Hall UK, 113-123.

Guha-Mazumder D. N., Haque R., Ghosh N., De B. K., Santra A., Chakaraborty D. and Smith A. H., 1998. Arsenic levels in drinking water and the prevalence of skin lesions in West Bengal, Inda. Int. Epidemiol. Assoc., 17:871-877.

Guo X., Fujino Y., Keneko S., Wu K., Xia Y. and Yosimura T., 2001. Arsenic contamination of ground water and prevalence of arsenical dermatosis in the hetao plain area, Inner Mongolia, China. Mol. Cell. Biochem., 222: 137-140.

Gupta N., Gupta D.K., Verma V.K. and Krishna G., 2006. Zinc-induced changes on chromosomes of a freshwater teleost, *Heteropneustes fossilis* (Bloch). Asian J. Exp. Sci., 20: 281-288.

Hamada K., Goldsmith C. A., Suzaki Y., Goldman A. and Kobzik, L., 2002. Airway hyper-responsiveness caused by aerosol exposure to residual oily fly-ash leachate in mice. J. Toxicol. Environ. Health, 65: 1351-1365.

Hammerschmidt C.R., Sandheinrich M.B., Weiner J.G. and Rada R.G., 2002. Effects of dietary methylmercury on reproduction of flathead minnows. Environ. Sci. and Technol., 36: 877-883.

Hayashi Y., Kondo T., Zhao Quang-Li, Ogawa R. and 4 others., 2004. Singnal transduction of p53-independent apoptotic pathway induced by hexavalent chromium in U937 cells. Toxicol. Appl. Pharmacol., 197: 96-106.

IARC, 1990. IARC monographs on the evaluation of carcinogenesis risks to humans, chromium, nickel and welding, Vol 49, IARC, Lyon, France.

I.O.S. and H.I. Center, 1999. International Occupational Safety and Health Information Center. Heavy metal toxicity. São Paulo, Brazil. (www.lifeextensionvitamins.com/hemeto)

Ireleaven J., Meller S., Farmer P., Birchal D., Goldman J. and Piller G., 1993. Lymphoma, 10: 343-345.

Ishihara Y., Kyono H., Serita F., Toya T., Kawashima H. and Miyasaka M., 2002. Inflammatory responses and mucous secretion in rats with acute bronchilitis induced by nickel chloride. Inhal. Toxicol., 14: 417-430.

Ishizaki A., 1965. Observation on urinary and faecal excretion of heavy metals in the patient of the so called Itai-itai disease. Japan J. Hyg., 20, 261-267.

Jahed Khaniki, Gholam Reza, Alli Inteaz, Nowroozi Ebrahim and Nabizadeh Ramin, 2005. Mercury Contamination in Fish and Public Health Aspects: A Review. Pakistan J. Nutrit., 4 (5): 276-281.

Jain K.L. and Mittal V., 2004. Heavy metal pollution in surface water bodies and its impact on fishes. Nat. Worksh. "Rational use of water resources for aquaculture", Deptt. Zool. and Aqua, CCS HAU,Hisar (March 18-19, 2004): 196-202.

Jiang H.X., Chen L.S., Zheng J.G., Han S., Tang N. and Smith B.R., 2008. Aluminium-induced effects on Photosystem II photochemistry in citrus leaves assessed by the chlorophyll a fluorescence transient. Tree Physiol., 28(12):1863-71.

Kalra V., 2004. Mercury in India: Toxic Pathways. Toxic Links.

Kang J., Zhang Y., Chen J., Chen H., Lin C., Wang Q. and Ou Y., 2003. Nickel-induced histone hypoacetylation: the role of reactive oxygen species. Toxicol. Sci., 74: 279-286.

Kasprzak A.A., Papas E.J. and Steenkamp D.J., 1983. Biochem. J., 211: 535-541.

Kasprzak K.S., 1991.The role of oxidative damage in metal carcinogenicity. *Chem. Res. Toxicol.*, 4 (6):604–615.

Kaur P. and Dani H.M., 2003. Carcinogenicity of nickel is the result of its binding to RNA and not to DNA. J. Environ. Pathol. Toxicol. Oncol., 22: 29-39.

Kennedy S., Crisler J.P., Smith E. and Bush M., 1977. Lead poisoning in sandhill cranes. J. Am. Vet. Med. Assoc., 171, 955–958.

Klassen C.D., 1991a. Heavy metals and heavy metal antagonists. The pharmacological basis of therapeutics. 8th ed. Vol. II. Pergamon Inc., Singapore. 66: 1592-1614.

Klassen C.D., 1991b. Principles of toxicology. In: Gilman, A.G., Tall, T.W., Nies, A.S., Taylor, P. (Eds.), Pharmacological Basis of Therapeutics, 8th ed. McGraw Hill, pp. 49–61.

Klinghardt D., 2010. On effective holistic heavy metal detoxification (Chelation). © Healing Cancer Naturally, September 2007.

Kodavanti U.P., Jaskot R.H., Costa D.L. and Dreher K.L., 1997. Pulmonary pro-inflammatory gene induction following acute exposure to residual oil fly ash: roles of particle-associated metals. 9(7):679-701.

Kumar Arun, Yang F., Goddard L. and Schubert S., 2004. Differing Trends in the Tropical Surface Temperatures and Precipitation over Land and Oceans. J. Climate, 17:653–664.

Languard S. and Norseth T., 1979. Chromium. In Handbook on the Toxicology of Metals. (L. Friberg, G. F. Nordberg and V. Vouk, eds.), pp. 383-397. Elsevier/ North-Holland Biomedical Press, Amsterdam.

Lansche A.M., 1965. Arsenic. In: Mineral facts and problems, Washington, DC, US, Department of the Interior (Bureau of Mines. Bulletin 630).

Larsson A., 1975. Some biochemical effects of Cd on fish: In Koeman, J.H. and JITWA Strik (eds). Sublethal effects of toxic chemicals on aquatic animals. Elsevier Scient. Pub. Co. Amsterdam., 3-13.

Larsson J.E., 1970. Environmental mercury research in Swedon, Stockholm. National Swedish Environmental protection board.

Laurier F.J.G., Mason R.P., Whalin L., Kato K., 2003. Reactive gaseous mercury formation in the North Pacific Ocean's marine boundary layer: A potential role of halogen chemistry. J. Geophys. Res.: Atmospheres (1984–2012),108 (D17): 16

Lenvik K., Steinnes E. and Pappas, A. C., 1978. Contents of some heavy metals in Norwegian rivers. Nord. Hydrol., 9: 197-206.

Liao C.M., Chen B.C., Singh S., Lin M.C., Liu C.W. and Han B.C., 2003. Acute toxicity and bioaccumulation of arsenic in tilapia (*Oreochromis mossambicus*) from a blackfoot disease area in Taiwan. Environ. Toxicol., 18, 252–259.

Lodenius M. and Tulisalo E., 1984. Environmental mercury contamination around a Finnish chlor-alkali plant. Bull. Environ. Contam. Toxicol., 32: 439-444.

Lunde G., 1970. Analysis of trace elements in seaweed. J. Sci. Food and Agri., 21(8): 416–418.

MacArthur J.M., Ravenscroft P., Safiullah S. and Thilwall M.F., 2001. Arsenic in groundwater; testing pollution mechanisms for sedimentary acquifers in Bangladesh. Water Res. Res., 37: 109-117.

Mahaffey K.R., Clickner R.P., Bodurow C.C., 2004. Blood organic mercury and dietary mercury intake: National Health and Nutrition Examination Survey, 1999 and 2000. Environ. Health Perspect., 112:562–570.

Maity S., Roy S., Chaudhury S., Bhattacharya S., 2008. Antioxidant responses of the earthworm Lampito mauritii exposed to Pb and Zn contaminated soil. Environ. Pollut., 151 : 1–7.

Mallick S. and Rajagopal N.R., 1996. Groundwater development in the arsenic affected alluvial belt of West Bengal. Curr. Sci., 70; 956-958.

Mandal B.K., Chaudhary G., Samanta G., and 3 others., 1998. Impact of safe water for drinking on five families for 2 years in West Bengal, India. Sci. Total Environ., 218: 185-201.

Mason R.P., Fitzgerald W.F. and Morel F.M.M., 1994. The biogeochemical cycling of elemental mercury. Geochimica et Cosmochimica Acta., 58(15); 3191-3198.

Merz W., 1993. Chromium in human nutrition: a review. J. Nutr., 123: 626-633.

Misra B.B., 1986. Role of blue green algae in bio-degradation of chlor-alkali waste (soil) with special reference to rice cultivation. Ph. D Thesis, Berhampur Univ., Orissa, India.

Mitra A.K., Bose B.K., Kabir H. Das B.K. and Hossain M., 2002. Arsenic-related health problems among hospital patients in southern Bangladesh. J. Health Popul Nutr., 20:198-204.

Mitra S., 1986. In: Mercury in the ecosystem. Trans. Tech. Publ. Ltd. Switzerland, 51.

NADP., 2006. National Atmospheric Deposition Programme- 2006. Effect of deposition in Coastal and Urban Environments, 24-26 Oct., Norfolk, Virginia.

Nagorski S.A., McKinnon T.E. and Moore J.N., 2003. Season and storm-scale variations in heavy metal concentration of two mining-contaminated strems, montana, USA. J. de Physique IV, 107: 909-912.

NAS., 1977. Medical and biologic effects of environmental pollutants: Arsenic, Washington, DC, National Academy of Sciences.

Needleman H.L., Schell A., Bellinger D., Leviton A. and Allred E.N., 1990. The long term effects of exposure to low doses of lead in childhood. J Med., 322:83.

Newman M.C. and Unger M.A., 2003. Fundamentals of ecotoxicology. New York: Levis.

Ninomiya T., Ohmori H., Hashimoto K., Tsuruta K. and Ekino S., 1995. Expansion of methylmercury poisoning outside of Minamata: An epidemiological study on chronic methylmercury poisoning outside Minamata. Environ. Res., 70: 47-50.

NRC., 2000. National Research Council, Toxicological Effects of Methylmercury, National Academy Press, Wash, DC.

Nriagu J.O., 1990. "Global Metal Pollution- Poisoning the Biosphere", Environment, 32: 7-33.

Ogwuegbu M.O.C. and Muhanga W., 2005. Investigation of lead concentration in the blood of people in the copper belt province of Zambia. J. environ. (1): 66-75.

Omura Y., Shimotsuura Y., Fukuoka A., Fukuoka H., Nomoto T., 1996. Significant mercury deposits in internal organs following the removal of dental amalgam. Acupunct Electrother Res., 21(2):133-60.

Paikaray S. Banerjee S. and Mukherjee S., 2005. Sorption of arsenic onto Vidhyan shales: Role of pyrite and organic carbon. Curr. Sci., 88: 1580-83.

Pain D.J., 1995. Lead in the environment. In: Hoffman, D.J., Rattner, B.A., Burton, G.A., Jr., Cairns, J., Jr. (Eds.), Handbook of Ecotoxicology. CRC Press Inc., Boca Raton, pp. 356–391.

Palaniappan P. and Vijayasundaram V., 2008. FTIR study of arsenic induced biochemical changes on the liver tissue of freshwater fingerlings, *Labeo rohita*. Romanian J. Biophy., 18; 135-144.

Pane E.F., Haque A., Goss G.G. and Wood C.M., 2004a. The physiological consequences of exposure to chronic, sub lethal water borne nickel in rainbow trout (*Oncorhynchus mykiss*); exercise vs. resting physiology. J. Exp. Biol., 207: 1249-1261.

Pane E.F., Haque A. and Wood C.M., 2004b. Mechanistic analysis of acute, Ni-induced respiratory toxicity in the rainbow trout (*Oncorhynchus mykiss*): an exclusively branchial phenomenon. Aquat. Toxicol., 69(1):11-24.

Pane E.F., Richards J.G. and Wood C.M., 2003. Acute waterborne nickel toxicity in the rainbow trout (*Oncorhynchus mykiss*) occurs by a respiratory rather than iono-regulatory mechanism. Aquat. Toxicol., 63: 65-82.

Patierno S.R., Sugiyama M., Basilion J.P. and Costa M., 1985. Preferential DNA-protein crosslinking by $NiCl_2$ in magnesium-insoluble regions of fractioned Chinese Hamster ovary cell chromatin. Cancer Res., 45: 5787-5794.

Pattee O.H. and Pain D.J., 2003. Lead in the environment. In: Hoffman, D.J., Rattner, B.A., Burton, G.A., Jr., Cairns, J., Jr. (Eds.), Handbook of Ecotoxicology. CRC Press Inc., Boca Raton: 373–408.

Penrose W.R., Conacher H.B., Black R., Méranger J.C., Miles W., Cunningham H.M. and Squires R.W., 1977. Implications of inorganic/organic interconversion on fluxes of arsenic in marine food webs, Environ. Health Perspect., 19: 53-59.

Peyster A. and Silvers J.A., 1995. Arsenic levels in hair of workers in a semiconductor fabrication faculty. Am. Ind. Hyg. Assoc. J., 56: 377-383.

Pope C.A. III., Burnett R.T., Thun M.J., Calle E.E., Krewski D., Ito K. and others., 2002. Lung cancer, cardiopulmonary mortality, and long-term exposure to fine particulate air pollution. JAMA., 287:1132–1141.

Powers K.M., Smith-Weller T., Franklin G.M., Longstreth, W.T. Jr., Swanson P.D. and Checkoway H., 2003. Parkinson's disease risks associated with dietary iron, manganese, and other nutrient intakes. Neurology, 60(11):1761-1766.

Quentin K.E. and Winkler H.A., 1974. Occurrence and determination of inorganic polluting agents. Zentralbl. Bakteriol. (Orig. B),158: 514-523.

Rahman M., 2002. Arsenic and Contamination of Drinking-water in Bangladesh: A Public-health Perspective. J. Health. Population Nutrition, 20 (3):193 – 197.

Rahman M., Wingren G. and Axelson O., 1996. Diabetes mellitus among Swedish art glass workers- An effect of arsenic exposure. Scand. J. Work, Environ. Health, 22: 146-149.

Ravichandran M., Aiken G. R., Ryan J. N. and Reddy M. M., 1999. Enhanced dissolution of cinnabar (Mercuric sulfide) by dissolved organic matter isolated from the Florida, Everglades. Environ. Sci. and Technol., *33*: 1418.

Reiser S., Powell A., Yang C.Y. and Canary J.J., 1987. Effect of Cu intake on blood cholesterol and in lipoprotein distribution in men. Nutr. Rep. Int., 36: 641-647.

Rencher A.C., Carter M.W. and McKee D.W., 1977. A retrospective epidemiological study of mortality at a large western copper smelter. J. occup. Med., 19: 754-758.

Roberts 1999. Reduced Cadmium Content in Cocoa beans.,31: 635-636.

Roulet M., Lucotte M., Canuel R., Rheualt I., Tran S. Goch Y.G.D. and 6 others., 1998. Distribuition and partition of total mercury in waters of the Tapajós river Basin, Brazilian Amazon. Sci. Total Environ., 213:203-211.

Round M., Marin A., Alter L., Tatsutani M., 1998. Mercury deposition in the Northeast. In: M Tatsutani (ed.) Proceedings, Northeast states and eastern Canadian provinces mercury study. NESCAUM *et al*.: VI1 - VI42.

Saas O., Grosshans E. and Simonart J.M., 1993. Chronic Arsenicism: Criminal Poisoning or Drug-Intoxication? Dermatology, 186: 303-305.

Saha K.C., 1995. Chronic arsenical dermatoses from tubewell water in West Bengal during 1983-87. Indian J. Dermatol., 40:1-12.

Saffi S.A., 1981. Distillery waste toxicity of metabolic dysfunctioning in nine freshwater teleost. Toxicol. Lett. 8:179-186.

Salinikow K., Davidson T., Zhang D., Chen L.C., Su W., Costa M., 2003. The involvement of hypoxia-inducible pathway in nickel carcinogenesis. Cancer Res., 63: 3524-3530.

Salnikow K., Li X. and Lippmann M., 2004. Effect of nickel and iron co-exposure on human lung cells. Toxicol. Appl. Pharmacol., 196: 258-263.

Schober S.E., Sinks T.H., Jones R.L., Bolger P.M., and others., 2003. Blood mercury levels in U.S. children and women of childbearing age, 1999–2000. JAMA, 289:1667–1674.

Seigneur C.K., Karamchandani P., Vijayaraghavan K., Lohman K. and Yelluru G., 2003. Scoping study for mercury deposition in the upper Midwest. San Ramon, CA: Atmospheric and Environmental Research, Inc.

Seller P., Kelly C.A., Rudd J.W.M. and MacHutchon A.R., 1996. Photodegradation of methylmercury in lakes. [Abstract] Nature 380: 694-697.

Sen Promila, Conway K. and Costa Max., 1987. Comparison of the Localization of Chromosome Damage Induced by Calcium Chromate and Nickel Compounds. Cancer Res., 47: 2142-214.

Sharma V. K. and Sohn M., 2009. Aquatic arsenic: Toxicity, speciation, transformations, and remediation. Environ. Internat., 35: 743–759.

Silva-Forsberg M.C., Forsberg, B.R., Zeideman, V.K., 1999. Mercury contamination in Humans Linked to River Chemistry in The Amazon basin. Royal Swedish Acad. Sci., 28: 519-521.

Singh A.K., 2006. Chemistry of Arsenic in Groundwater of Gange-Brahmaputra river basin. Curr. Sci., 91: 599-.

Singh S., 2001. A physiologically based pharmacokinetic and pharmacodynamic model for arsenic accumulation in aquacultural fish from black foot disease area in Taiwan. Unpublished PhD dissertation, National Taiwan University.

Smith A. H., Arroyo A. P., Mazumder D.N.G., Konsnett M. J. and 4 others., 2000. Arsenic-induced skin lesions among Alacameno people in Northern Chile despite good nutrition and centuries of exposure. Environ. Helth Prospect., 108: 617-620.

Suzuki Y., Suzuki,Y., Fujii N and Mouri T., 1974. Environmental contamination around a smelter by arsenic. Shikoku Igaku Zasshi, 30: 213-218.

Takeuchi T., 1968. In: Pathology of Minamata diseases: Study group of Minamata diseases, Japan, 11: 141.

Temme C., Slemer F., Ebinghaus R. and Einax W., 2003. Distribution of mercury over the Atlantic ocean in 1996 and 1999-2001. Atmos. Environ., 37(14): 1889-1807.

Thomas L.D., Hodgson S., Nieuwenhuijsen M. and Jarup L., 2009. Early kidney damage in a population exposed to cadmium and other heavy metals. Environ. Health Perspect., 117; 181-184.

Tondel M., Rahman M., Magnuson A., Chaudhary L. A., Faraqee M.H. and Ahmed S.A., 1999. The relationship of arsenic levels of drinking water and the prevalence rate of skin lesions in Bangladesh. Environ Health Propect., 107: 727-729.

Tseng C.H., 2002. An overview on peripheral vascular disease in blackfoot disease-hyperendemic villages in Taiwan. Angiology, 53: 529-537.

Tseng C.H., 2004. The potential biological mechanisms of arsenic-induced diabetes mellitus. Toxicol. Appl. Pharma., 197: 67-83.

Tseng W.P., 1977. Effects and dose-response relationships of skin cancer and Blackfoot disease with arsenic. Environ. Health Perspect., 19: 109-119.

Tsuda T., Babazono A., Yamamoto E., Kurumatani N., Mino Y., and 3 others., 1995. Ingested arsenic and internal cancer: a historical cohort study followed for 33 years. Am. J. Epidemol., 141: 198-209.

Underwood E.J., 1971. Trace elements in human and animal nutrition. 3rd ed. Acad. Press. NY. 208 pp.

UNEP., 2002. United Nations Environment Programme, Environment for development. Global Mercury Assessment Report, December 2002. *www.unep.org › Mercury.*

U.S.EPA., 1987. Interim practices for estimating risks associated with exposers to mixtures of chlorinated dibenzo-p-dioxins and dibenzofuranes (CDDs and CDFs). EPA/625/3-87/012. Risk assessment Forum, Washington DC.

U.S.EPA., 1989. Interim practices for estimating risks associated with exposers to mixtures of chlorinated dibenzo-p-dioxins and dibenzofuranes (CDDs and CDFs) and 1989 update. EPA/625/3-87/016. Risk assessment Forum, Washington DC.

U.S.EPA., 1995. U.S. Environmental Protection Agency, Hazardous Air Pollutant Hazard Summary Fact Sheets, EPA: In Risk Information System.

U.S.EPA., 1997. United States Environmental Protection Agency. Mercury study report to Congress. Volume VI: An ecological assessment for anthropogenic mercury emissions in the United States. http://www.epa.gov/ttn/oarpg/t3/reports/volume6.pdf

U.S.EPA., 1999. Guidelines for Carcinogen Risk Assessment Review draft. NCEA-F-0644, Jul., 1999.

U.S.EPA., 2001. United States Environmental Protection Agency. Water quality criterion for the protection of human health: methylmercury. http://www.epa.gov/waterscience/criteria/methylmercury/merctitl.pdf

Valentine J.L., He S.Y., Reishoro L.S., Lachenbruch P.A., 1992. Health response by questionnaire in arsenic-exposed populations. J. Clin. Epidemiol., 45: 487-494.

Venkataramana P. and Radhakrishnaiah K., 1987. Lethal and sub-lethal effects of copper on the protein metabolism of the freshwater fish, *Labeo rohita* (Ham.). Trends Life. Sci., 2(2): 82-86.

Venkataramani S.H., 1987. Slow poisoning. India Today, 144-149.

Verstraeten S.V., Aimo L. and Oteiza P.I., 2008. Aluminium and lead: molecular mechanisms of brain toxicity. Arch Toxicol., 82(11):789-802.

Viessman W. and Hammer M., 1985. Water Supply and Pollution Control. 4th Edn., Harper and Row Publishers, New York: 797pp.

Ward M.K., Feest T.G., Ellis H.A., Parkinson I.S. and Kerr D.N., 1978. Osteomalacic dialysis osteodystrophy: Evidence for a water-borne aetiological agent, probably aluminium. Lancet, 1(8069):841-5.

Weis P. and Weis J.S., 1999. Accumulation of metals in consumers associated with chromatic copper arsenate-treated wood penals. Mar. Environ. Res., 48: 73-81.

WHO., 1977. Environmental Health Criteria. 3- Lead. World Health Organization, Geneva.

WHO., 1981. Arsenic. World Health Org. Geneva.

WHO., 1993. Guidelines for drinking water quality. WHO Geneva.

WHO., 1998. Aluminium. Guide lines for drinking-water quality, 2nd addition, Health criteria and other supporting information (1998), Ganeva,:3-13.

Wiener J.G., 1996. Mercury in fish and aquatic food webs. In: Abstracts for presentations given at the USGS workshop on mercury cycling in the environment. (http://toxics.usgs.gov/pubs/hg/abstracts.html).

Wiener J.G., Drabbenhoft K.P., Heinz G.H., Scheuhammer A.M., 2002. Ecotoxicology of mercury. In: DJ Hoffmann, BA Rattner, GA Burnton Jr., and J Cairns Jr., eds., Handbook of ecotoxicology. Boca Raton, FL: CRC Press: 61 p. Chapter 16: 409-464.

Windham B., 1999. Annotated Bibliography: Health Effects Related to Mercury from Amalgam Fillings and Documented Clinical Results of Replacement of Amalgam Fillings" 1999.

Yan Y., Khuz T., Zhang P., Chen H. and Costa M., 2003. Analysis of specific lysine histone H3 and H$ acetylation and methylation status in clones of cells with a gene silenced by nickel exposure. Toxicol. Appl. Pharmacol., 190: 272-277.

Yan-Chu H., 1994. Arsenic distribution in soils. In: Arsenic in the environment. Part 1 Cycling and characterization.(Ed.J.O. Nriagu). John Wiley and Sons, INC. New York.

Yokel R.A. and McNamara P.J., 2001. Aluminium toxicokinetics: an updated mini review. Pharmacol Toxicol., 88(4):159-67.

Yoshida T., Yamauchi H. and Sun G. F., 2004. Chronic health effects in people exposed to arsenic via the drinking water: dose-response relationships in review. Toxicol. Appl. Pharmacol., 198: 243-252.

Younger P.L., Banwart S.A. and Hedin R.S., 2002. Mine water: Hydrology, pollution, remediation, Kluwer Acad. Publ., Netherlands.

Zhao J., Yan Y., Salnikow K., Kluz T. and Coata M. 2004a. Nickel induced down-regulation of serpin by hypoxic signaling. Toxicol. Appl. Pharmacol., 194910: 60-8.

Zhao J., Chen H., Davidson T., Thomas K., Zhang Q. and Costa M., 2004b. Nickel induced 1-4 alpha-glucan branching enzyme I up-regulation via the hypoxic signaling pathway. Toxicol. Appl. Pharmacol., 196: 404-409.

Zhavaronkov N.I., Makhno P.M. and Titova I.M., 1987. Epithelioma disease of salmon associated with a high content of heavy metals in tissues. Veterin. Moscow., 9: 56-59.

Zima T., · Mestek O., Tesar V., Tesarová P., Nmecek K., Zák A. and Zeman M., 1998. Chromium levels in patients with internal diseases. Biochem. and Molecular Boil. Internat., 46(2):365-74.

Chapter 4

Heavy Metal Contaminations in River Waters and Sediments

The vulnerability of aquatic habitats to heavy metal contamination with the increasingly discharge of industrial effluents and industrial wastes is now well established. Expeditious rate of industrialization and urbanization during the last few decades has left an indelible mark of destruction of many of our vital aquatic ecosystems (Walters, 1981). The Himalaya Rivers, such as 'The Gange and Yamuna' the major contributor to inland capture fisheries of the country, are showing declining trend in fish production and also a shift in diversity of the native fish population. Das *et al.* (2007) have stated that the 14 major rivers including 'The Gange and The Godawari basins' cover about more than 85 per cent of the surface flow, and harbour one of the richest fish genetic resources in the world. The Gangetic system alone harbours arround 265 species of fish. The fish catch statistics, however, had revealed highly disturbing trends in this riverine sector over the past few decades. The total fish catch in 'The Gange basin' alone has declined from 85.2 tones in 1959 to 65.5 tones in 2004; attributing to water contamination due to fast industrilization, agricultural run-off and discharge of domestic sewage and other wastes into these rivers. According to Tandon and Dhaman (1967) who had earlier reported 56 species of fish occurring in Budha Nallah, a tributary of the river Satluj in the Ludhiana city of Punjab in India, have revealed today 7 species missing in these waters; and not even single species of fish was said to exist in this Nallah in the vicinity heavily polluted with heavy metals (Kaur *et al.*, 2000). Ludhiana is known for large number of industries for manufacturing bicycle parts, Nickel–Chrome plating, dyeing and woolen hosiery *etc.*, and releasing the heavy metal chromium in the effluents especially from chrome plating and chrome tanning industries thus, polluting river water heavily with such heavy metal rich wastes and effluents.

Crandall and Goodnight (1962) have suggested that heavy metals in low doses produce chronic intoxication and occurrence of this chronic toxicity enable determination of "safe pollution level" for fish difficult. Koushik *et al.* (1999) collected water samples from three lentic water bodies (Motijlhil, Surajajkund and Ranital) of Gwalior region and detected some heavy metals like Cu (0.017 to 0.09 mg/l), Zn (0.065 to 0.120 mg/l), Ni (below detection limit 0.001 mg/l to 0.004 mg/l), Co (0.003 to 0.009 mg/l) Pb (0.002 to 0.009 mg/l) Mn (0.009 to 0.016 mg/l) Cr (0.020 to 0.048 mg/l), Cd (0.009 to 0.019 mg/l) and As (below the detection limit 0.001 mg/l), indicating presence of Cd and Zn in higher levels in these waters. It is generally the ionic form of a metal that produces immediate fish mortalities, while complex metal compounds tend to act by accumulation in the body tissues over a considerably long time. Table 4.1 represents the levels of various major metals found in the air, sludge, drinking water and the aquatic organisms as per USEPA report.

Table 4.1

Heavy Metal	*Max. Metal Conc. in Water Supporting Aquatic Life (ppm)*	*Max. Metal Conc. in Sludge (ppm)*
Cd	0.008	85
Pb	0.0058	420
Zn	0.0766	7500
Hg	0.05	<1

1. Metal Contamination in Rivers Waters

Many workers till date have reported physico-chemical status of industrial effluents and of lake/river waters in India and abroad receiving the industrial wastes (Sikander and Tripathi, 1984; Mukhopadhyay *et al.*1994). Several research reports have been published on the concentration of heavy metals and their toxicity from different corners of the globe (Kureishy *et al.*, 1983; Mastala *et al.*, 1992; Okamura *et al.*, 1996; Kaviraj and Ghosal, 1997; Mendez *et al.*, 1998). Ray and David, 1962 reported fish mortality on a large scale in river Daha in North Bihar due to the deposition of ferric ions in water. Since then the effects of metals like lead, chromium, arsenic, cadmium, mercury and copper, have been under studies. A survey of major rivers in India indicates that all the rivers are polluted to varying degree (Jhingran, 1974; Banerjee *et al.*, 1992). Sahay *et al.* (2002) have enlisted some of the major industries polluting some Indian rivers, such as Bhadra river in Karnatka, being polluted with effluents from pulp, paper and steel industries; Cauvery water from the wastes of Tanneries, distilleries, paper and Rayon industries; Godavary river in Central India from the out flow of paper mills; and Gomti, Gange and Kosi rivers in UP State by the effluents from paper and pulp mills and a variety of chemicals from sugar industries, tanneries and textile industries. Nanda *et al.* (2000) recorded 96-hour LC_{30} values for air-breathing fishes exposed to paper mill effluent as 6.09, 80.35 and 81.28 per cent in *Anabas testudineus, Channa punctatus* and *Clarias batrachus*, respectively.

Saifi and Singh, 2011 analyzed the effluents of the sugar mill in Agra and recorded metals in general below the standard levels except that of Manganese and

Iron. The metal contamination values were: Cr 0.22-0.31; Mn 0.14-1.10; Fe 3.21-7.14; Ni < 0.20; Cu 0.00-0.17; Zn 0.05- 1.47; Cd 0.01-0.11 mg/L and Hg was only in traces. Vinod and Chopra (2010) have recorded enrichment of soil with some heavy metals in the order of Cr>Pb>Cd>Cu>Zn after irrigating soil with sugar mill effluents at Haridwar in Uttarakhand State (India). Mohammed *et al.* (1988) determined distribution of heavy metals in the water and the sediments collected from Kalinadi, U.P. (India). The river Yamuna in Delhi and Damodar in Jharkhand are perhaps the most heavily polluted, as these receive effluents and wastes from large number of industries. Likewise, a 72 km long small river Cooum, flowing through Chennai in Tamil Nadu and draining into the Bay of Bengal, has been found to be polluted to the extent that one liter of its water is reported to contain 900 mg of iron, 275 mg of lead, 1313 mg of nickel and 32 mg of zinc (Sukumarm and Sivakumar, 1999). Hooghly estuary in Bengal is reported to receive about 0.43 mm^3/100ml of effluents from various industries located in around Calcutta (Ghosh and Bagchi, 1980). Sahoo *et al.*, 2002 have reported that this Hooghly river receive around 50 thousand liters of industrial effluents, containing 28.5 ppm of Cr, mainly from tanneries and 127. 7 ppm of Hg. Chromium is one of the important constituent in tannery wastes as well as in the wastes of electroplating, textile, dyeing and printing industries and pharmaceuticals *etc.* Considerable variations have been shown in the concentration of heavy metal from one sampling station to the other, which may be due to the variation in quality of industrial and sewage waters added to the river at different places.

The upper stretch of the river Gange although is free from any anthropogenic activities and appear contaminated with heavy metals in very small amounts, which may be attributed to the geo-chemical source. Dwivedi and Tiwari (1997) reported that the levels of these heavy metals were within the maximum permissible limits at all the river ghats of the Gange. Saikia *et al.* (1988) studied the water and wet sediments of 'The Gange' over a stretch of 480 kms extending from Badrinath to Narora and reported presence of Mn, Fe, Co, Cu and Zn below the toxic limits thereby indicating the water to be unpolluted. As it passes through the UP state, a variety of effluents from Paper and pulp mills, sugar industries, tanneries and textile industries show a progressive increase in concentration of variety of heavy metals (Sahay *et al.*, 2002). The river Yamuna is one of the most important tributaries of river Gange in India. It takes its origin in the Himalayas and flows through the North West and southern Gangetic plains and ultimately joins the main river at Allahabad, popularly known as the 'Sangam. This river passes through the highly polluted and industrial cities of Delhi, Mathura, Agra and Allahabad. Though a major source for fishes, drinking water and for agricultural irrigation, it is heavily contaminated due to severe discharge of industrial pollutants and city sewerage wastes. There have been many attempts in the past few decades to study the ecology of this river in relation to the changing environment (Bhargava *et al.*, 1988; CPCB, 2000); Jhingran and Joshi (1987) recorded six heavy metals of eco-toxicological importance *viz.*, Zn, Cu, Cr, Cd, Pb, and Hg in its water at seven different locations in between Delhi and Allahabad (Table 4.2). They have mentioned maximum metal levels in water and sediments at Delhi, followed by at Agra, Mathura and Allahabad. Renu *et al.* (1991) studied water quality and metal speciation of the Yamuna river from Dakpathar to Agra and reported the metal pollution load increased along its route downstream the river. A large percentage of cadmium, copper and zinc were found in the bound form, whereas lead was more bio-available.

Table 4.2: Detection of Metals in Yamuna river Water as Reported by Jhingran and Joshi, 1987

*Metal/µg/l**	*Metal Conc. in Water*	*Metal Conc. in Sediments (Dry wt)*	*Maximum Tolerance Limit**
Zn	22.0-54.76	43.04-284.8	5.0
Cu	2.34-18.03	10.54-51.35	1.0
Cr	Nd-1.32	3,09-59.45	0.05
Cd	Nd-0.38	Nd-3.85	0.01
Pb	0.82-6.90	16.04-105.12	0.05
Hg	Nd-0.17	0.06-1.06	0.05

Textile and dyeing industries contribute much to water and soil pollution (Mani, 2002). These toxins end up in lakes, rivers, and oceans causing horrific damage. Heavy metals like copper (Cu), chrom (Cr), and zinc (Zn) are basically used in textile processing, especially in dyeing and printing (Smith, 1988). About 10-15 per cent of the dyes and chemicals are directly wasted in wastewater (Banat *et al.*, 1996). Textile dyeing industrial effluents containing hazardous chemicals like Azo- dyes, concentrated mineral organic acids, strong alkalis, bleaching agents appear highly toxic to aquatic flora and fauna, as they cause widespread itching on the skin. Toxic chemicals from dyes and pigments mainly contain Cr 5-20 mg, Pb >1 mg and Fe, Ni, Zn, and Cu > 0.1 mg/l. Textile effluents were also found to be toxic in decreasing RBC count and hemoglobin contents in animals. Due to mutagenicity of some of these synthetic dyes, disposal of textile printing wastewater is further a serious concern for humans and animal health. Some heavy metals contained in these effluents (either in free form or adsorbed in the suspended solid from the industries have been found to be carcinogenic (Tamburlini *et al.*, 2002), while other chemicals equally present are poisonous depending on the dose and exposure duration (Kupchella and Hyland, 1989). Analysis of effluents from five major textile industries in Kaduna (Nigeria) revealed presence of Al, Mn, and Zn in 80 per cent samples but within the limit, while Fe was detected in 60 per cent and Cu in 80 per cent samples with limits exceeding about 3 folds on the average (Yusuff and Sonibare, 2004).

Tannery effluents are today a major source of aquatic pollution in most South Asian countries like the Bangladesh, as the leather industries being most hazardous to the environment, have been transplanted by the Western ones mostly in these developing countries, due to enforced environmental protection legislation in their countries. There are nearly 300 tanneries in Hazaribagh area in Dhaka alone (Rusal *et al.*, 2006; Mohanta *et al.*, 2010). Mohanta *et al.*, 2010 reported that the effects of tannery effluents on snake head fish, *C. punctatus* were so toxic that fishes could not survive in it even for two hours. Even dilution of the effluents by freshwater, had the same effects. The tannery industry in Ethopia is also among the country's largest external earner (Eskindir, 2002). Central Inland Fisheries Research Institute and National Environmental Engineering Research Institute reported in 1989 that Hooghly estuary received 28.5 ppm of chromium from tannery and 0.007 ppm to 129.7 ppm of mercury from other various industries. Kaviraj (1989) has reported accumulation of 381.6 ppm of zinc and 9.6 ppm of chromium in the sediments, 120.4 ppm of zinc, 127.8 ppm of copper and 20 ppm chromium in fishes and 112.2 ppm zinc, 10 ppm copper and 0.3 ppm chromium in shrimps. In India, in Palar river basin around 700 tanneries are located and more than 50 per cent of these tanneries are located in Tamil Nadu (TN), and 80 per cent of these are operating only in the Dindigul town in upper Kodaganar river basin of the TN. Nearly 2 lakh tons of Cr rich effluents are released from various tanneries annually (Mondal *et al.*, 2005). These alone contribute up to 30 ppm of Cr contamination in Indian rivers. Cr posses an exceptional tendency to accumulate in the organism and is transported through serum proteins and is well stored in fish liver, muscles, kidneys and gills.

The Bhilai steel plant which is a major industry of Durg district in Chhattisgarh region in India, is said to dispose its effluents in surrounding area. Shrivastava and Pande, 2002 analyzed the underground waters near Somni stream receiving Bhilai steel plant effluents. The pre-monsoon ground water samples collected from the adjacent areas of Somni Nala course showed high concentrations of Cd and Pb exceeding the maximum permissible limits for drinking. 58 per cent samples showed maximum concentration of Cd (0.05 ppm) and 75 per cent samples containing lead up to 0.014 ppm. Water pollution with iron and chromium was found to reach up to 1.25 and 0.10 ppm, respectively due to Bharat Heavy Electrical Limited in Trichy in Tamil Nadu state (Sahayaraj *et al.*, 2002). Gupta *et al.*, 2002 in the Betwa river receiving effluents from the Mandideep industrial township (Bhopal) showed presence of Cd, Cu and Pb in concentrations in river water above the permissible limits. Bhattacharrya and Chakraborty, 2002 made an attempt to investigate distributional pattern of heavy metals *viz.*, Cd, Cu and Pb in different structural components of two wetlands located in the vicinity of an iron factory involved in extraction from different ores in West Bengal, India. They found the concentration of Cd, Cu, and Pb up to 0.23, 0.07 and 3.03 ppm, respectively in polluted water. Bharti and Katyal (2011) have provided a comparative account of study of eight major water quality indices (WQI) perceived as basic and most important indices for water quality assessment.

Paper mill effluents are reported to contain metals like Pb, Cu, Zn, Ni, Co and Cd, and the levels were said to be higher than the limits, as stated by World Health Organization (Hakeem and Bhatnagar, 2010). Pandey *et al.* (2004) estimated varying concentrations of heavy metals *i.e.* Ni (0.06-0.08 mg/l), Zn (0.02-0.8 mg/l), Ca (nil-0.70 mg/l), Pb (nil-0.26 mg/l), Cd (nil-0.84 mg/l) and As (nil-0.16 mg/l) in the different parts of river Pandu at Kanpur. Naik and Wanganeo (2008) determined five heavy metals *viz.*, Cu, Cd, Fe, Pb and Zn in surface water at different sites of upper basin of Bhoj wet land in Bhopal. Vashney (2008) found groundwater in Ankleshwar Industrial Estate in Bharuch district in Gujarat highly contaminated, as a result of more than 3,000 industrial units in the estate. Analysis of water samples of Chenab, Ramban area for Ca, Mg, K, Na, Fe, Mn, Cu, Ni, Zn and Pb (Fotedar *et al.*, 2008 revealed these metals within the permissible limits, except Fe, Mn and Ni which were slightly in higher concentrations. Lake ecosystems are said to be in particular, vulnerable to heavy metal pollution (Papagiannis *et al.*, 2004). Recording seasonal variations in Zn, Cu and Pb levels in the aquatic phase and underlying surface sediment from three stations (*viz.*, Shankarpur, Canning and Bali Islands) of the coastal zone of West Bengal during different seasons showed the order of the heavy metal level in the ambient media of the selected stations as Zn> Cu> Pb (Chakraborty *et al.*, 2009). Highest concentrations of heavy metals were recorded in the surface water during monsoon, the period characterized by lowest salinity and pH of the ambient aquatic phase. Lead was also detected in considerable amounts in the water and sediments of sewage fed aquaculture ponds in Kolkata, Increase in heavy metal concentration was reported in waters of Vasai Creek in Maharasta (Lokhande and Kelker, 1999), as well as in the surface and groundwater of Delhi (Dixit *et al.*, 2003). Roy *et al.* (2007) studied in Wardha district, Maharashtra the impact of effluents on soil and recorded heavy metals (Cd, Co, Cr, Ni and Pb), but within permissible limits.

The Canadian Global Emission Interpretation Center (CGEIC) has shown that India is one of the identified hotspots of Hg pollution, as 0.1 to 0.5 ton of Hg is released into the atmosphere every year. India exports around 200 tons of Hg per year for use in chlor alkali industries and the effluents of the chlor industry located in Gunjam area of Orissa in India was said to contain Hg residues up to 3.02 mg/l against the national standards fixed at 0.01 mg/L (Sahoo *et al.*, 2002). Maximum concentration (2.26mg/kg ww) was observed for *Caranx affinis* (*Mean:* 0.67 mg/kg ww). Sahoo *et al.* (2002) have further stated that water in river estuary contained Hg contamination level of 0.031 mg/L due to inflow of the chlor-alkali effluents. Hooghly river water in Kolkata in India is reported to receive 127.7 ppm of Hg from various industrial effluents. Hg pollution in India has been reported in many water bodies, especially near chlor alkali plants (Chandra, 1980). WHO (1976) had established a correlation between high Hg concentration and increasing industrialization. Hg in trace quantities is also released from coal and crude oil into the atmosphere continuously. Trivedy and Dubey, 1978 reported Hg contamination up to 15 mg/l in the effluents of a caustic-chlorine industry of Birlagram Nagda, MP. Sahu (1987) and Shaw (1987) analyzed the physico-chemical characteristics of the effluents from these industries and have reported that these wastes possess high alkalinity (212.5-256.2 mg/l), chlorinity (1739-1997 mg/l), hardness (465.2-487.6 mg/l) and high BOD value ranging between 22.5-39.3 mg/l, and Hg (2.85-3.02 mg/l) as the major pollutants. Shaw, 1987 stated the sediments of the effluent channel contained Hg between 457.7- 2053.3 mg/kg dry wt. These industries produce caustic soda and soda ash containing Hg, which are used in paper and rayon industries. These industries in majority use mercury as the cathode. India imports about 200 tons of Hg each year, of which about 180 tons are used by the chlor-alkali industries alone (CSE report, 1985). Most of this Hg comes as effluents or as sludge into water. According to Misra, 1986, the grayish white solid wastes of a chlor alkali industry contain Hg residues up to 984 µg/g of dry wt. Concentrations of Hg, total organic carbon (TOC), Al, Fe and Mn were determined in sediment of the Amba Estuary between the mouth and the head over a distance of 24 km in December and May during 1997-2002 (Anirudh *et al.*, 2009). Temporal and spatial changes in metal concentrations appear to be due to sediment movement associated with tidal movements. Elevated quantity of mercury and other heavy metals were reported in various environmental compartments of Cochin backwaters like in water, sediments and the biota including fish (Toms *et al.*, 2009; Shylesh *et al.*, 2010). On average water contamination with Hg was up to 0.2 ppm. A random survey conducted to evaluate human health risks associated with fish consumption contaminated with Hg near the banks of Cochin backwaters (Jayasooryan *et al.*, 2011) revealed total mercury in fish in the range of 0.093 to 2.321 mg/kg wet weight.

2. Metal Accumulation in River Sediments

Sediments represent an important sink for heavy metals and other contaminants in aquatic systems. Sediments may be deposited into lakes and streams through soil erosion caused by the clearing, excavating, grading, transporting, and filling of land. Mining is the principle source of various metals being added to the sediments. Levels of heavy metals in sediments are often several orders of magnitude greater than those in overlying water. Because of their close association with sediments, benthic invertebrates readily accumulate metals and other materials from contaminated sediments (Hare *et al.*, 1989) and represent an important link to higher trophic levels. Although most metals show little tendency to biomagnify up food chains, concentrations in fish can reach harmful levels owing to reduced prey diversity and increased consumption of contaminated prey (Dallinger *et al.*, 1987). Mercury in the Gange river ecosystem at Varanasi, followed a trend of: water<sediment<benthic macro vertebrates< fish (Sinha *et al.*, 2007), while Duzzin *et al.*, 1988 had stated the order as: water< fish< sediment< large aquatic invertebrates in the order of their increasing affinity. The bio-concentration factors (ratio between Hg concentration in biota and that in water) in sediment, benthic macro-invertebrates and fish (Sinha *et al.*, 2007) as 300, 500 and 10000 show successive accumulation and magnification in the bio-physical environment of the river. A study by Indian Toxicological Research Center (ITRC) Lukhnow during 1986-92, showed maximum annual concentration at Rishikesh as 0.081 ppb and at Allahabad as 0.043 ppb; whereas, over all mercury amount ranged in the Cauvery river in between 0.001 to 0.0136 ppm, which was much lower than that in Gange river at Varanasi. Gange river at Varanasi was found well within the maximum permissible standard of 0.001 ppm, as prescribed by the World Health Organization (WHO, 1985). Cauvery water is polluted by the chlor-alkali industries (Raj, 1986). Mondal *et al.*, 1999 have reported mercury concentration from below detection level to 0.0009ppm, and Mittal *et al.*, 1998 below detection levels to 7.2 ppm. Damoder river is polluted primarily by the coal mines and the coal based industries. Kannan *et al.*, 1998 determined concentrations of Hg and MeHg in sediment and fish collected from estuarine waters of Florida, and found total Hg concentration in sediments ranging from 1 to 219 ng/g dry wt.; and MeHg accounted for on average 0.77 per cent of total Hg in sediments. The concentrations of Hg in fish muscle were between 0.03 to 2.22 ug/g(wet wt.), and MeHg contributing to 83 per cent.

Previous investigations on the upper Arkansas river have demonstrated significant effects of heavy metals on benthic macro-invertebrate and fish populations. Roline (1988) reported reduced diversity and increased abundance of tolerant macro-invertebrates, particularly caddisflies, downstream of California Gulch. Certain benthic invertebrates and abiotic sediment components were also shown to accumulate heavy metals. This metal accumulation persisted even when metal concentrations in the water were diminishing. It was observed that the metal-tolerant (accumulating) organisms tended to predominate and served as the major food source for the brown trout (*Salmo trutta*). The diet of brown trout at the Arkansas river was dominated by benthic invertebrates. In general, Ephemeroptera, Plecoptera, Trichoptera, and Chironomidae (primarily Orthocladiinae) accounted for between 40-95 per cent of the diet of these organisms. Mathis and Kevern (1975) had stated that sediments have shown to act as sinks for many dissolved radionuclides including trace metals. Magnitude and spatial distribution of heavy metals like copper and cadmium in the river Yare, Norfolk was investigated by Bubb *et al.* (1991). The bottom sediments of the river Yare contained elevated amounts of copper and cadmium with mean transect loadings of 27-192 mg of Cu/kg and 0.78-4.09 mg of Cd/kg. Turnpenny, 1985 stated that fisheries status in upland streams of UK was positively correlated with pH, Ca concentration and the alkalinity, but negatively with the concentration of Al, Cu, Pb and Zn. Samanidou *et al.*, 1991 attributed Cd enrichment in river sediments to high anthropogenic activities. Accumulation of 381.6 ppm of zinc and 9.6 ppm of chromium in the sediments, 120.4 ppm of zinc, 127.8 ppm of copper and 20 ppm chromium in fishes and 112.2 ppm zinc, 10 ppm copper and 0.3 ppm chromium in shrimps was reported by Kaviraj (1989). Enrichment of Pb, Fe, Cd, Zn, Cr, Co and Cu were observed in water and sediments near Trichy in Tamil Nadu (Sahayaraj *et al.*, 2002). Water pollution with iron and chromium was found to reach up to 1.25 and 0.10 ppm due to a Heavy Electrical Limited industry (Sahayaraj *et al.*, 2002). Galhoom *et al.*, 2000 revealed presence of various metals (Pb, Hg, Cu, Cd and Zn) in the range 0.0006-0.910 in contaminated water in Bahr El-Bakar drain in Egypt.

Evaluation of the environmental impact of arsenic and certain other heavy metals on a 105 km area of the mining sites in Mexico (Razo *et al.*, 2004) showed soil samples to contain various metals in concentrations ranging between 19–17384 mg kg^{-1} of As,15–7200 mg kg^{-1} of Cu, 31–3450 mg kg^{-1} of Pb and 26–6270 mg kg^{-1} of Zn; these concentrations in dry stream sediment samples were found to vary between 29–28 600 mg kg^{-1} of As, 50–2160 mg kg^{-1} of Pb, 71–2190 mg kg^{-1} of Cu, and 98–5940 mg kg^{-1} of Zn. Metal concentration in water bodies like the Cikijing river in Indonesia, after receiving the main industrial effluent pollutants from PT X were highly increased (Roosmini *et al.*, 2010). PT X is the largest textile industry on those areas which discharge its effluents into Cikijing river. Metal concentration in water bodies after receiving the main effluent of PT X (Point-8) were found highly increased, the range before the effluents receiving point were 0.017 ppm to 0.033 ppm (Cu) and <0.01 ppm to 0.04 ppm (Cr); while the range after receiving point were 0.214 ppm to 0.54 ppm (Cu) and 2.745 ppm to 3.335 ppm (Cr). Meanwhile in sediments, the metal concentrations at segment before receiving the effluent varied from 22.125 to 46.575 ppm for Cu and 15.625 to 34.375 ppm for Cr; after receiving the effluent, the concentrations were on range from 37.5 ppm to 62.238 ppm for Cu, 68.125 ppm to 87.5 ppm for Cr.

There are further evidences of presence of metal pollutants in water and sediments in various other regions of the world. According to Singh *et al.* (2005), metals in an aquatic system also become a part of the water-sediment system and their distribution is controlled by a dynamic set of physical-chemical interactions and equilibrium, such as in sediments largely governed by pH, concentration and type of ligand, the chelating agent, oxidation state of the mineral components and the redox conditions of the system. Suwondo *et al.* (2005) stated that Zn solubility in water is relatively low and Zn with chloride and sulphate group is easily absorbed by sediments, consequently Zn in water bodies is highly precipitated in the bottom of a water body. Likewise, metals such as chromium and copper may interact with organic matter in the aqueous phase and settle, resulting in a high concentration of these metals in the sediment (Begum *et al.*, 2009b). Semwal and Alkolkar (2006) said that major contamination of river water is due to an increase in dissolved solids, total suspended solids and total alkalinity; ranging in between 90-121, 126-237 and 37-96 mg/l, respectively in these waters. Religious places further contribute lot of sulphates (1.66-20.0 mg/l) and traces of some heavy metals like Cu, Zn and Fe in the river waters. Lokeshwari and Chandrappa (2006) showed presence of some of the heavy metals like Zn, Cu, Pb and Cd, beyond the limits of Indian standards in rice and vegetables in the vicinity of Bellandur lake in Bangalore. Bellandur lake is the largest lake in Bangalore and is 130 years old, spreading across an area of 892 acres. Sewage from residential areas near the Bangalore international airport is directly allowed into the lake through the main drain. Analysis of heavy metals (Begum *et al.*, 2009a) in water sediments in Madivala lake of Banglore in Karnatka, also revealed an appreciable increase in metal concentrations in the sediment samples than in water. The heavy metal concentration, in water was in the order Pb > Cr > Cd > Ni, in sediments Pb > Cr > Cd > Ni. Meena *et al.*, 2010 while making heavy metal analysis of the groundwater quality in Ajmer city (Rajasthan) indicated that except Cd, all other heavy metals studied were under the permissible limit and Ni, Cu and Cr were below detectable limit at all sites. The general order of decreasing metals concentration in the ground water was Zn > Cd > Pb > Fe. Achi *et al.*, 2011 analyzed the fractionation of heavy metals in soils around the vicinity of automobile mechanic workshops in Kaduna Metropolis in Nigeria and reported that the residual fraction was the most abundant pool for all the metals examined. A significant amount (2.00 to 73.47, 1.77 to 17.78, 9.76 to 58.54 and 1.71 to 54.11 per cent, respectively) of Cu, Cd, Cr and Pb was present in the potentially available fraction: non residual fraction. Contamination of Zn, Ni and Fe in these soils was not as severe as Cu, Cd, Cr and Pb. Overall, the order of contamination was Cu > Cr > Pb > Cd > Ni > Zn > Fe.

3. Metal Contamination in Estuaries

Estuaries play a major role in the sustenance of commercial fisheries; these form the biotopes wherein many commercial fishes and shell fishes spend part of their life cycle. Estuarine biotopes owe their existence to the interaction of sea, land and freshwater and are the regions with steep and variable gradients in environmental conditions. The combination of marine conditions at the mouth and riverine conditions at the head of the estuary establishes a dynamic interaction between salt and freshwater. These estuaries are characterized by high population densities of microbes, plankton, benthic flora and fauna, and nekton; however, these organisms tend to be highly vulnerable to human activities in coastal watersheds and adjoining embayment. Pollution inputs and the loss and alteration of estuarine habitat due to the anthropogenic stress indicates that water quality in estuaries particularly urbanized systems, is often compromised by the overloading of nutrients and organic matter, the influx of pathogens, and the accumulation of chemical contaminants. Habitat destruction has far reaching ecological consequences, modifying the structure, function, and controls of estuarine ecosystems and contributing to the decline of biodiversity. Trends suggest that by 2025 estuaries will be most significantly impacted by habitat loss and alteration associated with a burgeoning coastal population, which is expected to approach six billion people (Kennish, 2002).

Estuaries around the world are not only the year round habitats for a variety of living organisms and as feeding grounds of many transient organisms, these are also the migration path for many catadromous and anadromous fish species as well as nurseries for many other vertebrate and invertebrate. India has an estuarine area of about 2.2 lakh ha including lagoons and lakes (Jhingran and Gopalkrishnan, 1973). Roughly two-third of the global harvest of sea-food is from the species whose existence depends upon the estuarine zone (Ketchum, 1967). For decades it has been obvious that pollution of estuaries has profoundly affected the food resources that are dependent on these organisms. Because of the multi use demands and stress placed on estuarine and coastal systems as a result of proximity to dense human populations, these systems have been the subject of much study in the past (Kennedy, 1990). Roy and Laha, 1981 reported declined fisheries in industrial zones of the Hooghly estuary. Kaviraj (1989) surveyed Hooghly estuary during 1977-81 and reported 112.2 ppm Zn, 10 ppm Cu and 0.3 ppm Cr in shrimps. In 1989, Central Inland Fisheries Research Institute and National Environmental Engineering Research Institute have reported that Hooghly estuary alone receive 28.5 ppm of chromium from tannery and 0.007 ppm to 129.7 ppm of mercury from other various industries. The Thengappatanam estuary at the tip of the Indian Peninsula, located in the southwestern part of Kanyakumari is today posed with the threat of eutrophication. Coir ret effluents from the connected canal and adjacent retting grounds and drainage from paddy fields gain entry in to the estuary (Vareethiah *et al.*, 1999). Tropical estuaries are further expected to undergo violent fluctuations in salinity following monsoon showers, resulting in pre-monsoon phase of high salinity, monsoon phase of low salinity and post monsoon phase of moderate salinity (Chandran and Ramamoorthy, 1984). Sand mining is other very important factor in the estuarine areas which induce changes to the geography structure of the river and significantly affect fish habit and abundance, and the habitats of benthic micro-invertebrates which cumulative negative effects on the biodiversity of the estuary (Vareethiah, 2002). Mukhopadhyay *et al.* (1994) concluded that heavy metals form important group of pollutants and these play adverse role on productive potentiality of the estuary. They reported the toxicity of industrial heavy metals, Zn, Cu, Cd and Cr in estuarine fishes : *Liza parsia, Appocryptis bato, Therapon jarbua, Mystus gulio* and Prawn, *Macrobrachium rude* and also the fish food organism such as *Cyclops* sp.,

Daphnia carinata, *Chironomus* sp. Amongst the estuarine organisms, crustaceans were most sensitive, hence least resistant to the toxicity of heavy metals. This is likely to affect food chain system, *i.e.* plankton density in the water system and the primary, secondary and the tertiary consumers. According to Williams *et al.*, 1994 salt marshes act as very efficient sinks for metal contaminants although metal concentrations in halophytes do not generally reflect environmental contamination levels.

4. Metal Contamination in Coastal Waters

Coastal fish are often subjected to pollution, especially near industrial or populated areas. Heavy metals such as copper, zinc and lead were found as normal constituents of marine and estuarine environments (Bryan, 1971). The coastal areas of India have some of the best developed industrial areas, metros and ports. These are the major contributor to the country's financial growth. While striving for excellence in development, however, we have failed to check the pollution of the rivers and oceans. The Indian coastal waters are under considerable stress from the effluents discharged by the industries and the municipalities located in the coastal areas. Worst sufferers are the marine biota. The industrialized centers along the coasts release a horde of chemical pollutants through their effluents in the streams and/or the local drainages, draining wastes into the sea. Most dreaded are the municipal wastes, sewage sludge and dredged spill dumping, oil spills and leakages. A survey of 12 heavy metals in the skeleton of the coral, *Siderastrea siderea* and reef sediments in 23 reefs along the Caribbean coast of Costa Rica and Panama indicated high levels of metal pollution in the region (Guzman and Jimenez, 1992). Many marine areas in the world have been reported to be contaminated by heavy metals and the organisms living in contaminated waters have reflected the high metal concentrations (Bryan and Langston, 1992). The entire coast is influenced by hundreds of rivers increasingly loaded with suspended sediments (associated with deforestation) which carry most of the metals several kilometers from the sources to the sea, where they are probably transported and distributed by currents through the entire region, even to pristine offshore reefs. Selvaraj (1999a,b) while monitoring the sea waters at Kalpakkam coast since 1979 implicated the condenser cooling water discharge from Madras Atomic Power Plant Station (MAPS) into the coastal waters as the sole source for elevated levels of Cu and Hg. The effluents and sewage that is discharged in to the sea at Mumbai alone are a gargantuan, *i.e.* about 2200 million litres a day. Situation is no better at Kolkata, Chennai and Visakhapatnam ports. Sea water is used in MAPS as the condenser cooling water (Iyengar *et al.* (1990) and the average value for Cu was reported as 1.7 mg/l. The values of total dissolved Cu reported by others (Rangarajan and Narasimhan, 1987) ranged from 0.8 to 4.5 mg/l, and the average value was 2.6 mg/l over a period of one year. Nayak *et al.* (2004) reported heavy metal contamination in the coastal sea of Orrisa, India. Satpathy *et al.* (2006) recorded it in the range from 34 to 52 mg/l, and the average value was 44.5 mg/l. The industrial hubs in Gujarat, Pondicherry and Orissa add to the misery of the marine biota, where the coastal and estuarine waters remain constantly in degraded conditions. Storelli *et al.* (2005) has stated that arsenic and Hg are nonessential microelements in a marine environment, where these are widely distributed as various inorganic and organic compounds of different toxicity to aquatic life. Goldberg, 1976 found that the concentration of Cd in sea water on an average was 0.15 mg/l. Beaupre *et al.*, 1983 studied the concentration of Cd in Ontario in Canada and recorded it in the range of 0.66 to 0.9 ppm. The National Academy of Environmental Sciences, 1972 considered 3.0 mg/l of Cd as a threat to the aquatic life. Cadmium is widely used either as the pure metal or alloy with Ag in nuclear reactors control rods and shielding. One such radio nucleotide is Cd^{113m} which is produced via neutron activation of Cd in reactor control rods and shielding. $Cadmium^{113m}$ enter the environment as a consequence of testing of nuclear weapons in the atmosphere (Salter, 1965; Palagyi and Larsen, 1983; Warwick and Croudace, 2008). Scientists (Dunn and Tisue., 1989) had demonstrated that Cd^{113m} can be determined accurately in water, soils, and sediments. In lake Michigan, the present concentrations of Cd^{113m} are 350 ± 150 μBq l^{-1} ($n = 16$) in the water, and 10.0 ± 5.5 mBq g^{-1} ($n = 12$) in the surface sediments. In undisturbed soils near lake Michigan, the mean $Cd^{113m}/^{137}Cs$ activity ratio is $2.0 \pm 1.4 \times 10^{-2}$ ($n = 4$), implying a total Cd^{113m} input to the lake of 2.7 ± 1.3 TBq.

Mercury pollution in the coastal waters has been a point of major concern throughout the world. Hg pollution was first detected in Japan (Irukayama, 1966), and thereafter in many other countries including Sweden (Jhonels *et al.*, 1967), the United States (Anon, 1970) and in Arctic Canada (El-Hayek, 2007); which were the result of such a magnification of Hg residues in large fishes in Hg contaminated waters. El-Hayek (2007) has reported methylmercury as a potent neurotoxic found at elevated concentrations in both the Arctic ecosystem and tissues of the local aboriginal inhabitants in Canada. In Japan, the Hg levels discharged in MinaMata continued to magnify until reaching values of 10-100 µg/g in sediments, 2000 µg/g at the discharge channl of the Chiss-o Corporation that used and dumped it, and 5-40 µg/g in fish (Aizpura *et al.*, 1997; Villanueva and Botello, 1998). MinaMata Bay, Japan has been a very well known case of Hg release to a marine environment in the 1950s but it typically involved direct industrial release of MeHg (Takizawa and Osame, 2001). McAlpine and Araki 1958 linked this unusual neurological disease with fish consumption from MinaMata Bay exposed to methyl Hg. Likewise, there has been Hg release in marine fish in Duch Wadden Sea and Ems Estuary (Essink, 1988), Liverpool Bay, UK (Leah *et al.*, 1993), in Princess Royal Harbour, Australia (Francesconi *et al.*, 1997) and the Haifa Bay, Israel (Herut *et al.*, 1993, 1995, 2003). Not surprisingly, the Qishon river has been identified as the main source of Hg, Cd, and other heavy metal pollution in the Haifa ports. Herut *et al.*, 2003 reported medium degree of pollution in the Qishon estuary by heavy metals (Hg, Cu, Zn, Cd and Ni) and high degree of pollution upstream by international environmental criteria. Moreover, tissue analysis of fishes, crustaceans and mollusks from th shallow waters of Haifa Bay at the mouth of Qishon river revealed high levels of heavy metals and deviations in oxidizing enzyme activities (Kress *et al.*, 1999; Herut *et al.*, 1999). The mercury levels in the sediments of the MinaMata Bay (Japan) in 1953 were recorded to vary between 28-73 mg/kg dry wt. (Fujuki, 1973). The contamination of Hg in seafood such as fish, shellfish, oyster and other types of seafood thus, is one of great concern particularly in the coastal areas (Bartoli *et al.*, 1995; Stein *et al.*, 1996).

Aquatic eco-systems in India have significant amount of Hg (Govindsamy *et al.*, 1998). The coastal waters of Mumbai were found to have a high content of mercury between 0.12 to 1.4µg/l and in sediments from 0.08 to 0.36µg/gram. At Karwar coast in Karnataka the mercury contents are as high as 2.68 µg/l in water and 1.32 µg/gram dry weights in sediments. Anirudh *et al.* (2009) measured concentrations of Hg, total organic carbon (TOC), Al, Fe and Mn in sediment of the Amba Estuary between the mouth and the head over a distance of 24 km in December and May during 1997-2002. Temporal and spatial changes in metal concentrations appear to be due to sediment movement associated with tidal movements. The Hg concentration varied in 0.05 -2.66 µg g^{-1} range and the profiles of its variation indicated PatalGange river that opens in the Amba Estuary is a major source of anthropogenic metal to the estuary.

All heavy metals exist in surface waters in colloidal, particulate, and dissolved phases, although dissolved concentrations are generally low (Kennish, 1992). Adsorbents like humic acids, sediments or the suspended solids greatly reduce bioavailability of a water contaminant to a fish, as the metals have a high affinity for humic acids, organic clays, and oxides coated with organic matter (Connell *et al.*, 1984). Heavy metals being non-degradable remain in water as suspended colloids forming complex compounds with inorganic and organic legends. White *et al.*, 1997 reported about the microbial solubilization and immobilization of heavy metals. The soluble forms are generally ions or unionized organo-metallic chelates or complexes. Bury *et al.* (1999) said that during acute exposure to silver, natural ligand such as dissolved organic matter (DOM), water Cl^- and hardness offer a great degree of protection to fish against silver toxicity. Higher water Cl^- levels (300 and 3000 mM), dissolved organic carbon (12 mg carbon/L) and hardness (150 and 400 mg/L as $CaCO_3$) offered some degree of protection against the silver induced ion-regulatory disturbance. In general, these factors appear to be less protective during chronic than during acute silver toxicity. The Biotic Ligand Model (BLM) is a relatively new computational approach for the generation of ambient water quality criteria (AWQC) that incorporates these influences in a quantitative manner, thereby replacing insensitive and inefficient "blanket" regulations. The BLM procedure has been provisionally accepted by the US EPA for the generation of acute freshwater AWQC for some metals (copper, silver), and is under various stages of review by other regulatory agencies (EU, Canada, Australia/New Zealand, Chile). Historically, AWQC for metals are described simply based on total metal levels in the water, or more recently on dissolved metal levels. BLMs for many other metals are currently under development, and the ultimate goal is to generate chronic and estuarine-marine AWQC as well. The BLM approach integrates basic principles in physiology, aquatic geochemistry, and toxicology, using the known water chemistry and the experimentally determined binding constants of key "toxic" sites (physiological processes) on the receptor surface (biotic ligand) of the organism to predict the metal load causing a defined toxic endpoint (*e.g.*, 96-h mortality). Metallic legends in the sediments increase metal transport in fish. The historical framework of the BLM has been comprehensively reviewed in a treatise by Paquin *et al.* (2002).The BLM assumes that the free cationic forms of metals are the source of acute toxicity, that these do their toxic damage by binding on or in the gill surface, and that the extent of short-term gill metal binding (*i.e.*, before pathology develops at 3 h or 24 h) is directly predictive of longer term mortality (*e.g.*, at 96 h). The greater the binding affinity of a metal for the gill (*i.e.*, the higher the log K value), the greater is the toxicity. Based on fish gill research, these "toxic" sites appear to fall into three categories: active Na+ uptake sites targeted by silver and copper, resulting in hyponatremia and hypochloremia; active Ca^{2+} uptake sites targeted by zinc, cadmium, lead, and cobalt, resulting in hypocalcemia; and allergic reaction sites targeted by nickel, resulting in gill edema and associated inhibition of respiratory gas exchange.

REFERENCES

Achi M.M., Uzairu A., Gimba C.E. and Okunola O.J., 2011. Chemical fractionation of heavy metals in soils around the vicinity of automobile mechanic workshops in Kaduna Metropolis, Nigeria. J. Environ. Chem. and Ecotoxicol., 3(7): 184-194.

Aizpura I.C.M., Tenuta-Filho A., Skuma A.M. and Zenebon O., 1997. Int. J. Food Sci., 32: 333-337.

Anonymous, 1970. Chem Eng. News, 48: 14

Anirudh Ram, Rokade M. A. and Zingde M. D. 2009. Mercury enrichment in sediments of Amba estuary. Ind. J. Mar. Sci., 38(1): 89-96.

Banat I.M., Nigam P. and Singh D. and Marchant R., 1996. Microbial decolorization of textile dye containing effluents: A Revew. Bio. Res. Technol., 58: 217-227.

Banerjee A.K., Verma A.K. and Pathak Vandna, 1992. The Heavy metal pollution, Ad Bios., 11(1): 61-66.

Bartoli A., Gerotto M., Marchiori M., Palonta M. and Troncon A., 1995. Analytical problems in mercury analysis of seafood. Annali dell Istituto Superiore di Sanita, 31: 359-62.

Beaupre P.W., Holland W.J. and Mckenney W.J., 1983. Mikrochim. Acta., 11: 415-420.

Begum Abida, Krishna Hari, Khan Irfanulla, 2009a. Analysis of Heavy metals in Water, Sediments and Fish samples of Madivala Lakes of Bangalore, Karnataka. Intl J. Chem. Tech. Res., 1(2): 245-249.

Begum A., Ramaiah M., Harkrishna S., Khan I. and Veena K., 2009b, Heavy Metal Pollution and Chemical Profile of Cauvery River Water, E-Journal of Chemistry, 6 (1): 47-52.

Bhargava D.S., Sexena B.S. and Dewakar, 1988. A study of geo-pollution in the Godawary river basin in India. Asian Environment, IOS press: 36-59.

Bharti N. and Katyal D., 2011. Water quality indices used for surface water vulnerability assessment. Int. J. Environ. Sci., 2 (1): 154-174.

Bhattacharrya N. and Chakraborty S.K., 2002. Distributional pattern of some heavy metals in the structural components of two contrasting wetlands in the vicinity of an iron extraction factory of Midnapore district, West Bengal, India. In Kumar A. (Ed) Ecology of polluted waters Vol. 1. APH Publ. Corp., New Delhi: 287-294.

Bryan G.W., 1971. The Effects of Heavy Metals (other than Mercury) on Marine and Estuarine Organisms. Proc Royal Soc Lond., B 177: 389-410.

Bryan G. and Langston W.J., 1992. Bioavailability, accumulation and effects of heavy metals in sediments with special reference to United Kingdom estuaries: a review. Environ. Pollut., 76, 89-131.

Bubb J.M., Rudd T. and Lester J.N. 1991. Distribution of heavy metals in the river yare and its associated broads II. Copper and Cadmium. Sci. Total Environ., 102: 169-188.

Bury N.R., Grosell M., Grover A.K. and Wood C.M.,1999. ATP-dependent silver transport across the basolateral membrane of rainbow trout gills. Toxicol. Appl. Pharmacol.,159: 1-8.

Chakraborty R., Zaman S., MukhopadhyayN., Banerjee K. And Mitra A., 2009. Seasonal variation of Zn, Cu and Pb in the estuarine stretch of West Bengal. Ind. J. Mar. Sci., 38(1):104-109.

Chandra S.V., 1980. Toxic metals in Environment. A survey report of R and D work done in India. Ind. Toxicol. Res. Center, Lukhnow.

Chandran R. and Ramamoorthy K., 1984. Hydrobiological studies in the gradient zone of Vellar estuary. 1. Physico-chemical parameters. Mahasagar, 17: 69-77.

Clark R., Frid C., and Attrill M., 1997. Marine pollution (4th ed.). New York: Oxford University Press. Eisler, R. (1988). Zink hazards to fish, Wildlife and Invertebrates: A synoptic review. US Fish Wildlife Serv. Biol. Rep., 85.

Connell B.S., Cox M. and Singer I., 1984. Nickel and chromium. In: Brunner F. and Coburn J.W. (Eds). Disorders of minerals metabolism. Academic Press, NY. Pp 472-532.

CPCB, 2000. Water quality status of Yamuna River, assessment and Development of River basin. CPC report, ADSORBS/32: 1999-2000.

Crandall C.A. and Goodnight C.J., 1962. Effects of sublethal concentrations of several toxicants on growth of the common guppy, *Lebistes reticulates*. Limnol. and Oecano., 7: 233-239.

CSE., 1985. Center for Science and Environment, New Delhi. The State of India's environment 1984-85. The 2nd Citiens Report.

Dallinger R., Prosi F., Segner H., and Back H., 1987. Contaminated food and uptake of heavy metals by fish: a review and a proposal for further research. Oecologia (Berl), 73:91-98.

Das M.K., Samantha S. and Saha P.K., 2007. Riverine Health and Impact on Fisheries in India. Policy paper o. 01, Central Inland Fisheries Res. Inst., Barrackpur, Kolkatta:49.

Dixit R.C., Verma S.R., Nitnaware V. and Thacker N.P., 2003. Heavy metals contamination in surface and ground water supply of than urban city. Ind. J. Environ. Health, 45: 107-112.

Dunn D., Tisue t., 1989. Cadmium^{113m} in the water, sediment and adjacent soil of lake Michigan. Sci. Total Environ., 87/88: 305-14.

Duzzin B., Parvoni B. and Donazzolo R., 1988. Water Res., 22:1353-63.

Dwivedi S. and Tiwari I.C., 1997. A study on the heavy metals in the Gange water at Varanasi. Polln. Res., 16(4): 265-270.

Eisler R., 1988. Lead hazards to fish, wildlife, and invertebrates: a synoptic review. US Fish Wildlife Serv. Biol. Rep., 85: 1–4.

El-Hayek Youssef H., 2007. Mercury contamination in Arctic Canada: possible implications for Aboriginal Health. J. dev. Disabilities, 13(1): 67-91.

Eskindir Zinabu, 2002. Assessment of the Impact of Industrial effluents on the quality of irrigation water and changes on soil characteristics, a case of Kombolcha town.

Essink K., 1988. Decreasing mercury pollution in the Dutch Wadden Sea and Ems Estuary. Mar. Poll. Bull., 19:317–319.

Fotedar Amita, Verma R.K. and Tikoo Veerji, 2008. Chemical status of waters of Chenab River, Ramban area, Jammu and Kashmir state. J. Soil and Water Conser., 7: 52-59.

Francesconi K.A., Lenanton R.C.J., Caputi N. and Jones S., 1997. Long term study of mercury concentrations in fish following cessation of a mercury containing discharge. Mar. Environ. Res., 43: 27-40.

Fujuki M., 1973. The transitional condition of Minamata Bay and the neighbouring sea polluted by factory waste containing mercury. In: Advances in Water Pollution Res. Ed. J.H. Jenkason. Pergamon Press, Oxford: 905-920.

Galhoom K.L., Laila G., Rizk G. and El-ezzaway M.H., 2000. Some biochemical and hematological parameters in Mugil fish (*Mugil caphalcus*) reared in Bahr El-Bakar drain. Egypt. J. Agril. Res., 78: 1-13.

Ghosh B.B. and Bagchi M.M., 1980. The pollution problem in Hoogly estury. In Summer inst. On Brachishwater Capture and Culture Fisheries, CIFRI Barrachpur, (Jul-Aug, 1980): 1-10.

Goldberg E.W., 1976. The Wealth of the Oceans. The UNESCO Press: 172.

Govindsamy C., Roy Vij A.G., and Jayapal A., 1998. Int J. Ecol. Environ. Sci. 24:141-146.

Gupta D.C., Peters E., and Yadava R.N., 2002. Concentration of heavy metals in waters of the area around Mandideep Industrial Complex and its impact on Human Health. In Kumar A. (Ed) Ecology of polluted waters Vol. 1. APH Publ. Corp., New Delhi: 515-536.

Guzman H.M. and Jimenez C.E., 1992. Contamination of coral reefs by heavy metals along the Caribbean Coast of Central America. (Costa Rica and Panama). Marine pollution bulletin, 24: 554-561.

Hakeem Arti Suri and Bhatnagar, 2010. Heavy metal reduction of pulp and paper mill effluents by indogenous microbes. Asian J. Exp. Biol. Sci., 1(1): 201-203.

Hare L, Saouter E., Campbell P.G.C., Tessier A., Ribeyre F., and Boudou A., 1989. Dynamics of cadmium, lead, and zinc exchange between nymphs of the burrowing mayfly *Heaenia riaida* (Ephemeroptera) and the environment. Can. J. Fish. Aquat. Sci., 48:39-47.

Herut B., Hornung H., Krom M.D., Kress N. and Cohen Y., 1993. Trace metals in shallow sediments from the Mediterranean coastal region of Israel. Mar. Pollut. Bull., 26: 675-682.

Herut B., Hornung H., Kress N., Krom M.D. and Shirav M., 1995. Trace metals in sediments at the lower reaches of Mediterranean coastal rivers, Israel. Wat. Sci. Tech., 32: 239-246.

Herut B., Kress N., Shefer E. and Hornung H. 1999. Trace element levels in mollusks from clean and polluted coastal marine sites in the Mediterranean, red and North seas. Helgoland Mar. Res., 53; 154-162.

Herut B., Shefer E. and Cohen Y., 2003. Environmental quality of Israel's Mediterranean Coastalwaters in 2002. Ann. Rep. Nat. Mr. Environ. Monit. Prog., IOLR Rep. H39/2003a.

Irukayama K., 1966. Advanc. Water Pollut. Res. Proc. Ist Conf.:153p.

Iyengar M.A.R., Nair K.V.K., Ganapathy S., Viswanathan E.K., Kannan V., Rajan M.P. and Subratanam T., 1990.Progress report of Environmental Survey Laboratory, BARC-1/796: 1–50.

Jayasooryan K.K., Shylesh C.M.S., Mahesh M. and Ramasamy E.V., 2011.Mercury contaminated fish consumption at Cochin backwaters. Disaster, Risk and Vulnerablity Conference 2011, School of Environmental Sciences, Mahatma Gandhi University, India.

Jhingran V.G. and Gopalkrishnan V., 1973. Estuarine fisheries resources in India in relation to adjacent waters. J. Mar. Biol. Assoc. Ind., 15: 323-334.

Jhingran Arun G. and Joshi H.C., 1987. Heavy metals in water sediments and fish in the river Yamuna. J. Inland Fish. Soc. India, 19 (1): 13-23.

Jhonels A.G., Westermark T., Berg W., Persson P.I. and Sjostrand, 1967. Oikos 18: 323.

Kannan K., Smith R.G. Jr., Lee R.F., and 4 others., 1998. Distribution of mercury and methyl mercury in water, sediments, and fish from South Florida Estuaries. Arch. Environ. Contam. Toxicol., 34; 109-118.

Kaur Tejinder, Sexena P.K. and Jior R.S., 2000. Impact of pollution in Budha Nalla brook on the occurrence of fishes in river Satluj. Ind. J. Fish, 47(1): 49-53.

Kaur H., Sexena P.K. and Jior R.S., 2013. Occurrence of heavy metals in the water and sediment of the river Satluj in Punjab. In: Pollution and Bio-monitoring of Indian Rivers. Ed. Dr. R.K. Trivedy, pp. 176-180.

Kaushik S., Sahu B.K., Lawania R.K. and Tiwari R.K. 1999. Occurrence of Heavy Metals in Lentic Water of Gwalior Region. Poll. Res. 18 (2): 137-140.

Kaviraj A., 1989. Heavy metal concentration in mullet and prawn from the Hoogly Estuary. Sci. and Cult., 55: 695-699.

Kaviraj A. and Ghosal T.K. 1997. Partioning and bioacculmulation of cadmium from the sediments of brachishwater ponds in Sunderbans, India. Asian Fisheries Sci., 9: 269-274.

Kennedy V.S. 1990. Anticipated effects of climate change on estuarine and coastal fisheries. In: Fisheries. Bull. AFS, 15: 16-24.

Kennish M.J., 1992. Ecology of Estuaries: Anthropogenic Effects. Boca Raton, USA: CRC Press: 494 pp.

Kennish M.J., 2002. Environmental threats and environmental future of Estuaries. Environ.Conser., 29, 78-107.

Ketchum B.N., 1967. Phytoplankton nutrients in estuaries. In: Estuaries (Ed. G.H. Lauff). Pub. No. 83, Am Assoc. Advanc. Sci., Washington D.C.: 329-335.

Kress N., Herut B., Shefer E. and Hornung H., 1999. Trace elements in fish from clean and polluted coastal marine sites in the Mediterranean Sea, Red sea and North Sea. Helgoland Mar. Res., 53: 163-170.

Kumar Vinod, Chopra A.K., 2010.Influence of sugar mill effluent on physico-chemical characteristics of soil at Haridwar (Uttarakhand), India. J. Appl. and Natural Sci., 2:269-279.

Kupechella C.E. and Hyland M.C., 1989, Environmental Science, Allyn and Baron, London.

Kureishy T.W., Sanzgiary S., George M.D. and Braganca A., 1983. Mercury, cadmium and lead in different tissues of fishes and in zooplanktons from the Andeman sea. Ind. J. Mar. Sci., 12: 60-63.

Leah R.T., Collings S.E., Johnson M.S. and Evans S.J., 1993. Mercury in plaice (*Pleuronectes platessa*) from the sludge disposal ground of Liverpool Bay. Mar. Poll. Bull., 26: 436–439.

Lokeshwari H. and Chandrappa G.T., 2006. Impact of heavy metal contamination of Bellandur lake in soil and cultivated vegetables. Curr. Sci., 91(5): 622-627.

Lokhande R.S., and Kelkar N., 1999. Studies on heavy metals in water of Vasai Creek, Maharashtra. Ind. J. Environ. Protect., 19(9): 664-668.

Mani P. Jothy, 2002. Impact of textile industry waste water on soil and water ecosystem- a brief revew. In Kumar A. (Ed) Ecology of polluted waters, Vol. 1. APH PUbl. Corp., New Delhi. pp 67-77.

Mastala Z., Balogh K.V. and Solanki J., 1992. Reliability of heavy metal pollution monitoring utilizing aquatic animals versus statistical evaluation methods. Arch. Environ. Contam. Toxicol., 23: 476-483.

Mathis B.J. and Kevern N.R., 1975. Distribution of mercury, cadmium, lead and rhallium in a eutrophic lake. Hydrobiologia, 46; 207-222.

McAlpine D. and Araki S., 1958. Mimamana disease. Arch. Neurol., 1: 78-86.

Meena Manoj Kumar, Dutta Subroto and Pradhan Rashmi, 2010. Environmental impacts of wastewater irrigation on groundwater quality – a case study of Ajmer city, Rajasthan (India). EJEAF Che., 9 (4): 760-766.

Mendez L., Acosta B., Alverez-Castaneda S.T. and Lechuga-Deveze C.H., 1998. Trace metals distribution along the Southern cCoast of Bahia de La Paz (Gulf of California), Mexico. Bull. Environ. Contam. Toxicol., 61: 616-622.

Misra B.B., 1986. Role of blue green algae in bio-degradation of chlor-alkali waste (soil) with special reference to rice cultivation. Ph. D Thesis, Berhampur Univ., Orissa, India.

Mittal B.K., Gupta M.S. and Prasad A., 1998. Proc. Int. Specialized Conf. on water quality and its management, New Delhi, pp 107-114.

Mohammad A., Uddin R. and Khan V.A., 1988. Heavy metals in water, sediments, plants and fish of Kalinadi, U.P. (INDIA). Env.Int., 14: 515-525.

Mohanta M.K., Salam, M.A., Saha A.K., Hasan A. and Roy A.K., 2010. Effects of tannery effluents on survival and histopathological Changes in Different Organs of Channa punctatus. Asian J. Exp. Biol. Sci., 1 (2): 294 – 302.

Mondal G.C., Tiwary R.K., Chakraborty M.K., Tewary B.K and Singh R.S., 1999. Proc. Nat. Sem. On Socio-econy, morbidy pattern, demographical changes and cultural heritage in Damodar Basin, Giridih.

Mondal N.C., Sexena V.K. and Singh V.S., 2005. Impact of pollution due to tanneries on ground water, Regime. Curr. Sci., 88: 1988-1993.

Mukhopadhyay M.K., Ghosh B.B. and Bagchi M.M., 1994. Toxicity of heavy metals to fish, prawn and fish food organisms of Hoogly estuarine system. Geobios., 21 : 13-17.

Naik A. A. and Wanganeo A., 2008. Seasonal variation of heavy metals in surface water samples of Upper Basin of Bhoj wetland, Bhopal. Poll. Res., 27(4): 649-653.

Nanda P, Panigrahi S, Nanda B, Behera M.K., 2000. Toxicity of paper mill effluent to fishes. Env. Eco., 18(1): 220-222.

Nayak B.B., Acharya B.C., Panigrahy P.K. and Panda U.C., 2004. Assessment of heavy metals contamination in the coastal sea of Orissa, India. Polln. Res., 23(4): 791-803.

Okamura H., Luo R. and Aoyama, 1996. Ecotoxicity assessment of the aquatic environment around Lake Kojima, Japan. Environ. Toxicol. and Water Qual., 11: 213-221.

Palagyi S. and Larsen R.P., 1983. Determination of ^{113m}Cadmium in natural water. J. Radio-analytical Chem., 80; 141-52.

Pandey R.K., Arya R.K., Verma A. and Misra S.K., 2004. Distribution of heavy metals in the waste water of river Pandu at Kanpur. Poll. Res., 23 (2): 335-342.

Papagiannis I., Kagalou I., Leonardos J., Petridis D. and Kalfakakou V., 2004. Copper and zinc in four freshwater fish species from Lake Pamvotis (Greece). Environ Int., 30(3):357-62.

Paquin P.R., Gorsuch J.W., Apte S., Batley G.E., Bowles K.C., CampbellP.G.C. and 16 others, 2002. The biotic ligand model: a historical overview.Comp. Biochem. Physiol., C 133: 3-35.

Pethkar A.V., Gaikaiwari R.P. and Paknikar K.M., 2001. Biosorptive removal of contaminating heavy metals from plant extracts of medicinal value. Curr. Sci., 80: 1216-1219.

Raj S.P., 1986. Pollut. Res., 5: 39-43.

Rangarajan S. and Narasimhan S.V., In Proceedings of International seminar on instrumental methods of electro-analytical techniques, Mysore University, 1987, pp. 109–119.

Ray P. and David A., 1962. A case of fish mortality caused by precipitation of ferric iron in the river Daha at. Siwan (North Bihar). Indian J. Fish., 9(1): 117 – 122

Razo Israel, Carrizales Leticia, Castro Javier, Díaz-Barriga Fernando and Monroy Marcos, 2004. Arsenic and Heavy Metal Pollution of Soil, Water and Sediments in a Semi-Arid Climate Mining Area in Mexico. Water, Air, and Soil Pollution, 152(1-4): 129-152.

Roline R.A., 1988. The effects of heavy metal pollution of the upper Arkansas River on the distribution of aquatic microinvertebrates. Hydrobiologia, 160; 3-8.

Roosmini D., Andarani P. and Nastiti A., 2010. Heavy metals (copper and chromium) pollution from textile industry in rivers (case study : Cikijing River, West Java). In proceeding of: 8th International Water Symposium South East Asia Water Environment (SEAWE), At Phuket, Thailand.

Roy P. and Laha G.C., 1981. Effect of industrial and municipal wastes on the trend of fisheries in the Hoogly estury. In; Inter. Symp. On water Resources, Conser. Pollut. And abatement, Roorkee, India. Dec., 1981: 121-130.

Roy R.P., Prasad J., Joshi A.P., 2007. Effect of sugar factory effluent on some physico-chemical properties of soils--a case study. J Environ Sci Eng., 49(4):277-82.

Rusal M.G., Faisal I. and Kamal Khan, M.M., 2006. Environmental pollution generated from process industries in Bangladesh. J. Environ. Poll., 28 (1/2): 44-161.

Saifi Mohd. Aslam and Singh H.B., 2011. Studies on physico-chemical parameters and concentrations of heavy metals in sugar mill effluents. Int. J. Chem. Sci., 9(2): 929-935.

Sahay U., Bakshi R., Sah H.C.P., Dan M.T. and Sahay S., 2002. Water pollution, laws and remedies. In Kumar A. (Ed) Ecology of polluted waters Vol. 1. APH PUbl. Corp., New Delhi. Pp 257-285.

Sahayaraj A., Arulsamy K.S. and Senthil Kumar, S., 2002. A study on the physico-chemical parameters of the canal water in Trichy. In Ecology of polluted waters. 1: 151-163.

Sahoo A., Panigrahi Minati K., Sahu S.K. and Panigrahi Ashok K., 2002. Aquatic mercury pollution by a chlor-alkali industry and its impact on the ecology of human health in Ganjam area- a case study. In Kumar A. (Ed) Ecology of polluted waters Vol.1. APH Publ. Corp., New Delhi.,pp 465-496.

Sahu A., 1987. Toxicological effects of a pesticide on a blue green alga: III. Effects of PMA on a blue green alg, Westiellopis prolifica Janet. And its ecological implications. Ph. D. Thesis, Berhampur Univ., Orissa, India.

Saikia D.K., Mathur R.P. and Srivastava S.K., 1988. Heavy metals in water and sediments of upper Gange. Ind. J. Environ., 1: 11-17.

Salter L.P., 1965. Radioactibe fallout from nuclear weapons tests. In: Klement A.W., Ed. CONF-765, AEC Symp. Series, 409-21: p694.

Samanidou V., Papadoyiannis I. and Vasilikiotis G., 1991. Verticle distribution of heavy metals in sediments from rivers in Norhern Greece. J. Environ. Sci. Health A, 26;(8); 1345-61.

Satpathy K. K., Natesan U. Kalaivani S., Mohanty A.K., Rajan M. and Baldev Raj., 2006. Total dissolved copper concentrations in coastal waters of Kalpakkam. Current Sci., 91 (NO. 8): 1008-10.

Selvaraj K., 1999a, Response: Measurement of mercury and copper, Curr. Sci., 77, 1387-1389.

Selvaraj K., 1999b, Total dissolvable copper and mercury concentrations in inner shelf waters, off Kalpakkam, Bay of Bengal, Curr. Sci., 77, 494-497.

Sembal N. and Akolkar P., 2006. Water quality assessment of sacred Himalayan rivers of Uttranchal. Curr. Sci., 91: 486-.

Shaw B.P., 1987. Eco-physiological studies of a waste of a chlor alkali factory on biosystems. PhD thesis, Berhampur University, Orisa, India.

Shrivastava P.K. and Pandey S.K., 2002. Impact of Bhilai steel plant effluent on groundwater quality of Somni Nala Basin, Durg District, Chatisgarh for drinking purpose. In Kumar A. (Ed) Ecology of polluted waters Vol. 1. APH Publ. Corp., New Delhi: pp89-98.

Shylesh C.M.S., Jayasooyan K.K., Subin K. Josh, Mahash Mohan and Ramaswamy E.V., 2010. Mercury in the core sediment of Vembanad backwater-An application towards anthropogenic contamination. LAKE 2010: Wetland, Biodiversity and Climate Change. School of Environmental Sciences, M G University, kottayam.

Sikander M. and Tripathi B.D., 1984. In: River Ecology and Human Health (Eds. Ambasht, R.S. and Tripathi, B.D.). Nat. Environ. Con. Ass., pp 53 – 61.

Singare Pravin U., Lokhande Ram S., Pathak P., 2010. Soil Pollution along Kalwa Bridge at Thane Creek of Maharashtra, India. J. Environ. Protect., 1(20: 121-128.

Singh K.P., Malik A., Mohan D. and Sinha S., 2005. Persistent organochlorine pesticide residues in alluvial ground water aquifers of Gangetic plains, India. Bull. Environ. Contam. Toxicol., 74(4): 275-280.

Sinha R.K., Sinha S.K., Kedia D.K., Kumari A., Rani N., Sharma G. and Prasad K., 2007. A holistic study on mercury pollution in the Ganga River system at Varanasi, India. Curr. Sci., 92(9): 1223.

Smith B., 1988. A Workbook for Pollution Prevention by Source Reduction in Textile Wet Processing, Pollution Prevention Pays Program of the North Carolina Division of Environmental Management.

Stein E.D., Cohen Y. and Winer A.M., 1996. Environmental distribution and transformation of mercury compounds. Crit. Rev. Environ. Sci. Technol., 26; 1-43.

Storelli M.M., Busco V.P. and Marcotrigiano G.O., 2005. Mercury and arsenic speciation in the muscle tissue of Scyllorhinus canicula from the Mediterranean Sea. Bull. Environ. Contam. Toxicol., 75: 81-88.

Sukumarm G.B. and Sivakumar E.K.T., 1999. Hydrogeochemistry of River Cooum, Tamil Nadu, South India. Proc. Inst. Semin. Appl. Hydrogeochem. Annamalai Univ., pp 238-249.

Suwondo Fauziah Y., Syafrianti dan Wariyanti S., 2005, Akumulasi Logam Cupprung (Cu) dan Zincum (Zn) di Perairan Sungai Siak dengan Menggunakan Bioakumulator Eceng Gondok (*Eichhornia crassipes*), J. Biogen., 1(2): 51-56.

Takizawa Y. and Osame M. (Eds), 2001. Understanding Minamata disease: Methylmercury Poisoning in Minamata and Niigata Japan. Japan Pub. Helth. Assoc., pp 27-32.

Tamburlini G., Ehrenstein O.V. and Bertollini R., 2002. Children's Health and Environment: A Review of Evidence, In: Environmental Issue Report No. 129, WHO/European Environment Agency, WHO Geneva, pp 223.

Tandon K.K. and Dhaman S.K., 1967. An annotated list of fishes of Budha Nallah. Res. Bull. Punjab Univ., Chandigarh, 18: 53-59.

Toms L.M., Bartkow M.E., Symons R. and Paeller J.F., 2009. Assessment of polybrominated diphenyl ethers (PBDEs) in samples collected from indoor environments in South East Queensland, Australia.Chemosphere, 76: 173-8.

Trivedy R.C. and Dubey P. S., 1978. Environ. Pollut., 17:75.

Tulasi S.J., Reddy P.U.M. and Ramano Rao J.V., 1989. Effects of lead on the spawning potential of the freshwater Fish, *Anabas testudineus*. Bull. Environ. Contam. Toxicol., 43: 858-863.

Turnpenny A.W.H. 1985. The fish populations and water quality of some Welsh and Pennine strems. Cent. Elect. Res. Lab. Res. Note TPRD/L/2859/N85. Leatherhead, Survey.

Vareethiah K., Narayanan M. and Arochiasamy S., 1999. Zooplankton ecology in the retting zone of A.V.M. Canal. Ind. J. Environ. Ecoplan., 2: 141-145.

Vareethiah K., 2002. Sand mining in a bar-built estuary. In Kumar A. (Ed) Ecology of polluted waters Vol.1. APH Publ. Corp., New Delhi. Pp.563-575.

Villanueva F.S. and Botello A.V., 1998. Metal pollution in coastal areas of Mexico. Rev, Environ. Contam. Toxicol., 157: 53-94.

Vashney V., 2008., Ankleshwar ground water high on heavy metals. Down to Earth, dt. 31.05.2008.

Walters J.K., 1981. Water resources: Persective and effects: In Industrial Effluent Treatment. Vol. I, Water and solid wastes. Eds. J.K. Walters and A. Wint. pp 1-19.

Warwick P.E. and Croudace I.W., 2008. Measurement of Cd ^{113m} in irradiated cadmum shielding using liquid scintillation counting. In: Advances in liquid Scintillation Spectrometry; Eds., J. Eikenberg, M. Beer and H. Baehrle, pp 429-434.

White C., Sayer J.A. and Gadd G.M.., 1997. Microbial solubilization and immobilization of toxic metals: key biogeochemical processes for treatment of contamination. Microbiol. Rev., 20(3-4): 503-516.

Williams T.P., Bubb J.M. and Lester J.N., 1994. Metal accumulation with in salt marsh environments: a review. Marine pollution bulletin 28, : 277-290.

WHO., 1976. World Hlth. Org. Environ. Hlth criteria I- Mercury., p 121.

WHO., 1985. World Health Organization, 1985: Guidelines for drinking water Quality, Geneva, Vol 3.

Yusuff I.O. and Sonibare J.A., 2004. Characterization of textile industries' effluents in Kaduna, Nigeria and pollution implications. Global Nest: The Int. J., 6 (3): 212-22.

Chapter 5

Biological Concentration and Metal Availability to Fishes

Aquatic environment particularly the lakes and rivers constitute major biological productivity systems, which depend on high density and diversity of biotic components, inhabiting water bodies. The fish being an excellent source of digestible proteins, vitamins, minerals and polyunsaturated fatty acids, like Omega-3 fatty acid and its high digestibility makes it a valuable food source. Fish and sea food are further a good source of essential heavy metals (WHO, 1999; Irwandi and Farida, 2009). In the quest of increasing fish production, there has been considerable ignorance to water quality. Water pollutants exert stress on aquatic productivity including fish vis-a-vis the consumers. Even the sewage fed ponds are used in many areas for fish production, where recycling of organic loads provide nutrients to the growing fish. These also contain metals and much of various other kinds of toxicants, posing severe health hazards both to the fish and the consumers of these contaminated fish (Maiti and Banerjee, 2002). Many other workers have also contended and proved these metals to be highly injurious to fishes (Hoo *et al.*, 2004; Ayas *et al.*, 2007; Kumar and Achyuthan, 2007, Shukla *et al.*, 2007, Verma and Srivastava, 2008a,b and Srivastava and Verma, 2009). Rafia *et al.*, 2008, have contended that concentration of heavy metals in a fish has been a matter of serious concern because of the fact that these ultimately enter and accumulate in humans and cause damage to their organs. They analyzed metals in water samples from marine and freshwater resources and showed that these were in higher concentrations in marine water as compared to freshwater, leading to integrated effects on metabolic functions of the fish, vis-a-viz fish behavior, growth, reproduction and survival; thus, altering fish populations and community structure. As the heavy metal concentration in fish tissues reflects their exposures via water or food, it also demonstrates its physiological status and the ecological imbalance in a fish population in a metal contaminated aquatic environment. They suggested assessing potential toxicity of metals/or the contaminated effluents at the point of their discharge is important to avoid such ecological disasters and the detrimental effects of toxic metals in such a high quality food for millions of people throughout the world.

Metals in general does not occur in soluble form for long in natural waters and in the course of time these tend to accumulate in bottom sediments or these enter the food chain and subsequently endanger human and animal health (Nriagu, 1968; Valler and Ulmer, 1972). Broadly, metal uptake is described as a function of: a)- body size, higher flow of water over gills apparently results in greater metal uptake in smaller fish; b)- affinity of gill surface: depending upon the epithelial lining and the mucus layer; c)- presence of the metal binding legend in mucus: the legends, glycoproteins and mucopolysaccharides change with starvation and the water conditions like hardness, pH, temperature and salinity, later is known to reduce accumulation. Metals, which are present as the free ions or as a weak metal complex are more bio-available than in strong complexes or adsorbed to colloidal and/or particulate matter. Connell *et al.*, 1984 opined that intake of food was an important parameter affecting metal intake particularly in young fishes, and they further mentioned that:

1. Free metal ions that are absorbed through respiratory surface (*e.g.*, gills) are readily diffused into the blood stream.
2. Free metal ions that are adsorbed onto body surfaces are passively diffused into the blood stream.
3. Metals that are sorbed onto food and particulates may be ingested, as well as free ions ingested with water.

Sprague, 1985 has further concluded that tendency of a chemical to form complexes with organic and inorganic legends (primarily chlorides, carbonates, and hydroxides) also varies with the type of a metal absorbed. For example, Cu binds to organics much more readily than does either Cd or silver (Engel *et al.*, 1981). Khaled (1997) made a conclusive study of various factors affecting metal accumulations in fish tissues and he concluded that the impact of heavy metal wastes into aquatic environment may result in physical, chemical and a biological response, which can be categorized as effects of the environment on the metal, and likewise the effects of the metal on the environment.

1. Biological Concentration Factor (BCF)

For making an accurate prediction of a chemical toxicity effects, it becomes necessarily important to keep in mind the relative degree of chemical effects in all the fish organs, as well as the organs in which a chemical gets accumulated. The organization of a metal in an ecosystem is important as it plays a major role in its transfer to the higher trophic levels by affecting metal assimilation capability of a consumer (Munger *et al.*, 1999). Biological biomagnifications and bio-accumulation are the major characteristic features of metal toxicity in living organisms. Heavy metals are particularly severe in their action due to the tendency of biomagnification in the food chain. Monitoring heavy metal contamination in river systems using fish tissues further helps not only to assess the quality of an aquatic ecosystem (Adams, 2002), it also assays their bioavailability to higher consumers.

There are two modes of contaminant availability to fishes, one is water borne which are better absorbed through skin and gills; the others are diet born which appear absorbed via the gut. Different metals appear absorbed differently either via gills or gut. The process of accumulation of waterborne metals by fish and other aquatic animals through non-dietary routes is defined as **Bioconcentration** (Hemond and Fechner-Levy, 2000). This bioconcentration factor, which relates the concentration of metals in water to their concentration in an aquatic animal at equilibrium, is generally used to estimate the propensity to accumulate metals in an organism. Veith *et al.*, 1979 had suggested the term **bio-concentration** be limited to the accumulation of a chemical directly from the water, excluding that obtained via the food; while **bioaccumulation** would refer to accumulation from both food and water. Bioaccumulation of heavy metals in aquatic life, especially in a fish is of imperative interest owing to their direct toxic effects to aquatic animals and their detrimental effects on human health as the consumers (Fergusson, 1990). The biological concentration factor (BCF) describes the extent to which something accumulates in an aquatic organism. BCF is a unit less value obtained by dividing the concentration in one or more of the fish tissues by the average concentration of the pollutant in water (Heath, 1995). The BCF represents the capacity for a species to accumulate a compound to an extent that is greater than the background level. The plants and animals act as the natural reservoirs of metals as the metals being non-biodegradable remain stored in biological tissues un-metabolized and un-depurated. Measured or predicted BCFs are a requisite component for both aqua-cultural ecosystems and human health risk assessment (Liao *et al.*, 2003). Veith *et al.*, 1979 determined bio-concentration factor for 30 common organic chemicals in fathead minnows by exposing them for 32 days in the laboratory.

2. Metal Availability and the Water Conditions

Physico-chemical variables, such as hardness, pH, alkalinity and DOC may influence the speciation and bioavailability of metals (Markich *et al.*, 2002). Both bioavailability and toxicity of metals to aquatic organisms is said to depend on site-specific water chemistry factors, such as hardness, salinity, specific ion levels, pH, alkalinity, and dissolved organic matter (DOM), referred earlier as dissolved organic carbon. It is further postulated that knowledge of the relationship between water chemistry variables, including hardness and alkalinity, and metal toxicity is useful for predicting the potential biological detriments to an aquatic system, and can be of much use to modify 'National Water Quality' guidelines on site-specific basis. Markich and Camilleri (1998), in case of tropical Australian waters have reported Al, Cd, Cu, Ni, Mn, Pb, U, V and Zn as the metals of greatest concern largely as a result of mining activities, and the fact that their bio-availability appeared reduced as a function of increasing water hardness. The water chemistry of an aquatic system further controls the rate of adsorption and desorption of metals to and from sediments. Adsorption removes metal from the water column and stores in the substrate. Desorption returns metal to the water column, where recirculation and bio-assimilation may take place.

The solubility of trace metals in surface water is further said to be predominately controlled by the water pH, the type and concentration of a ligand on which the metal could adsorb, and the oxidation state of the mineral components and the redox potential of the system. Many data indicate that the concentrations of cadmium and lead, but not zinc, are considerably higher in the fish from acidified lakes (Grieb *et al.*, 1990; Haines and Brumbaugh, 1994; Wiener *et al.*, 1990; Horwitz *et al.*, 1995). Campbel and Stokes (1985) have reported Cd uptake in direct correlation to pH changes, but Pb exhibited just the opposite *i.e.* lowering the pH increased its uptake in fish exposed to metals. Water pH, means water acidification, a lower pH increases the competition between metal and hydrogen ions for binding sites. A decrease in pH may also dissolve metal-carbonate complexes, releasing free metal ions into the water column. Bradley and Sprague (1985) reported lowering pH to decrease zinc uptake into rainbow trout fish gills in soft water. On the contrary, Bentley (1992) observed increased zinc uptake at lower pH in channel catfish. Accumulation of copper was found also higher at lower pH (Cogun and Kargin 2004). There is, however, some contradiction about the effects of pH change in water upon uptake of methyl-Hg. Drummond *et al.* (1974) found a faster rate of accumulation of methyl-Hg into the gills and blood cells of brook trout at pH 6 than at pH 9. On the contrary, Rodgers *et al.* (1987) stated that pH over 5-7 had no effects on methyl-Hg accumulation in walleye and rainbow trout. Ponce and Bloom (1991) recorded methyl-Hg absorption in rainbow trout almost twice as fast at pH 5.8 as at 6.3. Playle and Wood (1991) specified that an increase in pH due to excretion of ammonia and carbon dioxide in the water around a fish causes precipitation and sloughing off of aluminium along with mucous from the gill surface. This change in microclimate near a fish, thus, allow little of aluminium to get into the fish (Spry and Wiener, 1991). Jezierska and Witeska (2006) concluded that water acidification affects bioaccumulation of metals by the fish in an indirect way, by changing solubility of metal compounds or directly, due to damage of epithelia which become more permeable to metals. They also suggested that competitive uptake of H^+ ions may also inhibit metal absorption.

Water hardness (presence of Ca cations in water) reduces metal absorption in a way that it affects membrane permeability to water and ions due to the increased positive charge outside the gill surface (Hunn, 1985). Eisler (1993) have shown that hardness (*i.e.* water [Ca^{2+}]) offers a protective effect against waterborne zinc toxicity. There are evidences that have shown that calcium inhibits branchial zinc uptake (Spry and Wood, 1989; Bentley, 1992; Hogstrand *et al.*, 1996), and correspondingly, that zinc competes with calcium uptake. This tends to repel the positively charged metal ions and reduces the metal uptake (Everall *et al.*, 1989; Wicklund and Runn, 1988). According to Playle *et al.* (1992), enrichment of water with calcium reduced copper accumulation in the gills. Likewise, Barron and Albeke (2000) indicated that calcium reduces zinc uptake in *Oncorhynchus mykiss.* Baldisserotto *et al.* (2005) reported that elevated dietary Ca^{2+} protect a fish against both dietary and waterborne Cd uptake. Kock *et al.* (1996) described the role of temperature in metal accumulation. They ascertained that higher uptake rates of cadmium and lead levels in *Salvelinus alpinus* liver and kidneys in summer was the result of high water temperature, as it increased metabolic rate. The data obtained by Douben (1989) also indicated increased rate of uptake and elimination of cadmium by *Noemacheilus barbatulus* with increased water temperature; the author has further suggested stronger effect of temperature on metal absorption than on elimination.

Metal accumulation is said also to be inversely related to the salinity (Thomson *et al.*, 1980; Mukhopadhyay and Konar, 1985). Metals may be desorbed from the sediment if the water experiences increase in salinity, decrease in redox potential, or decrease in pH (Connell *et al.*, 1984). According to Stagg and Shuttleworth (1982), seawater adapted *Platichthys flessus,* showed lower copper concentration than freshwater adapted individuals. The rate of lead accumulation by *Gillichthys mirabilis* was also reported as inversely proportional to the salinity of the medium (Somero *et al.*, 1977). Salinity increase *i.e.* elevated salt concentration creates an increased competition between the cations and metal ions for binding sites. Often metals will be driven off into the overlying water. Estuaries are prone to this phenomenon because of fluctuating river flow inputs. A decreased redox potential, as is often seen under oxygen deficient conditions, will change the composition of metal complexes and release metal ions into the overlying water. Sediment composed of fine sand and silt will generally have higher levels of adsorbed metal than that of a quartz, feldspar, and detritus carbonate-rich sediment.

Khaled (1997) concluded that natural conditions show interspecies differences on the accumulation of metals in fish tissues. The levels of metals in tissues of the benthic fishes are higher than in pelagic fishes, due to variations in their feeding habits, habitats and the behavior of the two species. The factors like salinity, DO, pH and inorganic and organic ligand affect speciation of metals. Lowering pH *i.e.* increasing the acidity of a solution increases solubility of metals and increases the concentrations of free metal ions in solution. This is due to the competition between H^+ and the metal ions for the binding sites on a ligand. For example, binding of the Cu ions to organic carbon decreases by increasing H+, as the hydrogen ion competes with Cu for binding sites on a carbon molecule. Due to this competitive relationship, there would be greater amounts of free Cu ion in solution that could precipitate gill secretions, causing thus, fish mortality by asphyxiation. Likewise, lead was stated to be more toxic at a pH value of 8.5 than at pH 6.5, it appear precipitated at a pH of 8.5. Same way at pH > 7, Zn begins to hydrolysis and form stable $Zn(OH)_2$ molecules.

3. Metal Absorption

Metals like other toxic chemicals have an accelerated release into the aquatic environment (Tulasi *et al.*, 1989), and their bio-magnification in the food chain (Eisler 1988; Clark *et al.*, 1997; Asuquo *et al.*, 1999; Asuquo *et al.*, 2004). Their persistence and bioaccumulation in food chains make them further toxic (Rishi and Jain, 1998) and highly deleterious in an aquatic environment. In an aquatic ecosystem, different organisms constitute different levels in a food chain and every food level contributes to about 100 to 10000 times multiplication in the metal residue level in the organisms at higher food chain level. Say, a water concentration of just less than 1ppm or even 1 ppb may reach up to 100ppm or even more after passing through planktons, herbivores, primary carnivores and ultimately in secondary or tertiary carnivores in an aquatic body. Among various metals, Hg is biomagnified easily, whereas, Cd and Pb seemingly do not. Shaw *et al.*, 1991 evidenced changes in the aquatic primary productivity in a estuary contaminated with Hg in water and estimated that as a result of biological magnification, freshwater macrophytes and planktons may concentrate Hg about 1000 times; a fish about 1000-3000 times and a shell fish up to 10 lac times. In a recent sample survey carried out in India, Pethkar *et al.* (2001) revealed that out of three thousand plant samples analyzed for heavy metals, about 20 per cent samples were recorded to contain lead, cadmium and chromium in concentrations exceeding I.0 ppm, which is the normal permissible limit for most of the heavy metals in food and drugs. Though, the World Health Organization has put a limit on the occurrence of the heavy metal in usable waters, but ecological magnification can further their hazards to the living world.

Heavy metals when present in the water source enter the food chain and accumulate in the fish that are often on the top of food chain and have the tendency to accumulate heavy metals (Mansour and Sidky, 2002). Chemical uptake, absorption, accumulation and excretion are fundamental in helping to understand the effects of an inorganic or organic chemical on a specific organ or a system in a fish. Gills, skin and kidneys, however, are the principal excretory routs in fish. In general, complexes of suspended inorganic and organic matter with metals decrease their bioavailability to aquatic biota. A short review concerning the role of organic matter on metal toxicity and bio-availability in aqueous systems was carried out by Calace and Petronio, 2004. In fact, the binding capacity and affinity is dependent on the number and type of ligands, on their position in the structures, on the ligand/metal ratio. In particular, concentrations of metals and their impact can be greatly modified due to interaction with chemical species present both in water and in sediments. Speciation studies have shown that metal affinity for different chemical species constitutes an important factor influencing metal distribution in the environment. For example, in particular 98-99 per cent of the Zn in oceanic surface water is present as organic complexes, whereas the Zn- organic forms in estuarine and freshwaters are 50-90 per cent of total.

High accumulation of various metals has been noted in the kidney of fish by Kumada *et al.* (1973), Dallinger *et al.* (1987), Mohammed Al- Mohanna (1994), Gupta and Sharma (1994) and Thiruvalluvan *et al.* (1997). Dallinger *et al.* (1987) recorded a high accumulation of several heavy metals in liver and kidney and suggested that these organs are target organs for final deposition of various heavy metals. There have been many other studies on metal accumulation in fish in different regions, like in Egypt (Saeed and Shaker, 2008); Finland (Voigt, 2007); India (Abid *et al.*, 2009); Jordan (Al-Weher, 2008); Malaysia (Irwandi and Farida, 2009); Nigeria (Olowu *et al.*, 2010); Pakistan (Rauf *et al.*, 2009); Saudi Arabia (Mohammed, 2009) and Turkey (Öztürk *et al.*, 2009). Since, metal concentrations found in fish tissues reflect the external concentrations; they become ideal tools for monitoring environmental levels. Investigations on quantitative accumulation of heavy metals in the fish, *Oreochromis nilotica* exposed to waste water showed lead and cadmium are accumulated nearly in a similar pattern in all the organs, whreas, copper is reported to accumulate more in liver and ovary (Maiti and Banerjee, 1999). Fish in general interact to water mainly through gills, and the gills are said to respire about 20 times faster than any terrestrial animal. Kumada *et al.* (1972) expressed that when the fish is exposed to heavy metals in an ambient medium, gill tissue rapidly accumulates the metal, since gills are always the main site of water movement (Motais *et al.*, 1969; Evans, 1993). According to Bryan, 1979 the mechanism of metal uptake through the gills is assumed mainly a simple diffusion. The gill tissue thus, serves as the main point of entery for any dissolved chemicals (Reid and McDonald, 1991). Several physico-chemical mechanisms such as the chemical concentration in water, molecular weight, molecular volume, solubility, rate of absorption or even presence of other adsorbents in the medium, influence the movement of chemicals across the biological membranes. Fish gill permeability is further affected by the water hardness, pH and temperature, thus, metal uptake by the gills. Metal accumulation in the tissues of fish, however, may vary according to the rates of uptake, storage, and elimination. This means that metals which have high uptake and low elimination rates in the tissues of fish are expected to be accumulated at higher levels.

The pattern of metal accumulation in fish tissue is further said to be not only metal specific, but also species specific and organ specific and the pattern of metal accumulation (Sultana and Rao, 1998). Each fish species possess a particular way to accumulate (and/or to eliminate) metal, when exposed to such a contaminant and analysis of edible part of any such fish (muscle) thus, will be important to investigate levels of metal contamination and direct transference of these heavy metals to a human population via fish as food. For example, metal concentrations in fish skin samples is said to decrease in the order of Cr>Pb>Ni>Co. In case of tilapia, Clements (1992) reported BCFs for arsenic in the range from 2 to 3.5 log units and the average arsenic concentrations in fish tissues as: 29.3, 10.9, 5.37, 5.04, and 3.55 μg g,$^{-1}$ respectively in intestine, stomach, liver, gill, and muscle. The order of tendency to accumulation of arsenic in tilapia tissues was reported as: intestine >stomach >liver =gill > muscle. The overall BCF's of arsenic was found to be highest in the intestine (1394). The other BCFs, of the stomach, liver, gill, and muscle, were 421, 180, 163, and 143, respectively. Takatsu and Uchiumi (1998) and Mason *et al.* (2000) also showed that arsenic concentrations are lower in muscle than in other tissues in big-scaled redfin, *Tribolodon bakonensis*. Fish gonads further play an important role in transferring contamination to the next generations, resulting in many chemical induced diseases or other kind of physiological ailments. These gonads are said to exhibit almost 1000 times biomagnifications of a chemical (Jhingran, 1991). The toxic chemicals may further tend to accumulate in a particular organ, which may not necessarily be the one with its highest concentration. For example, water pollutants like DDT tend to accumulate in fat where it produces no known effects; it acts instead on nerves and on the process of electrolyte and amino acid uptake in the gut.

A chemical after coming in contact with a fish either via gills, skin or gut is usually circulated in body tissues or organs via blood, where these come in contact with a variety of proteins or enzymes which metabolize these further for assimilation or accumulation into any storage organ like liver, or may be excreted via the kidneys. Proteins like metallothionein, plasma ceruplasmin or glutathione in blood after binding to these contaminants help in their transportation, accumulation or excretion. Cu is said to bind to ceruplasmin and Hg to hemoglobin. Laboratory studies by various workers have revealed that the industrial effluents and/or their constituents affect biochemical composition of the fish flesh leading to decline in its nutritive value (Palanichamy, *et al.*, 1990; Virk and Dhawan, 1997; Nanda *et al.*, 2000; Virk and Sharma 2003). The metals like Pb, Hg, Cd, Cr, Ni and As, appear further toxic to various fish functions because of their bioaccumulation in liver, kidney, gills and intestine. 'Uptake efficiency' is the most commonly applied term to how readily a fish removes a contaminant or a metal from the water via gills (Boddington *et al.*, 1979; McKim and Goeden, 1982). Khaled (1997) further concluded that the age of an organism can increase or reduce the uptake of heavy metals; cadmium and mercury accumulation was said to increase with age, whereas accumulation of Mn, Cr, Cu and Zn decrease with age. Due the similarity of cadmium with zinc, Cd has the ability to replace zinc from Zn/Zn-containing metalloenzymes in an irreversible reaction, which results in the alteration of the stereo structure of an enzyme and impairing its catalytic activity. Rainbow trout accumulates 70 per cent of its mercury from its food; a bivalve, *Macoma balthic* takes up about 82 per cent of cobalt from its food; the uptake of copper from the food was nearly similar to its uptake from water. Philips and Buhler (1978) also recorded 70 per cent absorption of Hg in rainbow trout, *Salmo gairdneri* via ingested food and only 10 per cent by the gills.

Dietary exposures of a fish to metals are said to be less toxic. Once a metal is absorbed via skin, gill or gut, is usually bound to the liver for transformation and/or storage, or excretion through bile or the urine. Storage however, may take place in liver, kidney, muscles or fats. Metal concentration acquired through bioaccumulation have been reported to be firstly a function of the body length and weight of the fish species (Barghigiani and Ranieri, 1992); secondly, it appear proportional to the levels in the aquatic environment in which the fauna reside (Akueshi *et al.*, 2003) and thirdly, variations in feeding habits of fish species (Kargin, 1996). Fishes in general, are said to accumulate metals in different tissues according to their nutritional level and body weight, as well as with increase in the metal levels in water. Measurements of bioaccumulation of iron, manganese, zinc, copper, nickel and lead by *Pseudocrenilabrus philander* from a mine-polluted impoundment revealed that there was an inverse relationship between metal concentrations and body mass of fish (De Wet *et al.* (1994). Ney and Van Hassel (1983) found that lead and zinc concentrations were higher in benthic fish. The results obtained by Campbell (1994) indicate that predators accumulated more zinc and nickel than benthivores, while the latter contained more cadmium. In fish, concentrations of most metals (except for mercury) are usually inversely related to the age and size. Kidwell *et al.* (1995) observed that predatory fish species accumulated more mercury, but the benthivores contained more cadmium and zinc. Higher concentrations of mercury in the predatory fishes comparing to the non-predatory ones was also reported by Voigt (2004).

Norstrom *et al.* (1976) developed a pollutant accumulation model, which had successfully predicted concentrations of PCBs and methylmercury in tissues of yellow perch (*P.flavescens*) from the Ottawa River, Canada. The model is based on pollutant biokinetics, coupled to fish energetic. These workers considered that uptake of pollutant from food is based on caloric requirements for respiration and growth coupled to concentrations of pollutant in food and its assimilation efficiency from the diet; and uptake of pollutant from water is based on flow of water past the gills for respiration coupled with concentration of pollutant in water and the efficiency of its assimilation by gills. Dallinger and Kautzky (1985) demonstrated that rainbow trout primarily accumulated metals through the diet when metal levels in the water were low. Harrison and Klaverkamp (1989) in rainbow trout and lake white fish exposed to cadmium in a continuous water flowing system found that Cd accumulated significantly in greater amounts through food rather than water. Many other workers (Miller *et al.*, 1993; Handy, 1996) have also emphasized upon diet as the major source of copper intake in a fish under optimal growth conditions. The liver accumulates a large proportion of the copper absorbed from the diet or water, and is the site for synthesis of the most abundant copper-containing protein in the body *i.e.* ceruloplasmin. Copper may also circulate in the body as Cu-albumin complex and other low-molecular mass proteins (Harris, 2000). The main site for secretion of excess copper in teleost fish is also ***via*** the bile (Grosell *et al.*, 1997, 2000).

Pollutant clearance is related to the body weight, raised to the power of -0.58, but is independent of metabolic rate. Under steady state conditions of chronic exposure, the predicted ratio of uptake to clearance is roughly constant at all weights, and the slope of a curve of log pollutant concentration in tissues vs. log body weight can be used to establish the exponent of body weight for clearance. Mackay (1991), Sijm *et al.* (1992), Landrum *et al.* (1994), and Wang *et al.* (1999) all reported that the process that cause a decrease in metal concentration in a larger fish may result from a growth dilution effect, suggesting that the key factor determining metal accumulation in fish seemed to be metabolism, such as (a) metabolic rate being linked with uptake/depuration rates and (b) small fish having a more rapid short-term uptake of metals and the like. Patrick and Loulit (1978) found younger fish to take zinc via food more rapidly than older ones as former have higher rates of metabolism. Mosquito fish (*Gambusia* sp.) was reported to show decreasing Zn accumulation with increasing body size (Newman and Mitz, 1988).

The mercury concentrations in various fish species (US EPA, 2001) are generally from about 0.05 to 1.4 mg/kg depending on factors such as pH and redox potential of the water, and species, age and size of the fish. Since mercury biomagnifies in the aquatic food web, fish higher on the food chain (or of higher trophic level) tend to have higher levels of mercury. Hence, large predatory fish, such as king mackerel, pike, shark, swordfish, walleye, barracuda, large tuna (as opposed to the small tuna usually used for canned tuna), scabbard and marlin, as well as seals and toothed whales, contain the highest concentrations. Salmon is a kind of fish which has good nutritive value to humans but it is reported to be contaminated with heavy metals, such as the arsenic (Budiati, 2010). Several other studies have suggested that stress and condition of salmon species are adversely affected by increases in heavy metal concentrations (Spry *et al.*, 1988). Reduced populations and poor survival of brown trout (*Salmo trutta*) has been attributed to heavy metal contamination of the Arkansas river in California (Lewis, 1987).

Moriarty *et al.* (1984) attributed most of the variations in metal contents of fish to body size are closely related to fish growth and metabolism. Most metabolic processes in animals itself are inversely proportional to body size; weight specific rate functions generally exhibit an exponent of 0.75 (Schmidt-Nielson, 1984). On the contrary, Murphy *et al.*, 1978 opined that metal accumulation in some fish is not related to body size. It was further stated that not all metals are equally absorbed, biomagnified or accumulated (Spry and Wiener, 1991). For example, Hg biomagnifies easily, whereas Cd and Pb seemingly do not, nor do the body concentrations of the later two metals increase with age or body size of the fish, as does Hg. According to Allen-Gill and Martynov (1995), an inverse correlation occurred between the age and lead content in *Coregonus clupeaformis*, and a similar relationship was found between accumulation of zinc, lead, cadmium and nickel and age in *Catostomus commersoni* (Ney and Van Hassel, 1983). The youngest fish showed the highest concentrations of metals, most distinct differences occurred for zinc. Negative relationships between fish length and metal concentrations (for Cr, Pb, and Cu) were also reported by Canli and Atli (2003). Field data show that mercury concentration in fish increases with age and size. Increase in body mercury level with fish age and size is probably related to the affinity of this metal to the muscle tissue (Goldstein *et al.*, 1996; Voigt, 2004).

Deshmukh and Marathe (1979) while recording 96 h LC_{50} values of Cu and Hg in the fry and the fingerlings of *L. rohita* and *C. carpio*, observed that the fish susceptibility varied with the size, and the fry being more susceptible. Similarly, the 96 h LC_{50} values of 3 size groups of *L. raticulatus* were found directly significantly proportionate to the body weight. This means to say that small animals of the same species operate at a higher rate than large ones. Concentrations of most trace metals in tissues of small- or medium-sized fish species usually decrease with increase in fish size (Al-Yousuf *et al.*, 2000). A significant correlation of arsnic concentrations was also demonstrated in tissues (muscle) with body weight in fish species, *Liza macrolepis* (Lin *et al.*, 2001), *Mugil cephalus* (Maher *et al.*, 1999), *A. forsteri* (Francesconi *et al.*, 1989), *Thunnus thynnus* and *T. toggle* (Ashraf and Jaffer, 1988), *S. maculata* (Edmonds and Francesconi, 1981), *Boreogadus saida* (Bohn and McElroy, 1976), *Reinhardtius hippoglossoides* (Bohn, 1975), *Anarchichas minor* (Bohn, 1975), and *Myoxocephalus scorpius* (Bohn, 1975; Bohn and Fallis, 1978). Maher *et al.* (1999) reported that arsenic concentrations in the intestine, stomach, gill, and muscle of *M. cephalus* decreased with fish body weight. They also pointed out that no significant differences in the As concentrations in fish were found by gender or by age.

These metals being non-degradable in soil and water systems as well as due to their lipophillic nature, these are known to bio-accumulate in animal tissues, thus, persisting longer in the environment and are not destroyed biologically; such metals, however, are merely transformed from one state to another in various organic complexes. Reid and McDonald, 1991 reported that metals are absorbed by fish in the ionic form, except in case of methyl-Hg, and according to Julshamm *et al.*, 1982 dietary methyl Hg is about 10 times more readily accumulated than dietary inorganic Hg into the tissues. Shaw (1987) has reported that while most Hg in the environment is inorganic, some is converted to the highly toxic mono-methyl which bioaccumulate in aquatic biotic forms; whereas some other forms like methoxy-methyl Hg, ethyl- and phenyl Hg were said to be converted into inorganic Hg. Anoxic conditions are said to favor net methylation in an ecosystem. Inorganic Hg, however, is not methylated directly by the fish, except that some of it may be methylated by the bacteria present in its gut, and methylation in gut however, make Hg easily bioavailable for absorption. MeHg contamination poses a particular challenge to public health because this toxicant is mainly contained in fish, which is a highly nutritious food with known benefits for human health. MeHg is bound to proteins and amino acids that are major components of muscles and are not removed by any cooking process. McKim *et al.*, 1976 stated that fish normally eliminate methyl Hg very slowly compared to its uptake rate, whereas inorganic Hg is excreted at a faster rate than that of the methyl Hg (Spry and Wiener, 1991), and overt toxicity due to methyl Hg does not occur until whole body concentration exceeds 10-30 μg/l ww of fish. Further studies on the Hg contents in some species of marine fish and mollusks (Caviglia and Cugurra, 1978) also revealed presence of Hg in organic form as methyl-Hg, which infact are biologically methylated by some microorganisms; present in sea and lake bottom (Jensen and Jernelov, 1969). Organic matter particularly higher molecular weight can limit Hg availability to bacteria such as the sulphate reducing bacteria or the Fe reducing bacteria, which have demonstrated methylation in pure culture. The Hg–S cycles are intimately linked, thus, linking acid rain to the Hg cycle. Sulphate and acidity stimulate methylation (Kamman *et al.*, 2004). Net methylation in the salt marshes can also be high (King *et al.*, 2001). MeHg concentrations are generally maximal at or near the sediment surface (Hines *et al.*, 2004). Hg availability for methylation is further reduced with Hg aging. Methyl mercury uptake is impacted by nature (pelagic vs benthic) and structure of the food web relative to the location of MeHg production (Gorski *et al.*, 2003) such as in the sediments or the lower portion of food web *i.e.* planktons as well as the fish population. Panday *et al.* (2005) in air-breathing teleost fish, *Channa punctatus* (Bloch) indicated that mercuric chloride is more toxic than malathion.

Because of the presence of negative charge on the outer surface of the gill tissue, it frequently attracts metallic ions. The affinity of the gill for metals tested differed as per the microenvironment of the gill surface and gill surface being further a complex structure because of the epithelial membrane, mucous, and a mixture of glycoproteins, mucopolysaccharides, assorted low molecular weight compounds and water. Altered gill structure due to metal toxicity further affects metal uptake, as observed in rainbow trout fish. Increased uptake of methyl-Hg was said to occur due to the damaged gill surface by the inorganic Hg (Rodgers and Beamish, 1983). Metal absorption at gill surface is said to be inversely proportional to its metal affinity, *i.e.* La> Ca= Cd> Cu. Interestingly contrary to this, absorption of metals into the gill epithelial cells is not as per their affinity to gill surface, rather it is said to be inversely proportional to the surface binding affinity. Thus, Cu, which has a low affinity, is absorbed into cells much more readily than either Ca or Cd (Read and Mc Donald 1991).

Mucous on the gill surface further play an important role in the accumulation of metals by the gills (Handy and Eddy, 1990a), *i.e.* a starved fish generally accumulate more of zinc on their gill surface than a well fed fish, probably due to the changed composition *i.e.* presence of more of metal binding legend in the mucous during starvation. Acute levels of certain metals like Cd, Pb, Zn and Cu, stimulate mucus secretion. Thus, it could serve as a way of reducing metal uptake rate by Chelation of metals and inhibition of diffusion (Part and Lock, 1983; Handy and Eddy, 1990b). Dai *et al.* (2009) indicated that the content of Fe, Cu, and Zn in liver and kidney decreased with the increasing dietary Pb concentrations.

Fish exposed to high concentrations of trace metals in water may take up substantial quantities of these metals. It is generally known that many freshwater fish organism are capable of concentrating trace elements within their tissues, although their concentration is at very low in water (Dallas and Day, 1993). Their bioaccumulation is further said to be species specific and their analysis can also serve as an indicator of pollution (Hellawell, 1988). Zn in certain concentration is desirable for the growth of freshwater animals but it's over accumulation is hazardous to an exposed organism as well as to those who consume them directly or indirectly through food chain. Zn damages gills hence lesser intake of Zn, but anoxia causes more intake of Cu, Cr and Pb under increased ventilation. (Freeman, 1980) Zn at 0.1 mg/l causes complete kill of rainbow trout but complete survival of brook trout. Holcombe and Andrew (1978) reported that brook trout was approximately 2.7 times more resistant to zinc toxicity than rainbow trout. Zinc toxicity in both the species was, however, found increased with the increase in pH, but decreased with the increasing hardness and alkalinity. Salmon are more sensitive to Zn than rainbow trout, and minnow more sensitive than trout fathead minnow or even rainbow trout usually survive at 8 ppm of Zn or 0.2 ppm of Cu in soft water * 8 hr exposure*, but succumb at even 8 times less concentration of Zn + Cu. Hogstrand *et al.* (1994) suggested an adaptation to water borne Zn by a change of *Km* of a mutual Ca^{2+}/Zn^{2+} carrier which may have reduced Zn influx. When exposure to high Zn level occurs and the liver's capacity may be removed by the exceeded level of Zn and the more toxic type of Zn (Zn^{2+}) may be transported through blood stream to other organs. In various tissues of *Channa punctatus* exposed to two treatments of zinc (6.62 and 13.24 mg L^{-1}) for a period of 45 days by Murugan *et al.* (2008), the pattern of zinc residue was found in the order of: liver > kidney > intestine > gill > muscle. Zinc accumulated significantly ($P<0.05$) in all the tissues except in muscle revealing further that liver and kidney are the prime sites of Zn accumulation. A preliminary biochemical investigation on the nuclei/cell debris fraction of the digestive tract tissue of common carp suggests that the main Zn binding substance(s) were probably protein(s). Certain amino acids like histidine and cysteine with high affinity for zinc also enhance its bioavailability (Glover and Hogstrand, 2002).

4. Metal Depuration

Although metal elimination routes are more abundant than the uptake routes, yet the metal accumulation occurs more rapidly than their removal, which is probably due to the presence of metal binding proteins in the tissues of the aquatic organisms (Kargin and Cogun, 1999). The term depuration is often used to refer riddling of some chemical from the body of the fish after the animal is put in non-contaminated water, or after the animal is fed the test chemical. Whole body depuration is inversely proportional to the body size, so small fish are able to do so more rapidly than the large ones (Sharpe *et al.*, 1977). Tawari-Fufeyin and Ekaye (2007) attributed metal bioaccumulation largely to differences in uptake as well as the depuration period for various metals in different fish species. The depuration of Cd is reported to be mediated principally by an active mechanism, possibly involving the accumulation in lysosome with a subsequent excretion of metal rich residual bodies (George and Viarengo, 1985). Mucus is also reported to bind with Pb for accumulation and subsequent depuration (Part and Lock, 1983) It in general reduce uptake of metals into fish, and *in vitro* preparations it was noticed to slow the diffusion rate of metals like Hg and Cd, but not of Ca. Part and Lock (1983) also found rainbow trout to exhibit higher tendency of gill to accumulate Cd, also rapid depuration from gut and gill. Its accumulation is 10 times more via food than direct from water.

According to Neumann and Mitz (1988) fish with higher body weights, have lower capacity to depurate metals. They showed in mosquito fish (*Gambusia* sp.) decrease in Zn accumulation with increasing body size (Newman and Mitz, 1988) as these being very small in size, depurated Zn more rapidly. They further recorded not many variations in the pattern of accumulation of lead and cadmium in these fish tissues. Fish can excrete at least to some extent, Cu, Cr, Cd, Hg, Se, As, Pb, and Zn (Bryan, 1976; Hodson *et al.*, 1980; Somero *et al.*, 1977). Depuration of Cd was faster for gut and gills, but not for kidneys. Gills, bile, kidney, and skin are the principle routes of excretion of metals in fish. Under continuous exposure to water borne chemical, the excretory process may take several hours or days to become activated, increasing thus, the body burden rapidly and actually declining somewhat thereafter. Like mammals, fish liver is a principle organ participating in the excretion of some metals via bile (Klassen, 1976), such as Cr (Buhler *et al.*, 1977), As (Sorenson *et al.*, 1979), and Cu (Felts and Heath, 1984). Hg and Zn, however, are excreted in a small amount in fish bile (Massaro, 1974; Hardy *et al.*, 1987). Excretion of metal via bile in fact lead to recycling of metal as it is again mixed up with the food in upper part of the intestine in which food is digested, and metals along with the digested food again absorbed into the blood. This process is called entero-hepatic circulation' (Klassen, 1976) and a chemical may be circulated more than once in the liver via bile and blood. Many workers have evidenced excretion of Zn, (Hardy *et al.*, 1987), Hg (Olson *et al.*, 1978), and As (Oladimeji *et al.*, 1984b) via the gills. In case of excretion of Hg methyl-Hg is first dimethylated and then excreted via gill. Kidneys are reported as poor excretory organs in fish (Rodgers and Beamuth, 1982); these are said to store some metals only for sequestering them such as Pb and Hg in salmonids (Reichert *et al.*, 1979).

The marine fish kidney since lack glomeruli thus, has little ability to excrete metals (Smith, 1982). Oronsaye and Brafield, 1984 suggested that exposure to some waterborne metals like Zn, Cu and Cd has been shown to stimulate the development of more chloride cells in the gills, which might be playing an important role in the excretion of these metals. This might be an osmoregulatory response. Loss of metals by skin and gills may involve mucous, as this proteinaceous material is constantly secreted and sloughed off by these tissues (Varanasi and Markey, 1978). Vijayram *et al.* (1989) determined the uptake and loss of Cd and Zn in various tissues of freshwater prawn, *Macrophthalamus malcolmsonii* and they reported different levels of depuration of both these metals in prawns. El-Shahawi and Al-Yousuf, 1998 reported sloughing of the epidermis and also of the dermis at a few places in the skin of most of the metal treated fishes. They further observed Ni, Co, Cr and Pb concentrations in the skin tissues of the fish, *Lethrinus lentjus* collected from the coastal waters of the United Arab Emirates between May and June 1993, decreased in the order of: Cr>Pb>Ni>Co. Ahmed and Al-Ghais (1996) made a comparison of Cd, Ni, Pb, Cu, Mn, and Zn in the muscle, liver and skin of *Epinepliehus tauvina* and *L. fulviflamma*. It was observed that the muscle tissue had a higher concentration of metals than the skin but markedly lower concentration as compared to the liver. Norrgren *et al.* (1991) have reported a decrease in aluminium of the kidney in *Phoxinus phoxinus* during recovery period at pH-7 in aluminium free water after pre-exposure to aluminium in an acidic medium for 12 days.

REFERENCES

Abid A.B., Ramalah M., Harikrishna A., Irfanulla K. and Veena K., 2009. Heavy metal pollution and chemical profile of Cauvery River Water. E-Journal of Chem., 6(1): 47-52.

Adams S. M. (Ed.), 2002. Biological Indicators of Aquatic Ecosystem Stress. (ed. S. M. Adams), Bethesda, USA: American Fisheries Society.

Ahmad S. and Al–Ghasis S.M., 1996. Metal contents in the tissues of *Lutjanus fulviflamma* (Smith, 1949) and *Epinephelus tauvina* collected from the Arabian Gulf. Bull. Environ. Contam. Toxicol., 57: 957-963.

Akueshi E.U., Oriegie E., Ocheskiti N. and Okunsebor S., 2003. Levels of some heavy metals in fish from mining lakes on the Jos Plateau, Nigeria. Afr. J. Nat. Sci., 6:82-86.

Allen-Gill S.M. and Martynov V.G., 1995. Heavy metal burdens in nine species of freshwater and anadromous fish from the Pechora River, northern Russia, Sci. Total Envron., 160/161:653–659.Al-Weher, 2008);

Al-Weher S.M., 2008. Levels of Heavy Metal Cd, Cu and Zn in three fish species collected from the Northern Jordan Valley, Jordan. Jordan J.Biol. Sci., 1(1): 41-46.

Al-Yousuf M.H., El-Shahawi M.S. and Al-Ghais S.M., 2000. Trace metals in liver, skin and muscle of *Lethrinus lentjan* fish species in relation to body length and sex. Sci. Total Environ., 256:87–94.

Ashraf M. and Jaffer M., 1988. Weight dependence of arsenic concentration in the Arabian Sea tuna fish. Bull. Environ. Contam. Toxicol., 40:219 –225.

Asuquo F.E., Ewa-Oboho I., Asuquo E.F. and Udo P.J., 2004. Fish species used as biomarker for heavy metal and hydrocarbon contamination for Cross river, Nigeria. The Environmentalist, 2: 29–37.

Asuquo F.E., Ogri O.R., and Bassey, E.S., 1999. Distribution of heavy metals and total hydrocarbons in coastal waters and sediments of cross River State, South Eastern Nigeria. Intl. J. Trop. Environ., 2: 229–242.

Ayas Z., Ekmerkci G., Yerli S.V. and Ozmen M., 2007. Heavy metal accumulation in water, sediments and fishes of Nallihan Bird Paradise, Turkey. J. Environ. Biol., 28, 545-549.

Baldisserottoa B.; Chowdhury M.J.; Wood Chris M., 2005. Effects of dietary calcium and cadmium on cadmium accumulation, calcium and cadmium uptake from the water, and their interactions in juvenile rainbow trout. Aquat. Toxicol., 72: 99–117.

Barghigiani C. and Ranieri De.S., 1992. Mercury content in different sizes of important edible species of the northern Tyrrhenian Sea. Mar. Poll. Bull., 24: 114-116.

Barron M.G. and Albeke S., 2000. Calcium control of zinc uptake in rainbow trout. Aquat. Toxicol., 50:257–264.

Bentley P.J., 1992. Influx of zinc by channel catfish (*Ictalurus punctatus*): uptake from external environmental solutions. Comp. Biochem. Physiol., 101C: 215-217.

Boddington M.J., MacKenzie B.A. and DeFreitas A.S.W., 1979. A respirometer to measure the uptake efficiency of water borne contaminants in fish. Ecotoxicol. Environment. Saft. 3: 383.

Bohn A., 1975. Arsenic in marine organisms. Mar. Pollut. Bull., 6: 87–89.

Bohn A. and Fallis B.W., 1978. Metals concentration (As, Cd, Cu, Fe and Zn) in shorthorn sculpins, *Myoxocephalus scorpius* (Linnaeus) and arctic char *Salvelinus alpinus* (Linnaeus) from the vicinity of Strathcona Sound, Northwest Territories. Wat. Res., 12: 659–663.

Bohn A. and McElroy R.O., 1976. Trace metals (As, Cd, Cu, Fe and Zn) in arctic cod *Boreogadus saida* and selected zooplankton from Strathcona Sound, Northern Baffin Island. J. Fish Res. Board Can., 33: 2836 –2840.

Bradley R.W. and Sprague J.B., 1985. Accumulation of zinc by rainbow trout as influenced by pH, water hardness and fish size. Envirn. Toxicol. Chem. 4: 685.

Bryan G.W., 1976. Heavy metal contamination in the sea. In R. Johnston, ed., Marine Pollution, Academic Press,London:185 - 302.

Bryan G.W., 1979. Bioaccumulation of marine pollutants. Philos. Trans. Res. Soc. London B., 286: 483.

Budiati T., 2010. The presence of arsenic as heavy metal contaminant on salmon: a risk assessment. Internat. J. Basic and Appl. Sci., 10: 6-12.

Buhler D.R., Stokes R.M. and Caldwell R.S., 1977. Tissue accumulation and enzymatic effects of hexavalent chromium in rainbow trout (*Salmo gairdneri*). J. Fish Res. Bd. Can., 34: 9.

Campbel P.G.C. and Stokes P.M., 1985. Acidification and toxicity of metals to aquatic biota. Can. J. Fish Aquatic Sci., 42: 2034.

Campbell K. R., 1994. Concentrations of heavy metals associated with urban runoff in fish living in storm water treatment ponds, Arch. Environ. Contam. Toxicol., 27: 352–356.

Canli M. and Atli G., 2003.The relationships between heavy metal (Cd, Cr, Cu, Fe, Pb, Zn) levels and the size of six Mediterranean fish species. Environ. Pollut., 121(1):129-136.

Caviglia A. and Cugurra F., 1978. Further studies on the mercury contents in some species of marine fish and mollusks. In Springer-Verlag, (0007-4861/78/0019-0528),1978, NY

Clark R., Frid C., and Attrill M., 1997. Marine pollution (4thed.). New York: Oxford University Press.

Clements William H., 1992. Bioaccumulation and food chain transfer of polycyclic aromatic hydrocarbons and heavy metals: a laboratory and field investigation. Ph.D. Thesis, Department of Fishery and Wildlife Biology, Colorado State University, Fort Collins., CO 80523.

Cogun H.Y. and Kargin F., 2004, Effects of pH on the mortality and accumulation of copper in tissues of *Oreochromis niloticus*, Chemosphere, 55: 277–282.

Connell B.S., Cox M. and Singer I., 1984. Nickel and chromium. In: Brunner F. and Coburn J.W. (Eds). Disorders of minerals metabolism. Academic Press, NY. Pp 472-532.

Dai Wei, Fu Linglin, Du Huahua, Jin Chengguan and Xu Zirong, 2009. Changes in growth performance, metabolic enzyme activities, and content of Fe, Cu, and Zn in liver and kidney of tilapia (*Oreochromis niloticus*) exposed to dietary Pb. Biol. Trace Elem. Res., 128(2):176-183.

Dallas H.F. and Day J.A., 1993. The effect of water quality variables on riverine ecosystems: A review. Water Research Commission Project No. 351.Pretoria, South Africa. pp 240.

Dallinger R., and Kautzky H., 1985. The importance of contaminated food uptake for the heavy metals by rainbow trout (*Salmo gairdneni*: a field study. Oecologia (Berl.), 67: 82-89.

Dallinger R., Prosi F., Segner H. and Back H., 1987. Contaminated food and uptake of heavy metals by fish: a review and a proposal for further research. Oecologia (Berl), 73: 91-98.

De Wet L.M., Schoonbee H. J., De Wet L.P.D., and Wiid A.J.B., 1994. Bioaccumulation of metals by the southern mouthbrooder, *Pseudocrenilabrus philander* (Weber, 1897) from a mine-polluted impoundment. Water SA., 20:119–126.

Deshmukh S.S. and Marathe V.B., 1979. Size related toxicity of copper and mercury to *Lepisteus reticulatus*, *Labeo rohita* and *Cyprinus carpio*. Ind. J. Exp. Biol., 18: 421-423.

Douben P. E. T., 1989. Uptake and elimination of waterborne cadmium by the fish *Noemacheilus barbatulus* L. (stone loach), Arch. Environ. Contam. Toxicol., 18: 576–586.

Drummond R.A., Olson G.F. and Batterman A.R., 1974. Cough response and uptake of mercury by brook trout, Salvelinus fontinalis, exposed to mercuric compounds at different hydrogen-ion-concentrations. Trans. Am. Fish Soc., 103: 244.

Edmonds J.S. and Francesconi K.A., 1981. The origin and chemical form of arsenic in the school whiting. Mar. Pollut. Bull., 12: 92–96.

Eisler R., 1988. Lead hazards to fish, wildlife, and invertebrates: a synoptic review. US Fish Wildlife Serv. Biol. Rep., 85: 1–4.

Eisler R. 1993. Zinc hazards to fish, wildlife, and invertebrate: a synoptic review. In Biological Report 10. Washington, DC: US Department of Interior, Fish and Wildlife Service.

El-Shahawi M.S. and Al-Yousuf M.H., 1998. Heavy metal (Ni, Co, Cr and Pb) contamination in liver and skin tissues of *Lethrinus lentjan* fish. Family: Lethrinidae (Teleost) from the Arabian Gulf. Int. J. Food Sci. Nutr., 49 (6): 447-451.

Engel D.W., Sunda W.G. and Fowler B.A., 1981. Factors affecting trace metal uptake and toxicity to estuarine organisms. 1. Environmental parameters. In: Biological monitoring of marine pollutants. Eds. J. Vemberg, A. Calabrese, F. Thurberg and Vemberg W.B. Acad. Press, NY: pp127.

Evans D., 1993. Osmotic and ionic regulation. In: The Physiology of Fishes, CRC Press, Boca Raton, FL. (Ch 11).

Everall N.C., Macfarlane N.A.A. and Sedgwick R.W., 1989. The interaction of water hardness and pH with the acute toxicity of Zinc to the brown trout *Salmo trutta* L. J. Fish Biol., 35: 27-36.

Felts P.A. and Heath A.G., 1984. Interactions of temperature and sublethal environmental copper exposure on the energy metabolism of bluegill, *Lepomis macrochrus*. J. Fish Biol., 25: 445.

Fergusson J.E., 1990. The heavy elements: Chemistry, Environmental Impact and Health, Pergamon Press, Oxford. p 614.

Francesconi K.A., Edmonds J.S. and Stick R.V., 1989. Accumulation of arsenic in yellow eye mullet (*Aldrichetta forsteri*) following oral administration of organoarsenic compounds and arsenate. Sci. Total Environ., 79:59–67.

Freeman B.J., 1980. Accumulation of cadmium, chromium, and lead by bluegill sunfish (*Lepomis macrochirus*) under temperature and oxygen stress, Ph.D. dissertation, university of Georgia, Athens.

George S.G and Viarengo A., 1985. An integration of current knowledge of uptake, metabolism and intracellular control of heavy metals in Mussels. In: Vernberg F.J., Thurberg F.P., Calabrese A. and Vernberg W.B. (Eds). Marine Pollution and Physiology: Recent Advances University of South Carolina Press, Columbia:125 – 144.

Glover C. N. and Hogstrand, C. 2002a. *In vivo* characterisation of intestinal zinc uptake in freshwater rainbow trout. J. Exp. Biol., 205: 141 -150.

Goldstein R.M., Brigham M.E., and Stauffer J.C., 1996, Comparison of mercury concentrations in liver, muscle, whole bodies, and composites of fish from Red River of the North, Can. J. Fish. Aquat. Sci., 53: 244–252.

Gorski P.R., Cleckner L.B., Hurley J.P., Sierszen M.E. and Armstrong D.E., 2003. Factors afffecting enhanced mercury bioaccumulation in inland lakes of Isle Royale National Park, USA. Sci Total Environ., 304: 327–348.

Grieb T.M., Driscoll C.T., Gloss S.P., Schofield C.L., Bowie G.L., and Porcella D.B.,1990. Factors affecting mercury accumulation in fish in the upper Michigan Peninsula, Environ. Toxicol. Chem., 9:919–930.

Grosell M.H., Hogstrand C. and Wood C.M., 1997. *Cu uptake and turnover in both Cu-acclimated and non-acclimated rainbow trout (Oncorhynchus mykiss).* Aquat. Toxicol., 38*:* 257 *-276.*

Grosell M., O'Donnell M.J. and Wood C.M., 2000. *Hepatic versus gallbladder bile composition: in vivo transport physiology of the gallbladder in rainbow trout.* Am. J. Physiol. Reg., I. 278*:* R1674 *-R1684.*

Gupta A.K. and Sharma S.K., 1994. Bioaccumulation of zinc in Cirrihinus mrigala (Hamilton) fingerlings during short-term static bioassay. J. Environ. Biol., 15(3): 231-237.

Haines T.A., and Brumbaugh W.G., 1994. Metal concentration in the gill, gastrointestinal tract, and carcass of white suckers (*Catostomus commersoni*) in relation to lake acidity. Water Air Soil Pollut., 73:265–274.

Handy R.D. and Eddy F.B., 1990a. The interaction between the surface of rainbow trout, *Onchorhynchus mykiss* and waterborne metal toxicants. Funct. Ecol., 4: 385.

Handy R.D. and Eddy F.B., 1990b. Influence of starvation on water borne zinc accumulation by rainbow trout, *Salmo gairdneri,* at the onset of episodic exposure in natural soft water. Wat. Res., 24(4): 521-527.

Handy R.D., 1996. Dietary Exposure to Toxic Metals in Fish. In: Toxicology of Aquatic Pollution: Physiological, Cellular and Molecular Approaches, Taylor, E.W. (Ed.). Cambridge University Press, Cambridge, England, ISBN: 0521455243, pp: 29-60.

Hardy R.W., Sullivan C.V. and Koziol A.M., 1987. Adsorption, body distribution and excretion of dietary zinc by rainbow trout. FDish Physiol. Biochem., 3: 133.

Harris E.D., 2000. Cellular copper transport and metabolism. Annu. Rev. Nutr., 20: 291 -310.

Harrison S.E. and Klaverkamp J.F., 1989. Uptake, elimination and tissue distribution of dietary and aquous cadmium by rainbow trout (*Salmo gairdneri* Richardson) and lake white fish (*Coregonus clupeaformis* Mitchell). Environ. Toxicol. Chem., 8: 87-97.

Heath A.G., 1995. Water pollution and Fish physiology. Second Ed. Lewis Publishers: 359 pages.

Hellawell J.M., 1988. Toxic substances in rivers and streams. Environ. Pollut., 50: 61-85.

Hemond H.F. and Fechner-Levy E.J., 2000. Chemical fate and transport in the environment. 2nd ed. New York: Academic Press. 433 p.

Hines N.A., Brezonik P.L. and Engstrom D.R., 2004. Sediments and porewater profiles and fluxes of mercury and methylmercury in a small seepage lake in northern Minnesota. Environ. Sci. Tech., 38; 6610-6617.

Hodson P.V., Spry D.J. and Blunt B.R., 1980. Effects on rainbow trout, Salmo gairdneri) of a chronic exposure to waterborne selenium. Can. J. Fish Aquat. Sci., 37: 233.

Hogstrand C., Wilson R.W., Polgar D. and Wood C.M., 1994. Effects of zinc on the kinetics of branchial calcium uptake in freshwater rainbow trout during adaptation to waterborne zinc. J. Exp. Biol., 186:55–73.

Hogstrand, C., Verbost, P. M., Wendelaar Bonga, S. E. and Wood, C. M. 1996. Mechanisms of zinc uptake in gills of freshwater rainbow trout: interplay with calcium transport. Am. J. Physiol. 270, R1141 -R1147.

Holcombe G.W. and Andrew R.W., 1978. The acute toxicity of zinc to rainbow and brook trout- comparison in hard and soft water. US Deptt. Commerce, Nat. Tech. Inf. Serv., PB-289 939.

Hoo L.S., Samat A. and Othman M.R., 2004. The level of selected heavy metals (Cd, Cu, Fe, Pb, Mn and Zn) at residential area nearby Labs river system riverbank, Malaysia. Res. J. Chem. Environ., 8: 24-29.

Horwitz R. J., Ruppel B., Wisniewski S., Kiry P., Hermanson M. and Gilmour C., 1995, Mercury concentrations in freshwater fishes in New Jersey, Water Air Soil Pollut., 80: 885–888.

Hunn J.B. 1985. Role of calcium in gill function in freshwater fishes. Comp. Biochem. Physiol. 82A: 543.

Irwand J. and Farida O., 2009. Mineral and heavy metal contents of marine fin fish in Langkawi island, Malaysia. Internat. Food Res. J., 16: 105-112.

Jensen S. and Jernelov A., 1969. In: The biochemistry of mercury in the environment (Ed. Nriagu J.O.). Elsevier Pub. Amsterdam: pp 203-210.

Jezierska B., and Witeska M., 2001, Metal Toxicity to Fish, Wydawnictwo Akademii Podlaskiej, Siedlce, 318 pp., 2006?

Jhingran V.G., 1991a. Fish and Fisheries of India. Hindustan publishing Corporation, New Delhi, India.

Julshamm K., Ringdal O. and Brackkan O.R., 1982. Mercury concentration in liver and muscle of cod (*Gadus morhua*) as an evidence of migration between waters with different levels of mercury. Bull. Environ. Contam. Toxicol., 29: 544.

Kamman N.C. and Burgess N.M., 2004. Mercury in freshwater fish tissues of northeast North America: A geographic perspective based on fish tissue monitoring databases. Presentation to U.S.EPA. Mercury Round Table Conference cell.

Kargin F. and Cogun H.Y., 1999. Metal interactions during accumulation and elimination of zinc and cadmium in tissues of the freshwater fish, *Tilapia nilotica*. Bull. Environ. Contam. Toxicol., 63: 511-519.

Kargin F., 1996. Seasonal changes in levels of heavy metals in tissues of Mullus barbatus and Trachurus mediterraneus collected from Iskenderua gulf (Turkey). Water Air Soil Pollut., 90: 557-562.

Khaled Azza M.M., 1997. A comparative study for distribution of some heavy metal in Aquatic organisma fished from Alexandria Region.

Kidwell J.M., Phillips L.J. and Birchar G.F., 1995, Comparative analyses ofcontaminant levels in bottom feeding and predatory fish using the national contaminant biomonitoring program data, Bull. Environ. Contam. Toxicol., 54:919–923.

King J.K., Kostka J.E., Frischer M.E., Saunders F.M. and Jahnke R.A., 2001. A quantitative relationship that demonstrates merury methylation rates in marine sediments are based on the community ccomposition and activity of sulphate-reducing bacteria. Environ. Sci. Technol., 35: 2491-2496.

Klassen C.D., 1976. Biliary excretion of metals. Drug. Metab. Rev., 5: 165.

Kock G., Triendl M., and Hofer R., 1996, Seasonal patterns of metal accumulation in Arctic char (Salvelinus alpinus) from an oligotrophic Alpine lake related to temperature, Can. J. Fish. Aquat. Sci., 53: 780–786.

Kumada H.S., Kimura S., Yokote M. and Matida Y., 1973. Acute and chronic toxicity, uptake and retention of cadmium in freshwater organisms. Bull. Freshwater Fish Res. Lab. (Tokyo), 22: 157-165.

Kumada H., Kimura M.Y. and Matida Y., 1972. Acute and chronic toxicity, uptake and retention of cadmium in freshwater organisms. Bull. Environ. Contam. Toxicol., 22: 157-165.

Kumar A.K. and Achyuthan H., 2007. Heavy metal accumulation in certain marine animals along the East Cost of Chennai, Tamil Nadu, India. J. Environ. Biol., 28: 637-643.

Landrum P.F., Hayton W.L., Lee H., McCarty L.S., Mackay D. and Mckim J.M., 1994. Bioavailability: physical, chemical, and biological interactions. Boca Raton, FL: CRC Press: 203–215.

Lewis W.S., 1987. An evaluation of mining related metals pollution in the upper Arkansas, River basin. M.S. Thesis, Colorado School of Mines.

Liao C.M., Chen B.C., Singh S., Lin M.C., Liu C.W. and Han B.C., 2003. Acute toxicity and bioaccumulation of arsenic in tilapia (*Oreochromis mossambicus*) from a blackfoot disease area in Taiwan. Environ. Toxicol., 18, 252–259.

Lin M.C., Liao C.M., Liu C.W. and Singh S., 2001. Bioaccumulation of arsenic in aquacultural large-scale mullet *Liza macrolepis* from blackfoot disease area in Taiwan. Bull. Environ. Contam. Toxicol., 67:91–97.

Mackay D., 1991. Multimedia environmental models: the fugacity approach. Ann Arbor, MI: Lewis Publishers: 211–214.

Maher W., Goessler W., Kirby J. and Raber G., 1999. Arsenic concentrations and speciation in the tissues and blood of sea mullet (*Mugil cephalus*) from Lake Macquarie NSW, Australia. Mar. Chem., 68:169 –182.

Maiti P. and Banerjee S., 1999. Accumulation of heavy metals in different tissues of fish, *Oreochromis nitoticus* exposed to waste water. Environ. Ecol., 17(4): 895-898.

Maiti P. and Banerjee S., 2002. Bioaccumulation of metals in different food fishes in wastewater fed wetlands. In: Kumar A. (ed), Ecology of Polluted Waters, vol. 1. APH Publ. New Delhi: 217-243.

Mansour S.A. and Sidky M.M., 2002. Ecotoxicological studies. 3: Heavy metals contaminating water and fish from Fayoum Governorate, Egypt. Food Chem., 78: 15-22.

Markich S.J., Brown P.L., Batley G.E., Apte S.C. and Stauber J.L., 2002. Incorporating metal speciation and bioavailability into water quality guidelines for protecting aquatic ecosystems. Aust. J. Ecotox., 7: 109-122.

Markich S.J. and Brown P.L., 1998. Relative importance of natural and anthropogenic influences on the freshwater chemistry of the Hawkesbury-Nepean River, South-eastern Australia. Sci. Total Environ., 117: 201-230.

Mason R.P., Laporte J.M. and Andres S., 2000. Factors controlling the bioaccumulation of mercury, arsenic, selenium, and cadmium by freshwater invertebrates and fish. Arch. Environ. Contam. Toxicol., 38: 283–297.

Massaro E.J., 1974. Pharmacokinetics of toxic elements in rainbow trout, USEPA. Ecol. Res. Ser. No. EPA-660/3-74-027. 30p. Washington, DC.

McKim J.M. and Goeden H.M., 1982. A direct measure of the uptake efficiency of a xenobiotic chemical across the gills of brook trout (*Salvelinus fontinalis*) under normoxic and hypoxic conditions. Comp. Biochem. Physiol. 72C: 65.

Miller P.A., Lanno R.P., McMaster M.E. and Dixon D.G., 1993. Relative contribution of dietary and waterborne copper to tissue copper burdens and waterborne copper uptake in rainbow trout (*Oncorhynchus mykiss).* Can. J. Fish. Aquat. Sci. 50, 1683.

Mohammed M. Al–Mohanna, 1994. Residues of some heavy metals in fishes collected from (Red Sea coast) Jizan, Saudi Arabia. J. Environ. Biol., 15(2): 149-157.

Mohammed A. A., 2009. Accumulation of heavy metals in Tilapia fish (*Oreochromis niloticus*) from Al-Khadoud Spring, Al-Hassa, Saudi Arabia. Am. J. Appl. Sci., 6 (12): 2024-2029.

Moriarty F., Hanson H.M. and Freestone P., 1984. Limitation of body burden as an index of environmental contamination heavy metal in fish, *Cotus gobio* L. from the river Ecclesbourne, Derbyshire. Environ. Pollut. Ser., A34: 297-320.

Motais R., Isaia J., Rankin J.C. and Maetz J., 1969. Adaptive changes of water permeability of the teleost gill epithelium in relation to external salinity. J. Exp. Biol., 51: 529-546.

Mukhopadhyay M.K. and Konar S.K., 1985. Toxic effects of metals mixture on fish and aquatic ecosystem. Environ and Ecol., 3: 22-29.

Munger C., Hare L., Craig A. and Charest P-M., 1999. Influence of exposure time on the distribution of cadmium within the Cladoceran Ceriodaphnia dubia. Aquatic Toxicol. 44: 195-200.

Murphy B.R., Atchson G.J. Mcintosh A.W. and Kolar D.J. 1978. Cadmium and zinc content of fish from an industrially contaminated lake. J. Fish Biol. 13: 327-335.

Murugan S.S., Karuppasamy R., Poongodi1 K. and Puvaneswari1 S., 2008. Bioaccumulation pattern of zinc in freshwater fish, *Channa punctatus* (Bloch.) after chronic exposure. Turk. J. Fisher. Aqua. Sci., 8: 55-59.

Nanda P, Panigrahi S, Nanda B, Behera M.K., 2000. Toxicity of paper mill effluent to fishes. Env. Eco., 18(1): 220-222.

Newman M.C. and Mitz S.V., 1988. Size dependence of Zn elimination and uptake from water by mosquito fish Gambusia affinis. Aquat. Toxicol., 12: 25-59.

Ney J.J., and Van Hassel J.H., 1983, Sources of variability in accumulation of heavy metals by fishes in a roadside stream, Arch. Environ. Contam. Toxicol., 12: 701–706.

Norrgren L., Wicklund Glynn A. and Malmborg O., 1991. Accumulation and effects of aluminium in the minnow (*Phoxinus phoxinus* L.) at different pH levels. J. Fish. Biol., 39: 833-847.

Norstrom R.J., McKinnon A.E., and DeFreitas A.S.W., 1976. A bioenergetics-based model for pollutant accumulation by fish. Simulation of PCB and methylmercury residue levels in Ottawa river yellow perch (*Perca flavescens*). J. Fish. Res. Bd. Can., 33 :248-267.

Nriagu J.O. 1968. A silent eoidemic of environmental metal poisoning. Environ. Pollut., 50: 139-161.

Oladimeji A.A., Quadri S.U. and deFreitas A.S.W. 1984b. Measuring the elimination of As by the gills of rainbow trout (*Salmo gairdneri*) by using a two compartment respirometer. Bull. Environ. Contam. Toxicol. 32: 661.

Olowu R.A., Ayejuyo O.O., Adewuyi G.O., Adejoro I.A., Denloye A.A.B., Babatunde A.O. and Ogundajo A.L., 2010. Determination of heavy metals in fish tissues, water and sediment from Epe and Badagry lagoons, Lagos, Nigeria. E Journal of Chem., 7(1), 215-221.

Olson K.R., Squibb K.S. and Cousins R.J., 1978. Tissue uptake, subcellular distribution and metabolism of (CH_3HgCl)-Cl_4 and (CH_3HgCl)-Hg_2O_3 by rainbow trout, *Salmo gairdneri*. J. Fish Res. Bd. Can., 35: 381-390.

Oronsaye J.A. and Brafield A.E., 1984. The effect of dissolved cadmium on the chloride cells of the gills of the stickleback, *Gasterasteus aculeatus* L. J. Fish Biol., 25:253.

Öztürk M., Özözen G., Minareci O. and Minareci E., 2009. determination of heavy meatals in fish, water and sediments of Avsar Dam Lake Turkey. Iran. J. Environ. Health. Sci. Eng., 6(2): 73-80.

Palanichamy S., Arunchalam S., Ali S.M. and Baksaran P., 1990. The effect of chemical factory effluent on the feeding energetics and body composition in the freshwater catfish *Mystus vittatus*. Proc. 2nd Asian Fisheries, Forum, Philippines. pp 935-937.

Pandey S., Kumar R., Sharma S.,Nagpure N.S., Srivastava S.K. and Verma M.S., 2005. Acute toxicity bioassays of mercuric chloride and malathion on air-breathing fish, *Channa punctatus* (Bloch). Ecotoxicol. and Environ. Safety, 61(1): 114-120.

Part P. and Lock R., 1983. Diffusion of calcium, cadmium and mercury in a mucus solution from rainbow trout. Comp. Biochem. Physiol., 76C: 254.

Patrick F.M. and Loulit M.W., 1978. Passage of metals to freshwater fish from their foods. Wat. Res., 12: 395.

Pethkar A.V., Gaikaiwari R.P. and Paknikar K.M., 2001. Biosorptive removal of contaminating heavy metals from plant extracts of medicinal value. Curr. Sci., 80: 1216-1219.

Philips G.R. and Buhler D.R., 1978. The relative contributions of methylmercury from food or water to rainbow trout, *Salmo gairdneri* in a controlled laboratory environment. Trans. Am. Fish Soc., 107: 853-861.

Playle R.C. and Wood C.M., 1991. Mechanisms of aluminium extraction and accumulation at the gills of rainbow trout, *Oncorhynchus mykiss* in acidic soft water. J. Fish Biol., 38: 23.

Playle R.C., Gensemer R.W., and Dixon D.G., 1992, Copper accumulation on gills of fathead minnows: influence of water hardness, complexation and pH of the gill microenvironment, Environ. Toxicol. Chem., 11: 381–391.

Ponce R. and Bloom N., 1991. Effect of pH on the bioaccumulation of low level methylmercury by rainbow trout, Oncorhynchus mykiss. Wat. Air Soil Pollut., 56: 631.

Rafia Azmat, Farha Aziz and Madiha Yousfi, 2008. Monitoring the Effect of Water Pollution on four Bioindicators of Aquatic Resources of Sindh Pakistan. Res. J. Environ. Sci., 2 (6): 465-473.

Rauf A., Javed M. and Ubaidullah M., 2009. Heavy metal levels in three major carps (Catla catla, Labeo rohita and Cirrhina mrigila) from the river Ravi, Pakistan. Pak. Vet. J., 29 (1): 24-26.

Reichert W.L., David A.F. and Malins C.D., 1979. Uptake and metabolism of lead and cadmium in coho salmon, *Onchorhynchus kisutch*. Comp. Biochem. Physiol. Pharmacol., 63: 229-234.

Reid S. and McDonald D., 1991. Metal binding activity of the gills of raibow trout, Onchorhynchus mykiss. Can. J. Fish Auat. Sci., 48: 1061.

Rishi K.K. and Jain M., 1998. Effect of toxicity of cadmium on scale morphology in Cyprinus carpio. Bull. Environ. contam. Toxicol.< 60: 323-328.

Rodgers D.W. and Beamish F.W.H., 1982. Dynamics of dietary methyl mercury in rainbow trout *Salmo gairdneri*. Aquat. Toxicol., 2: 271-290.

Rodgers D.W. and Beamish F.N.H., 1983. Water quality modifies uptake of waterborne methylmercury by rainbow trout, *Salmo gairdneri*. Can. J. Fish Aquat. Sci., 40: 824.

Rodgers D.W., Watson T., Langan J. and Wheaton E., 1987. Effect of pH and feeding regime on methylmercury accumulation within aquatic microcosms. Environ. Pollut., 45: 261.

Saeed S.M. and Shaker I.M., 2008. Assessment of heavy metals pollution in water and sediments and their effect on *Oreochromis niloticus* in the Northern Delta Lakes, Egypt. 8th Intern. Symp. Tilapia in Aqua., pp 475-490.

Schmidt-Nielsen K. 1984. Scaling: Why is Animal Size So Important? Cambridge University Press, 27-Jul-1984 - Medical: 241 pp

Sharpe M.A., deFreitas A.S.W., and McKinnon A.E., 1977. The effect of body size on methylmercury clearance by goldfish (*Carassius auratus*). Environ. Biol. Fish., 2: 37-43

Shaw B.P., 1987. Eco-physiological syudies of the waste of a chlor-alkali industry on bio-systems. PhD thesis, Berhampur Univ., Orrisa, Ind.

Shaw B.P., Dash S. and Panigrahi A.K., 1991. Effect of methyl mercuric chloride treatment on haematological characteristics and erythrocyte morphology of swiss mice. Environ. Poll., 73: 43-52.

Shukla V., Dhankhar M., Prakash J. and Sastry K.V., 2007. Bioaccumulation of Zn, Cu and Cd in *Channa punctatus*. J. Environ. Biol., 28: 395-397.

Sijm D.T.H.M., Selnen W. and Opperhulzen A., 1992. Life-cycle biomagnifications study in fish. Environ Sci Tech., 26: 2162–2174.

Smith L.S., 1982. Introduction to fish Physiology. TFH Publ. Neptune, NY.

Somero G.N., Chow T.J., Yancey P.H. and Snyder C.B., 1977. Lead accumulation rates in tissues of the esturine teleost fish Gillichthys mirabilis: salinity and temperature effects. Bull. Environ. Contam. Toxicol., 6: 337.

Sorenson E.M.B., Henry R.E., Yancey P.H. and Sayder C.B., 1979. Arsenic accumulation, tissue distribution and cytotoxicity in teleostsfollowing aidirect aqueous exposure. Bull. Environ. Contam. Toxicol., 21:162.

Sprague J.B., 1985. Factors that modify toxicity. In: Fundamentals of Aquatic toxicicology. Rand G.M. and Petrocelli S.R. (Eds.) Hemisphere Publ., Washington DC.

Spry D.J., Hodson P.V., and Wood C.M., 1988. Relative contributions of dietary and waterborne zinc in the rainbow trout, Salmo gairdneri. Can J. Fish Aquat. Sci., 45:32-41

Spry D.J. and Wiener J., 1991. Metal bioavailability and toxicity to fish in low alkalinity lakes; a critical review. Environ. Pollut., 71: 243.

Spry, D.J. and Wood, C.M. 1989. A kinetic method for the measurement of zinc influx *in vivo* in the rainbow trout, and the effects of waterborne calcium on flux rates. J. Exp. Biol. 142, 425 -446.

Srivastava N. and Verma H., 2009. Alterations in biochemical profile of liver and ovary of zinc-exposed freshwater murrel, *Channa punctatus*., J. Environ. Biol., 30(3): 413-6.

Stagg R.M. and Shuttleworth T.J., 1982. The accumulation of copper in *Platichthyes nesus* L. and its effects on plasma electrolyte concentrations. J.Fish Biol., 20: 491-500.

Sultana R. and Rao D.P., 1998. Bioaccumulation pattern of zinc, copper, lead and cadmium in grey mullet, *Mugil cephalus* (L.), from harbour waters of Visakhapatnam, India. Bull. Environ. Contam. Toxicol., 60: 949–955.

Takatsu A. and Uchiumi A., 1998. Abnormal arsenic accumulation by fish living in a naturally acidified lake. Analyst., 123:73–75.

Tawari-Fufeyin P. and Ekaye S.A., 2007. Fish species diversity as indicator of pollution in Ikpoba River, Benin City, Nigeria. Rev. Fish Biol. Fisheries, 17: 21-30.

Thiruvalluvan M., Nagendran N. and Mahoharan A. Charles, 1997. Bioaccumulation of cadmium and methyl parathion in *Cyprinus carpio var. communis* (Linn). J. Environ. Pollut., 4(3): 221-224.

Thomson K. W., Hendricks A.C. and Crains J. Jr., 1980. Acute toxicity of Zn and Cu singly and in combination to blue gill (*Lepomis macrochrus*). Bull. Environ. Contam. Toxicol., 25: 122-129.

Tulasi S.J., Reddy P.U.M. and Ramana Rao J.V., 1992. Accumulation of lead and effects on total lipid derivatives in the freshwater fish, *Anabas testudineus* (Bloch). Ecotoxicol. Environ. Saf., 23: 33-38.

U.S.EPA., 2001. United States Environmental Protection Agency. Water quality criterion for the protection of human health: methylmercury (Executive summary). EPA-832-R-01-001. 2001. (http://www.epa.gov/waterscience/criteria/methylmercury/merctitl.pdf)

Valler B.I. and Ulmer D.D., 1972. Biochemical effects of mercury, cadmium and lead. Annu. Rev. Biochem., 41: 91-128.

Varanasi U. and Markey D., 1978. Uptake and release of lead and cadmium in skin and mucous of coho salmon (*Onchorhynchus kisutch*). Comp. Biochem. Physiol., 60C: 187.

Veith G.D., DeFoe D.L. and Bergstedt B.V., 1979. Measuring and estimating the bioconcentration factor of chemicals in fish. J. Fish Res. Bd. Can., 36: 1040.

Verma H. and Srivastava N., 2008. Effects of sublethal concentrations of zinc on bioaccumulation and architectural alterations in the liver of fish, *Channa punctatus*. J. Environ. Sociobiol., 5 : 135-140.

Vijayram K., Geraldine P., Varadarajan T.S., George J. and Loganathan P., 1989. Cadmium induced changes in the biochemistry of an air-breathing fish, *Anabas testudineus*. J. Ecobiol., 1 (4): 245-251.

Virk S. and Dhawan A., 1997. Effect of nickel and chromium on flesh quality of *Cyprinus carpio*. Ind. J. Ecol., 24: 1-9.

Virk S. and Sharma A., 2003. Changes in the biochemical constituents of gills of *Cirrhinus mrigala* following exposure to metals. Ind. J. Fish. 50: 113-117.

Voigt H.R., 2004. Concentrations of mercury (Hg) and cadmium (Cd), and the condition of some coastal Baltic fishes, Environmentalica Fennica., 21: 26 pp.

Voigt H.R., 2007. Heavy metal (Hg, Cd, Zn) concentrations and condition of Eelpout (*Zoarces viviparus* L.), around Baltic Sea. Polish J. Environ. Stud., 16(6): 909-917.

Wang P.P., Chu K.L.M. and Wong C.K., 1999. Study of toxicity and bioaccumulation of copper in the silver sea bream *Sparus sarba*. Environ. Int., 25: 417– 422.

WHO., 1999. Food safety issues associated with products from aquaculture. Report of a Joint World Health Organ. Tech. Rep., 883, i-vii: 1-55.

Wicklund A. and Runn P., 1988. Calcium effects on cadmium uptake, redistribution and elimination in minnows, *Phoxinus phoxinus* acclimated to different calcium concentrations. Aquatic. Toxicol., 13: 109.

Wiener J.G., Martini R.E., Sheffy T.B., and Glass G.E., 1990, Factors influencing mercury concentrations in walleyes in northern Wisconsin lakes, Trans. Am. Fish Soc., 119:862–870.

Chapter 6

Metal Uptake and their Distribution in Fish Tissues

Metal availibilty and their uptake in fish is an index of metal pollution in the aquatic bodies (Karadede-Akin and Ünlü, 2007; Tawari-Fufeyin and Ekaye, 2007) and its toxicity status when present at higher concentrations (Dural *et al.*, 2007). Fish often being at the top of an aquatic food chain particularly the predatory fish, may concentrate not only large amounts of some metals from the water (Mansour and Sidky, 2002), these unlike organic contaminants that loose toxicity inside a fish tissue due to their natural biodegradation, heavy metals concentrations might be increased due to bioaccumulation (Aksoy, 2008). Mentioned below are the status of major toxic metals in fish tissues and their biological significance.

1. Mercury

Mercury (Hg), a very important water pollutant has been reported to accumulate in fishes in different regions of the world, and the rate of accumulation is said to depend majorly upon fish age (Johnels *et al.*, 1967), fish weight (Olson and Jensen, 1975), fish metabolism (Norstron *et al.*, 1976) and the type of fish species (Matsunaga, 1978). Romeo *et al.*, 1999, showed that Hg concentration in edible muscles of pelagic fish species was lower than those of benthic fish species. Presence of unacceptable levels of Hg along with lead (Pb) in the tissues of the African catfish, *C. gariepinus* from river Niger has also been reported by Lawani and Alawode (1996). Considerable information is available about the accumulation of Hg in liver, gill and spleen. Fish may absorb Hg from water through the digestive tract (Jernelov and Lann, 1971; Olson *et al.*, 1973, Phillips, 1977) or through gills and skin (Norstron *et al.*, 1976). Neumann *et al.* (1997) showed presence of total Hg residues in fish tissues beyond the US-EPA's health screening values of 0.6 mg/kg. As per UNEP data (2002) the Hg concentrations in various fish species are generally from about 0.05 to

1.4 mg/kg, depending on factors such as pH and redox potential of the water, and species, age and size of the fish. As per US FDA 1990-2007 data, sharks contained 0.979 ppm and swordfish up to 0.995 ppm of MeHg. Because of Hg biomagnifiction in the aquatic food web, fish being at higher trophic level in the food chain tend to have higher levels of mercury. Hence, large predatory fish, such as king mackerel, pike, shark, swordfish, walleye, barracuda, large tuna, scabbard and marlin, as well as seals and toothed whales, contain the highest Hg concentrations. Some other workers (Hattula *et al.*, 1978; Claisse *et al.*, 2001) also reported highest Hg concentration in predatory fish species. Rimmer *et al.* (2009) have ascertained that food web structure and composition also impact metal bioaccumulation through food quality and gut chemistry. Jhonels *et al.*, 1967 reported much low levels of Hg in fishes in Swedish lake. Caviglia and Cugurra (1978) analyzed different fish species in Italy and found about 53 species with accumulation of Hg beyond the tolerance limit Of 0.7ppm (up to 0.20 to 2 50 ppm) in their body tissue. They could find Hg traces up to 1.7 ppm in some of the marine fish species and majority of the species contained in between 0.02 to 0.30 ppm of Hg. Later tend to accumulate in tissues because of its low excretion.

Survey of The Gange river for Hg pollution Near Varanasi (Sinha *et al.*, 2007) revealed 0.00023 ppm of Hg present in water. They found highest amounts of Hg accumulation in the fish, *Rhinomugil corsula* (0,141-0.762 ppm) and *Macrognathus pancalus* (up to 91ppm). Haines (1996) reported mercury concentration in wild freshwater fish populations in 125 randomly selected lakes in Maine, commonly exceeding 1.0 µg/g wet weight, the level deemed unfit for human consumption (U.S.DHHS. and U.S.EPA, 2004). Half the samples from the Haines study exceeded 0.5 µg/g. U.S.EPA.'s two-year study of the National Fish Tissues found concentrations exceeding 1.0 µg/g in only 8 out of 282 composite samples of predatory and freshwater game fish distributed across the United States, while in the 237 composite samples of bottom dwelling fish the highest concentration was 0.531 µg/g (U.S.EPA., 2002,2003). However, more than 28 per cent of the total samples exceeded 0.1µg/g, with over 86 per cent of predatory fish exceeding this level. Evidence suggests that more than a quarter of all the mature fish contained methyl mercury concentrations above 1 µg/g. Neumann *et al.* (1997) reported levels of mercury in fish tissues exceeding the US Environmental Protection Agency's (EPA) health screening value of 0.6 mg/kg. FDA (USA) established a safe limit of 0.5 ppm of Hg in fish. Kamman and Burgess (2004) in the northeastern United States, suggested widespread occurrence and mercury uptake characteristics of brook trout (*Salvelinus fontinalis*) and yellow perch (*Perca flavescens*), which may make these species as the preferred subjects for monitoring mercury impact on natural systems.

Metal uptake in fishes is reported usually as weight specific (smaller > larger), but not in predaceous fish. Anderson and Spear (1980) stated that Hg in fish bioaccumulate faster; whereas, Cd and Pb do not, the later metals also do not increase with age as in predaceous fish. The fish muscle in Thane Creek (Govindsamy *et al.*, 1998) was reported to contain mercury ranging from 0.217 to 0.512µg/g. They attributed this due to a combination of up-food chain biomagnifications as well as larger fish size. It is reported to range between 10-30 per cent of the total Hg in marine plants, 20-80 per cent in invertebrates, while in fish and higher predators generally accounting for more than 80 per cent of total Hg. Tariq *et al.* (1979) have reported Hg contamination in some Indian fishes. Sahoo *et al.* (2002) reported that the fish in estuary water receiving chlor-alkali industry effluents near Rushikulya river in India were found to accumulate Hg between 0.26 to 2.70 mg per kg wt in 1986 and 0.28 to 5.45 in 1996 in muscles. They found Hg levels as high as 5 mg/kg fresh wt in different fish species. They further observed *Mugil oligolepis* collected from river estuary, having mean Hg residues as 1.03, 1.42 and 0.39 in their brain, liver and muscles, when Hg contamination level in the river water was of 0.031 mg/L, due to inflow of the chlor-alkali effluents. Fish in the Western coast of Bombay was reported to contain up to 82ppm of Hg (Anon., 2003).

The metal concentration factor in a given tissue changes over time of exposure due to its course of movement within tissues, so it is not an absolute value for any metal or organ, evidently one would observe different metal concentrations in different organs at different times. Similar time course changes in the ratio of liver-muscle mercury concentrations were reported in rainbow trout by Olson *et al.*, 1978. Greig *et al.* (1978) reported increased accumulation of Hg in muscles and liver, but with higher levels in muscles in majority of the cases than in liver. Julshamm *et al.*, 1982 further suggested that Hg after absorption into blood first move into liver, thereafter from liver to muscles. Hornung *et al.* (1980) investigated total mercury concentration in the edible portion (muscle tissue) of twelve commercial trawl fishes along the Mediterranean coast of Israel. It showed an increase in mercury concentration with length for *Saurida undosquamis, Trachurus mediterraneus, Upeneus moluccensis* and *Pagellus erythrinus,* except in *Mullus barbatus* in which the correlation between mercury concentration and total length was low. Amundsen *et al.* (1997), however, stated that Hg accumulation in fish takes place less in muscles and more in liver and gill. The mercury concentration of this species appeared to increase with depth.

Storelli *et al.* (2005) have mentioned a positive correlation between the fish body weight and the Hg contents. Recent studies suggest trophic transfer of Hg also occur in terrestrial food chains (Rimmer *et al.*, 2005). Benson *et al.*, 2007 investigated Hg accumulation in a variety of fishes in tropical waters of Nigeria. They recorded maximum accumulation of Hg in *Pomadasys jubelini* (0.063-+0.03 mg kg^{-1}), followed by *O. nilotica, Brycinus nurse, Hemichromis faciatus, Chrysichthys nigrodigitatus and Lutianus ava.* Rafia *et al.*, 2008 analyzed Hg concentration in muscles of two freshwater species and found it to be higher as compared to marine water fish. *Pomodasy argyrew* showed highest averaged value (5 µg kg^{-1}) whereas *Liza subviridus* showed lowest (3.2 µg kg^{-1}). In addition there was a significant difference among the average concentration of Hg in two stations. This data indicate that different species have various capabilities to accumulate and store water contaminates independent of their level in water. Same phenomena were observed by De la Torre *et al.* (2000). Endo *et al.* (2008) reported Hg concentrations in the muscle of tiger sharks (*Galeocerdo cuvier*) from the coast of Ishigaki Island, Japan to increase proportionally with body length in the tiger sharks, whereas that in the liver increased rapidly after maturity. Muscle Hg levels were higher than liver concentrations in immature sharks, with the inverse trend observed in mature sharks. This rapid increase in hepatic Hg concentration concurrent with the onset of maturity in sharks may result from the continuous intake of Hg via food and the slower growth of mature sharks.

Concern over mercury in the environment arises from its extremely toxic forms, in which mercury can occur. Keeping in view the highly toxic effects of Hg in food particularly through the contaminated fish, many industrialized countries have established procedures and policies to assess minimize and prevent humans exposure to Hg. Presence of Hg in contaminated fish and mollusks is mostly in organic form and these appear dangerous to human health. Hg in fish tissues from wild catch mostly appears in methylated form (Rudd *et al.,* 1980). Histological observations of liver and kidney showed that MeHg accumulates mostly in the glomeruli of the kidney and in the nucleus of the endoplasmic reticulum of liver cells (Baatrup *et al.,* 1986). The EPA's reference dose for ingested methyl-Hg is 0.1 µg/kg BW/day and it has been suggested that food preparation factor be taken into consideration as it changes the metal concentration level (US EPA, 2001); Burger *et al.,* 2003). Based on an average daily intake of 17.5 gram of fish, the US EPA also calculated a Tissue Residue Criterion of 0.3 mg methyl-mercury per kg of fish (0.3 mg/kg). This limit is weighted on all fish and shellfish consumed (US EPA, 2001). The process of deep fry increases Hg concentration, whereas, no significant change occurs when the fish is cooked simply. Data from US Food and Drug Administration (Center for Food Safety and Applied Nutrition, Office of Seafood, 2001) suggested that pregnant women or even those expecting to become pregnant to avoid eating the long-lived larger fish that feed on other fish, such as tilefish, king mackerel, swordfish and shark, as these fish species were found to have high methyl-Hg concentrations in their tissues. These fish contained Hg levels up to 1.30- 1.67 ppm in tuna fish, king mackerel, groups and lobster fish; but as much as up to 4.54ppm in shark, 3.73 ppm in tilefish and 3.22 ppm in swordfish. Wiener *et al.* (2003) have reported Hg residues >5 mg per kg ww in brain and >20 mg in liver in a variety of these predatory species. Storelli *et al.,* 2003a stated the range in between 72-83 per cent for eagle ray and glass-shark fishes, and for sardines the range was in between >85 per cent (Joiris *et al.,* 1999). Storelli *et al.,* 2005 in the muscles of small spotted sharks along Italian coast in the Adriatic Sea, reported total Hg concentration ranging between 0.26 to 2.06 mg/l dominated with the methyl-Hg.

As a result of biomagnifications of Methyl Hg (MeHg), long lived piscivorous or other top predators feeding on aquatic food chains, such as pikes and trouts, birds (common loom, osprey, eagles, kingfishers) and even mammals like minks, polar bears and seals are at a greater risk for elevated dietary Hg exposures, accumulation and toxicity. Walker, 1976 mentioned methyl Hg between 70-92 per cent in Elasmobranch fish muscles. Bull *et al.* (1981) studied growth rates and Hg content of Roach, Perch and Pike living in sewage effluent and recorded slightly elevated Hg content in Roach and Perch. MeHg represented a high proportion of Hg and was found in all the three species. Elevated MeHg concentrations may impair the long term sustainability of aquatic ecosystems and deleterious effects on the behavior and health of associated wild forms. MeHg is largely responsible for the accumulation of Hg in organisms and its transfer from one trophic level to another, and accumulation of this potent toxicant has also been shown to vary within different phyla, due to the distinct routes of uptake of the element (Anderson and Depledge, 1997). Study of time course changes in metal accumulation in tissues revealed that methyl mercury (MeHg) first move to liver than to muscles (Olson *et al.*, 1973). Storelli *et al.*, 2003b reported presence of Hg as methyl mercury ranging between 55-100 per cent in the muscles of a variety of fish species.

2. Arsenic

Arsenic (As) another major water pollutant in most of the south-east countries has been further a matter of serious concern for the fish consumers. Oladimeji *et al.* (1984) studied the long term effects of arsenic accumulation in rainbow trout, *Salmo gairdneri*. Turoczy *et al.* (2000) made a study on the deep sea dog sharks. The concentrations of As along with Hg were found to exceed the normal limits in all the available fish species. Storelli *et al.*, 2005 in the muscles of small spotted sharks along Italian coast in the Adriatic Sea, reported total As in the range of 4.96 to 14.27 mg/l, dominated as the organic As among various metal species. Accumulation of arsenic in the tissues of rainbow trout was reported to increase with the increasing levels of arsenic in the diet and the liver contained the highest level of arsenic, which has been associated with the functional change and it may cause histological changes as well. The muscles of this fish were seen accumulating arsenic up to 135 µg/kg of BW within 4 weeks after their exposure at the lowest concentration of 10 mg/kg of diet. The gill did not seem to accumulate any substantial amount of arsenic during continued exposure. This might be due to fast elimination of arsenic via the gills. Arsenic concentrations in all tissues of tilapia (Liao *et al.*, 2003) were found to be allometric *i.e.* negatively correlating with fish body weight, where as in muscle tissue these were positively correlated with As accumulation in the viscera. Significantly higher concentrations of As were obtained in the viscera of tilapia (12.65 ± 10.17 µg g^{-1} dry wt.) than in the muscle tissue (3.55 ± 0.42 µg g^{-1} dry wt.). They suggested that a simple way of reducing the health risk associated with consuming tilapia is to trim and cook the fish properly, *i.e.* removing the viscera of tilapia can greatly reduce the amount of As ingested and consequently reduce the health risks.

Liao *et al.* (2003) in the black foot disease (BFD) area of Taiwan have reported the average concentration of As in pond water ranging from 17.8 to 49 µg L^{-1}. Acute toxicity tests showed that the As concentration that caused toxicity to tilapia ranged from 69 060 µg L^{-1}, in the 24-h toxicity test, to 28 680 µg L^{-1}, in the 96-h toxicity test. The highest bioconcentration factor (BCF) was found in the intestine (maximum value: 2270). The order of BCFs was: intestine > stomach > liver= gill >muscle. Allen *et al.* (2004) studied the biochemical toxicity of arsenic trioxide in a freshwater edible fish, *Channa punctatus* on its exposure for 7 to 90 days. The arsenic concentration increased exponentially in liver, kidney, gills and muscles of fish up to 60 days of exposure of arsenic. However, arsenic concentration in these tissues declined at 90 days of exposure. As concentration in muscles of the four species (Rafia *et al.*, 2008) showed variation in two freshwater and two marine water species. Results showed that interaction of metal pollutants vary from species to species. High concentrations of contaminants were found in tissues of fishes collected from marine water as compared to freshwater fishes. *Liza subviridus* showed highest (9 μg kg^{-1}) accumulation of As, while *Cyprius carpio* showed lowest value (3 μg kg^{-1}). In addition to this *Pomodasy argyrew* (5 μg kg^{-1}) and *Johnius belengerii* (5 μg kg^{-1}) showed the same pattern of accumulation of As, regardless the marine or freshwater sites. Eisler (2010) reviewed As contents in 10 species of Mediterrarian Sea sharks and stated highest As contents in the muscles of ghost shark, *Chimaera monstrosa* (52.4 mg/kg FW) and in the liver of the longnose spurdog, *Squalus blainvillei* (14.2 mg/kg FW). Most of the shark species in general possessed As maximally in their liver as compared to that in muscles and other tissues (*e.g.* spottrail shark, 23.2; silky shark, 20.0 mg; sandbar shark, 11.2 mg; dusky shark, 10.0 mg; cownose ray, 17.0 mg; atlantic guitar fish, 16.0 mg mg/kg DW).

3. Cadmium

Bonnell *et al.* (1960) observed that cadmim (Cd) is accumulated initially both in the liver and kidney in comparable concentrations while on prolonged exposure, it accumulates preferentially in the fish kidney. The kidneys did not accumulate Cd as much as the liver. Cross *et al.* (1973) showed cadmium concentration in tissue of fish was in decreasing order of kidney > liver > gills > muscle. It is clear that certain fish species like rainbow trout (*Salmo gairdneri*) are especially sensitive to the presence of Cd in the water, others such as roach, perch, rudd and stone loach can tolerate much higher concentrations of the dissolved metal (Solbe and Flook, 1975). Cd toxicity differs significantly from Hg in its distribution within body of a fish as per fish species (Hawkins *et al.*, 1980). In contrast to Hg, they recorded neither skeleton muscles nor the brain accumulated Cd in a spot fish, *Leiostomus xanthurus* (marine fish). In a spot fish Cd accumulation was observed after 48 hrs in the order of: liver> gut> kidney> gill and 99 per cent accumulation occurred only in liver, kidney and gill (Thomas *et al.*, 1983). Same was the case with another marine fish rainbow trout, except that there was no accumulation of Cd in its gut. Anderson and Spear (1980) stated that Hg in fish bioaccumulate faster, whereas Cd and Pb do not, the later metals also do not increase with age as in predaceous fish. Stromberg *et al.* (1983) studied the pathology of lethal and sub-lethal exposure of fathead minnow to the Cd. The radiotracer studies using Cd^{115} showed a rapid uptake and two-phase elimination of Cd by the fish. Brown *et al.* (1986) compared the distribution of Cd accumulated in the organs and the sensitivities to Cd among several species of fish and concluded that the susceptible species, rain-bow trout (*Salmo gairdneri*), retained a much greater proportion of the notional dose of Cd than did the resistant species, roach (*Rutilus rutilius*) or stone loach (*Noemacheilus barbatulus*). They made an analysis of accumulation of Cd in three fish species. They evidenced the total body loads of Cd were found not correlated with the Cd concentration to which individuals were exposed. They devised an alternative comparator which as the quotient of the total body cadmium accumulation (μg/100g BW) and the notional Cd dose (μg/l) x weeks, was described as a 'fractional retention coefficient (FRC) for Cd'. The coefficient was constant for each species at different periods of exposure to Cd alone, the values were, however, much lower for roach (*R. rutilus*) and stone loach (*N. barbatulus*), as compared to that of rainbow trout. Interestingly, when rainbow trouts were pre-exposed to Zn (100μg/l) for 5 days before being exposed to Cd, the FRC value for Cd fell to a value similar to those of roach and stone loach fishes exposed to Cd alone.

The toxicity of cadmium to adult fish, its accumulation within the tissues, and its impact on the physiological mechanisms was well studied by Heath (1987). Gupta (1998) observed that in case of cadmium accumulation in the tissues of *Heteropneustes fossilis* and *Channa punctatus* after 96 hours exposure, gills accumulated the highest concentration of cadmium, followed by kidney and liver. *Heteropneustes fossilis* accumulated relatively more cadmium in tissues than *Channa punctatus.* Olasson *et al.* (1988) exposed rainbow trout to cadmium. They observed that kidney accumulated the maximum amount as compared to the other tissues. The accumulation of cadmium in different tissues of *Heteropneustes fossilis* was also reported by Das and Kaviraj (1990) after exposure to different concentration of cadmium (2.5, 6.4 and 55.5 mg/l), exhibiting accumulation of maximum amount of Cd in liver upon longest perod of exposure, followed by the gills and the kidneys. Kraal *et al.* (1995) reported that there was a decreasing order in the Cd accumulation in tissues of the 8 weeks old common carp *C. carpio* and the order was found to be gut > kidney > liver = gill > muscle. Whereas, in the Cd contaminated water, the corresponding values in the decreasing order were as: gut > gills > kidney > liver > muscle. Kaviraj and Ghosal (1997) studied the bioassays lasting 180 days with common carp, *C. carpio* exposed to eight different treatment of the Cd (2.5 mg/l) alone or with compost manure. Harrison and Klaverkamp, 1989 assessed the accumulation of Cd in various tissues in rainbow trout and Lake Whitefish (*Coregonus clupeaformis*) exposed to either water borne or food born cadmium for 72 days and subsequent 56 days depuration. Most notable is the tendency for gills to accumulate the most Cd, even in those exposed to Cd via food. Considerable amounts of food born Cd, however, found its way to gut and kidneys, but not when it was water born. There was about 10 fold greater accumulation of Cd from the food than from water. De-Smet and Blust (2001) in common carp, *Cyprinus carpio* after exposer to 0, 0.8, 4 and 20 mM cadmium over a 29 day period, reported Cd accumulation in the tissues in the following order: kidney>liver>gills.

Cd alone did not show any adverse effect on the growth of the fish but compost manure, irrespective of its dose and its combination with Cd markedly increased the growth of the fish and primary productivity of water. Lin *et al.* (2000) studied Cd tolerance in the larva of *Oreochromis mossambicus*. It was revealed that exposing the female *Tilapia* to the Cd did not affect its reproduction, the egg quality and the survival of their offspring. De-Smet and Blust-Ronny (2001) exposed common carp *C. carpio* to Cd over a period 29 days and observed Cd accumulation in the tissue in the following order: kidney > liver > gills. Effects of cadmium compounds on common carp, *Cyprinus carpio* exposed to cadmium via an intra peritoneal injection were assessed by Brucka-Jastrzebska and Protasowicki (2004). Metal accumulated in different organ in the series of decreasing order: kidney> liver > skin> gills>mid posterior part of the alimentary tract> muscles. Vijayram *et al.* (1989) determined the uptake and loss of Cd in various tissues of freshwater prawn, *Macrophthalamus malcolmsonii*. A dose and time dependent uptake of Cd was noted in all the tissues.

The prawns were, however, able to regulate the tissue concentrations of Cd up to a threshold level of dissolved metal exposure. Since the concentration of the Cd found in the tissues reflect the external medium concentration, these prawns are ideal tools for monitoring the environmental Cd level. Harrison and Klaverkamp (1989) assessed accumulation of Cd in various tissues in rainbow trout and the lake whitefish, *Coregonus clupeaformis* and noticed a tendency of the gills to accumulate most of the Cd, even when exposed to Cd given through the food. Gut and kidney also accumulate considerable amounts of the metal when it is in the food, but the accumulation is not there in the water.

Cd toxicity increases with increase in water temperature, pH and hardness, with hyperactivity especially in males, and increased mortality at spawning time (Alabaster and Lloyd, 1982). Finally, depuration appears most rapid for the gut and the gills. The kidney, however, shows little depuration of Cd, perhaps due to transfer of Cd from other tissues to the kidney during excretion. The percentage of Cd accumulated because of the dose was almost 10 times greater in the food than from the water. Even in the present studies depuration of metals from the gills was more than from muscles, probably due to its more accumulation over there. Jastrzêbska and Protasowick (2006) assessed comparatively cadmium elimination dynamics by various organs of common carp, *Cyprinus carpio* L. after transferring them to a laboratory environment. Average cadmium contents in the examined in fish organs and tissues (liver, kidneys, skin, gills, alimentary tract and muscles) ranged from 0.004 to 0.053 mg g^{-1} wet weight. The highest values of Cd content were recorded in the liver, the mid posterior section of the alimentary tract, and the gills, while the lowest value was noted in the muscles. The radiotracer studies using Cd^{115} showed a rapid uptake and two-phase elimination of Cd by the fish. Rainbow trout are considerably more sensitive to Cd than roach (*Rutilus rutilus*) and some loach (*Noemacheilus barbatulus*), as reflected in lower LC_{50} values (Norey *et al.*, 1990). They showed that these differences are correlated with the relative rate of accumulation of Cd into tissues.

4. Copper

With copper (Cu), liver was reported to have the highest concentration factor (Buckley *et al.*, 1982; Felts and Heath, 1984), since dietary Cu is directly taken into liver from the intestine, while water borne Cu is initially accumulated in various tissues. Low levels of Cu in the kidney suggest low capacity of this organ for detoxification and elimination of this metal. Increase in Cu accumulation in the liver and fall in the muscles in these fish, could be attributed to the fact that on chronic exposure Cu is rapidly taken into the liver, where it is finally stored (Buckley *et al.*, 1982). Toxicological studies further revealed that elevated copper concentrations in water can lead to increased copper levels, particularly in the gills and the liver (McCarter and Roch, 1984; Laurén and McDonald, 1987a,b), suggesting that the gills can serve as a route of copper uptake. In the light of the high volume of water passing through the gills (18 l kg^{-1} h^{-1}) (Wood and Jackson, 1980), even the low ambient copper concentrations normally present in freshwater (8–80 nmol l^{-1}=0.5–5.0 mg l^{-1}) (Spry *et al.*, 1981) offer a potential source for normal copper assimilation by the gills. In a time course study involving exposure of Coho salmon to Cu, the plasma copper rose on day-1; after 2 weeks or more it increased steadily only in liver, indicating that Cu was being constantly removed from plasma by the liver. Segner (1987) after exposing yearling roach, *Rutilus rutilus* to 80 μg Cu l^{-1} and keeping it starved for 7 days found that the nutritional status of a fish is of great importance in modifying its response to sublethal copper contamination. They evidenced that the gill tissues had significant uptake of copper in both fed and starved fishes, while the liver tissues of only starved specimens showed significant accumulation. Re-feeding roach after 7 days of starvation along with continuous Cu contamination, however, resulted in a significant decrease of liver copper content. No copper release from the liver occurred after cessation of exposure but starvation continued. Exposure of fed common carp (*Cyprinus carpio*) to sublethal (1 μM) copper levels for 28 d (Hashemi *et al.*, 2007) also showed effects of rate of feeding on copper accumulation in liver and gills. It showed Cu accumulations in the gills, liver, and kidney significantly higher than in high fed fish (HFR, @5 per cent of BW) and in low fed fish (LFR, 0.5 per cent). Copper accumulation in the liver of both feeding treatments occurred in a time-dependent manner and did not reach steady state in any treatment. On the contrary, copper concentration in the gills reached a steady state for both HFR and LFR fish within the first week of exposure. No copper accumulation was found in muscle tissues of either treatment. Fýrat and Kargýn (2010) recorded increasing level of Cu in the *C. carpio* gill with increasing concentrations of metal in the exposure medium, and with increasing duration of exposure.

The metals in general are readily absorbed in the living organisms and show a greater affinity with sulfhydryl (-SH) group and appear to exert toxic effects largely by combining with such groups on proteins, thus, disrupting cellular functions. Radha Krishnaiah (1988) studied the degree of accumulation of copper in the gills, liver, brain and muscles of *Labeo rohita*. Mazon and Fernandes (1999) studied the toxicity and differential tissue accumulation of copper in tropical freshwater fish *Prochilodus sarofa*. Zhou *et al.* (2003) demonstrated that the accumulation of Cu in the tissues of fish was related to the metallic species (Zn, Pb or Cd), their concentration and the type of tissue. The interaction became more significant with an increase in the concentration of metals. Accumulation of Cu was said to decrease in the gills in an interaction of a combination of metal, but increased in the liver and brain. The metal mixture could not affect the muscle for accumulation of Cu. The order of Cu accumulation in the tissues of the fish was liver>gill>brain>muscles and the other metals did not change this order of Cu accumulation in the tissues. Subathra and Karuppasamy (2008) analyzed the bioaccumulation pattern of copper in different sizes (fingerlings and adult age) of healthy *Mystus vittatus* when exposed to Cu-water, containing one-third 96-hr LC_{50} level (6.20 and 15.95 mg L^{-1}) for short-term (120 hr) and one-eighth 96-hr LC_{50} level (2.33 and 5.98 mg L^{-1}) for long-term experimentation, respectively. It showed a maximum deposition ($p < 0.01$) of Cu in the liver (82.12 and 70.65 μg/g), followed by gill (74.35 and 63.69 μg/g) and kidney (61.52 and 54.09 μg/g) both in fingerlings and adult fish, respectively, during 28 days of exposure. The lowest deposition of Cu was found to be 0.83 and 0.93 μg/g in fingerlings and 0.79 and 0.86 μg/g in adult muscle tissue during short-term (120 hr) and long-term (28 days) exposure periods, respectively. Comparing accumulation of Cu on the two size groups at both exposure levels, it is obvious that the fingerlings showed higher Cu concentration in all tissues than those of adult fish. Another equally important finding was that the depuration of Cu by maintaining the Cu exposed fish (long-term exposed group) of both size groups in quality dechlorinated ground water, it revealed a significant ($p < 0.05$) reduction in Cu concentration in different tissues as the days passed. A comparison of the performance of the two size groups in respect of depuration clearly indicates that the fingerlings have taken 24–43 days (gill-kidney), whereas, in mature fish it is 21–39 days (gill-kidney) to reach the level of control fish. Among the various tissues in both size groups, gill took the minimum number of days for complete recovery, whereas the muscle tissue did not significantly eliminate Cu even after 30 days of depuration. These data constitute a reference for future studies on the evolution of Cu accumulation and elimination tendency in relation to different size groups of fish in the ecotoxicological testing scheme for hazard assessment.

5. Lead

Lead accumulates in freshwater fish mostly in scales, bone, kidney, gill and liver, but in an estuarine fish it was more in spleen, followed in decreasing order in gills, fin and intestine (Spry and Wiener, 1991). Holcombe *et al.*, 1976 recorded lead accumulation in brook trout in the decreasing order as kidney, gill, liver, and spleen with virtually none in the muscle. Surprisingly, there are reports that liver and muscles did not accumulate any lead (Somero *et al.*, 1977). They further suggested lead binding to mucous is an especially important mechanism for its accumulation and subsequent depuration, and gills and intestine have an abundant mucous. On the contrary, Kumar and Mathur (1991) on exposure of Indian freshwater fish, *Colisa fasciatus* to lead for 24 days observed almost same concentration factor for lead in gill and muscle, whereas the liver had much less accumulation. According to Rafia *et al.*, 2008, Pb concentration in muscles of two biomarkers from freshwater, *Cyprius carpio* (0.6 $\mu g\ kg^{-1}$) and Pomodasy argyrew (5.8 $\mu g\ kg^{-1}$) and two from marine water (*Liza subviridus* (7.4 $\mu g\ kg^{-1}$) and *Johnius belengerii* (7.6 $\mu g\ kg^{-1}$) showed significant difference. *Johnius belengerii* showed averaged highest value (7.6 $\mu g\ kg^{-1}$), whereas, *Pomodasy argyrew* showed averaged lower value (5.8 $\mu g\ kg^{-1}$). It indicated that interaction of heavy toxic metals with biomarkers vary species to species and more prominent in marine water fish. Extensive clinical and experimental evidence support the significance of Pb-Ca interaction which is apparent in current study that *Liza subviridus* has lowest concentration of Ca (22 $mg\ kg^{-1}$) and that of Pb 7.4 $\mu g\ kg^{-1}$ and *Johnius belengerii* (7.6 $\mu g\ kg^{-1}$) and Ca 24 $mg\ kg^{-1}$. These interactions occur at the cellular and molecular level and are the abilities of Pb to displace Ca during specific physiological process. It is likely that Pb blocks Ca efflux from cells by substituting Ca in Ca^{++}/Na adenosine triphosphate (Simons, 1986).

6. Zinc

Accumulation of zinc (Zn) in various organs of fish has been described by a few workers (Handy and Eddy, 1990ab; Gupta and Sharma, 1994; Pandey *et al.*, 1995). Zinc shows greatest BCF in skin, and bone (Mount, 1964), and some accumulation in liver, gill and kidney (Holcombe *et al.*, 1979). Dietary Zn may further show a different pattern of distribution from that of the water borne (Hardy *et al.*, 1987). In rainbow trout they recorded higher accumulation level of Zn in blood than that in gills, liver, kidney and spleen. Ghosh and Bagchi (1980) had reported in resident fish, *Rita rita* of the Hooghly river maximum accumulation of Zn in gonads (170 µg/g), followed by in gills (52.5 µg/g), kidney (43.3 µg), liver (31.7 µg), and muscles (8.5 µg). Mirenda (1986) studied acute toxicity and accumulation of Zn in the Crayfish, *Orconectes virilis* and found that gills were the main site of Zn uptake from the water. Appreciable levels of Zn were found in the gills at all exposure concentrations (130 mg Zn/l, 63.3 mg Zn/l, 26.8 mg Zn/l, 12.2 mg Zn/l and 5.2 mg Zn/l) and the level increased further as the Zn concentration in the water increased. The abdominal muscles were the least variable tissue with regard to changes in the Zn concentration over the range of test concentrations used. Under normal conditions, the crayfish, however, appears to be capable of regulating the Zn content through mechanism for absorption across the body surface and its loss via fecal production. High zinc concentration in the kidney of treated fish is probably a result of the kidney being one of the major organs for detoxification and elimination of metallic pollutants (Dallinger *et al.*, 1987). Mohan and Chaudhary (1991) studied the uptake of Zn using Atomic Absorption Spectrophotometer in a few tissues of the fish, *Puntius sophare* by exposing to 7.5 mg/l of $ZnCl_2$ for 48 and 96 hr. The highest amount of Zn in the intestine accumulated on dry weight basis followed by gills, liver and the kidney. Gupta and Chakraborty (1993) studied toxicity of Zn in the freshwater teleost, *N. notopterus* (Pallas) and *Punctius javanicus* (Blkr). Gupta and Sharma (1994) observed high retention and slow elimination of zinc by the kidney of fingerlings of *C. mrigala* maintained in normal freshwater for a period of 96 hr after pre-exposure to zinc for 96 hr. The dose levels of zinc administerated in the fish, *C. punctatus* (Gupta and Srivastava, 2006) appear to affect the detoxication mechanism within the kidney thereby retarding metal elimination and enhancing its accumulation. Study of recovery effects in this fish further evidenced accumulation of zinc much quicker than its elimination as retention of zinc was noted up to 15 days, as there was significant retention of zinc by the kidneys in comparison to control.

The comparative account of the behavioral response against Zn toxicity exhibited that *P. javanicus* was more sensitive than the *N. notopterus*. Rema and Philip (1996) concluded that the levels of the Zn being an essential metal, are under the metabolic control. Jeng-Sen-Shyong, (1999) reported that the digestive tract tissue of the common carp, *C. carpio* had very high concentration of Zn (300-500 mg/g fresh tissue), while concentration of more than 100 mg/g fresh tissue were there in the kidney, gills, skeletal tissue and the spleen. 40 per cent of the Zn in the digestive tract tissue of the common carp could be extracted by water. Analysis of the digestive tract tissue of the common carp, *C. carpio* revealed a very high concentration of Zn (300-500 mg/g fresh tissue), while a concentration of more than 100 mg/g (fresh tissue) were found in the kidney, gills, skeletal tissue and the spleen (Jeng-Sen-Shyong, 1999). A preliminary biochemical investigation on the nuclei/cell debris fraction of the digestive tract tissue of common carp suggests that the main Zn binding substance(s) were probably protein(s). Liyaquat *et al.* (2003) assessed acute and chronic toxicity of zinc to different freshwater animals. Comparative freshwater animal species sensitivity pattern against zinc and zinc smelter effluent based on acute toxicity studies was recorded as: Moina> Limnaea>*Labeo* fry> frog tadpoles>

Notonecta. Moina turned out to be most sensitive to zinc with 48 hrs and 96 hrs LC_{50} values of 1.2 and 0.57 mg/l for Zn and 23 and 7 per cent of effluent, respectively. Gupta and Srivastava (2006) in *C. punctatus* on exposer to zinc (10 mg/l, 15 mg/l and 25 mg/l) for 15 days also showed significant dose dependent increase in zinc concentration in fish of all treated groups. Post exposure recovery in fish was noted after transferring these fish to normal tap water for another fortnight.

7. Metal Accumulations and Inter-metallic Interactions

Natural waters since are usually contaminated with mixtures of metals and other toxic compounds, inter-metallic interactions further play an important role in the toxicity effects of each metal, especially in terms of their uptake in presence of other metal or other water pollutant in fishes. Accumulation of certain metals in fish may be altered in the presence of the other compounds (Wicklund *et al.*, 1988; Allen, 1995; Pelgrom *et al.*, 1995). Significant concentration of heavy metals in natural waters lead to accumulation of majorly of metals like Cu, Fe, Pb, Hg, Ni, Cd and Cr in fish (James *et al.*, 1996). Presence of Zn in water is reported to enhance Hg uptake in soft water by trout (Ramamoorthy and Blumhagen, 1984), low levels of Cd were found to inhibit the uptake of Zn from the water, whereas Zn was seen to yield no effect on Cd absorption (Bentley, 1992). Likewise, presence of a detergent, linear alkylaryl sulphonate (LAS) and a emulsifier, nonylphenol ethoxylate (NP-10EO) in water affect uptake of Cd by the gills (Part *et al.*, 1985). They found that LAS stimulated Cd uptake due to its lower affinity for the metal. Since any other detergent did not cause such effects on Cd uptake, it was concluded that the increased uptake of Cd was not merely due to reduced surface tension by the detergent. The rate of metal uptake into the gill tissue further correlates with the weight-specific metabolic rate, thus, a small fish since have the higher flow rate of water over the gills apparently result in metal accumulation more rapidly than a large fish (Anderson and Spear, 1980). Freeman (1980) noticed fast flow of water over the gills under mild hypoxia also causes a greater intake of Cd, Cr and Pb by the fish gills. On the contrary, since Zn is known to cause histological damage to gill epithelium, thus, affecting Zn absorption into the gills and resulting in lesser intake of Zn into its gills (Hughes and Flos, 1978).

Brown *et al.* (1986) compared the distribution of Cd accumulated in the organs and the sensitivities to Cd among several species of fish like rain-bow trout (*Salmo gairdneri*), and roach (*Rutilus rutilius*) or stone loach (*Noemacheilus barbatulus*). They evidenced that when rainbow trouts were pre-exposed to Zn (100µg/l) for 5 days before being exposed to Cd, the FRC value 'fractional retention coefficient' for Cd fell to a value similar to those of roach and stone loach fishes exposed Jezierska and Witeska (2006) have stated metal levels in live fish usually follow the ranking: Fe > Zn > Pb > Cu > Cd > Hg. According to some authors both liver and kidney are the principal storage organs for these metals (Calamari *et al.*, 1982; Handy, 1992). While considering the action of a toxic chemical in a fish, the liver is considered as an important organ. It is the primary organ of biotransformation of the organic xenobiotics, and probably also for the excretion of the harmful trace metals. Since many of these metals and the organics tend to accumulate in the high concentrations in the liver, therefore the cells in liver are exposed to very high levels of these harmful chemicals, as compared to those present in the environment or in other organs of the fish. Metal accumulated in different organ is said to be in the decreasing order as: kidney> liver > skin > gills > mid posterior part of the alimentary tract> muscles. The behavior of metals like lead further changes with other metals. When bluegill were exposed for 7 days to 0.5 ppm of waterborne Pb, Cr, or Cd the gills had higher concentration of all the three metals as compared to the liver (Freeman, 1980). Moreover, the differences were large (*i.e.* order of magnitude greater). In gills, the Cr and Pb reached higher concentration as compared to Cd, while all the three metals were having the same concentration in the liver.

Liver and gills of fish species namely, *Sparus aurata, Dicentrarchus labrax, Mugil cephalus,* and *Scomberomorus cavalla,* have been reported to accumulate highest levels of cadmium, lead, copper, zinc, and iron (Dural *et al.*, 2007; Ploetz *et al.*, 2007). A study conducted on *Leuciscus cephalus* and *Lepomis gibbosus* (Yilmaz *et al.*, 2007) has reported maximum accumulations of cadmium, cobalt, and copper in the liver and gills, while these accumulations were least in the fish muscle. The higher levels of trace elements such as lead and chromium in liver relative to other tissues may be attributed to the affinity or strong coordination of metallothionein protein with these elements (Ikem *et al.*, 2003). Kumari *et al.*, 2014 has reported that study of the Cr bioaccumulation pattern in the selected tissues of the fish, *L.rohita* showed highest concentrations of Cr in the liver and gills and minimum concentration in muscle and brain. Amongst all the tissues studied (liver, muscle, gills, and brain), the highest concentration of chromium ($4.56 \pm 1.45 \mu g g^{-1}$) was observed in the liver of the fish *Labeo rohita* after 96hrs period of exposure which was followed by gills ($2.18 \pm 0.62 \mu g g^{-1}$), muscle ($2.0 \pm 0.60 \mu g g^{-1}$); and brain ($0.135 \pm 0.065 \mu g g^{-1}$). After 15 days of exposure the chromium accumulation, following trend: gills ($121.8 \pm 46.22 \mu g g^{-1}$); liver ($100.80 \pm 16.6 \mu g g^{-1}$); muscle ($35.65 \pm 4.25 \mu g g^{-1}$); and brain ($22.75 \pm 5.60 \mu g g^{-1}$). Significant concentration of heavy metals in natural waters lead to accumulation of metals majorly of Cu, Fe, Pb, Hg, Ni, Cd and Cr in fish (James *et al.*, 1996). Low levels of chromium in fish muscles and brain appear to be due to low levels of binding proteins. A study conducted on *Cyprinus carpio, Barbus capito,* and *Chondrostoma regium* caught at 5 stations on the Seyhan river system has also confirmed the maximum accumulation of cadmium and chromium in the liver and gills than muscles (Canli *et al.*, 1998).

Greig *et al.* (1983) studied the concentration of Ag, Cd, Cu and Pb in the liver of Window pane flounder and reported values which were typical for fin fish. Felts and Heath (1984) observed Cu to accumulate maximally in liver, much less in brain and kidney. Carpene *et al.* (1989) studied the uptake of Cd, Zn and Cu in liver and bile of gold fish, *Carassius auratus*, after exposing three metals individually. They reported higher accumulation of cadmium in liver. Whereas, zinc accumulation was less evident and copper was absent. The samples of muscles, gills, liver and the kidneys were collected from 30 adult carps and examined for Cu, Cd, Pb and Zn concentration by Dobicki *et al.* (1991). The samples of the muscles showed the lowest concentration of all the four elements tested (0.42-1.52, 0.44-0.72, 0.54-2.0, 4.67-9.13 ppm, respectively). The highest concentration of Cd and Pb were found in the gills (0.65-1.64 and 1.39-5.02 ppm, respectively), of Cu in the liver (0.80-3.23 ppm) and that of Zn in the kidneys (156.9-230.4 ppm). Jezierska and Witeska (2006) also stated that fish muscles, comparing to the other tissues, usually contain the lowest levels of metals. In rainbow trout Cr accumulation was faster in blue gill fish, but slower in trout. Natural waters are usually contaminated with mixtures of metals and other toxic compounds. Accumulation of certain metals in fish may be altered in the presence of the other compounds (Wicklund *et al.*, 1988; Allen, 1995; Pelgrom *et al.*, 1995). Jezierska and Witeska (2006) have stated metal levels in a live fish usually follow the ranking as: Fe > Zn > Pb > Cu > Cd > Hg. Tulasi *et al.* (1992) studied lead accumulation in inner organs *viz.*, kidney and liver of freshwater fish, *Anabas testudineus*. Abu and Weis (1993) studied concentration levels of Hg, Cd, Cu and Zn in two populations of *Fundulus heteroclitus*. Kureishy and Silva (1993) conducted experiments on the uptake and the loss of Hg, Cd and Pb in various tissues of the *T. mosambica*, *Perna viridus* and *Villorita cyprinoids*. They found different metals showing preference for a particular organ, such as Hg showed affinity towards muscles, Cd towards liver and the Pb towards gills. Mercury was further found to be highly toxic to the clams as compared to the fish and mussels as it appeared accumulated in the greater amounts as compared to Cd and Pb. Pb was found to be least toxic. Similar observations were made in the tissues of fishes collected from the Indian coast, except that Pb showed affinity towards gills as well as the kidneys.

Mukhopadhyay *et al.* (1994) found that metal accumulation is also a function of fish species involved, as *O niloticus* was said to have increased accumulation of copper in liver, than *L. rohita* and *C. catla*. Allen (1995), following acute exposure of *Oreochromis aureus* to cadmium or lead (for 24-h or 1-week exposure to 0.1 mg l^{-1} of lead as $PbCl_2$; and cadmium as $CdCl_2$) observed trunk kidney accumulating the highest lead concentrations of all tissues, highest accumulations of cadmium was in the intestine. Caudal muscles consistently accumulated the lowest levels of lead or cadmium after a 1-week exposure period. Vigh *et al.* (1996) in case of grass carp, *Ctenopharyngodon idella*, reported the highest metal concentration in kidney and liver and to be the lowest in muscle and the gut, while Ni concentration was lower in liver, muscle and the gills of the fish and Hg was lowest in the gut of grass carp. Biomagnification of Hg occurred only in predatory species. Ahmad and Al-Ghais (1996) made a comparison of Cd, Ni, Pb, Cu, Mn, and Zn in the muscle, liver and skin of *Epinepliehus tauvina* and *L. fulviflamma*. It was observed that the muscle tissue had a higher concentration of metals than the skin but markedly lower concentration as compared to the liver. Presence of unacceptable levels of Hg and Pb in the African catfish, from river Niger has been reported (Lawani and Alawode, 1996). Deb and Santra (1997) reported the bioaccumulation of Cu, Pb, Zn and Cr in the liver, intestine, ovary, muscle and brain in four fishes, *Channa punctatus, Tilapia mossambicus, Catla catla* and *Labeo rohita* in a sewage fed Jheel ecosystem. Virk and Dhawan (1997) exposed *C. carpio* to Ni (7.5 and 15.0 mg/l) and Cr (10 and 20 mg/l) and observed a significant decline in flesh quality during pre-spawning phase in males, but during the spawning phase in females.

Cobalt, manganese and zinc are essential trace metals which have vital role in normal body metabolic functions in animals, and Cr is said to involve in many enzymes responsible for sugar metabolism and occur in many foods up to 1300mg/kg. Co is an essential element in Vit.B_{12}, and Ni occurs naturally in water. Excess of Zn in water is reported to cause metabolic disorders, and Cu damaging gill surface due to gill lamellae rupturing and various other histopathological and enzymological dysfunctions. Accumulation of Co metal in marine fishes was studied by Udea and Nakashera (1983). Liver accumulated higher concentration of these metals as compared to in other parts of the body *i.e.* head, abdomen and tail. El-Shahawi and Al-Yousuf, (1998) reported concentrations of Ni, Co, Cr and Pb in the skin tissues of *Lethrinus lentjus* fish collected from the coastal waters of the United Arab Emirates. Metal concentrations in skin samples decreased in the following order: Cr>Pb>Ni>Co. *Tilapia* fish were reported to show higher retention of Hg in gills, than in liver, muscle and brain. The recovery study also showed higher per cent excretion of Hg in gills. Clear signs of recovery were also reported by a few other workers in fish on their transfer to metal free water, as in liver (Kaushik and Srivastava, 2003) and kidney (Gupta and Srivastava, 2006) of the fish *C. punctatus.* Recovery of histopathological condition was also inversely proportional to the exposure dose level.

Krishna Murti *et al.* (1999) studied concentration of Cu, Zn, Pb and Ni in different tissues of 10 species of fish from Thane and Bassein creeks of Bombay. The trends in bioaccumulation indicated that irrespective of a fish species, Cu and Zn contributed 42-58 per cent to the total metals accumulated. Maximum concentration of all metals was found either in liver or in alimentary canal with lowest level in muscle tissue. Maiti and Banerjee (1999) studied quantitative accumulation of heavy metals like Cu, Zn, Cd, Pb and Cr in muscle, kidney, ovary, liver, gills and fins of nile tilapia (*Oreochromis niloticus*) exposed to waste waters. They reviewed metal accumulation in three fish species: *C. catla, L. rohita* and *Tilapia* sp. Cu accumulated more in liver in *Tilapia* (up to 90 µg/g of body weight), whereas in *C. catla,* Cd was found to accumulate more in kidney (61.29 µg/g of body wt.) and Zn, more in liver (28.10 µg of body wt.). In *L. rohita,* Zn accumulation went up to 32.2µg in kidney and up to 51.10 µg in liver. In *C. catla,* Pb further enhanced the Cu uptake but Cd inhibited Zn uptake, thus, exhibiting their complex interaction within the body. Heavy metals may replace other essential metals too in the body tissues. Maiti and Banerjee (2002) further further recorded maximum Zn accumulation in the liver and kidney of *L. rohita* and *C. catla* fishes. A preliminary biochemical investigation on the nuclei/cell debris fraction of the digestive tract tissue of common carp suggests that the main Zn binding substance(s) were probably protein(s). They recorded Cu accumulation more in liver of *L. rohita* (61.68 µg/g), *C. Catla* (20.69 µg/g) and *O. nilotica* (90.0 µg/g) because these organs act as the major site for metal storage and metabolic transformations.

Sinha *et al.* (2002) analyzed gill, liver, kidney, intestine and muscle of some of the common edible fishes captured from Kharkai river for their Fe, Zn, Ni, Pb, Cu, Mn, Cr, and Co contents. Kharkai river is one of the principal tributaries of river Subarnarekha and is considered as one of the most polluted rivers in the Chotanagpur plateau. Misra *et al.* (2002) assessed the concentration of eight heavy metals *viz.*, Mn, Pb, Cu, Cr, Hg, Cd, Ni and Zn in different parts of grass carp,

Ctenopharyngodon idella. Comparatively higher concentration of Mn, Pb, Ni and Zn were observed. Liver accumulated higher concentration as compared to other parts of the body *i.e.* head, abdomen and tail. Rehulka (2002) observed highest level of Al (5.310 mg/kg) in the muscle tissue of Pike, while that of Zn (12.355 mg/kg) in *Rutilus rutilus*. The contamination of slightly above 0.5 mg/g of Pb was found in the muscle tissue in a few specimens only. Omeregie *et al.* (2002) also reported enhanced levels of Pb, Cu and Zn in *O. nilotica* (Nile Tilapia) from river Delini. Higher concentrations of many metals were also reported in *C. gariepinus* and *O. nilotica*, cultured in some disused mining lakes (Akueshi *et al.*, 2003).

Accumulation of some heavy metals such as mercury, lead, nickel and cadmium was compared in ten species of fish available in local markets of Shillong by Chakraborty *et al.* (2003). The levels of lead and mercury were alarmingly high in certain body parts of some species. Nickel was also detected in considerable amount in the muscles, head and bones of almost all the fish species investigated. Ni accumulated in the muscles of almost all the fish species except *C. batrachus*. The highest concentration of Ni was recorded in the muscle of *A. testudineus* (60.0 – 65.2µg/Kg dry weight). In other fish species the concentration ranged between 8.0 – 25.0 µg/Kg dry wt. Cd could not be detected in the muscle of any of the fish species under study.Pane *et al.* (2004) reported Ni accumulation in the gill, kidney and plasma, with the plasma as the main sink for Ni, in rainbow trout on its exposer to 384 and 2034 µg Ni/l. Sharma *et al.* (2003) estimated bioaccumulation of zinc and cadmium in freshwater fish species, *viz., Labeo rohita, Catla catla, Channa punctatus* and *Poecilia reticulata* after acute and sub lethal exposure. They revealed that liver and kidney are the prime sites of metal accumulation and heavy metal load in the muscles was far low as compared to these organs. Kebede and Wondimu (2004) in tilapia, *O. niloticus*, collected from Lakes Awassa and Ziway, revealed organ specific distribution of trace metals, such as Cd: 0.44-1.43, 4.58-4.93, 2.20- 2.85, and 1.08-1.90; Co: 2.47-3.59, 17.1-18.9, 8.28-10.1, and 10.2-13.0; Cu: 1.68-4.95, 6.65-7.58, 7.08-8.58, and 602-797 (mg/kg DW) in muscle, bones, gills and liver, respectively. Furthermore, significant correlations ($P < 0.05$) among the selected metals and order of metal accumulation in the fish muscle were determined between Fe-Cu, Fe-Ni, Fe-Cr, Mn-Ni, Mn-Cr, Mn-Cr, Cu-Ni, and Ni-Cr. Cr posses an exceptional tendency to accumulate in the organism and is transported through serum proteins and stored in liver, muscles and kidney and gills (Jyothrmayee *et al.*, 2006).

Young marine life stages have been reported more sensitive than adults (Mance, 1987), and the shorter diffusion distances in small fingerlings may anticipate faster toxicokinetics than in adults or larger fingerlings. Wang *et al.* (1999) showed that the order of concentration of copper in organs was intestine>liver>gonad>gills, skin and muscle. Peyghan *et al.* (2003), in their work with common carp *C. carpio* exposed to copper in shortterm and long-term experiments, showed Cu accumulation in liver and muscle after only 1 day exposure to 1 mg L^{-1}. Accumulation in liver and muscle follow osmoregulatory imbalance (Hansen *et al.*, 2006) by disturbance of the hydromineral equilibrium in gills (De Boeck *et al.*, 2003). Bashir *et al.* (2012) evaluated the trace metal levels in water and tissues of two commercial fish species: *Arius thalassinus* and *Pennahia anea* collected from Kapar and Mersing coastal waters. It showed the concentrations of Fe, Zn, Al, As, Cd and Pb in the muscle, liver and gills tissues of the fishes in the descending order: livers > gills >muscles. A positive association between the trace metal concentrations and weight and length of the sample fishes was investigated. The metal levels in fish, however, were below the permitted level of Malaysian and international standards. Yosef and Ghada (2011) determined some heavy metal contents in fresh and salted mullet fish collected from El-Burullus Lake in muscle, gills and liver. Heavy metal concentrations were found varied significantly depending on the type of the tissue and salt. Generally, the lowest concentrations of the studied metals were found in fresh fish, while the highest concentrations were recorded in salted fish in the order: liver>gills>muscle and gills>muscle>liver, respectively. The metal levels were ranked as follow: Fe>Zn>Cu>Pb>Cd. Fe and Cd concentrations exceeded the upper limit of standards while, Zn one recorded much below level in all deliberate fish organs. Cu content was slightly higher than the permissible limit in muscle and gills of salted fish by industrial salt, whereas Pb level was almost double the concentration of the permissible limit in the liver of all groups and muscle of salted fish by industrial salt, while it were triple the concentration of the permissible limit in gills of that group.

Accumulation of copper and lead in the tissues of a freshwater fish, *Tilapia zilli* was studied by Ay *et al.* (1999). Szkoda *et al.* (1999) collected samples of muscles and liver of 311 carps at fish farms in selected regions of Poland. The concentrations of the toxic elements (Pb, Cd, Hg, As) in carp tissues were relatively low, rather lead and cadmium were seldom detected in the muscle (15 per cent and 35 per cent, respectively). Rashed (2001b) determined heavy metals Co, Cr, Cu, Fe, Mn, Ni, Sr and Zn in different tissues of *T. nilotica* (ages 1, 1.5, 2, 2.5 and 3 years from the High Dam Lake, Aswan (Egypt). The results showed that of all the fish parts, liver accumulated the highest levels of Cu and Zn, whereas, cadmium and lead concentrations were higher in fish scales and vertebral column than in the other parts of the fish. Manganese was present in the intestine and stomach in the highest concentration. Scales exhibited the highest levels of Co, Cr, Ni and Sr; the gill and vertebral column contains the lowest level of these elements. Heavy metals in different parts of *T. nilotica* differ with the fish growth, but were in the safety permissible levels for human uses. The levels of six heavy metals (Cd, Cr, Cu, Ni, Pb and Zn) in different tissues of some cultured marine fishes collected from three fish culture sites in Hong Kong were evaluated by Wong *et al.* (2001). Faraq *et al.* (2002) studied the distribution of the metals; Cu, As, Cd, Pb and Zn during digestion by cut throat trout. Pb was the lowest of all the metals in the fluid portion of the intestinal contents. However, more than 80 per cent of all the metals in the hindgut were associated with the particulate fraction where they may still be available for the uptake. In another study various organs of the common carp, *C. carpio* were analyzed for the heavy metals. Significantly higher Hg contents were detected in the muscle, Cd and Zn in the kidney and Cu in the liver. A significantly higher content of Cu and Zn was found in eggs also (Svobodova *et al.*, 2002). Tissues of all fish species contained high concentration of Zn and Cu but much lower concentrations of Ni, Pb, Cd and Cr. Allinson *et al.* (2002) analyzed samples of the muscle and liver of tilapia (*Oreochromis mossambicus*) obtained from the five reservoirs in four catchments in southern Sri Lanka for 16 elements: As, Ca, Cd, Co, Cr, Cu, Fe, K, Mg, Mn, Na, Ni, Pb, Sr and Zn by inductively coupled plasma emission spectrophotometry and Hg by atomic absorption spectrophotometry. The concentrations of Cr, Ni, and Pb were below the detection limits of the instrumental techniques employed in all samples. The elements As, Ca, Co, Cu, Fe, Hg, K, Mg, Mn, Na, Sr and Zn were detected in the muscle and liver tissue, with Cd detected in some liver samples. The degree of contamination is said to depend on pollutant type (Asuquo *et al.*, 2004). Lake Pamvotis (NW Greece) is a typical Mediterranean ecosystem of great importance in regard to biodiversity and its aesthetic value. Most commonly found fish species in this lake are: *Cyprinus carpio, Silurus aristotelis, Rutilus ylikiensis, and Carassius gibelio*. Papagiannis *et al.*, 2004 showed that *C. carpio* and *R. ylikiensis* contained the highest metal content in their body. Tissues analysis revealed that liver and gonads accumulated the highest levels of Cu and Zn. Metal concentration in the edible part of the examined fish *i.e.* the muscles were however, within the safe permissible levels for human consumption.

Sultana and Rao (1998) studied the bioaccumulation of Zn, Cu, Pb and Cd in grey mullet, *Mugil cephalus* from harbor waters of Vishakhapatnam in India and reported that alimentary canal and gills can be considered as the interface of the organism and its ambience. The gills are directly exposed to the ambient conditions. Alimentary canal is also known for its excretory functions for some metals like Zn. Galhoom *et al.*, 2000 compared data for the presence of heavy metals in various tissues in mugil fish (*Mugil cephaleus*). Analysis revealed increased levels in their muscles of Pb and Hg, but decreased levels of Cu, Cd and Zn. Metals Pb, Hg, Cu, Cd and Zn in the fish muscles were present in the range as follows: Pb- 0.0226-0.9373, Hg- 0.488-3.628, Cu- 0.0569-0.875, Cd- 0.00416-0.0324 and Zn- 0.04989-0.3175 μg/g of the tissue. Chen and Chen (1999) studied accumulation of toxic metals in grey mullet, *Liza macrolepis*. After 190 days of sediment exposure, significant increase in Cu, Mn, Fe, Zn and Cd concentrations in whole fish were found 3-, 1.5-, 3-, 2- and 2- times, respectively, higher than the control group. The bioaccumulation factors (BAFs) in whole fish ranked as Cd>Zn>Cu>Mn>Ni>Fe. The accumulation behavior of Cu, Zn, Cd and Pb concentrations in flesh, gills, liver and gonads of the commercial fishes: *Mullus barbatus, Merluccius merliccius and Boops boops* from the Aegean sea in Turkey have been studied. Cu, Zn and Pb concentration in flesh were found low, while in gonad Cd was found high. Liver showed higher concentration of Cu than other fish organs. Gonads and liver recorded high concentration of Cd (Zyadah and Chouikhi, 1999). Canli and Atli 2003 revealed presence of heavy metal (Cd, Cr, Cu, Fe, Pb, Zn) concentrations in the muscle, gill and liver of fish species (*Sparus auratus, Atherina hepsetus, Mugil cephalus, Trigla cuculus, Sardina pilchardus* and *Scomberesox saurus*) from the northeast Mediterranean Sea. The relationships between fish size (length and weight) and metal concentrations in the tissues were investigated by linear regression analysis. Metal concentrations (as μg/g d.w.) were highest in the liver, except for iron in the gill of *Scomberesox saurus* and lowest in the muscle of all the fish species. Highest concentrations of Cd (4.50), Cr (17.1) and Pb (41.2) were measured in liver tissues of *T. cuculus, Sardina pilchardus* and *A. hepsetus*, respectively. The liver of *M. cephalus* showed strikingly high Cu concentrations (202.8). The gill of *Scomberesox saurus* was the only tissue that showed highest

(885.5) iron concentrations. The concentration of the different heavy metals *i.e.* As, Se, Cd, Co, Fe, Cu, Mn, Ni, Mb, Zn and Pb in the different fish species range from 0.004-80.36 mg/kg body weight of the fish having the maximum concentration of Zn. Total Hg was found to be high *i.e.* 0.61 mg/g. 12 out of 17 fishes exceeded the safe level of 0.01 mg/kg of different heavy metals (Gutleb *et al.*, 2002).

Recently, there has been an increasing interest over metal pollution in Bay of Poudre D'Or which has assured interest in edible fishes since high concentration of metals has been hazardous to health. There have been certain recent reports from Mauritius about the accumulation of high contents of Cu in the liver of flat head mullet, *Mugil cephalus* (Bhagwant, 2003a). Copper levels among the species were said to vary considerably (Table 6.1), ranging from 0.038 to 42.7 mg/kg w wt, highest amounts being in the liver of *Mugil cephalus*, which very much exceed the permissible limit of 30 ppm proposed by FAO (1983), and therefore need to be monitored in this fish. The trend in the distribution of Fe in various tissue/organs of *Mullioidicthys flavolineatus*, *Lutjanus notatus* and *Lutjanus fulvus* was liver > gut > gill > muscle. The levels of Cu amongst different fish species vary considerably ranging from 0.038 to 42.7 mg/kg wet weight. A high content of Cu recorded from the liver of *Mugil cephalus* far exceeded the prescribed permissible limits and thus, requires close monitoring. While the Cu levels in the muscle tissues (edible part) were below the permissible limits. Conversely, the bioconcentration order of Fe in *M. cephalus* was gill > gut > muscle > liver. The bioaccumulation of the Fe in the fish species show the following decreasing order: *L. notatus* > *L. fulvus* > *M. flavolineatus* > *M. cephalus*, thus, suggesting a possible use of *L. notatus* as the biological indicator for the monitoring of Fe. The extent of metal accumulation differed among species, possibly due to differences in behavior and their feeding habits. Bhagwant (2003b) also studied bioaccumulation of lead, copper and zinc levels in tissue of three edible lagoon fish species from Tombeau Bay, Mauritius, namely; *Scarus ghobban*, *Acanthurus mata* and *Mugil cephalus.* This study reports on the bioaccumulation of Pb, Zn and Cu in different organ tissues of fish species. Pb and Zn levels were higher in liver samples compared with muscle tissues in all three species. Cu was found in the liver of *M. cephalus* only.

Table 6.1: Accumulation of different Metals in some Marine Fish Species of the Bay of Poudre D'Or, as Determined by Bhagwant, 2003a

Mg/kg Wet wt.	*Liver/Cu*	*Muscle/Cu*	*Liver/Fe*	*Muscle/Fe*
Mugil cephalus	41.21-44.19	41.44.19	0.288-0.523	7.80-10.79
LutjanusFulvous	1.578-1.709	1.578-1.709	218.6-235.3	12.25-19.55
L. notatus	.067-.339	7.475-8.720	247.5-262.9	16.80-18.40
Mulloidicthys flavolineatus	1.286-2.140	1.286-2.140	217.7-267.3	7.653-13.41

*Cu in gills ND-6.028 and gut ND-12.06; Fe in gills 89.02-138.4 and in gut 53.61-228.4.

Shukla *et al.* (2005) reported maximum accumulation of the heavy metals in liver and minimum in muscles in the adult *C. punctatus* when exposed to Zn (18.62mg/l), Cd (11.8 mg/l) and Cu (0.56mg/l). The degree of accumulation varied in the five tissues and it was in the order: gill > liver > kidney > blood> muscle in case of Zn:

gills > kidney > liver > blood > muscle in case of Cd and gills > kidney > blood > liver > muscle in case of exposure with Cu. Shelly (2005) further recorded different metal concentrations (Mn, Zn, Fe, Cu, Cr, Ni, Pb, Hg, Cd and As) in muscle of the *C. carpio*. These were found to vary between 0.115 to 11.157, 1.875 to 50.650, 2.805 to 180.550, 0.193 to 7.200, 0.365 to 13.200, 0.628 to 38.800, 0.765 to 45.316, 0.020 to 26.800, 0.004 to 1.500 and 0.480 to 7.0 mg/kg weight basis, respectively. Yilmaz (2005) studied Fe, Cu, Ni, Cr, Pb and Zn contents in muscle, skin and gonads of two fish species (*Mugil cephalus and Sparatus aurata*). Cu, Zn, Fe, Pb, Ni, and Cr concentrations were found to be highest in *M. cephalus respectively* with 1.39, 47.25, 16.38, 10.02, 1.34 and 1.71 μg/g ww in muscle and 44.50, 269.06, 332.56, 90.97, 9.17 and 8.68 μg/g ww in gonads. Shah and Altindag (2005) studied the effects of already accumulated heavy metals (Hg, Cd, Pb) in the body of a tench on the 96-h LC_{50} values of the respective heavy metal. They observed that the fish with the lowest body concentration of the heavy metal showed the lowest 96-h LC_{50} value of the respective heavy metal and vice-versa. Andreji *et al.* (2005) assessed concentrations of selected metals (Fe, Mn, Zn, Cu, Ni, Co, Cr, Pb, Cd, and Hg) in the muscle of four common Slovak fish species (chubs: *Leuciscus cephalus*, barbel: *Barbus barbus*; roach: *Rutilus rutilus*, and perch: *Perca fluviatilis*).

The Paraíba do Sul river located in southeast Brazil, is one of the most suitable habitats for a variety of fish species; but is reported to be heavily polluted (Molisani *et al.* (1999), as it receives many organic and industrial effluents which directly affect the ichthyo-fauna. It drains one of the most industrial regions in the country. Terra *et al.* (2007) determined concentration of four heavy metals (Cu, Cr, Zn and Pb) in muscle and gonads of three abundant fish species (*Oligosarcus hepsetus*–carnivore, *Geophagus brasiliensis*–omnivore and *Hypostomus luetkeni*–detritivore) from different trophic levels of the Paraíba do Sul river. They showed that all the metals examined except Cu, exceeded the maximum permitted metal concentration in muscles and/or gonads of the all the three fish species (*O. hepsetus, G. brasiliensis and H. luetkeni*) examined. *O. hepsetus* (carnivore) showed the highest contamination levels, followed by *G. brasiliensis* (omnivore) and *H. luetkeni* (detritivore), confirming the hypothesis that carnivorous species are prone to incorporate more heavy metals than fish in other trophic levels. Generally, all the fish species showed higher metal concentration in gonads than in muscles, except for Cr. *O. hepsetus* further showed metal contamination levels in tissues as per the river segment, *i.e.* highest levels of Pb contamination in muscles (7.98±3.73 μg g^{-1}) in the middle-upper river segment and of Cr in the upper (5.53±0.05 μg g^{-1}) and middle-upper (4.20±0.85 μg g^{-1}) segments, indicating thus, non-consummation of these fish species by the humans from these segments of the Paraíba do Sul river. Analysis for Hg, Zn and Cu concentrations in the liver and muscle of tiger sharks (*Galeocerdo cuvier*) from the coast of Ishigaki Island, Japan (Endo *et al.*, 2008) evidenced that the Hg concentrations in the muscle increased proportionally with body length, whereas that in the liver increased rapidly after maturity. Notably, the Zn and Cu concentrations in the liver tended to decrease with increasing body length. The high concentrations of the essential metals Zn and Cu in immature sharks may be explained by the physiological demands related to rapid growth.

The order of metal concentrations in the fish muscles was: Fe > Zn > Hg > Pb > Cu > Mn > Cd > Cr > Ni > Co. Levels of Zn, Cu, Cd, As, and Pb in the kidney, liver, gills and heart of African cat fish (*Clarias gariepinus*) from the Ogun river in Ogun State located close to six major industries in the South Western part of Nigeria, were determined (Farombi *et al.*, 2007), using Bulk Scientific Atomic Absorption Spectrophotometer. Fishes were also collected from Government owned fish farm in Agodi, Ibadan which was considered a reference site. The trend of accumulation of the metals in the organs is as follows: heart - Zn > Cu > Pb > As > Cd; gills - Zn > Cu > Pb > Cd > As; kidney - Zn > Cu > Pb > As > Cd; liver - Zn > Cu > Pb > As > Cd. The order of concentration of the metals in the organs is as follows: arsenite - kidney > liver > gills > heart; zinc - gills > liver > kidney > heart; lead- liver > kidney > gills > heart; copper- kidney > liver > gills > heart; cadmium > liver > gills > kidney > heart. The levels of heavy metals ranged between 0.25- 8.96 ppm in the heart, 0.69-19.05 ppm in the kidneys, 2.10-19.75 ppm in the liver and 1.95-20.35 ppm in the gills. Begum *et al.* (2009a) analyzed metals in ten fish species: Catla, Silver carp, Common carp, Tilapia (*Oreochromis niloticus*), Mrigal (*Etroplus suratensis*), Murrels (*Channa marulius*), *Nandus nandus*, *Amblypharyngodon mola*, Catfishes (*Heteropneustes fossilis*) from a lake. She showed the presence of elevated levels of Pb and Cd in almost all the fishes and considered eating fish caught in this lake a serious matter, as these had the potential for human exposure to heavy metals. She recorded the order of heavy metal levels in various organs as; Muscles >Gills >liver >kidney. The order of heavy metal concentration in muscles was: Pb >Cd>Ni>Cr, in gills: Pb =Cd>Ni>Cr; in kidney: Pb>Cd>Ni>Cr and in liver: Pb>Cd>Ni>Cr. Naz *et al.* (2003) evidenced effects of lead in the fish diet in carps. Analysis for trace element concentrations (Hg, Cd, Pb, Cr, Ni, Cu, Zn) in the muscle, gonads, skin, and brain of smooth hound *Mustelus mustelus* (Storelli *et al.*, 2011) indicated that metal accumulation depended on the tissues probably as a consequence of metabolic needs, physiochemical properties, and detoxification processes specific for each element. They estimated higher concentration of metals in gonads (Hg 0.10–0.70 μg g^{-1}; Cd 0.02–0.10 μg g^{-1}; Pb 0.08–0.39 μg g^{-1}; Cr 0.06–0.36 μg g^{-1}; Ni 1.37–3.00 μg g^{-1}; Zn 9.15–16.30 μg g^{-1}; Cu 1.95–21.62 μg g^{-1}) and skin (Hg 0.16–0.66 μg g^{-1}; Cd 0.01–0.04 μg g^{-1}; Pb 0.10–0.62 μg g^{-1}; Cr 0.15–0.68 μg g^{-1}; Ni 1.60–7.20 μg g^{-1}; Zn 9.00–16.00 μg g^{-1}; Cu 0.78–6.80 μg g^{-1}) than brain (Hg 0.04–0.34 μg g^{-1}; Cd 0.01–0.05 μg g^{-1}; Pb 0.03–0.59 μg g^{-1}; Cr 0.08–0.48 μg g^{-1}; Ni 5.59–9.69 μg g^{-1}; Zn 5.90–7.35 μg g^{-1}; Cu 0.90–4.02 μg g^{-1}), while muscle always exhibited the lowest levels (Hg 1.03–2.58 μg g^{-1}; Cd 0.01–0.06 μg g^{-1}; Pb 0.02–0.16 μg g^{-1}; Cr 0.05–0.28 μg g^{-1}; Ni 1.13–2.48 μg g^{-1}; Zn 2.64–5.06 μg g^{-1}; Cu 0.33–2.23 μg g^{-1}). Ni and Hg took exception having the highest concentrations in brain and muscle, respectively. Mhadhbi *et al.*, 2012 stated that Ría de Vigo and coastal zone NE Atlantic (Spain) are in particular, vulnerable to heavy metals pollution. Concentrations of the heavy metals examined in fish in these waters were found to range as follows: Cd (0.01- 0.7), Pb (0.1- 2.5), Hg (0.01- 0.7), Cu (3.3- 46.7), Zn (15- 274), As (3- 151), Se (0.9- 18.2), Mn (0.9- 9.76) μg/g dwt. Kidney and liver showed the highest levels of Cd, Cu, Pb and Zn.

8. Metal Accumulation in other Aquatic Organisms

In aquatic animals, mollusks are endowed with the highest capacity for accumulation of heavy metals (Martin and Flegal, 1975). Janseen and Scholz (1979) observed that the Cd concentration after 21 days of exposure to 100 µg Cd l^{-1}, at 10°C and a salinity of 25 per cent in various organs of the mussels decreased in the order of muscles > gill > kidney > adductor muscle > mantle > foot. When the rates of uptake of Cd during the first five days of exposure were recorded, it was noted that the gills accumulate Cd at the rate of 21.4 µg g^{-1} day^{-1}. The loss-rates in the gills being 67.5 per cent of the total Cd released during the first depuration period are also considerable. In the gills of *M. edulis*, the Cd uptake occurred by diffusion, and both the uptake and the accumulation was facilitated by intracellular binding and sequestration (Carpene and George, 1981). The data on the loss of Cd in *M. edulis* corresponds to the results on the uptake and depuration of trace metals in the oysters, *Crassostrea gigas* and *C. virginica* (Okazaki and Panietz, 1981; Zaroogian and Johnson, 1983). The release of Cd did not continue until the original control level was reached. A field study on the accumulation of Cd from seawater by *M. edulis* indicated a significant equilibrium between the total recoverable Cd in seawater and its concentration in the mussel (Talbot, 1985). This was possibly due to Cd binding to metal binding proteins thereby hampering the excretion of the metal-rich residual bodies as described by George and Viarengo (1985). This depicts the importance of metallothionin induction in the reduction of the cytotoxic effects exerted by high levels of Cd accumulation. Strong and Luoma (1981) examined the variation in the correlation of body size with concentration of Cu and Ag in the bivalve *Macoma balthica*. Strong positive and strongly negative relationships between body size and metal concentration were observed. They concluded correlations appeared to be influenced by the degree of enrichment in tissues, size-dependent differences and seasonal variations in growth rate, and size dependent differences in uptake rates. Jhingran and Joshi (1987) reported maximum accumulation of Zn, Cu and Pb in fishes and mollusks (Table 6.2) in Yamuna river waters.

Table 6.2: Detection of Metals in Fishes and Mollusks in Yamuna River Water as Reported by Jhingran and Joshi, 1987

Metal	*Metal Conc. in Fish (Wet wt)*	*Metal Conc. in Molluscs (Wet wt)*
Zn	6,16-36.3	28.1-193.0
Cu	0.41-3.57	2.3-18.8
Cr	Nd-1.25	Nd-6.723
Cd	Nd-0.19	Nd-1.8
Pb	0.46-1.56	7.3-38.6
Hg	Nd-1.38	Nd-0.62

After an exposure of 50 µg/l at 15°C for 40 days, Cd accumulated linearly in kidney of mussel (*Mytilus edulis*) to 300 µg g^{-1} dry weight, in the gills to 130 µg g^{-1} dry weight and in the muscles to 170 µg g^{-1} dry weight (Everarts, 1990). Cd release, in terms of the absolute decrease of the Cd concentration, from the kidney was very fast during the first four days (300 µg g^{-1}) in clean seawater. This loss was about twice in the gills (150 µg g^{-1}) and three times higher as compared to that in the muscles (100 μg g^{-1}). After 17 days in clean seawater, Cd loss expressed as the per cent of total Cd concentration after 43 days exposure to 50 μg Cd l^{-1} in the various organs, shows the following sequence : kidney (65.1 per cent > foot (62.4 per cent) > gill (60.3 per cent) > muscles (42.9 per cent) > adductor muscle (40.1 per cent) > mantle (14.1 per cent). Tiwari *et al.* (2000) studied the growth and the accumulation of heavy metals in green mussel, *Mytilus viridis*. The specific growth rate of the mussel, 13.25 cm in length and 31.61 g in weight was 3.38 per cent per month during the study period and it was further revealed that the metal load increased with the increasing size. Bhattacharrya and Chakraborty, 2002 in the vicinity of an iron factory in West Bengal, India, also reported presence of metals in highest concentrations in mollusks and the macrophytes. The pattern of accumulation of Cd, Cu and Pb in the mollusks near the industry ranged in between 4.91-9.08, 29.6-156.96 and 99.69-155.31 μg/g, respectively in different seasons and the lowest values for these metals however, occurred during the monsoon period. The macrophytes were found to contain these metals in the range in between 1.01-1.387, 1.53-23.43 and 12.15-23.63 μg/g in this area.

Exposer of certain *Cladocerans* and *Copepodes* to Cu, Cd and Zn (Gautam, 1990) revealed the order of toxicity of these metals as Cu>Cd>Zn; and the *Cladocerans* were relatively much more sensitive than the *Copepodes* to these metals. Madigosky *et al.*, 1991 evaluated the concentration of Pb, Cd, and Al in tissues of crayfish, *Procambarus clarkii* from several wetland sites located adjacent to roadways and were compared to crayfish harvested from a commercial site free from roadside influences. Abdominal muscle, hepato-pancrease, alimentary tract, exoskeleton and blood were analyzed for metal content. Results indicated that levels of contamination obtained in almost all tissues; crayfish from roadside ditches contained significantly higher amounts of metals than those of the commercially harvested control crayfish. Detection limits of Pb, Cd, and Al ranged from 0.04 μg Pb/g to 16.15 μg Pb/g,.001 μg Cd/g to.13 μg Cd/g, and 1.22 μg Al/g to 981 μg Al/g, respectively. Concentrations of Pb, Cd, and Al were highest in the hepato-pancrease and alimentary tract. High levels of these elements were also detected in the exoskeleton. In contrast, muscle tissue was least affected. Likewise, Dobicki *et al.*, 1991 showed several significant correlations among concentrations of metals on comparing a variety of tissues in *Procambarus clarkii.* They examined samples of muscles, gills, liver and the kidneys collected from 30 adult carps for Cu, Cd, Pb and Zn concentration and showed in muscles the lowest concentration of all the four elements tested (0.42-1.52, 0.44-0.72, 0.54-2.0, 4.67-9.13 ppm, respectively). The highest concentration of Cd and Pb were found in the gills (0.65-1.64 and 1.39-5.02 ppm respectively), of Cu in the liver (0.80-3.23 ppm) and that of Zn in the kidneys (156.9-230.4 ppm). In the crayfish, *Astacus astacus,* Simon *et al.* (2000) reported that the trophic uptake of Cd leads to high concentrations in the hepato-pancreas and small accumulation in the muscle tissue; while highest levels of bioaccumulation of MeHg occurred in its green gland and in the tail muscle. Turoczy *et al.* (2001) in tissues of the King Crab (*Pseudocarcinus gigas*) from southeast Australian waters, observed significant differences between tissues for all metals, and the distribution of metal between the tissues was different for each metal. The concentrations of mercury and zinc in muscle tissue were found increased with crab size and were high compared to other crab species. The concentrations of cadmium and copper present in edible tissues were not especially high compared to other crab species, but the concentration of cadmium in the hepato-pancreas was of dietary concern.

REFERENCES

Abu T. and Weis J.S.,1993. Bioaccumulation of heavy metals in two populations of mummi chog (*Fundulus heteroclitus*). Bull. Environ. Contam. Toxicol. 51: 1-5.

Ahmad S. and Al–Ghasis S.M., 1996. Metal contents in the tissues of *Lutjanus fulviflamma* (Smith, 1949) and *Epinephelus tauvina* collected from the Arabian Gulf. Bull. Environ. Contam. Toxicol., 57: 957-963.

Aksoy A., 2008. Chicory (*Cichorium intybus* L.): a possible biomonitor of metal pollution. Pak. J. Bot., 40(2): 791-797.

Akueshi E.U., Oriegie E., Ocheskiti N. and Okunsebor S., 2003. Levels of some heavy metals in fish from mining lakes on the Jos Plateau, Nigeria. Afr. J. Nat. Sci., 6:82-86.

Alabaster J.S. and Lloyd R. (Eds.), 1982. Water quality criteria for freshwater fish. 2nd Ed. Published by arrangement with the Food and Agriculture Organization of the United Nations by Butterworths, London: 361 pp.

Allen T., Singhal R. and Rana S.V.S., 2004. Resistance to oxidative stress in a freshwater fish, *Channa punctatus* after exposure to inorganic arsenic. Biol. Trace Elem. Res., 98(7): 63-72.

Allen P., 1995. Accumulation profiles of lead and cadmium in the edible tissues of *Oreochromis aureus* during acute exposure. J. Fish Biol., 47(4): 559–568.

Allinson G., Nishikawa M., DeSilva S.S., Laurenson L.J. and De Silva K., 2002. Observations on metal concentrations in tilapia (*Oreochromis mossambicus*) in reservoirs of south Sri Lanka. Ecotoxicol. Environ. Saf., 51(3): 197-202.

Amundsen P.A., Staldvik F.J., Lukin A.A., Kashulin N.A. and others, 1997. Heavy metal contamination in freshwater fish from the border region between Norway and Russia. Sci. Total Environ., 201: 211-224.

Anderson J.L. and Depledge M.H., 1997. A survey of total mercury and methylmercury in edible fish and invertebrates fom Azorean waters. Mar. Environ. Res., 44: 331-350.

Anderson P.D. and Spear P.A., 1980. Copperpharmakinetics in fish gills.1. Kineticsin pumpkinseed sunfish, *Lepomis gibbosus* of different body sizes. Water Res., 14: 1101.

Andreji, J., Stranai, I., Massanyi, P. and Valent, M. (2005). Concentration of selected metals in muscle of various fish species. J. Environ. Sci. Health A Tox. Hazard Subst. Environ. Eng. 40(4): 899-912.

Anonymous, 2003. Multiresidue detection method by AINP.

Asuquo F.E., Ewa-Oboho I., Asuquo E.F. and Udo P.J., 2004. Fish species used as biomarker for heavy metal and hydrocarbon contamination for Cross river, Nigeria. The Environmentalist, 2: 29–37.

Ay Q., Kalay M., Tamer L. and Canli M., 1999. Copper and lead accumulation in tissues of freshwater fish Tilapia zilli and its effects on the branchial Na^+ K^+ ATPase activity. Bull. Environ. Contam. Toxicol., 62: 160-168.

Baatrup E., Nielsen E. and Danscher G., 1986. Histochemical demonstration of two mercury pools in trout tissues; mercury in kidney and liver after mercuric chloride exposure. Ecotoxicol. Environ. Safety, 12: 267.

Bashir F.A., Shuhaimi-Othman Md. and Mazlan A.G., 2012. Evaluation of trace metal levels in tissues of two commercial fish species in Kapar and Mersing Coastal Waters, Peninsular Malaysia. J. Environ. and Pub. Hlth., 10pp.

Begum Abida, Krishna Hari, Khan Irfanulla, 2009a. Analysis of Heavy metals in water, sediments and fish samples of Madivala Lakes of Bangalore, Karnataka. Intl J. Chem. Tech. Res., 1(2): 245-249.

Benson N.U., Essein J.P., Willium A.B. and Bassey D.E., 2007. Mercury accumulation in fishes from tropical aquatic ecosystems in the Niger Delta, Nigeria. Curr. Sci., 92; 781-785.

Bentley P.J., 1992. Influx of zinc by channel catfish (*Ictalurus punctatus*): uptake from external environmental solutions. Comp. Biochem. Physiol., 101C: 215-217.

Bhagwant S., 2003a. Bioaccumulation of copper and iron in edible fish species from Pourdre D' Or Bay, Mauritius. Polln. Res. 22(2): 293-297.

Bhagwant S., 2003b. Lead, copper and zinc levels in organ tissue samples of three edible lagoon fish species from Tombeau bay, Mauritius. Polln. Res., 22(2): 299-303.

Bhattacharrya N. and Chakraborty S.K., 2002. Distributional pattern of some heavy metals in the structural components of two contrasting wetlands in the vicinity of an iron extraction factory of Midnapore district West Bengal, India. In Kumar A. (Ed) Ecology of polluted waters Vol. 1. APH Publ. Corp., New Delhi: 287-294.

Bonnel J.A., Ross J.H. and King E., 1960. Renal lesions in experimental cadmium poisoning. Dist. J. Indian Med., 17: 69-80.

Brown M.W., Thomas D.G., Shurben D., and 3 others, 1986. A comparison of the differential accumulation of Cadmium in the tissues of three species of freshwater fish, Salmo gairdneri, Rutilus rutilus and Noemacheilus barbatulus. Comp. Biochem. Physiol. 84c: 213-217.

Brucka–Jastrzebska E. and Protasowicki M., 2004. Elimination dynamics of cadmium, administered by a single introperitoneal injection in common carp, *Cyprinus carpio* (L). Acta. Ichthyol. Piscat., 34(2): 167-179.

Buckley J.T., Roch M., McCarter J.A., Rendel C.A. and Mathieson A.T., 1982. Chronic exposure of coho salmon to sublethal concentration of copper. I. Effect on growth, accumulation and distribution of copper and on copper tolerance. Comp. Biochem. Physiol., 72C: 15-19.

Bull K.R., Dearsley A.F. and Jnskip M.H., 1981. Studies of growth rates and mercury content of roach, perch and pike living in sewage effluent. Environ. Poll., 25: 229-240.

Burger J. Dixon C., Boring C. S. and Gochfeld M., 2003. Effect of deep frying fish on risk from mercury. J. Toxicol. Environ. Hlth. Part A, 66: 817-828.

Calamari D., Gaggino F.F., and Pacchetti G., 1982. Toxico-kinetics of low levels of Cd, Ni and their mixure in long terms treatment on *Salmo gaidneiri*. Rich Chemosphere, 11: 59-70.

Canli M and Atli G., 2003.The relationships between heavy metal (Cd, Cr, Cu, Fe, Pb, Zn) levels and the size of six Mediterranean fish species. Environ. Pollut., 121(1):129-136.

Canli M., Ay Ö. and Kalay M., 1998. Levels of heavy metals (Cd, Pb, Cu, Cr and Ni) in tissue of *Cyprinus carpio, Barbus capito* and *Chondrostoma regium* from the Seyhan river, Turkey. Turk. J. Zool., 22(2):149–157.

Carpene E., Fedrizzi G. and Giani G., 1989. Concentrations of Cd, Zn and Cu in liver and bile of *Carassius aurratus* (Italian). Arch. Ital., 40(3): 197-202.

Carpene E. and George S.G., 1981. Absorption of cadmium by gills of *Mytilus edulis* (L.). Mol. Physiol., 1: 23-34.

Caviglia A. and Cugurra F., 1978. Further studies on the mercury contents in some species of marine fish and mollusks. In Springer-Verlag,

Chakraborty R., Dey S., Dkhar P.S., Ghosh D., Singh S., Sharma D.K. and Myrboh B., 2003. Accumulation of heavy metals in some freshwater fishes from eastern India and its possible impact on human health. Polln. Res., 22(3): 353-358.

Chen-Meng-Hsien and Chen Chiee-Young, 1999. Bioaccumulation of sediment bound heavy metals in grey mullet, *Liza macrolepis*. Marine Polln. Bull., 39(1-12): 239-244.

Claissae D., Cossa D., Bretaudeau-Sanjuan J., Touchard G. and Bombled B., 2001. Methylmercury in mollusks along the French coast. Mar. Pollut. Bull., 42: 329-332.

Cross F.A., Hardy L.H., Jones N.Y. and Barber R.T., 1973. Relation between total body weight and concentrations of manganese, iron, copper, zinc and mercury in the white muscle of blue fish (*Pohnatonus saltatri*) and a bathy demersal fish (*Antiwora rostrata*). J. Fish. Res. Bd. Can., 30: 1287-91.

Dallinger R., Prosi F., Segner H. and Back H., 1987. Contaminated food and uptake ofheavy metals by fish: a review and a proposal for further research. Oecologia(Berl), 73: 91-98.

Das S.K. and Kaviraj A., 1990. Accumulation of Heteropneustes fossilis and changes in haematological parameters. J. Nat. Conserv., 2: 25-30.

De Boeck G., De Wachter B., Vlaeminck A. and Blust R., 2003. Effect of cortisol treatment and/or sublethal copper exposure on copper uptake and heat shoch protein levels in common carp, *Cyprinus carpio*. Environ. Toxicol. Chem., 22: 1122-1126.

De la Torre F.R., Saliian A. and Ferrari L., 2000. Biomarker assessments in Juvenile Cyprinus carpio exposed to water born cadmium. Environ. Poll., 109: 277-282.

Deb S.C. and Santra S.C., 1997. Bioaccumulation of metals in fishes. An *in vivo* experimental study of sewage fed ecosystem. Journal Article, ASEA-Part3.

De-Smet-Hans and Blust-Ronny, 2001. Stress responses and changes in protein metabolism in carp, *Cyprinus carpio* during cadmium exposure. Ecotoxicol. Environ. Saf., 48(3): 255-262.

Dobicki W., Szulkowska W.E., Marck J. and Polechonski R., 1991. Concentration of some heavy metals in tissues of carp from selected ponds of lower Silesia. Zeszyty- Naukowe- Alkademii- Rolniczy., 34: 41-46.

Dural M., Göksu M.Z.L. and Özak A.A., 2007. Investigation of heavy metal levels in economically important fish species captured from the Tuzla lagoon. Food Chemistry., 102(1): 415–421.

Eisler Ronald, 2010. Compedium of trace metals and marine biota. Vertebrates, Vol.-2 (Ch-2: 7-13). Elsevier, UK.

El-Moselhy Kh. M., 2001. Toxicity of cadmium to the marine fish Mugil seheli and its accumulation in different tissues. J. Egypt Acad. Soc. Environ. Develop., 2 (1) :17–28.

El-Shahawi M.S. and Al-Yousuf M.H., 1998. Heavy metal (Ni, Co, Cr and Pb) contamination in liver and skin tissues of *Lethrinus lentjan* fish. Family: Lethrinidae (Teleost) from the Arabian Gulf. Int. J. Food Sci. Nutr., 49 (6): 447-451.

EndoTetsuya, Hisamichi Yohsuke, Haraguchi Koichi, Kato Yoshihisa, Ohta Chiho and Koga Nobuyuki, 2008. Hg, Zn and Cu levels in the muscle and liver of tiger sharks (*Galeocerdo cuvier*) from the coast of Ishigaki Island, Japan: Relationship between metal concentrations and body length. Mar. Poll. Bul., 56 (10): 1774-1780.

Everarts J.M., 1990. Uptake and release of Cadmium in various organs of Common Mussel, *Mytilus edulis* (L.). Bull. Environ. Contam. Toxicol., 45: 560 – 567.

FAO, 1983. Compilation of legal lmits for hazardous substances in fish and fishery products. FAO fishery circular no. 464: 5-100.

Faraq A.M., Suedkamp M.J., Meyer J.S., Banaws R. and Woodward D.F., 2002. Distribution of metals during digestion by cut-throat trout fed benthic invertebrates contaminated in the Clark Fork River, Montana and the Coeurd' Alene River, Idaho, U.S.A., and fed artificially contaminated Artemia. J. Fish Biol., 56: 173-190.

Farombi E.O., Adelowo O.A., Ajimoko Y.R., 2007. Biomarkers of oxidative stress and heavy metal levels as indicators of environmental pollution in African cat fish (*Clarias gariepinus*) from Nigeria Ogun River. Int. J. Environ. Res. Public Health, 4(2): 158-65.

Felts P.A. and Heath A.G., 1984. Interactions of temperature and sublethal environmental copper exposure on the energy metabolism of bluegill, *Lepomis macrochrus*. J. Fish Biol., 25: 445.

Fırat Özgür and Kargın Ferit , 2010. Response of *Carpus carpio* to copper exposure: alterations in reduced glutathione, catalase and proteins electrophoretic patterns. Fish Physiol. and Biochem., 36 (4): 1021-1028,

Freeman B.J., 1980. Accumulation of cadmium, chromium, and lead by bluegill sunfish (*Lepomis macrochirus*) under temperature and oxygen stress, Ph.D. dissertation, university of Georgia, Athens.

Galhoom K.L., Laila G., Rizk G. and El-ezzaway M.H., 2000. Some biochemical and hematological parameters in Mugil fish (*Mugil caphalcus*) reared in Bahr El-Bakar drain. Egypt. J. Agril. Res., 78: 1-13.

Gautam A.K. 1990. Toxic effect of copper, cadmium and zinc to *Daphanosoma brachyurum* and *Eucyclops soparatus*; 139-144. In: Environmental Pollution and Health Hazards, Env. Sr. III (Ed. Prakash and Choubey S.M.). Society of Biosciencs, Muzaffarnagar (UP).

George S.G and Viarengo A., 1985. An integration of current knowledge of uptake, metabolism and intracellular control of heavy metals in Mussels. In: Vernberg F.J., Thurberg F.P., Calabrese A. and Vernberg W.B. (Eds). Marine Pollution and Physiology: Recent Advances University of South Carolina Press, Columbia:125 – 144.

Ghosh B.B. and Bagchi M.M., 1980. The pollution problem in Hoogly estury. In: Summer Inst. on brachishwater capture and culture fisheries, CIFRI Barrachpur, (Jul-Aug, 1980): pp1-10.

Govindsamy C., Roy Vij A.G., and Jayapal A., 1998. Int J. Ecol. Environ. Sci., 24:141-146.

Greig R.A., Wenzloff D.R., Mckenzit C.L., Meril A.S. and Zdanowicez V.S., 1978. Trace metals in the sea Sallops Pacepedent magelanicus from Eastern United States. Bull. Environ. Contam. Toxicol., 19: 326-334.

Greig R.A., Schurman S., Preira J. and Patricia N., 1983. Metal concentration of silver, cadmium, copper and lead in livers of windopane flounder. Bull. Environ. Contam. Toxicol., 31 : 257-262.

Gupta A.K., 1998. Accumulation of the cadmium in the fishes *H. fossilis* and *Channa punctatus*. Environ. Ecol., 6: 577-580.

Gupta A.K. and Chakarbarti P., 1993. Toxicity of zinc to freshwater teleosts, *Notopterus notopterus* and *Puntius javanicus*. Freshwater Biol., 5 (4): 359-363

Gupta A.K. and Sharma S.K., 1994. Bioaccumulation of zinc in *Cirrihinus mrigala* (Hamilton) fingerlings during short-term static bioassay. J. Environ. Biol., 15(3): 231-237.

Gupta P. and Srivastava N., 2006. Effects of sublethal concentrations of on histological changes and bioaccumulation of zinc by kidney of fish, *Channa punctatus* (Bloch). J. Environ. Biol., 27: 211-215.

Gutleb A.C., Helsberg A. and Mitchell C., 2002. Heavy metal concentrations in fish from a pristine rainforest valley in Peru: a baseline study before starting of oil drilling activities. Bull. Environ. Contam. Toxicol., 69 (4): 523-529.

Haines, T.A., 1996. Mercury in fish and fish-eating birds in Maine [Abstract]. In: Abstracts for presentations given at the USGS workshop on mercury cycling in the environment. (http://toxics.usgs.gov/pubs/hg. html).

Hall L.W. Jr. and Burton D.T., 1982. Effect of power plant coal pile and coal waste runoff and leachate on aquatic biota: an overview with research recommendations. Boca Raton, FL: CRC Press: 287–297.

Handy R.D., 1992. The assessment of episodic metal pollution. II The effects of cadmium and copper enriched diets on tissue contaminant analysis in Rainbow trout (Onchorhynchus mykiss). Arch. Environ. Contam. Toxicol., 22: 82-87.

Handy R.D. and Eddy F.B., 1990a. The interaction between the surface of rainbow trout, *Onchorhynchus mykiss* and waterborne metal toxicants. Funct. Ecol., 4: 385.

Handy R.D. and Eddy F.B., 1990b. Influence of starvation on water borne zinc accumulation by rainbow trout, *Salmo gairdneri*, at the onset of episodic exposure in natural soft water. Wat. Res., 24(4): 521-527.

Hansen B.H., Rømma S., Garmo Ø.A., Olsvik P.A. and Andersen R.A., 2006. Antioxidative stress proteins and their gene expression in brown trout (Salmo trutta) from three rivers with different heavy metal levels. Comp. Biochem. Physiol. C Toxicol. Pharmacol., 143(3): 263-74.

Hardy R.W., Sullivan C.V. and Koziol A.M., 1987. Absorption, body distribution, and excretion of dietary zinc by rainbow trout,*Salmo gairdneri.*Fish Physiol. Biochem., 3: 133-143.

Harrison S.E. and Klaverkamp J.F., 1989. Uptake, elimination and tissue distribution of dietary and aquous cadmium by rainbow trout (*Salmo gairdneri* Richardson) and lake white fish (*Coregonus clupeaformis* Mitchell). Environ. Toxicol. Chem., 8: 87-97.

Hashemi Shodja, Blust Ronny, de Boeck Gudrun, 2007. The effect of food rations on tissue-specific copper accumulation patterns of sublethal waterborne exposure in *Cyprinus carpio*. Environ. Toxicol. and Chem., 26, Issue 7: 1507–1511.

Hattula M.L., Sarkka J., Janatuinen J., Passivirta J. and Roos A., 1978. Environ. Pollut., 17: 19.

Hawkins W.E., Tate L.G. and Sharpie T.G., 1980. Acute effects of cadmium on the spot, Leiostomus xanthurus (Teleostei): tissue distribution and ultrastructure. J. Toxicol. Environ. Hlth., 6: 283.

Heath A.G., 1987. Effects of waterborne copper or zinc on the osmoregulatory response of bluegill to a hypertonic NaCl challenge. Comp. Biochem. Physiol., 88: 307.

Holcombe G.W., Benoit D.A., Leonard E.N. and McKim J.M., 1976. Long term effects of lead exposure on three generations of brook trout, Salvelinus fentinalis. J. Fish Res. Bd. Canada, 33: 1731-1741.

Holcombe G.W., Benoit D.A. and Leonard E.N., 1979. Long-term effects of zinc exposures on brook trout (*Salvelinus fontinalis*). Trans. Am. Fisher. Soc., 108: 76-87.

Hornung Hava, Zismann Lyka and Oren O.H., 1980. Mercury in twelve Mediterranean trawl fishes of Israel. Environ. Intrl., 3: 243-248.

Hughes G.M. and Flos R. 1978. Zinc content of the gills of rainbow trout, *Salmo gairdneri* after treatment with zinc solutions under normoxic and hypoxic conditions. J. Fish Biol., 13: 717.

Ikem A., Egiebor N.O. and Nyavor K., 2003. Trace elements in water, fish and sediment from Tuskegee Lake, Southeastern USA. Water, Air, and Soil Pollution., 149(1–4): 51–75.

James R., Sampath K., Deakiamma and Pattu V.J., 1996. Recovery of dehydrogenase enzymes activities in *Heteropneustes fossilis* (Bloch) exposed to sublethal levels of mercury. Ind. J. Fish, 43: 293-297.

Janseen H.H. and Scholz N., 1979. Uptake and cellular distribution of cadmium in *Mytilus edulis*. Mar. Biol., 55: 133-141.

Jastrzêbska Ewa Brucka and Protasowick Miko³aj, 2006. Changes of cadmium content in various organs of common carp (*Cyprinus carpio* L.) during the fast growth period following initial rearing in contaminated water. Arch. Pol. Fish, 14: 183-194.

Jeng Sen Shyong, 1999. Zn and Zn binding substances in the tissue of Common Carp. Comp. Biochem. Physiol., 122 (4): 461-468.

Jernelov A. and Lann H.,1971. Mercury accumulation in food chains. Oikos., 22: 403–406

Jezierska B. and Witeska M., 2006. The metal uptake and accumulation in fish living in polluted waters. Soil and Water Pollution Monitoring, Protectionand Remediation: 3–23 pp

Jhingran Arun G. and Joshi H.C., 1987. Heavy metals in water sediments and fish in the river Yamuna. J. Inland Fish. Soc. India, 19 (1): 13-23.

Jhonels A.G., Westermark T., Berg W., Persson P.I. and Sjostrand, 1967. Oikos., 18: 323.

Joiris C.R., Das H.K. and Holsbeek L.,1999. Total and methylmercury in sardines, *Sardinella aurellia* and *Sardina pilchardus* from Tunisia. Mar. Pollut. Bull., 38: 188-192.

Jyothirmayee S., Janetheophillus, Padma Balaravi, Narender Reddy T. and Reddy P.U.M., 2006. Endosulfan induced changes in esterases of *Anabas testudineus* and *Clarias batrachus*. Ind. J. Comp. Animal Physiol., 24: 95-99.

Julshamm K., Ringdal O. and Brackkan O.R., 1982. Mercury concentration in liver and muscle of cod (*Gadus morhua*) as an evidence of migration between waters with different levels of mercury. Bull. Environ. Contam. Toxicol., 29: 544.

Kamman N.C. and Burgess N.M., 2004. Mercury in freshwater fish tissues of northeast North America: A geographic perspective based on fish tissue monitoring databases. Presentation to U.S.EPA. Mercury Round Table Conference cell.

Karadede-Akin H. and Ünlü E., 2007. Heavy metal concentrations in water, sediment, fish and some benthic organisms from Tigris River, Turkey. Environ. Monitor. and Assess., 131(1–3): 323–337.

Kaushik N. and Srivastava N., 2003. Recovery in zinc induced hepatotoxicity of *Channa punctatus*. J. Ecophysiol. Occup. Hlth., 3, 199-204.

Kaviraj A. and Ghosal T.K., 1997a. Effects of poultry litter on the chronic toxicity of cadmium to common carp, *Cyprinus carpio*. Biol. Res. Technol., 60: 239-243.

Kaviraj A. and Konar S.K., 1983. Lethal effects of mixtures of mercury, chromium and cadmium on fish. Environ. Ecol., 1: 153-158.

Kebede Aweke and Wondimu Taddese, 2004. Distribution of trace elements in muscle and organs of tilapia, Oreochromis niloticus, from lakes awassa and ziway, ethiopia. Bull. Chem. Soc. Ethiop., 18(2), 119-130.

Kraal M.H., Kraak M.H.S., Groot C.J., Davids C. and De–Groot C.J., 1995. Uptake and tissue distribution of dietery and aqueous Cd by Carp (Cyprinus carpio). Ecotoxicol. Environ. Safety., 31(2), 179-183.

Krishna Murti, Asha Jyoti and Nair V.R., 1999. Concentration of metals in fishes from Thane and Bassein creeks of Bombay, India. Ind. J. Mar. Sci., 28 (1): 39-44.

Kumar A. and Mathur R., 1991. Bioaccumulation kinetics and organ distribution of lead in a freshwater teleosts, *Colisa fasciatus*. Environ. Technol., 12: 731.

Kumari Kanchan, Khare Ankur, and Dange Swati, 2014. The applicability of oxidative stress biomarkers in assessing chromium induced toxicity in the Fish, *Labeo rohita*. Biomed. Res. Int., 2014: 782493.

Kureishy T.W. and D'Silva C., 1993. Uptake and loss of mercury, cadmium and lead in marine organisms. Ind. J. Exp. Biol., 31: 373-391.

Laurén D.J. and McDonald D.G., 1987a. Acclimation to copper by rainbow trout, *Salmo gairdneri*: physiology. Can. J. Fish. Aquat. Sci., 44: 99–104.

Laurén D.J. and McDonald D.G., 1987b. Acclimation to copper by rainbow trout, Salmo gairdneri: biochemistry. Can. J. Fish. Aquat. Sci., 44: 105–111.

Lawani S.A. and Alawode J.A., 1996. Concentration of lead nd Hg in river Niger and its fish at Jebba. Biosci. Res. Commun., 8: 47-49.

Liao C.M., Chen B.C., Singh S., Lin M.C., Liu C.W. and Han B.C., 2003. Acute toxicity and bioaccumulation of arsenic in tilapia (*Oreochromis mossambicus*) from a blackfoot disease area in Taiwan. Environ.Toxicol., 18, 252–259.

Lin H.C., Hsu S.C. and Hwang P.P., 2000. Maternal transfer of cadmium tolerance in larval *Oreochromis mossambicus*. J. Fish Biol., 57: 239 -249.

Liyaquat F., Sharma M.S. and Chishty N., 2003. Evaluation of zoo toxicity of zinc and zinc smelter effluent to selected freshwater animals. Pollun. Res., 22(4): 577-584.

Madigosky S.R., Alvarez-Hernandez X. and Glass J., 1991. Lead, cadmium, and aluminium accumulation in the red swamp crayfish *Procambarus clarkii* G. collected from roadside drainage ditches in Louisiana. Arch. Environ. Contam. Toxicol., 20(2): 253-8.

Maiti P. and Banerjee S., 1999. Accumulation of heavy metals in different tissues of fish *Oreochromis nitoticus* exposed to waste water. Environ. Ecol., 17(4): 895-898.

Maiti P. and Banerjee S., 2002. Bioaccumulation of metals in different food fishes in wastewater fed wetlands. In: Kumar A. (ed), Ecology of Polluted Waters, vol. 1. APH Publ. New Delhi: 217-243.

Mance G., 1987. Pollution threat of heavy metals in aquatic environments. London. Elsevier Applied Science.

Mansour S.A. and Sidky M.M., 2002. Ecotoxicological studies. 3: Heavy metals contaminating water and fish from Fayoum Governorate, Egypt. Food Chem., 78: 15-22.

Martin J.H. and Flegal A.R., 1975. High copper concentration in squid rivers in association with elevated levels.of Ag, Cd and Zn. Mar. Biol., 30: 51-55.

Matsunaga k., 1978. Concentration of mercury in marine animals. Bull. Fac. Fish Hokkaido Univ., 29 (1): 70-74.

Mazon A.F. and Fernandes M.N., 1999. Toxicity and differential tissue accumulation of copper in the tropical freshwater fish, *Prochilodus sarofa*. Bull. Environ. Contam. Toxicol., 63: 797-804.

McCarter J.A. and Roch M., 1984. Chronic exposure of coho salmon to sublethal concentrations of copper. III. Kinetics of metabolism of metallothionein. Comp. Biochem. Physiol., 74C, 133–137.

Mhadhbi L., Palanca A., Gharred T. and Boumaiza M., 2012. Bioaccumulation of metals in tissues of *Solea vulgaris* from the outer Coast and Ria de Vigo, NE Atlantic (Spain). Int. J. Environ. Res., 6(1):19-24.

Mirenda J.R., 1986. Acute toxicity and accumulation of zinc in the Crayfish, *Orconectes virilis* (Hagen). Bull. Environ. Contam. Toxicol., 37: 387-394.

Misra S.M., Borana K., Pani S., Bajpai A. and Bajpai A.K., 2002. Assessment of heavy metal concentration in grass carp (*Ctenopharyngodon idella*). Pollun. Res. 21 (1): 69-71.

Mohan D. and Choudhary A., 1991. Zinc accumulation in a few tissues of fish, Puntius sophare (Ham.) after sublethal exposures. J. Nat. Conserv., 3: 205-209.

Molisani M.M., Salomão M.S.M., Ovalle A.R.C., Rezende C.E., Lacerda L.D. and Carvalho C.E.V., 1999. Heavy Metals in Sediments of the Lower Paraíba do Sul river and Estuary, RJ, Brazil. Bull. Environ. Contam. Toxicol., 63, 682–690.

Mount, D. 1964. An autopsy technique for zinc – caused fish mortality, Trans. Am. Fish. Soc., 93: 174.

Mukhopadhyay M.K., Ghosh B.B. and Bagchi M.M., 1994. Toxicity of heavy metals to fish, prawn and fish food organisms of Hooghly estuarine system. Geobios., 21 : 13-17.

Naz S, Javed M, Hayat S, Abdullah S, Bilal M, Shaukat T., 2008. Long term effects of lead (Pb) toxicity on the growth performance, nitrogen conversion ratio and yield of major carps. Pak. J. Agri. Sci., 45(3):53–58.

Neumann C.M., Kauffman K.W. and Gilroy D.J., 1997. Methyl mercury in fish from Owyhee reservoir in southeast Oregon: Scientific uncertainty and fish advisories. Sci. Total Environ., 204: 205-214.

Norey C.G., Brown M.W., Cryer A. and Kay J., 1990. A comparison of the accumulation, tissue distribution and secretion of cadmium in different species of freshwater fish. Comp. Biochem. Physiol., 96C:181.

Norstrom R.J., McKinnon A.E.and DeFreitas A.S.W., 1976. A bioenergetics-based model for pollutant accumulation by fish – Simulation of PCB and methylmercury residue levels in Ottawa River yellow perch. J. Fish. Res. Board Can., 33: 248–267.

Okazaki R.K and Panietz M.H., 1981. Depuration of twelve trace metals in tissues of the *Oysters Crassostrea gigas* and *C. virginica*. Mar Biol., 63:113 – 120.

Oladimeji, A.A., Quadri, S.U. and Defreitas, A.S.W. 1984a. Long term effects of arsenic accumulation in rainbow trout, *Salmo gairdneri*. Bull. Environ. Contam. Toxicol. 32: 732-741.

Olasson P.E., Larson A. and House C., 1988. Metallothionein and heavy metals in Rainbow trout (*Salmo gairdneri*) during exposure to cadmium in water. Mar. Environ. Res., 24: 151-153.

Olson K.R., Bergman and From P.O., 1973. J.Fish Res. Bd. Can., 33: 248.

Olson M. and Jensen S., 1975. Rep. Inst. Preshw. Res. Drattningholm, 54: 83.

Omeregie E., Okoronkwo M.O., Eziashi A.C. and Zoakah A.I., 2002. Metal concentration in water column, benthic macroinvertebrates and tilapia from Delimi River, Nigeria. J. Aqua. Sci., 17:55-59.

Pandey B.K., Sarkar U.K., Bhowmik M.L. and Tripathi S.D., 1995. Accumulation of heavy metals in soil, water, aquatic weed and fish samples of sewage-fed ponds. J. Environ. Biol., 16(2): 97-103.

Pane E.F., Haque A., Goss G.G. and Wood C.M., 2004. The physiological consequences of exposure to chronic, sub lethal water borne nickel in rainbow trout (*Oncorhynchus mykiss*); exercise vs. resting physiology. J. Exp. Biol., 207: 1249-1261.

Papagiannis I., Kagalou I., Leonardos J., Petridis D., Kalfakakou V., 2004. Copper and zinc in four freshwater fish species from Lake Pamvotis (Greece). Environ Int., 30(3): 357-62.

Part P., Svanberg O. and Bergstrom E. 1985. The influence of surfactants on gill physiology and cadmium uptake in perfused rainbow trout gills. Ecotoxicol. Environ. Safety, 9:135.

Pelgrom S.M.G.J., Lamers L.P.M., Garritsen J.A.M. and 4 others, 1994. Interactions between copper and cadmium during single and combined exposures in juvenile tilapia *Oreochromis mossambicus*. Influence of feeding condition on whole body metal accumulation and the effect of the metals on tissue water and ion content. Aquat. Toxicol., 30: 117-135.

Pelgrom S.M.G.J., Lock R.A.C. Balm P.H.M. and Wendelaar Bonga S.E., 1995.Effects of combined waterborne Cd and Cu exposures on ionic composition and plasma cortisol in tilapia, *Oreochromis mossambicus*. Comp. Biochem. Physiol., 11: 227-235.

Peyghan R., Razijalaly M., Baiat M. Rasekh A., 2003. Study of bioaccumulation of copper in liver and muscle of common carp *Cyprinus carpio* after copper sulphate bath. Aquacult. Int., 11: 597-604.

Phillips D.J.H., 1977. The use of biological indicator organisms to monitor trace metal pollution in marine and eustuarine environments- a review. Environ. Pollut., 13: 281-317.

Ploetz D.M., Fitts B.E. and Rice T.M., 2007. Differential accumulation of heavy metals in muscle and liver of a marine fish, (king mackerel, *Scomberomorus cavalla* cuvier) from the Northern Gulf of Mexico, USA. Bull. Environ. Contam. and Toxicol., 78(2):134–137.

Radha Krishnaiah K., 1988. Accumulation of copper in the organs of freshwater fish, *Labeo rohita* (Ham.) on exposure to lethal and sublethal concentration of copper. J. Environ. Biol., 9: 319-326.

Rafia Azmat, Farha Aziz and Madiha Yousfi, 2008. Monitoring the effect of water pollution on four bioindicators of aquatic resources of Sindh Pakistan. Res. J. Environ. Sci., 2 (6): 465-473.

Ramamoorthy S. and Blumhagen K.,1984. Up-take of Zn, Cd and Hg by the fish in the presence of competing compartments. Can. J. Fish Aquat. Sci., 41: 750.

Rashed M.N., 2001b. Cadmium and lead levels in fish (*Tilapia nilotica*) tissues as biological indicator for lake water pollution. Environ. Monitor. Assess., 68(1): 75-89.

Rehulka J., 2002. Contents of organic and inorganic pollutants in the fish from the Slezska Harta reservoir. Czech. J. Anim. Sci., 47(1): 30-44.

Rema L.P. and Philip B., 1996. Metallothiolein on metallothionein like proteins and heavy metals toxicity in *Oreochromis mossambicus* (Peters). Ind. J. Exp. Biol., 34: 527-530.

Rimmer C.C., McFarland K.P., Evers D.C., and 4 others, 2005. Mercury contamination in Bicknell's thrush and other insectivorous passerines in montane forests of northestern north Ameica. Ecotoxicology 14: 223-240.

Rimmer C.C., Miller E.K., McFarland K.P., Taylor R.J. and Faccio S.D., 2009. Mercury bioaccumulation and trophic transfer in the terrestrial food web of a montane forest. Ecotoxicology, 13pp.

Romeo M., Siau Y., Sidoumon Z. and Gnassia-Barelli M., 1999. Heavy metal distribution in different fish species from the Mauritania coast. Sci. Total Environ., 232: 169-175.

Rudd J.W.M., Furutani A. and Turner M.A., 1980. Mercury methylation by fish intestinal contents. Appl. Environ. Microbiol., 40: 477.

Sahoo A., Panigrahi Minati K., Sahu S.K. and Panigrahi Ashok K., 2002. Aquatic mercury pollution by a chlor-alkali industry and its impact on the ecology of human health in Ganjam area- a case study. In: Kumar A. (Ed) Ecology of polluted waters Vol.1. APH Publ. Corp. New Delhi, pp 465-496.

Segner H., 1987. Response of fed and starved roah, *Rutilus rutilus* to sublethal Cu contamination. J. Fish Biol., 30: 423.

Shah S.L. and Altindag A., 2004. Haematological parameters of tench (*Tinca tinca*) after acute and sub chronic exposure to lethal and sublethal Hg treatments. Bull. Environ.Contam. Toxicol., 73: 911-918.

Sharma M., Liyaquat F. and Chishty N., 2003. Bioaccumulation of Zn and Cd in freshwater fishes. Ind. J. Fish, 50: 53-65.

Shelly Virmani, 2005. Effects of heavy metals on histopathological and biochemical changes in *Cyprinus carpio* (L) and *Cirrhinus mrigal* (Am). PhD thesis, Deptt. Zoology., CCS HAU., Hisar, Haryana: 195+xxxiii

Shukla V., Dhankar M. and Sastry K.V., 2005. Bioaccumulation of Zn, Cu and Cd in fish Coaser, Nat. Symp. Challenges and Opportunities Anim. Sci. Edu. and Res., Meerut, 17-19 March, 2005.

Simon O., Ribeyre F. and Boudou A., 2000. Comparative experimental study of cadmium and methylmercury trophic transfers between the Asiatic clam *Corbicula fluminea* and the crayfish *Astacus astacus.* Arch. Environ. Contam. and Toxicol., 38 (3): 317-326.

Simons T.J.B., 1986. Lead-calcium interaction in cellular lead toxicity. Neurotoxicol., 14: 77-85.

Sinha A.K., Dasgupta P., Chakraborty S., Bhattacharyya G. and Bhattacharjee S., 2002. Bio-accumulation of heavy metals in different organs of some of the common edible fishes of Kharkai river, Jamshedpur. Ind. J. Environ. Hlth., 44(2): 102-107.

Sinha R.K., Sinha S.K., Kedia D.K., Kumari A., Rani N., Sharma G. and Prasad K., 2007. A holistic study on mercury pollution in the Gange river system at Varanasi, India. Curr. Sci., 92(9): 1223.

Solbe J.F.De.L.G.and Flook V.A., 1975 Studies on the toxicity of zinc sulphate and of cadmium sulphate to stone loach Noemacheilus barbatulus (L.) in hard water. J. Fish Biol., 7(5): 631-637.

Spry D.J., Hodson P.V., and Wood C.M., 1988. Relative contributions of dietary and waterborne zinc in the rainbow trout, Salmo gairdneri. Can J. Fish Aquat. Sci., 45:32-41

Spry D.J., Wood C.M. and Hodson P.V., 1981. The effects of environmental acid on freshwater fish with particular reference to the softwater lakes in Ontario and the modifying effects of heavy metals: A literature review. Can. Tech. Rep. Fish. Aquat. Sci., 999, pp 1–144.

Storelli M.M., Busco V.P. and Marcotrigiano G.O., 2005. Mercury and arsenic speciation in the muscle tissue of *Scyllorhinus canicula* from the Mediterranean Sea. Bull. Environ. Contam. Toxicol., 75: 81-88.

Storelli M.M., Cuttone G. and Marcotrigiano G.O., 2011. Distribution of trace elements in the tissues of smooth hound *Mustelus mustelus* L. from the southern–eastern waters of Mediterranean Sea (Italy). Environ. Monit. Assess., 174 (1-4): 271-281.

Storelli M.M., Giacominelli S. R. and Marcotrigiano G. O. 2003a. Total and methylmercury residues in cartilaginous fish from the Mediterranean Sea. Mar. Pollut. Bull. 44: 1354-1358.

Storelli M.M., Giacominelli S.R., Strolli A. and Marcotrigiano G.O., 2003b. Total mercury and methylmercury content in edible fish from the Mediterranean Sea. J Food Prot., 66: 300-303.

Stromberg P.C., Ferrante J.G. and Carter S., 1983. Pathology of lethal and sublethal exposure of fathead minnow, *Pimephalus promelas* to cadmium. J. Toxicol. Environ. Health., 11 : 247-259.

Strong C.R. and Luoma S.N., 1981. Variations in the correlation of body size with concentrations of Cu and Ag in the bivalve *Macoma balthica*. Can. J. Fish Aquat. Sci., 38:1054-1064.

Subathra S. and Karuppasamy R., 2008.Bioaccumulation and depuration pattern of copper in different tissues of *Mystus vittatus*, related to various size group. Arch. Environ.Contam. Toxicol., 54(2): 236-244.

Sultana R. and Rao D.P., 1998. Bioaccumulation pattern of zinc, copper, lead and cadmium in grey mullet, *Mugilcephalus*(L.), from harbour waters of Visakhapatnam, India. Bull. Environ. Contam. Toxicol., 60: 949–955.

Svobodova Z., Zlabek V., Celechovska O. and Janouskova D., 2002. Metal content in tissues of marketable common carp and in bottom sediments of selected ponds of South and West Bohemia. Czech. J. Anim. Sci., 47(8): 339-350.

Szkoda J., Zmudzki J. and Zelazny J. 1999. Trace elements in carp tissues. Bromatologia Ichemia Toksykologiczna, 32: 311-316.

Talbot V., 1985. Relationship between cadmium concentrations in sea water and those in the mussel *Mytilus edulis*. Mar Biol., 85: 51 – 54.

Tariq W.K., George M.D. and Gupta R.S., 1979. Mar. Poll. Bull., 10: 357.

Tawari-Fufeyin P. and Ekaye S.A., 2007. Fish species diversity as indicator of pollution in Ikpoba River, Benin City, Nigeria. Rev. Fish Biol.Fisheries, 17: 21-30.

Terra B.F., Araújo F.G., Calza C.F., Lopes R.T. and Teixeira T.P., 2007. Heavy metal in tissues of three fish species from different trophic levels in a Tropical Brazilian River. Water Air Soil Pollut., 187: 275–284.

Thomas D.G., Solve J.F., Kay J. and Cryer A., 1983. Environmental cadmium is not sequestered by metallothionein in rainbow trout. Biochem. Biophys. Res. Commun., 110: 584.

Tiwari A., Joshi H.V., Raghunathan C., Kumar Sravan V.G. and Kotiwas Q.S., 2000. New record of *Mytilis virdis* L., its density, growth and accumulation of heavy metals on Saurashtra Coast, Arabian Sea. Current Sci., 78 (1): 97-101.

Tulasi S.J., Reddy P.U.M. and Ramana Rao J.V., 1992. Accumulation of lead and effects on total lipid derivatives in the fresh water fish, *Anabas testudineus* (Bloch). Ecotoxicol. Environ. Saf., 23: 33-38.

Turoczy N.J., Larenson L.J.B., Allinson G., Nishikawa M. and Smith C., 2000. Observations on metal concentrations in Sharks from South Eastern Australian waters. J. Agric. Food Chem., 48(9): 4357-4364.

Turoczy Nicholas J., Mitchell Bradley D., Levings Andrew H. and Rajendram Vijaya S., 2001. Cadmium, copper, mercury, and zinc concentrations in tissues of the King Crab (*Pseudocarcinus gigas*) from southeast Australian waters. Environ. Intl., 27: 327-334.

Udea T. and Nakashera M., 1983. Accumulation of Co by marine fish. Bull. Jap. Soc. Fish., 49 (4): 655-656.

UNEP., 2002. United Nations Environment Programme, Environment for development. Global Mercury Assessment Report, December 2002. *www.unep.org › Mercury.*

US DHHs., 1994. Report: Agency for Toxic substances and disease registry, Atlanta, GA, TP 93.10.

U.S.EPA., 2001. United States Environmental Protection Agency. Water quality criterion for the protection of human health: methylmercury (Executive summary). EPA-832-R-01-001. 2001.

U.S.EPA., 2002. United States Environmental Protection Agency. National study of chemical residues in lake fish tissues, year 1 data [CD-ROM]. http://cta.policy.net/reports/reel_danger/reel_danger_report.pdf

U.S.EPA., 2003. United States Environmental Protection Agency. National study of chemical residues in lake fish tissues, year 2 data [CD-ROM].

U.S.EPA., 2004. United States Environmental Protection Agency. Fish Advisories. http://www.epa.gov/ost/fish/states.htm.

US.FDA., 1997-2007. United States Food and Drug administration. Mercury Levels in commercial fish and shellfish. http://www.cfsan.fda.gov/~frf/sea-mehg.html.

Vigh P., Mastala Z. and Balogh K.V., 1996. Comparison of heavy metal concentration of grass carp in a shallow eutrophic lake and a fish pond. Chemosphere, 32 (4): 691-701.

Walker T.L., 1976. Effect of species, sex, length and locality on the mercury content of school shark *Galeorhinus australis* and gummy shark *Mustelus antatticus* from south-eastern Australia waters. Aust. J. Mar. and Freshwater Res., 27: 603-616.

Wang P.P., Chu K.L.M. and Wong C.K., 1999. Study of toxicity and bioaccumulation of copper in the silver sea bream *Sparus sarba*. Environ. Int., 25: 417– 422.

Wicklund A. and Runn P., 1988. Calcium effects on cadmium uptake, redistribution and elimination in minnows, *Phoxinus phoxinus* acclimated to different calcium concentrations. Aquatic. Toxicol., 13: 109.

Wiener J.G., Krabbenhoft D.P., Heinz G.H. and Scheuhammer A.M., 2003. Ecotoxicology of mercury.in: D.J.H. Hoffman, B.A. Rattner, G.A. Burton Jr. and J Cairns Jr. (Eds), Handbook of ecotoxicology. Second Ed., Lewis, Boca Raton, Florida, USA: 409-463 pp.

Wong C.K., Wong P.P.K and Chu L.M., 2001. Heavy metal concentrations in marine fishes collected from fish culture sites in Hong Kong. Arch. Environ. Contam. Toxicol. 40(1): 60-69.

Wood C.M. and Jackson, E.B., 1980. Blood acid–base regulation during during environmental hyperoxia in the rainbow trout (*Salmo gairdneri*). *Respir. Physiol.* 42, 351–372.

Yilmaz A.B., 2005. Comparison of heavy metal levels of grey mullet (*Mugil cephalus* L.) and Sea bream (*Sparus aurata* L.) caught in Ishkenderun bay (Turkey). Turk. J. Vet. Anim. Sci., 29: 257-262.

Yilmaz F., Özdemir N., Demirak A. and Tuna A.L., 2007. Heavy metal levels in two fish species *Leuciscus cephalus and Lepomis gibbosus*. Food Chemistry,100(2): 830–835.

Yosef T.A. and Ghada M.G., 2011. Assessment of some heavy metal contents in fresh and salted (Feseakh) mullet fish collected from El-Burullus Lake, Egypt. J. Am. Sci., 7(10): 137-144.

Zaroogian G.E and Johnson M., 1983. Chromium uptake and loss in the bivalve *Platichthys flexus* L. and its effects on plasma electrolyte concentrations. J. Fish. Biol., 20: 491-501.

Zhou X.W., Zhu G.N. and Sun J.H., 2003. Effects of the interaction of heavy metals on the accumulation of copper in the tissues the fish (*Carassius auratus*). J. Zhejiang-Univ., Agric. Life Sci. 28 (4): 427-430.

Zyadah M. and Chouikhi A., 1999. Heavy metal accumulation in *Mullus barbatus, Merluccius merluccius* and *Boops boops* fish from the Aegean Sea, Turkey. Internat. J. Food Sci. and Nutr., 50(6): 429-434.

Chapter 7

Health Risks of Heavy Metals in Fishes

1. General Metal Toxicity Parameters

Toxic effects of heavy metals on fish are multidirectional and are manifested chiefly by certain detectable changes in the biochemical processes of their body systems including interference with their normal metabolic processes. When ingested in an acid medium inside stomach, the metals are converted to their stable oxidation state (Zn^{2+}, Pb^{2+}, Cd^{2+}, As^{2+}, As^{3+}, Hg^{2+} and Ag^{2+}) and these combine with the body's bio-molecules such as proteins and enzymes to form strong and stable chemical bonds (Holum, 1983); in the process the hydrogen atoms or even the non-toxic metal groups are replaced by a toxic metal, thus, inhibiting the enzyme. Furthermore, a metal ion in the body's mettallo-enzyme complex can be conveniently replaced by another metal ion of similar size, such as the Cd^{2+} replacing Zn^{2+} in some dehydrogenating enzymes, leading to Cd^{2+} toxicity. In the process of inhibition, the structure of a protein molecule can be mutilated to a bio-inactive form resulting in completely destroyed enzyme. Metal interactions with proteins, leading to their denaturation and precipitation is termed as an allosteric effect or enzyme inhibition (Mizrahi and Achituv, 1989). Allosteric regulation (or allosteric control) is the regulation of a protein by binding an effector molecule at a site other than the protein's active site. Such changes lead to amorphous abnormalities in tissue forms and their functions, leading to the origin of various abnormalities in body functions such as the neurological, cardiological, pneumatical or the hematological diseases. They may also bind to nucleic acids, leading to irreversible conformational changes in DNA/RNA molecules.

A chemical's capacity to harm fish and other aquatic animals is a function of a variety of factors, largely its inherent toxicity, exposure time, dose rate, and its persistence in the environment; and its form, its concentration and the fish species affected. The inherent toxicity of a metal further depends upon its capacity to disturb the dynamic equilibrium of life processes in biological systems by combining with cell organelles, macromolecule and metabolites. The chief criterion regarding the toxicity of a chemical is the dose, *i.e.* the amount of exposure per unit of time of a substance. The relationship between chemical dose and effects on an exposed organism is of high significance in toxicology. Toxicity of these metals also depends upon their interaction within metals and hydrological parameters like temperature, pH, dissolved oxygen, conductivity as well as animal's status of acclimation to the toxicants. It also relates to biological half-life of a chemical in terms of its biodegradation (Jackson, 1976). Animal's physical and physiological condition, its milieu characteristics, age and sex also contribute to animal's tolerance and its response to any chemical. Chronic effects of metals in fishes or in other animals further depend upon their susceptibility, mode of administration and duration of their exposure.

Heavy metals affect many general body functions of the fish (Heath, 1987; Jain and Sharma, 2003). Slightly elevated metal levels in natural waters may cause overt latent or chronic effects, usually termed as the sub lethal effects; stated in general as the bio-chemical effects on cell chemical constituents, cell enzymes and the blood chemistry; leading to physiological changes. It is now well recognized that chronic exposure to sub lethal levels of toxicants including the heavy metals and the pesticides and the risk of their accumulation through continuous biomagnifications via food chain can yield severe effects on biotic components in an ecosystem. The down flow water in a river away from the point of contamination, even if it contain 0.001ppm of a chemical might increase in concentration in carnivorous fishes through planktons and herbivorous fishes, up to 100ppm, much beyond the tolerance limit of 0.01ppm for most of the metals. Heavy metal pollution assessment through comparison of different indices in sewage-fed fishery pond sediments at East Kolkata Wetland (Sarkar *et al.*, 2010) further revealed metals like lead, cadmium, and chromium are responsible for the contamination.

The distribution of toxic chemicals into various organs is influenced primarily by the route of its uptake. Metals find their way in an aquatic organism via three main pathways, *i.e.* as free metal ions that are absorbed through gills and are readily diffused into the blood stream; as free metal ions that are directly absorbed onto body surface and are passively diffused into the blood stream; and as metal ions that are absorbed onto food and particulates and may be digested along with the free ions taken in with the drinking water. Metal uptake rates also vary according to the organism and the type of metal. Smaller organisms like small fishes, phytoplanktons and zooplanktons often assimilate quickly the available metal, because of their high surface area to volume ratio. This envisages disturbances in digestion, osmo-regulations and respiration; reduced energetic resources and metabolism; tissue damage especially the skin, liver and the gills; suppression of growth and development; abnormality in behaviors and reproductive functions. Such xenobiotics that enter the circulation via gills can be rapidly distributed throughout the body to all organs, whereas, the xenobiotics that are absorbed via the intestine will on the other hand pass first to liver via the hepatic portal system, where even some part of the toxic chemical may even be metabolized or sequestered. The toxic effects produced in the digestive system may alter the activity of digestive enzyme and the metabolism. Many organisms are also able to regulate metal concentrations in their tissues by way of their detoxification, metabolism and excretion. Fishes and crustaceans can excrete metals, such as Cu, Zn, and Fe when present in excess and some species can excrete even non-essential metal like Hg and Ca, although this is usually less successful. These metals are known to exert toxic effects in an organism at tissue, cellular and molecular level, affecting adversely various physiological and biochemical functions in a fish. The physiological disorders related to heavy metal toxicity in fishes are reported to be associated with respiration, hematology, enzymology, immunology *etc.*, as well as the histological aberrations, leading to tissue damages and occurrence of a variety of diseases. Heavy metals also induce synthesis of new proteins in fishes (Boone and Vijayan, 2002; Ali *et al.*, 2003). The regulation of heavy metal induced gene expression was investigated in two fish cell lines: the Rainbow Trout Hepatoma (RTH) and Chinook Salmon Embryo (CHSE) cells (Price-Haughey and Gedamu, 1987). Throughout the world, millions of people are at risk of metal toxicity and suffer with cancers, heart disease and diabetes because of their chronic exposures. Metal interaction with DNA can lead to mutation or carcinogenesis (Cohen *et al.*, 1993).

There are, however, narrow and pre-determined boundaries of tolerance levels for various metals (Van Vuren *et al.*, 1999). Toxicity effects of cadmium and copper metals are well illustrated in fishes by many workers. Copper is reported in general as more toxic than cadmium for the larvae and adult fish species (Denton and Burdon-Jones, 1986; Cui-Keduo *et al.*, 1987 and Wu-Yulin *et al.*, 1990). A trace metal usually less than 0.01 per cent of the body mass of an organism is essential, and in the absence of that metal an organism either fails to grow or may have even retorted growth or an incomplete life cycle. When these trace metals occur in high concentrations inside a living cell, these alter cell physiological functions (Venugopal and Luckey, 1975; Heath, 1987). Zn though an essential element and it acts on the structural components having specific properties indispensable for life (Bengari and Patil, 1986). Zn deficient diet in fish is known to cause certain deficiency symptoms, and Zn levels of 10 and 20 mg/kg of diet had significant effects on fish growth in *C. mrigala* fingerlings (Gupta and Soni, 1998). Zn in higher levels may damage tissues by reacting with proteins that could affect respiratory and osmo-regulatory function of the gills (Lioyd, 1992). Likewise, feeds supplemented with cobalt chloride @ 1 mg/day though were found to enhance survival and growth of the fingerlings of Indian major carps, but at 7.5 ppm dose it was found to impair gonadosomatic index (GSI) and growth in goldfish (Ahilan and Jeyaseelan, 2001).

2. General Metal Exposures in Fishes

i. Mercury

The high toxicity of Hg compounds in biotic forms was known since a long time (Bidstrup 1964). There have been many such reports indicating Hg contamination in water due to discharge of effluents from the chlor-alkali industries (Turney, 1970;

Carlos, 1979; Panigrahi and Misra, 1979; Glass *et al.*, 1986; Shaw *et al.*, 1986; Sorensen *et al.*, 1990), and its impact on biota (Weiss *et al.*, 1971; Suckcharoen, 1980; Lodenius and Tulisalo, 1984; Shaw *et al.*, 1985, 1986). A study by The Indian toxicological Research Center, Luknow in 1986-87 showed maximum concentration of Hg in Rishikesh and Allahabad areas of the river. Cauvery and Damodar rivers are said to be more polluted with Hg due to chlor-alkali industries and coal mine sand coal based industries, respectively (Raj, 1986). Sahoo *et al.* (2002) reported the chlor-alkali industry effluents near Rushikulya river estuary in India to contain Hg up to 0.447 mg/l, and the estuary water contained Hg up to 1.55 mg/l in the month of maximum pollution; and they further recorded mean Hg residues as 1.03, 1.42 and 0.39 mg/l in their brain, liver and muscle tissues in fish, *Mugil oligolepis* collected from river estuary, as compared to Hg contamination level in the river water of 0.031 mg/L due to inflow of the chlor-alkali effluents. The distribution of mercury along the west coast of India (Kaladharan *et al.*, 1999) also showed a conspicuous pattern, showing low levels ranging up to 0.058 mg/l during pre-monsoon and monsoon seasons and an increase of 100 per cent during the post-monsoon season. The Central Board for the prevention and control of Water Pollution in India, has limited Hg pollution level maximally up to 0.01 mg/l. Mercury discharged in an aqueous system is generally lost to sediments by precipitation or by the death of the contaminated aquatic organisms (Clifton and Vivian, 1973).

Mercury pollution in water is caused by various forms of mercury *i.e.* methyl mercury, mercuric chloride, phenyl mercuric acetate *etc.* which are more dangerous because of their high toxicity, non-biodegradability and bio-magnification tendencies. The multifarious deleterious effects of $HgCl_2$ at different dose levels on various physiological parameters of different fish species during short or long term exposures has indicated that it is highly toxic even at low doses. These different forms of mercury including mercuric chloride ($HgCl_2$) adversely affect the reproductive processes at different levels in fishes (Kime, 1995; Jagadeesan and Pillai, 2007). Methyl mercury (MeHg) is one of the most widespread waterborne contaminants (U.S.EPA., 2004). Of the total Hg present in the aquatic organisms, 30 per cent (Ward *et al.*, 1979) to 98 per cent (Hattula *et al.*, 1978) has been found as methyl-Hg, which is many times more toxic and injurious to animal health including fishes (Wood *et al.*, 1968). Methylation is believed to involve a non-enzymatic reaction between Hg^{++} and methyl-cobalamine compound (an analogue of Vit. B_{12}) that is produced by bacteria (Wood and Wang, 1983). This reaction takes place primarily in the aquatic systems. The intestinal bacterial flora of various animal species including fish, are able to convert ionic mercury into methyl mercuric compounds, although to a much lower degree (Nielson, 1992). The generation, bioaccumulation, and biomagnification of MeHg within aquatic systems has been studied for decades, following the identification of severe neurological and teratogenic impacts to humans associated with the consumption of contaminated fish in Minamata Bay, Japan in the 1950s (Eisler 1987; Ninomiya *et al.*, 1995). MeHg has been linked to potential reproductive and immune system effects in humans and wildlife (Wiener *et al.*, 1996; U.S.EPA., 1997; Round *et al.*, 1998). Methyl-Hg is more toxic and an organism requires a considerably longer period of time to eliminate it. MeHg has been most commonly found in prey fish at concentrations toxic to piscivorous birds and mammals (Sams, 2007). In U.S.EPA.'s National Fish Tissue Survey, the Lowest Adverse Effect Concentration of 0.1 g/g wet weight was exceeded in 28 per cent of total samples and 86 per cent of predatory fish samples (U.S.EPA., 2002). Evidence suggests that more than a quarter of all mature fish contain MeHg concentrations above this level.

Hg is known to strongly bind to sulpher containing organic or inorganic particles (Walters and Wolrey, 1974) and to a lesser extent also to clay soils, fine sands and other minerals (Reimers and Krenkel, 1974). The immobilized Hg in sediments returns to water either by bottom stirring or majorly by the chemical process called methylation (Stopford and Goldwater, 1975). Unhealthful levels of MeHg in fish have led to the issuance of fish consumption advisories by at least 46 states (U.S.EPA., 2004). Mercury in *C. carpio* (Alam and Maughan, 1995) and $HgCl_2$ in *C. punctatus* (Patil and Dhande, 2000) and *N. notopterus* (Sindhe *et al.*, 2002) were reported to be toxic. Adverse effects of $HgCl_2$ on testicular recrudescence (Masud *et al.*, 2001), first ovarian maturity (Masud *et al.*, 2003), behavioral and hematological responses (Masud *et al.*, 2005) have also been reported in *C. carpio*. To protect humans from mercury poisoning, the U.S. Environmental Protection Agency (U.S.EPA.) has set a generalized, default fish tissue mercury residue criterion for freshwater and estuarine fish at 0.0175 mg per kg of fish per day (U.S.EPA., 2001).

ii. Arsenic

Arsenic (As) is another important heavy metal known as one of the most serious ground water pollutants in most of the south-east countries, including India and Bangladesh. The U.S. Food and Drug Administration (U.S.FDA, 1993), has indicated that fish and other seafood account for 90 per cent of total arsenic exposure. Trivalent form of arsenic may show an adverse effect on aquatic biota and is considered more toxic than its inorganic pentavalent form (Hall and Burton, 1982). Donohue and Abernathy (1999) reported that total arsenic in a marine shellfish and a freshwater fish tissue, ranged from 0.19 to 65 and from 0.2 to 125.9 (from 0.007 to 1.46 $\mu g\ g^{-1}$ dry wt.), respectively. Koch *et al.* (2001) demonstrated that total As in freshwater fish ranged from 0.28 to 3.1 $\mu g\ g^{-1}$ dry wt. for whitefish (*Coregonus clupeaformis*), from 0.98 to 1.24 $\mu g\ g^{-1}$dry wt. for sucker fish (*Catostomus commersoni*), from 0.46 to 0.85 $\mu g\ g^{-1}$ dry wt. for walleye (*Stizostedion vitreum*), and from 1.30 to 1.40 $\mu g\ g^{-1}$ dry wt. for pike (*Esox lucius*). Arsenic contents in several fish farming ponds were found even to exceed the water quality criteria for total arsenic amounts in a freshwater ecosystems of 150 $\mu g\ L^{-1}$ as per limits set under the criterion continuous concentration (U.S.EPA., 2002). Maximum arsenic concentration in pluvial water storage ponds (Razo *et al.*, 2004) was found to be up to 265 $\mu g\ L^{-1}$ which exceed by 5 times the Mexican drinking water quality guideline (50 $\mu g\ L^{-1}$).

Allen *et al.* (2004) studied the biochemical toxicity of arsenic trioxide in a freshwater edible fish, *Channa punctatus* on exposure ranging from 7 to 90 days. The arsenic concentration increased exponentially in liver, kidney, gills and muscles of fish up to 60 days of exposure of arsenic. However, arsenic concentration in these tissues declined at 90 days of exposure. Storelli *et al.*, 2004 recorded As residues ranging between 5.75 to 14.49 ppm in muscles of slotted shark along

Italian coast in the Adriatic sea, 92.3- 99.1 per cent being the organic arsenic. Storelli and Marcotrigiano (2004) showed that total amounts of As vary greatly among individuals and species, may be due to differences in diet. Fish feeding on crustaceans and small molluscs usually have more As since these invertebrates generally contained more As in their tissue. In tilapia, waterborne arsenic is reported to significantly accumulate in fish organs like gill, liver, and intestine; and was further reported to manipulate growth (Liao *et al.*, 2004; Tsai and Liao, 2006). Maher (1985) in the muscle tissues of marine animals observed diet independent low inorganic arsenic concentrations in all muscle tissues as compared to the total arsenic concentrations. Arsenic was found to be present in all tissues predominately as a methanol-water soluble form. The ratio of lipid soluble and un-extractable arsenic to total arsenic was higher in plankton feeders than in herbivores and carnivores. Hwang and Tsai (1993) suggested that different tolerances to arsenic between freshwater and seawater-adapted tilapia (*Oreochromis mossambicus*) may be ascribed to their different membrane permeability for arsenic. Suhendrayatna *et al.* (2002) indicated that around 14–42 per cent accumulated arsenic were transformed to methylated-As in tilapia under chronic exposure conditions; methylation, however, render arsenic to be lesser toxic (Newman and Unger 2003; Sharma and Sohn, 2009). Usually organo-metallic compounds are reported as much more toxic than ions of the corresponding inorganic compounds, except in case of arsenic which represents an exception because most organo-arsenicals are less toxic than inorganic arsenic species.

3. Cadmium

Cadmium (Cd), being an important constituent in industrial effluents and municipal waste discharge, is also a potentially hazardous pollutant in the environment and is an extremely toxic water pollutant to which fish are very sensitive. Cadmium derives its toxicological properties from its chemical similarity to zinc an essential micronutrient for plants, animals and humans. Cadmium toxicity in fish has been further reported to depend on water quality criteria (Calamari *et al.*, 1980). Acute and chronic effects of cadmium have been widely described for different aquatic organisms and exposure routes. Main target organs for cadmium toxicity are the kidneys, lungs, bones and the immune system (Alloway, 1990; ATSDR, 1992; Allen, 1994). Zn and Cu are known to enhance Cd toxicity. The absorption of Cd is said to continue irrespective of the body burden of this element (Cotzias *et al.*, 1961), but it behaves similar to zinc with respect to its excretory metabolism and it imply that Cd may function, in part, as an anti-metabolite of Zn; Cd, however, show significant physiological discrimination in comparison *i.e.* at the level of excretory processes its excretion is extremely slow (Lucis *et al.*, 1969). It thus, enhances its retention time in the body of a fish (Neathery and Miller, 1976). Toxic effects of cadmium with different concentrations and duration of exposure are well documented in teleosts (Singh and Sivalingam, 1982). Chemical symptoms of cadmium toxicity are further well documented (Browning, 1969). Absorption and accumulation of cadmium in the digestive tract have been described in the fingerlings of perch (*Perca fluviatilis*) (Edgren and Notter, 1980). Cd accumulation in fish occur in bones and muscles, more in gills, kidneys, liver and gut in fish. Lethal

Cd levels in fish are said to be up to 1000 ppb in soft water and up to 70000 ppb in hard water, chronic toxicity limits are respectively 4 and 60 ppb. The acute toxicity of Cd is due to its action as a calcium antagonist, and its pathological effects, thus, tend to be less severe at higher water calcium levels *i.e.* water hardness (Wood 2001). Scott *et al.*, 2003 found water borne 2µg Cd l^{-1} in 7 days eliminated normal anti predatory behaviors in juvenile rainbow trout, *O. mykiss* exhibited in response to alarm substances.

4. Chromium

Chromium is also a compound of biological interest, probably having an essential role in glucose and lipid metabolism (Langard and Norseth, 1979). Chromium often accumulates in aquatic life, adding to the danger of eating a fish contaminated with high levels of chromium. Chromium, one of the most common ubiquitous pollutants in the environment, and is classified as a carcinogen possessing both mutagenic and teratogenic properties. It is present as divalent [Cr(II)], trivalent [Cr(III)], and hexavalent [Cr(VI)] oxidation states, with Cr(VI) and Cr(III) being the most stable forms. The Cr hazards are dependent on its oxidation state, ranging from the low toxicity of the metal form to the high toxicity of the hexavalent form. Trivalent Cr plays an important role in glucose metabolism by serving as a cofactor for insulin action. Recently, naturally occurring Cr(VI) has been found in ground and surface waters at values exceeding the World Health Organization limit for drinking water of 50 µg of Cr(VI) per liter (Venkatramreddy *et al.*, 2009). The aquatic toxicology of Cr depends on both biotic and abiotic factors, particularly the water pH (Abbasi *et al.*, 1995). Various workers have reported toxic effect of hexavalent chromium on survival and physiology of fishes (Sastry and Sunitha 1984; Ambrose *et al*, 1994; Sornaraj *et al*, 1995; Abbasi *et al*, 1995; Singh, 1995; Vutukuru, 2003), gills of roach (Strike *et al.*, 1975), inhibition of various metabolic processes (Nath and Kumar, 1987), osmoregulatory ability and respiration in fish (Artillo and Melodio, 1988) and the physiological processes of mudskipper, *B. detanus* (Kundu *et al.*,1992). Short-term studies on the acute effects of Cr revealed that the metal exerts its toxicity at various functional levels across fish species. Further, individual variations exist across species in terms of susceptibility. For instance, recent studies designed to test the sensitivity of five fish species: rainbow trout, three-spinned stickleback, roach perch and dace, exposed to acute concentrations of Cr(VI) revealed that rainbow trout is 1.16 to 2.52 times more sensitive than the other test species to the metal (Svecevicius, 2006). The early life stage of Chinook salmon (*Oncorhynchus tshawytscha*) exposed to Cr(VI) (0 to 260 µ L^{-1}) from a contaminated groundwater source was evaluated for 98 days (Farag *et al.*, 2006). The study revealed no significant changes in the survival rate, development, and behavior. Chromium accumulation, however, was increased in tissues with respect to concentration and the time factor. Even the growth and survival of fish appears to depend on the dose and duration of exposure to Cr (Farag *et al.*, 2006). Chinook salmon exposed to Cr concentrations from 24-120 µg and 54-266 µg L^{-1} for 105 to 134 days significantly affected both survival and growth rate.

Trivalent chromium is said to alter the osmoregulatory function of various fish. Loss of osmoregulatory and respiratory functions was reported in rainbow trout by

Van Der Putte *et al.*, 1982. Studies on *L. rohita* exposed to Cr(VI) (39.4 mg L^{-1}) revealed significant decreases in the per cent of hemoglobin (Hb) and the total erythrocyte count at the end of both 24 h and 96 h (Vutukuru, 2005). The study further reported that the results reflect the anemic state of the fish and could be due to iron deficiency and its consequent decreased utilization for Hb synthesis. A dose-dependent decrease in erythrocyte counts, Hb content, and packed cell volume indicated anemia. According to a study, Na_2CrO_4 known as highly cytotoxic to medaka (*Oryzias latipes*) fin cells in a dose-dependent manner, at very low concentrations (1 µM) was found to induce chromosomal aberrations, causing chromatic lesions and exchanges that increased with concentration (Goodale *et al.*, 2006). Prabakaran *et al.* (2007) studied the immune response and non-specific immunity in the tilapia (*Oreochromis mossambicus*) exposed to sublethal concentrations of tannery effluent (TE) containing Cr (88.2 mg L^{-1}) and significant amounts of calcium carbonate, and sodium sulphate. The chronic exposure of fish to 0.53 per cent of TE significantly suppressed the antibody response, nonspecific serum lysozyme activity, and ROS (reactive oxygen species) and RNI (reactive nitrogen intermediates) production. Tannery wastes were also reported to induce chromosomal aberrations in plants and changes in mitotic indices. Singh *et al.*, 2005 reported Cr up to 150 ppm in tannery waste at Warangal. Cr is also said to be carcinogenic.

5. Copper

Copper being a trace metal and essential for living organisms also becomes toxic to fish at high concentrations. According to the studies conducted by Richey and Roseboom (1978), bluegill fish in soluble copper concentrations of 3.6 mg/l or greater becomes very aggressive, as dominant individuals chase and nip their tank-mates. Such aggressiveness occurs after every about 8 min and lasts for about 5 to 45 sec. Duarte *et al.* (2009) analyzed copper sensitivity ($CuCl_2{\cdot}2H_2O$) of 10 wild ornamental fish from Rio Negro, an ion-poor river in the Amazon, in the absence of dissolved organic carbon (DOC) and found no mortality at concentrations up to 700 µg of Cu l^{-1} for two species exposed to $CuCl_2{\cdot}2H_2O$ dissolved in standard EPA-USA water.

Copper concentration in water is further affected by the prevailing water conditions, as high alkalinity, pH, and organic substances will have low cupric ion concentrations. Many other workers have observed variability in fish response both as per its stage of development as well as the species. Sauter *et al.*, 1976 found that the eggs of brown trout, channel catfish, and walleye were more resistant to copper than other life stages, king salmon eggs also hatched in copper concentrations where the fry did not survive (Hazel and Stephen, 1970). Ezeonyejikaku *et al.* (2011) also observed different mortality responses in fish species to the varying concentrations of copper and copper was recorded as more toxic to *O. niloticus* than the catfish. 96 hrs LC_{50} values for *O. niloticus and C. gariepinus* were revealed as 58.837 and 70.135 mg/L, respectively. Holland *et al.* (1960) reported that copper salts combine with proteins present in the mucus of the fish's mouth, gills, and skin prevent aeration of the blood; death sometimes result due to suffocation. Benoit (1975) observed periodic muscle spasm for several weeks in bluegill fish following exposure to copper concentrations. Baker (1969) working with flounders, observed neuromuscular disorders just prior to death. Brook trout that were exposed to copper had increased cough frequencies (Drummbod, 1973). Loss of appetite was also noticed in brook trout. Perhaps it was the same response in fathead minnows that prevented its fry in copper solutions from growing as rapidly as the control fry (Mount, 1968).

Early in 1959, Vogel showed that gold fish treated with 100mg/l of Cu subjected to severe nephrotoxic and neurotoxic effects. To name a few effects, such as on chemosensory functions (Carreau and Pyle, 2005), immunological parameters (Dethloff *et al.*, 2001), antioxidant stress (Dautremepuits *et al.*, 2004; Sanchez *et al.*, 2005) and gill and intestine homeostasis (Monteiro *et al., 2005*; Carvalho and Fernandes, 2006) in a fish have been reported. Besides, toxicity can vary with the age of fish (Sellin *et al.*, 2005). In seabream, Antognelli *et al.* (2003) reported (organ-specific) alterations of the metabolic pathway of glyoxalase of adult fish exposed to sublethal concentrations of copper sulphate. Copper sulphate, however, is commonly used under the form of baths as a chemotherapeutic in aquaculture (Schlotfeldt, 1992; Carvalho and Fernandes, 2006) to treat various external diseases as a disinfectant treatment against protozoan ectoparasites of the genus *Cryptocarion* and *Amyloodinium*, as well as against bacteria (Rigos *et al.*, 2001; Chen *et al.*, 2006). Chen *et al.* (2006) found that copper sulphate baths (1–2 h, 0.1–1 mg L^{-1}) reduced bacterial pathologies on white seabass, *Atractoscion nobilis* and increased larval and juvenile survival, without microscopic evidences of toxicity due to the treatments.

6. Lead

James *et al.*, 1993 reported increased fish mortality with increase in lead concentration in *O. mossambicus*. In major carps, lead interacts with carboxyl and phosphoryl groups and interferes with the heme synthesis. Lead poisoning is said to causes breakage of the villi of intestine, cell swellings, shrinkage of glomeruli and the degeneration of goblet cells. In the fish exposed to the highest concentrations of lead and also of zinc, produced degenerations in the cytoplasm of tubular epithelial cells, it was so drastic to the extent that the tubule as a whole appeared as if these were ghosts of tubules (Al-Zahaby *et al.* (1998). The homeostatic tissue also appeared crowed with necrotic cells in case of fishes exposed to the LC_{50} of lead.

7. Zinc

Zn though an essential element, is needed in small quantities for the normal development and metabolism (Gupta and Sharma, 1994; Srivastava and Sharma, 1996; Srivastava and Kaushik, 2001; Shukla *et al.*, 2002), exposures of fishes to Zn to higher dose levels is reported to cause enfeeblement, retardation of growth and may bring about metabolic and pathological changes in various organs in fishes (Sharma and Sharma, 1994; Singh and Gaur, 1997). Like copper, the water quality of a water body, like hardness, temperature, dissolved oxygen, pH has an important effect on the toxicity of zinc to fish (Reed *et al.*, 1980). Zinc is reported to cause reduction in growth and delay in sexual maturity in the common guppy, when subjected to a 90-day exposure of 1.15 mg/L of zinc in water containing 80 mg/l total hardness (Crandall and Goodnigh, 1962). Bengtsson (1974a,b) also observed reduced growth in minnow yearlings at 0.13 mg/l zinc in freshwater over 150 days.

Adult minnows exposed to different concentrations of zinc nitrate were also found to develop hemorrhage and lesions.

Toxic concentrations of zinc compounds are reported to cause adverse changes in the morphology and physiology of fish. Jones (1938) reported gill damage and heavy secretions of mucus in sticklebacks, resulting in death of a fish. Lloyd (1960), however, contradicted the role of mucous as a cause of fish death, rather it was probably the zinc induced damage to the gill epithelium. Crandall and Goodnigh (1962) had further recorded abnormalities among the internal organs, like degenerated and undersized liver and pancreas, distorted and hemorrhaged kidneys, and under developed and vacuolated skeletal muscles. Acutely toxic concentrations of zinc were also shown to adversely affect appetite, swimming and equilibrium, and caused severe cytological damage in gills along with coagulation of large amounts of mucous over body. Zinc in low dose levels *i.e.* chronic toxicity though caused enfceblement and wide spread histological in many organs, but not in gills. Mattthiessen and Brafield (1973) also observed detachment of epithelial cells upon exposure to zinc and Burton *et al.*, 1972a also noticed hypoxia as the major physiological change preceding death in acute toxicity studies with zinc. Sparks *et al.*, 1972 used breathing rate to detect zinc levels in water. The zinc concentration as low as 2,55 mg/l could be detected in bluegills by an breathing rate or a change in breathing rate variance.

Zn also acts on the structural components and having specific properties indispensable for life (Bengari and Patil, 1986) such as the synthesis of nucleic acid RNA and DNA polymerases and metallothionein. Small quantities of zinc are required for the normal development and metabolism (Srivastava and Sharma, 1996). Zn toxicity to most organisms above certain concentrations results in general enfeeblement, growth retardation and may bring about metabolic and pathological changes in various fish organs (Ambrose *et al.*, 1994; Sharma and Sharma, 1994; Singh and Gaur, 1997) due to its accumulation. The dangers of even such essential metals however, appear aggravated due to their persistence almost for an indefinite period in the environment. Zn deficient diet in fish is known to cause certain deficiency symptoms, and Zn levels of 10 and 20 mg/kg of diet had significant effects on fish growth in *C. mrigala* fingerlings (Gupta and Soni, 1998). Low dietary Zn was also found to increase whole body tissue protein and carbohydrate levels, but moderately the fat values. Everall *et al.*, 1989 reported Zn as a potential toxicant to fish; known to cause disturbances in fish acid-base and ion regulations and disruptions in the gill tissues; result in symptoms of hypoxia (Hogstrand *et al.*, 1994). In case of Zn poisoning fish die from suffocation due to deposition of precipitated Zn salts onto the gills. Zn may damage tissues by reacting with proteins that could affect respiratory and osmoregulatory functions of the gills.

8. Cumulative Metal Effects

Metals produce cumulative toxicity in small doses over long periods of time and acute toxicity in higher doses. Metals that are ingested may pose great risks to animal health. Even the trace metals like lead and Cd will interfere with essential nutrients of similar appearance, such as calcium and zinc. Copper acts as a cofactor for a number of key proteins (*i.e.* superoxide dismutase, ceruloplasmin and its flexible redox state means it plays a vital role in cellular respiration, with cytochrome-**c** oxidase being an important copper protein (Watanabe *et al.*, 1997). However, in excess, copper is toxic and the primary toxic action is predominantly the production of free radicals in tissues where copper accumulates. In addition, dietary copper toxicity can occur at several other loci in the gut and includes inhibition of digestive enzymes and reduced gut motility (Woodward *et al.*, 1995). Conversely, high concentrations of waterborne copper affect branchial function, the main toxic action being a perturbation of sodium homeostasis (Wood, 2001). Likewise, zinc is essential due to its vital structural and/or catalytic importance in more than 300 proteins that play important roles in piscine growth, reproduction, development, vision and immune function (Watanabe *et al.*, 1997). Consequently for fish, of the essential metals, zinc is second in quantitative importance only to iron. But like Cu and Fe, it is redox inert, enabling the formation of relatively stable associations within the cellular environment (Vallee and Falchuk, 1993). Consequently, in contrast to copper and iron, zinc does not form free radical ions, and in fact has antioxidant properties (Powell, 2000). Zinc may, however, generate toxicity to fish by interfering with calcium homeostasis (Spry and Wood, 1985; Hogstrand and Wood, 1996). Surplus zinc is either excreted **via** the bile, intestinal sloughing (Handy, 1996) or the gills (Hardy *et al.*, 1987), whilst urinary loss of zinc in fish is minimal (Spry and Wood, 1985). The major routes of zinc assimilation in fish are the gills and the gut. Shears and Fletcher (1983) determined that the anterior intestine was the most important region for zinc absorption in winter flounder *Pseudopleuronectes americanus* and plaice *Pleuronectes platessa*, respectively.

9. Metallothionein

Metallothionein (MT) is a protein that binds with the excess of metal ions to render them unavailable and Ca induces synthesis of MT which makes the essential metals unavailable by binding with them. A role for metallothionein (Shears and Fletcher, 1984) and glutathione (Shears and Fletcher, 1984; Jiang *et al.*, 1998), in cellular zinc uptake and metabolism has been proposed. Ca also interferes with the ability of metallothionein to regulate Zn and Cu concentration in the tissues. The interaction among various chemical pollutants in the aquatic ecosystem may further be synergistic, antagonistic or additive. In synergism, toxicological manifestation of the individual chemical, when in combination often gets increased. Cd and Cu are generally antagonistic in action to each other. The antagonistic behavior of Cd to Cu absorption and iron was demonstrated by Weigel *et al.* (1984). Roch and McCarter (1986) reported moderate antagonistic effect for the mixture of Zn, Cu and Cd in *Salmo gairdneri*. Sprague and Ramsay (1965) however, stated that toxicity of a mixture of Zn and Cu to the rainbow trout *Salmo gairdneri* and atlantic salmon *Salmo salar* was additive. Cu is said to be antagonistic to methyl Hg (Roales and Perlmutter, 1977) in *Trichagaster trichopterus*. Biney *et al.* (1994), stressed upon the need to identify the sources of the discharge of heavy metals into aquatic systems nationwide and appropriate measures to control them. Khunyakari *et al.* (2001) studied toxic effects of three heavy metals *viz.*, Nickel, copper and zinc in *Poecilia reticulata*. Among these, copper was found to be most active followed by zinc and nickel.

3. LC_{50} Values of Metals for Fishes

In an aquatic system, accurate prediction of chemical composition of the medium and their accumulation by a fish are needed for human health and the ecological risk assessment. Most important aspect of chemical toxicity studies is thus, assessment of their relative toxicity values, doses or their effects in humans or the other test animals. Assessment of chemical toxicity in the past used to rely heavily on mortality-based acute tests, but a chemical that does not cause death within a specific duration, may still have long-term deleterious impact on a test organism (Wo *et al.*, 1999), such effects are termed commonly as sub lethal or chronic toxicity effects. The sub lethal toxicity has received insufficient attention, probably due to the inherent difficulty; first in predicting what compounds are to be examined and secondly a long period of exposures required to assess their effects sufficiently. Most appropriate methods are the determination of LD_{50} values of the toxic chemicals. The term LD_{50} refers to the dose of a toxic substance that kills 50 per cent of a test population. LD_{50} estimations in animals are pertinent only in the context of acute toxicity assessments involving 1-3 day treatment tests, but in reference of chronic toxicity (usually >30 days), it is the sub lethal dose ($<LD_{50}$), may be LD_{10} or even lower than that is important in such clinical chronic tests.

Khanee *et al.* (1991) in *O. mossambicus* determined the median lethal concentration of mercury for 96 hr as 0.06 ppm. The 96h LC_{50} values for silver (Ag) in freshwater teleost fish are situated in the 5–70 $\mu g\ l^{-1}$ range, whereas these values for seawater teleost fish vary between 183 and 1200 $\mu g\ l^{-1}$ (Shaw *et al.*, 1997, 1998; Ferguson and Hogstrand, 1998; Hogstrand and Wood, 1998). Deshmukh and Marathe (1979) reported 96 h LC_{50} values of Cu and Hg in the fry and the fingerlings of *L. rohita* and *C. carpio*. It was observed that the susceptibility varied with the size of the fish and the fry being very small is more susceptible. Similarly, the 96 h LC_{50} values of 3 size groups of *L. raticulatus* were found to be directly proportionate to their body weight with a significant correlation coefficient value. Wong *et al.* (1999) in their work with silver sea bream (*Sparus sarba*) reported a 24 h LC_{50} value of 2.01 mg L^{-1} for 9.4 ± 2.1 g fingerlings. Kaviraj and Konar (1983) determined the LC_{50} and LC_{95} values of Hg-Cr-Cd mixture for the fish, *Tilapia mossambicus*, which was in the range of 90.0 to 190.5 ppm. Sehgal and Saxena (1986) examined acute toxicity of Cd and Zn to *Puntius ticto*. They reported the LC_{50} for 24, 48, 72 and 96 h as 35, 30, 28 and 26 ppm for Cd and 75, 70, 66 and 60 ppm for Zn, respectively. The 96 h LC_{50} value of copper was 1.83 ppm for fish, *Etroplus maculaus* as reported by Gaikwad (1989). 96 h LC_{50} of Cu was said to range from 0.2 to 3 mg/l for various marine fish and crustaceans (Bryan, 1976). Taylor (1981) reported LC_{50} values of Cd about 0.3 to 50 mg/l. The variation in the LC_{50} values for the same metal may be due to; species type, chemical structure of metal compound, and the conditions of the experiment (water temperature, salinity, oxygen content and pH). Mukhopadhyay *et al.* (1994) reported the LC_{50} values for Zn, Cu, Cr and Cd as: 9.09, 34.75, 21.05 and 242.31 mg/L, respectively, and for Zn, Cu and Cr in combination as 10.80. For rainbow trout (*Oncorhynchus mykiss*), the 96-h LC_{50} for Co was reported as 1,406 mg/L, and the incipient lethal value (ILL) as 346 mg/L (Marr *et al.*, 1998). Ezeonyejikaku *et al.* (2011) also observed copper as more toxic to *O. niloticus* than the catfish; and 96 hrs LC_{50} values for *O. niloticus and C. gariepinus* were revealed as 58.837 and 70.135 mg/L, respectively.

The 96-h LC_{50} values of cadmium on *Salmo gairdneri* was reported to be 80 mg/l by Woodal *et al.* (1988); while Muley *et al.* (2000) reported the 96-h LC_{50} value of cadmium and lead in *C. carpio* as 121.8 and 594 ppm, respectively. The LC_{50} values of cadmium on rainbow trout (*O. mykiss*) for 24, 48, 72 and 96 h were found to be 7.76, 1.95, 0.5, and 0.45 mg/l, respectively by Oryan and Nejatkhah (1997). Chambers (1995), investigated the effect of acute cadmium toxicity on marron, *Cherax tenuimanus*. He found 96-h LC_{50} value with 95 per cent confidence limits as 17.9 (13.4–23.9) mg/l. Asato and Reish (1988) found the LC_{50} value of cadmium on *Holmesimysis costata* (Crustacea: Mysidacea) as 0.008 mg/l. Sehgal and Pandey (1984) observed that LC_{100} and LC_{50} values for guppy fish (*Lebistes reticulatus*) were 300 and 250 ppm, respectively of $CdCl_2$. Yýlmaz *et al.* (2004) estimated a 96-h LC_{50} value for the guppy fish, *L. reticulatus* to Cd as 30.4 mg/l in a static bioassay test system and 30.6 mg/l with Behrens–Karber's method. Murugappan *et al.* (1999) determined the LC_{50} value of cadmium in *C. carpio* for 96 hr duration as 3 ppm. El-Moselhy (2001) stated that toxicity of Cd to *Mugil seheli* decreased with the increase in the exposure time and the LC_{50} values were recorded as 12.34, 8.92, 6.01 and 3.45 mg/l for 24, 48, 72 and 96 hours, respectively. Hamed (1992) found that the 72 h LC_{50} of Cd for *Mugil seheli* as 4.87 mg/l. The study of acute toxicity of cadmium in *Channa punctatus* (Subathra and Karuppasamy, 2003) showed the LC_{50} values as 158.49, 125.89, 110.92, 102.00 and 89.13 mg/l for 24, 48, 72, 96 and 120 hrs, respectively. The observed results indicate that the mortality of the test fish due to Cd was dose-time dependent.

The U.S. EPA established water quality criterion for protection of aquatic biota for arsenic as 100 $\mu g\ L^{-1}$ (Hellawell, 1988). Clements (1992) reported 96-h LC_{50} for As 28.68 mg L^{-1}, close to the range of the 96-h LC_{50} of As for the seawater tilapia (26.5 mg L^{-1}), rainbow trout, *Oncorhynchus mykiss* (26.6 mg L^{-1}, Spehar *et al.*,1980), the bluegill, *Lepomis macrochirus* (29–35 mg L^{-1}), and the stonefly, *Pteronarcys californica* (38 mg L^{-1}, Johnson and Finley, 1980). 96-h LC_{50} of As for the freshwater tilapia was reported as 71.7 mg L^{-1} by Hwang and Tsai (1993). The median lethal concentration (LC_{50}) of Cu was reported to vary with the time of exposure as 0.397 after 24 hr, 0.341 after 48 hrs and 0.322 after an exposure of 96 hrs. The 96 h LC_{50} bioassays carried out in the blue tilapia, *Oreochromis aureus* (Abhas *et al.*, 2003) indicated LC_{50} values as 8.17 and 25.86 mg/l, respectively for Cu and Pb. Emad *et al.*, 2005 showed high toxicity of copper to cadmium in marine fish, *Mugil seheli*, They estimated the 96-h LC_{50} values of Cd and Cu as: 5.36 and 1.64 ppm, respectively. Shah and Altindag (2005) while studying the effects of already accumulated heavy metals (Hg, Cd, Pb) in the body of a tench on the 96-h LC_{50} values of the respective heavy metal, observed that the fish with the lowest body concentration of the heavy metal showed the lowest 96-h LC_{50} value of the respective heavy metal and vice-versa. Anandhan and Hemalatha (2009) recorded toxic effects of aluminium on a freshwater fish, *Brachydanioreri*. They found 61.66, 59.57, 57.94 and 56.92ppm as 24h, 48h, 72h and 96 hr LC_{50} values for Al, whereas LC_0 and LC_{100} values were 52 ppm and 65ppm, respectively.

REFERENCES

Abbasi S.S., Kunahmed T. and others, 1995. Influence of the acidity on chromium toxicity –A study with the teleost *Nuria dannicus* as model. Poll. Res., 14(3): 317-323.

Abhas H.H., Zaghloul K.H. and Mousa M.A., 2003. Effect of some heavy metal pollutants on some biochemical and histopathological changes in blue Tilapia, *Oreochromes aureus*. Egyptian J. Agric. Res., 80 (3): 1395-1410.

Ahilan B. and Jeyaseelan M.J., 2001. Effect of cobalt chloride and vitamin B_{12} on the growth and gonadal maturation of goldfish *Carassius auratus*. Ind. J. Fish, 48 (4): 369-374.

Alam M.K. and Maughan O.E., 1995. Acute toxicity of heavy metals to common carp (*Cyprinus carpio*). J. Environ. Sci. Hlth., 30, 1807-1816.

Ali A., Al-Ogaily S.M., Al-Asgah N.A. and Gropp J., 2003. Effect of sublethal concentrations of copper on the growth performance of *Oreochromis niloticus*. J. Appl. Ichthyol., 19(4): 183–188.

Allen D., 1994. Mercury accumulation profiles and their modification by interaction with cadmium and lead in the soft tissues of the *Cichlid areochromis* during chronic exposure. Bull. Environ. Contam. Toxicol., 53: 684–692.

Allen T., Singhal R., Rana S.V.S., 2004. Resistance to oxidative stress in a freshwater fish, *Channa punctatus* after exposure to inorganic arsenic. Biol. Trace Elem. Res., 98: 63-72.

Alloway B.J., 1990. Cadmium. In: Alloway, B.J. (Ed.), Heavy Metals in Soils, Blackie, Glasgow and London. John Wiley and Sons, New York, pp. 100–124.

Al-Zahaby A.S., Hemmaid K.Z., Gamal A.M. and Ghonemy O.I., 198. The pollutant effects of copper, zinc, and lead on the histological patterns of fish kidney. Egypt. J. quat. Biol. and Fish, 2(3): 15-41.

Ambrose T., Vincent S. and Cyril L., 1994. Susceptibility of the freshwater fish, *Gambusia affinis* (Baird and Giard), *Sarotherodon mossambicus* (Peters) and *Cirrhinus mrigala* (Ham.) to zinc toxicity. J. Environ. Toxicol., 4: 29-31.

Anandhan R. and Hemalatha S., 2009. Acute toxicity of aluminium to zebra fish, *Brachydanio rerio* (Ham.). The Internet J. Veterin. Med., 7: 1-5.

Ambrose T., Cyril Arun, Kumar L. and others.1994. Biochemical responses of *Cyprinus carpio communis* to toxicity of tannery effluent. J. Ecobiol., 6(3): 213-216.

Antognelli C., Romani R., Baldracchini F., De Santis A., Andreani G., Talesa V., 2003. Different activity of glyoxalase system enzymes in specimens of *Sparus auratus* exposed to sublethal copper concentrations. Chem-Biol. Interact., 142: 297-305.

Artillo A. and Melodio F., 1986. Effect of hexavalent chromium on the trout mitochondria. Toxicol. lett., 44: 71-76.

Asato S.L., Reish D.J., 1988. Effects of heavy metals on the survival and feeding of *Holmesimysis costata* (Crustacea: Mysidacea). In: Proceedings of the 7th Symposium of Marine Biology (Memorias Del 7 Simposium de Biologia Marina), pp 113–120.

ATSDR., 1992. Agency for Toxic Compounds and Disease Registry. Toxicological profile for cadmium. US Public Health Service, US Department of Health and Human Services, Atlanta, GA.

Baker Jeremy Y., 1969. Histological and electron microscopical observations on copper poisoning in the winter flounder (*Pseudopleuronectes americanus*) J. Fisheries Res. Brd. Canada v. 26; 2785-2793.

Bengeri K.V. and Patil H.S., 1986. Respiration, liver glycogen and bioaccumulation in Labeo rohita exposed to zinc. Indian J. Comp. Anim. Physiol., 4: 79–84.

Bengtsson B.E., 1974a. Effect of zinc on growth of the minnow, *Phoxinus phoxinus*, Oikos., 25: 370-373.

Bengtsson B.E., 1974b. The effect of zinc on the mortality and reproduction of the minnow, *Phoxinus phoxinus* L. Arh. Environ. Contam. Toxicol., 2: 342-355.

Benoit D.A., 1975. Chronic effects of copper on survival growth and reproduction of the bluegill (*Lepomis macrochirus*). Trans. Am. Fisheries Soc. v., 104(2): 353-358.

Bidstrup P.L., 1964. Toxicity of mercury and its compounds. Elsevier Publ. NY.

Biney C., Amazy A.T., Calamari D., Kaba N., Mbome I.L., Naeve H., Ochumba P.B.O., Osibanju O., Radegonde V. and Saad M.A.H. 1994. Review of heavy metals in the African aquatic environment. Ecotoxicol. Environ. Safety, 31: 134-159.

Boone A.N. and Vijayan M.M., 2002. Constitutive heat shock protein 70 (HSC 70) expression in rainbow trout hepatocytes: effect of heat shock and heavy metal exposure. Comp. Biochem. Physiol. C. Toxicol. Pharmacol., 132 (2 : 223-233.

Browning E., 1969. Toxicity of Industrial Metals. Butterworth and Co., London, pp119-131.

Bryan G.W. 1976. Some aspects of heavy metal tolerance in aquatic organisms. In: Effects of pollutants on Aquatic organisms, A.P.M. Lockwood (Ed). Cambridge Univ. Press, NY. p7

Burton D.T., Jones A.H. and Cairns J., 1972a. Acute zinc toxicity to rainbow trout (*Salmo girdneri*): Confirmation of the hypothesis that death is related to tissue hypoxia. J. Fisher. Res. Brd. Canada, 29: 1463-1466.

Calamari D., Marchetti R. and Vailati G., 1980. Influence of water hardness on cadmium toxicity on *Salmo giardneri*. Richardson. Water Res., 14: 1421.

Carlos S., 1979 Aust. J. Mar. Fish Wat. Res., 30: 623.

Carreau N.D., Pyle G.G., 2005. Effect of copper exposure during embryonic development on chemosensory function of juvenile fathead minnows (*Pimephales promelas*). Ecotoxicol Environ. Safe., 61: 1-6.

Carvalho C.S., Fernandes M.N., 2006. Effect of temperature on copper toxicity and hematological responses in the neotropical fish, *Prochilodus scrofa* at low and high pH. Aquaculture, 251: 109-117.

Chambers M.G., 1995. The effect of acute cadmium toxicity on marron, *Cherax tenuimanus* (Smith, 1912; Parastacidae), Freshwater Crayfish 10. In: 10th Intern. Sym. Astacol., 10–15, pp. 209–220.

Chen M.F., Apperson J.A., Marty G.D., Cheng Y.W., 2006. Copper sulfate treatment decreases hatchery mortality of larval white sea bass, *Atractoscion nobilis*. Aquaculture, 254: 102-114.

Clements William H. 1992. Bioaccumulation and food chain transfer of polycyclic aromatic hydrocarbons and heavy metals: a laboratory and field investigation. Ph D Thesis, Department of Fishery and Wildlife Biology, Colorado State University, Fort Collins, CO 80523.

Clifton A.D. and Vivian C.M.G. 1973. Retention of mercury from an industrial source in Swansea Bay sediments. Nature, 253: 621-622.

Cohen A., Nugegoda D. and others, 2001. Metabolic responses of fish following exposure to two different oil spill remediation techniques. Ecotoxicol. Environ. Saf., 48(3): 306-10.

Cotzias G.C., Borg D.C. and Selleck B. 1961. Virtual absence of turnover in cadmium metabolism: Cd^{109}studies in the mouse. Am. J. Physiol., 201: 927-930.

Crandall C.A. and Goodnight C.J., 1962. Effects of sublethal concentrations of several toxicants on growth of the common guppy, *Lebistes reticulates*. Limnol. and Oecano., 7: 233-239.

Cui-Keduo; Liu-Yumei and Hou-Lanying, 1987. Effects of six heavy metals on hatching eggs and survival of larval of marine fish. Oceanological. Limnology, 18 (2): 138 - 144.

Dautremepuits C., Paris-Palacios S., Betoulle S., Vernet G., 2004. Modulation in hepatic and head kidney parameters of carp (*Cyprinus carpio* L.) induced by copper and chitosan. Comp. Biochem. Physiol., C 137: 325-333.

Denton G. R. W. and Burdon-Jones C., 1986. Environmental effects on toxicity of heavy metals to two species of tropical marine fish from Northern Australia. Chem. Ecol., 2 (3): 233 - 249.

Deshmukh S.S. and Marathe V.B., 1979. Size related toxicity of copper and mercury to *Lepisteus reticulates, Labeo rohita* and *Cyprinus carpio*. Ind. J. Exp. Biol., 18: 421-423.

Dethloff G.M., Bailey H.C., Maier K.J., 2001. Effects of dissolved copper on select hematological, biochemical, and immunological parameters of wild rainbow trout (*Oncorhynchus mykiss*). Arch. Environ. Contam. Toxicol., 40: 371-380.

Donohue J.M. and Abernathy C.O., 1999. Exposure to inorganic arsenic from fish and shell fish.In: Chappell W.R., Abernathy C.O. and Calderon R.L. (Eds). Arsenic Exposure and Health effects, Elsevier, Amsterdam: pp 89-98.

Drummond R.A., 1973. Some short term indicators of sublethal effects of copper on brook trout, *Salvelinus fontinalis*. J. Fisheries Res. Brd. v. 30(5): 698-701.

Duarte R.M., Menezes A.C.L., Rodrigues l.da S., Maria V., de Almeida-Val F., Val A.L.2009. Copper sensitivity of wild ornamental fish of the Amazon. Ecotoxicol. and Environ. Safety, 72 (3):693–698.

Edgren M. and Notter M., 1980. Cadmium uptake by fingerlings of perch (*Perca fluviatilis*) studied by Cd-115m at two different temperature. Bull. Environ. Contamin. Toxicol., 24: 647.

Eisler R. 1987. Mercury hazards to fish, wildlife, and invertebrates: a synoptic review. U.S. Fish and Wildlife Service Biological Report 85.

El-Moselhy Kh. M., 2001. Toxicity of cadmium to the marine fish Mugil seheli and its accumulation in different tissues. J. Egypt Acad. Soc. Environ. Develop., 2 (1) :17–28.

Emad H., Abou El-Naga, Khalid M., El-Moselhy and Mohamed A. Hamed., 2005. Toxicity of cadmium and copper and their effect on some biochemical parameters of marine fish, *Mugil sehel*. Egyptian J. Aqua. Res., 31: 60-71.

Everall N. C., Macfarlane N.A.A. and Sedgwick R.W., 1989a. The interaction of water hardness and pH with the acute toxicity of Zinc to the brown trout *Salmo trutta* L. J. Fish Biol., 35; 27-36.

Everall N. C., Macfarlane N.A.A. and Sedgwick R.W. 1989b. The effects of water hardness upon the uptake, accumulation and excretion of zinc in the brown trout, *Salmo trutta* L. J. Fish Biol. 35: 881.

Ezeonyejikaku C.D., Obiakor m.O. and Ezenwelu C.O., 2011.Toxicity of copper sulphate and behavioral locomotor response of Tilapia (*Oreochromis niloticus*) and catfish (*Clarius gariepinus*) species. Online J. Anim. and Feed Res., 1(4): 130-134.

Ferguson, E.A., Hogstrand, C., 1998. Acute silver toxicity to seawater acclimated rainbow trout: influence of salinity on toxicity and silver speciation. Environ. Toxicol. Chem. 17, 589–593.

Farag A.M., May T., Marty G.D., Easton M., Harper D.D., Little E.E. and Cleveland L. 2006. The effect of chronic chromium exposure on the health of Chinook salmon (*Oncorhynchus tshawytscha*) Aquat. Toxicol.,76(3-4):246–257.

Gaikwad S. A., 1989. Acute toxicity of mercury, copper and selenium to the fish, *Etroplus maculatus*. Environmental. Ecology, 7 (3): 694 - 696.

Glass G.E., Leonmard Edward N., Chan W.H. and Orr D.B., 1986. Great lakes research, Int. Assoc. Great Lakes Res., 12: 37.

Goodale B.C., Walter R., Pelsue S.R., Thompson W.D., Wise S.S., Winn R.N., and others., 2008. The cytotoxicity and genotoxicity of hexavalent chromium in medaka (*Oryzias latipes*) cells. Aquat. Toxicol., 87(1):60–7.

Gupta A.K. and Sharma S.K., 1994. Bioaccumulation of zinc in *Cirrihinus mrigala* (Hamilton) fingerlings during short-term static bioassay. J. Environ. Biol., 15(3): 231-237.

Gupta A.K., and Soni, M. K., 1998. Dietary requirements of Zinc for the fingerlings of *Cirrhinus mrigala*. Geobios., 25: 261-265.

Hall L.W. Jr. and Burton D.T. 1982. Effect of power plant coal pile and coal waste runoff and leachate on aquatic biota: an overview with research recommendations. Boca Raton, FL: CRC Press: 287–297.

Hamed, M. A-F., 1992. Seawater quality at the northern part of the Gulf of Suez and the nearby area of the Suez Canal. Master Science Thesis, Faculty of Science, El-Mansoura University.

Handy, R.D., 1996. Dietary Exposure to Toxic Metals in Fish. In: Toxicology of Aquatic Pollution: Physiological, Cellular and Molecular Approaches, Taylor, E.W. (Ed.). Cambridge University Press, Cambridge, England, ISBN: 0521455243, pp: 29-60.

Hardy, R.W., Sullivan, C.V. and Koziol, A.M. 1987. Absorption, body distribution, and excretion of dietary zinc by rainbow trout, *Salmo gairdneri*. Fish Physiol. Biochem. 3: 133-143.

Hattula M.L., Sarkka J., Janatuinen J., Passivirta J. and Roos A. 1978. Environ. Pollut. 17: 19.

Hazel Charles and Stephen Meith, 1970. Bioassay of king Salmon eggs and sac fry in copper solutions. Calif. Fish and Game v. 56(2): 121-124.

Heath A.G., 1987. Water pollution and fish physiology. CRC Press, Inc., Florida, pp 245.

Hellawell J.M., 1988. Toxic substances in rivers and streams. Environ. Pollut., 50: 61-85.

Hogstrand C., Wilson R.W., Polgar D. and Wood C.M. 1994. Effects of zinc on the kinetics of branchial calcium uptake in freshwater rainbow trout during adaptation to waterborne zinc. J. Exp. Biol., 186:55–73.

Hogstrand C. and Wood C.M., 1996. The physiology and toxicology of zinc in fish. In Toxicology of Aqu. Pollut. (ed. E. W. Taylor), pp. 61-84. Cambridge: Cambridge University Press.

Hogstrand C., Wood C.M., 1998. Toward a better understanding of the bioavailability, physiology, and toxicity of silver in fish: implications for water quality criteria. Environ. Toxicol. Chem., 17, 547–561.

Holland G.A., Laster J.E., Neuman E.D. and Eldridge W.E., 1960. Toxic effects of organic pollutants on young salmon and trout. Washington State Deptt. Fisheries, Res. Bull. No. 5.

Holum J.R., 1983. Elements of general and biological chemistry, 6th Ed., John Wiley and Sons, NY., pp324, 326, 353 and 469.

Hwang P.P. and Tsai Y.N., 1993. Effects of arsenic on osmoregulation in the tilapia *Orechromis mossambicus* reared in seawater. Marine Biol 117:551–558.

Jackson, G.A., 1976. Biological half life of eldrin in channel cat fish tissues. Bull. Environ. Contam. Toxicol. 16 : 505-507.

Jagadeesan G. and S. Sankarsami Pillai. 2007. Hepatoprotective effect of taurine against mercury induced toxicity in rat J. Environ. Biol., 28, 753-756.

Jain K.L., Sharma M., 2003. Toxic effects of mercury and cobalt on biochemical composition of freshwater fish *C. mrigala*. Proc. 3rd Interact. Workshop, December 17-18, 2003. pp 203-207.

James R.V., Alagurathinam S. and Sampath K., 1993. Hematological changes and oxygen consumption in *O. mossambicus*, exposed to sublethal concentration of lead. Ind. J. Fisheri., 38: 49-54.

Jiang L.J., Maret W. and Vallee B.L. 1998. The glutathione redox couple modulates zinc transfer from metallothionein to zinc-depleted sorbitol dehydrogenase. Proc. Natl. Acad. Sci. USA, 95: 3483 -3488.

Johnson W.W., Finley M.T., 1980. Handbook of acute toxicity of chemicals to fish and aquatic invertebrates, US Department of the Interior Fish and Wildlife Services, Publication no: 137: 1-98.

Jones J.R.E., 1938. The relative toxicity of salts of lead, zinc, and copper to the stickleback (*Gasterosteus aculeatus* L.) and the effect of calcium on the toxicity of lead and zinc salts. J. Exp. Biol., 15: 394-407.

Kaladharan P, Pillai V.K., Nandakumar A, Krishnakumar P.K. 1999. Mercury in seawater along the west coast of India. Ind. J. Mar. Sci, 28(3): 338-340.

Kaviraj A. and Konar S.K. 1983. Lethal effects of mixtures of mercury, chromium and cadmium on fish. Environ. Ecol., 1: 153-158.

Khanee K.M.A., Nazeemul James R., Sivakumar K. and Muthuramalakshmi M. 1991. Sublethal effects of mercury on growth and biochemical composition and their recovery in *Oreochromis mossambicus*. Ind. J. Fish. 38(2): 111-118.

Khunyakari- Rupesh P., Tare-Vrushali and Sharma R.N., 2001. Effects of some trace heavy metals on *Poecilia reticulates* (Peters). J. Environ. Biol. 22(2): 141-144.

Kime D.E., 1995. The effects of pollution on reproduction in fish. Rev. Fish Biol. Fisheries, 5: 52-96.

Koch I., Reimer K.J., Beach A., Cullen W.R., Gosden A. and Lai V.W.M., 2001. Arsenic speciation in fresh-water fish and bivalves. In: Chappell W.R., Abernathy C.O., Calderon R.L. (Eds). Arsenic exposure and health effects IV. Oxford, UK: Elsevier: 115–123.

Kundu R., Lakshmi R. and others, 1992. The entry of mercury through the membrane: an enzymological study using a tolerant fish, *Boleophthalmus dentatus*. Proc. Acad. Environ. Biol.,1: 1-6.

Langard S. and Norseth T., 1979. Chromium. In: Hand book on the Toxicology of metals. Eds- Friberg L., Gunner F.N. and Velimir B.V., 1979. Elsevier- North Holland, Biochemical Press, Netherlands, pp 383-394.

Liao C.M., Tsai J.W., Ling M.P., Liang H.M., Chou Y.H. and Yang P.T., 2004. Organ-specific toxicokinetics and dose-response of arsenic in tilapia *Oreochromis mossambicus*. Arch. Environ. Contam. and Toxicol., 47, 502–510.

Lloyd Richard, 1960. The toxicity of zinc sulphate to rainbow trout. Ann. Appl. Biol., 48(1): 84-94.

Lloyd R., 1992. Pollution and fresh water fish. Fishing-News books, Div. Blackwell Sci. Publ. Ltd., UK: 176pp.

Lodenius M. and Tulisalo E., 1984. Environmental mercury contamination around a Finnish chlor-alkali plant.- Bull. Environ. Contam. Toxicol., 32:439-444.;

Lucis O.J., Lynk, M.E and Lucis R., 1969, Turnover of Cd^{109} in rat. Arch. Environ. Health. 18, 307-310.

Maher W.A., 1985. Distribution of arsenic in marine animals: Relationship to diet. Comp. Biochem.Physiol., 82C: 433-434.

Marr J.C.A., Hansen J.A., Meyer J.S., Cacela D, Podrabsky T, Lipton J. and Bergman H.L., 1998. Toxicity of cobalt and copper to rainbow trout: Application of a mechanistic model for predicting survival. Aquat Toxicol., 4:225–237.

Masud S., Singh I.J and Ram R.N., 2001. Testicular recrudescence and related changes in *Cyprinus carpio* after a long term exposure to mercurial compound. J. Ecophysiol. Occup. Hlth., 1, 109-120.

Masud S., Singh I.J and Ram R.N., 2003. First maturity and related changes in female *Cyprinus carpio* in response to long-term exposure to a mercurial compound. J. Ecophysiol. Occup. Hlth, 3, 1-14.

Masud S., Singh I.J and Ram R.N., 2005. Behavioural and hematological responses of *Cyprinus carpio* exposed to mercurial chloride. J. Environ. Biol., 26, 393-397.

Mathiessen P. and Brafield A.E., 1973. The effects of dissoved zinc on the gills of the stickleback, *Gasterosteus aculeatus* (L). J. Fish Biol., 5: 607-613.

Mizrahi L. and Achituv Y., 1989. Effect of heavy metals ions on enzyme activity in the Mediterranean mussel, *Donax trunculus*. Bull. Environ. Contam. Toxicol., 42(6):854-9.

Monteiro SM, Mancera J.M, Fontaínhas-Fernandes A. and Sousa M., 2005. Copper induced alterations of biochemical parameters in the gill and plasma of *Oreochromis niloticus*. Comp. Biochem. Physiol. C Toxicol. Pharmacol., 141(4):375-383.

Mount D.I., 1968. Chronic toxicity of copper to fathead minnows (*Pimephales promellas*). Water Res. v., 2: 215-223.

Mukhopadhyay M.K., Ghosh B.B. and Bagchi M.M., 1994. Toxicity of heavy metals to fish, prawn and fish food organisms of Hooghly estuarine system. Geobios., 21 : 13-17.

Muley D.V., Kamble G.B. and Bhilave M.P., 2000. Effect of heavy metals on nucleic acids in *Cyprinus carpio*. J. Environ. Biol., 21, 367–370.

Murugappan R.M., Muthusamy M. and Gunasundari M., 1999. Genotoxicity and bioaccumulation of BHC and cadmium on the freshwater fish, *Cyprinus carpio*. J. Environ.Biol., 20: 313-316.

Nath K. and Kumar N., 1987. Effects of hexavalent chromium on the carbohydrate metabolism of a freshwater tropical teleost, Colisa fasciatus. Bull. Inst. Zool. Acad. Sin.,(Taipei) 26: 245-248.

Neathery M. W. and Miller W. J., 1976. Metabolism and toxicity of Cd Hg and lead in animals. A review. J. Diary Sci., 58: 1767-1781.

Newman, M.C. and Unger M.A., 2003. Fundamentals of Ecotoxicology. New York: Levis.

Nielson J.B., 1992. Toxicokinetics of mercuric chloride and methylmercuric chloride in mice. J. Toxicol. And Environ. Helth., 37: 85-122.

Ninomiya T., Ohmori H., Hashimoto K., Tsuruta K. and Ekino S., 1995. Expansion of methylmercury poisoning outside of Minamata: An epidemiological study on chronic methylmercury poisoning outside Minamata. Environ. Res., 70: 47-50.

Oryan, S., Nejatkhah, P., 1997. Effects of cadmium on plasma cortisol and prolactin levels in the rainbow trout (*Oncorhynchus mykiss*). J. Fac. Vet. Med. Univ. Tehran., 51: 39–53.

Panigrahi A.K. and Misra B.N., 1979. Bull. Environ. Contam. Toxicol., 23: 784.

Patil G.P. and Dhande R.R., 2000. Effect of HgCl2 and CdCl2 on haematobiochemical parameters of the freshwater fish, *Channa punctatus* (Bloch). J. Ecotoxicol. Environ. Monit., 10: 177-181.

Powell, S. R., 2000. The antioxidant properties of zinc. J. Nutr., 130: 1447S -1454S.

Prabakaran M., Binuramesh C., Steinhagen D. and Dinakaran Michael R., 2007. Immune response in the tilapia, *Oreochromis mossambicus* on exposure to tannery effluent. Ecotoxicol. Environ. Saf., 68(3):372–8.

Price-Haughey J. and Gedamu L., 1987. Heavy metal induced protein synthesis in fish cell lines. Experientia Suppl., 52: 465-469.

Raj S.P., 1986. Pollut. Res., 5: 39-43.

Randall C.A. and Goodnight C.J., 1962. Effects of sublethal concentrations of several toxicants on growth of the common guppy, *Lebistes reticulates*. Limnol. and Oceano., 7(2): 233-239.

Razo Israel, Carrizales Leticia, Castro Javier, Díaz-Barriga Fernando and Monroy Marcos, 2004. Arsenic and Heavy Metal Pollution of Soil, Water and Sediments in a Semi-Arid Climate Mining Area in Mexico. Water, Air, and Soil Pollution, 152(1-4): 129-152.

Reed P., Richey D. and Roseboom D., 1980. Acute toxicity of zinc to ome fishes in high alkalinity water. Circular no. 142, Illinois state Water Survey, Urbana. State of Illinois.

Reimers R.S. and Krenkel P.A., 1974. CRC Grit Revs. Environ. Contam., 4: 265.

Richey D. and Roseboom D., 1978. Acute toxicity of copper to some fishes in high alkalinity water. Circular no. 131, Illinois state Water Survey, Urbana. State of Illinois, 27pp.

Rigos G., Pavlidis M., Divanach P., 2001, Host susceptibility to Cryptocaryon sp. infection of Mediterranean marine broodfish held under intensive culture conditions: a case report. Bull. Eur. Assoc. Fish Pathol., 21, 33-36.

Roales R.R., and Perlmutter A.,1977. The effect of sub lethal doses of methylmercury and copper applied singly and jointly on the immune response of the blue gourami to viral and bacterial antigen, Arch. Environ. Contam. Toxicol., 38:325-331.

Roch M. and McCarter J.A., 1986. Survival and hepatic metallothein in developing rainbow trout exposed to a mixture of Zn, Cu and Cd. Bull. Environ. Contam. Toxicol., 36: 168-175.

Round M., Marin A., Alter L., Tatsutani M., 1998. Mercury deposition in the Northeast. In: M. Tatsutani (ed.) Proc. Northeast states and eastern Canadian Provinces, mercury study. NESCAUM *et al.*: VI1 - VI42.

Sahoo A., Panigrahi Minati K., Sahu S.K. and Panigrahi Ashok K., 2002. Aquatic mercury pollution by a chlor-alkali industry and its impact on the ecology of human health in Ganjam area- a case study. In Kumar A. (Ed) Ecology of polluted waters Vol.1. APH Publ. Corp., New Delhi, pp 465-496.

Sams Charles E., 2007. Impacts on Aquatic Systems and Terrestrial Species, and Insights for Abatement. In: Advancing the Fundamental Sciences: Proceedings of the Forest Service National Earth Sciences Conference, San Diego, CA, 18-22 October 2004, PNWGTR- 689, Portland: 11pp.

Sanchez W., Palluel O., Meunier L., Coquery M., Porcher J.M., Ait- Aissa S., 2005, Copper-induced oxidative stress in three-spined stickleback: relationship with hepatic metal levels. Environ. Toxicol. Pharmacol., 19, 177-183.

Sarkar S., Ghosh P.B., Sil A.K. and Saha T., 2010. Heavy metal pollution assessment through comparison of different indices in sewage-fed fishery pond sediments at East Kolkata Wetland, India.Environ. Earth Sci., 63 (5): 915-924.

Sastry K.V. and Sunitha K., 1984. Chronic toxic effects of chromium in *Channa punctatus* Biochemical studies. J. Environ. Biol., 5 (1):53-56.

Sauter S., Buxton K.S., Macek K.J. and Petrocel S.R., 1976. Effects of exposure to heavy metals on selected freshwater fish: Toxicity of copper, cadmium, chromium and lead to eggs of seven species. Environ. Res. Lab., US Environmental Protect. Agency, Duluth, Minnesota.

Schlotfeldt H.J., 1992. Current practices of chemotherapy in fish culture. In: Michel, C., Alderman, D.J. (Eds.). Chemotherapy in Aquaculture: from theory to reality. Office International des Epizooties, Paris, pp 25-38. 270 I.

Scott G.R., Sloman K.A., Rouleau C. and Wood C.M. 2003. Cadmium disrupts behavioural and physiological responses to alarm substances in juvenile rainbow trout (*Oncorhynchus mykiss*). J. Exp. Biol., 11: 1779-90.

Sehgal R., Pandey A.K., 1984. Effect of cadmium chloride on testicular activities in guppy, *Lebistes reticulatus*. Comp. Physiol. Ecol., 9, 225–230.

Sehgal R. and Sexena A.B.N., 1986. Studies on the toxicity of cadmium and zinc to freshwater teleost *Puntius ticto*. J. Hydrobiol., 2: 33-35.

Sellin M.K., Tate-Boldt E. and Kolok A.S., 2005. Acclimation to Cu in fathead minnows: Does age influence the response? Aquat. Toxicol., 74: 97-109.

Shah S.L. and Altindag A., 2005. Effects of heavy metals accumulation on the 96-h LC_{50} values in Tench, *Tinca tinca* L. Turk. J. Vet. Anim. Sci., 29: 139-144.

Sharma A. and Sharma M.S., 1994. Toxic effect of zinc smelter effluent to some developmental stages of freshwater fish, *Cyprinus carpio* (L.). J. Environ. Biol., 15(3): 221-229.

Sharma V.K. and Sohn Mary, 2009. Aquatic arsenic: Toxicity, speciation, transformations, and remediation. Environ. Internat., 35 (4): 743–759

Shaw B.P., Sahu A. and Panigrahi A.K., 1985. Residual mercury concentration in brain, liver and muscle and in contaminated fish collected from an estuary near a caustic-chlor industry. Curr. Sci., 54: 810-812.

Shaw B.P, Sahu A. and Panigrahi A.K., 1986. Mercury in plant, soil and water from a caustic chlorine industry. Bull. Environ. Contam. Toxicol., 36; 229-305.

Shaw J.R., Hogstrand C., Price D.J. and Birge W.J., 1997. Acute and chronic toxicity of silver to marine fish. Proceedings of the Fifth International Argentum Conference, Transport, Fate, and Effects of Silver in the Environment. Hamilton, Ontario, Canada, September 28–October 1, 1997, pp. 317–324.

Shaw J.R., Wood C.M., Birge W.J. and Hogstrand C., 1998. Toxicity of silver to the marine teleost (*Oligocuttus maculosus*): effects of salinity and ammonia. Environ. Toxicol. Chem., 17: 594–600.

Shears M.A. and Fletcher G.L., 1983. Regulation of Zn^{2+} uptake from the gastrointestinal tract of a marine teleost, the winter flounder (Pseudopleuronectes americanus). Can. J. Fish. Aquat. Sci., 40 (Suppl. 2): 197 -205.

Shears M.A. and Fletcher G.L., 1984. The relationship between metallothionein and intestinal zinc absorption in the winter flounder. Can. J. Zool., 62: 2211 -2220.

Shukla V., Rathi P. and Shastry K.V., 2002. Effect of cadmium individually and in combination with other metals on the nutritive value of freshwater fish, *Channa punctatus*. J. Environ. Biol., 23(2): 105-110.

Sindhe V.R., Veeresh M.V. and Kulkarni R.S., 2002. Ovarian changes in response to heavy metal exposure to the fish, *Notopterus notopterus*. J. Environ. Biol., 23(2): 137-141.

Singh, K.P., Malik, A., Sinha, S., Singh, K. and Murthy, R.C., 2005. Estimation of source of heavy metal contamination in sediments of Gomti river (India) using principal component analysis, Water, Air, and Soil Poll. (Springer), Vol 166: 321-341.

Singh Mahipal, 1995. Hematological responses in a freshwater teleost, *Channa punctatus* to experimental copper and chromium poisoning. J. Environ. Biol.,16: 339-341.

Singh M, Gaur K.K., 1997. Effects of mercury, zinc and cadmium on the proteinic value and their accumulation in trunk muscle of *Channa punctatus* (Bloch.). Adv. Bios., 16(11): 109-114.

Singh M.S. and Sivalingam P.M., 1982. *In vitro* study on the active effects of heavy metals on catalase activity of *Sarotherodon mossambicus* (Peters). J. Fish. Biol., 29: 623.

Sorensen J.A., Glass G.E., Schmidt K.W., Huber J.K. and Rapp G.R.Jr., 1990. Environ. Sci. Technol., 24: 1716.

Sornaraj R, Baskaran P. and others, 1995. Effects of heavy metals on some physiological responses of air breathing fish, *Channa punctatus* (Bloch). Environ. Ecol., 13(1): 202-207.

Sparks R.E., Cairns J. and Heath A.G., 1972. The use of bluegill breathing rates to detect zinc. Water Res., 6: 895-911.

Spehar R.L., Fiandt J.T., Anderson R.L. and Defoe D.L.,. 1980. Comparative toxicity of arsenic compounds and their accumulation in invertebrates and fish. Arch. Environ. Contam. Toxicol., 9: 53–63.

Springue J.B., and Ramsay B.A., 1965. Lethal levels of mixed Cu and Zn solutions for juvenile salmon (*Salmo salar*). J. fish Res. Bd. Canada, 22: 245-355.

Spry D.J. and Wood C.M., 1985. Ion flux rates, acid–base status, and blood gases in rainbow trout, Salmo gairdneri, exposed to toxic zinc in natural soft water. Can. J. Fish. Aquat. Sci., 42: 1332 -1341.

Srivastava N. and Kaushik N., 2001. Use of fish as bioindicator of aquatic pollution. In: Abstracts, Intl. Cong. Chem. and Environ. (16 - 18th Dec. 2001), Indore, India.

Srivastava N. and Sharma R., 1996. Toxicity of zinc in fish (*Channa punctatus* Bloch.), as influenced by temperature and pH of water. Ind. J. Anim. Nutr., 13(2): 87-90.

Stopford W. and Goldwater L.J., 1975. Environ. Hlth. Persp., 12: 115.

Strike JJTWA., De longh H.H. and others, 1975. Toxicity of chromium VI in fish, with special reference to organ weights, liver and plasma enzyme activities, blood parameters and histological alterations. In: J.H. Koeman and J.J.T.W.A. Strike (Eds), Sublethal effects of Toxic chemicals on aquatic animals. Elsevier, New York, pp 31-41.

Storelli M.M. and Marcotrigiano G. O. 2004. Interspecific variation in total arsenic body oncentration in elasmobranch fish from the Mediterranean Sea. Mar. Pollut. Bull. 48: 1145-1167.

Subathra S. and Karuppasamy R., 2003. Bioassay evaluation of acute toxicity levels of cadmium on mortality and behavioural responses of an air breathing fish, *Channa punctatus* (Bloch). J. Exp. Zoo. India, 6(2): 245-250.

Suckcharoen S. 1980. Mercury contamination of terrestrial vegetation near a caustic soda factory in Thailand. Bull. Environ. Contam. Toxicol., 24: 463-466.

Suhendrayatna Ohki A., Nakajima T. and Maeda S., 2002. Studies on the accumulation and transformation of arsenic in fresh organisms II. Accumulation and transformation of arsenic compounds by *Tilapia mossambica*. Chemosphere, 46: 325–331.

Svecevicius G., 2006. Acute toxicity of hexavalent chromium to European freshwater fish. Bull Environ. Contamn. Toxicol, 77(5):741–747.

Taylor D., 1981. A summary of the data on the toxicity of various materials to aquatic life. 2. Cadmium. Report Bl/A/2073, ICI Brixham Laboratory, Freshwater quarry, Overgang, Brixham, Devon.

Tsai J.W. and Liao C.M., 2006. Mode of action and growth toxicity of arsenic to tilapia *Oreochromis mossambicus* can be determined bio-energetically. Arch. Environ. Contam. and Toxicol., 50, 144–152.

Turney W.G., 1970. The mercury pollution problem in Michigan. In: Environmental Mercury contamination. Eds R. Hartung and B.D. Dinman. Ann. Arbor. Sci. Publ. Michigan, p 29.

U.S.EPA., 1997. United States Environmental Protection Agency. Mercury study report to Congress. Volume VI: An ecological assessment for anthropogenic mercury emissions in the United States. http://www.epa.gov/ttn/oarpg/t3/reports/volume6.pdf

U.S.EPA., 2001. United States Environmental Protection Agency. Water quality criterion for the protection of human health: methylmercury (Executive summary). EPA-832-R-01-001. 2001.

U.S.EPA., 2002. United States Environmental Protection Agency. National recommended water quality criteria. EPA-822-R-02–047. Website:http://www. epa. gov/ost/pc/revcom.pdf.

U.S.EPA., 2004. United States Environmental Protection Agency. Fish Advisories. (http://www.epa.gov/ost/fish/states.htm).

US FDA. 1993. U.S. Food and Drug Administration. Guidance document for arsenic in shellfish. Washington, DC: U.S. Food and Drug Administration: 25–27.

Vallee B.L. and Falchuk K.H., 1993. The biochemical basis of zinc physiology. Physiol. Rev., 73: 79-118.

Van Der Putte I., Laurier MBHM. and Van Eijk GJM., 1982. Respiration and osmoregulation in rainbrow trout (*Salmo gairdneri*) exposed to hexavalent chromium at different pH values. Aquat. Toxicol., 2:99–112.

Van Vuren J.H.J., Van Merwe M. and others, 1994. The effect of copper on the blood chemistry of *Clarias gariepinus*. Ecotoxicol.Environ.Saf, 29: 187-199.

Venkatramreddy Velma, Vutukuru S.S. and Tchounwou Paul B., 2009. Ecotoxicology of hexavalent chromium in freshwater fish: A critical review. Rev. Environ. Health., 24(2): 129–145.

Venugopal B. and Luckey T.D., 1975. Toxicology of non-radioactive heavy metals and their salts. In: Heavy metal toxicity safety and hoemology. Eds. Luckey T.D., Venugopal B., Hutchenson D. Stuttgart, Thieme: 4-73.

Vogal F.S. 1959. The deposition of exogenous copper under experiments conditions with observations on its neurotoxic and nephrotoxic properties in relation to Wilsons disease. J. Exp. Med., J, Exp. Med., 110: 801-809.

Vutukuru S.S., 2003. Chromium induced alterations in some biochemical profiles of the Indian major carp, *Labeo rohita* (Hamliton). Bull. Environ. Toxicol., 70: 118-123.

Vutukuru S.S., 2005. Acute effects of hexavalent chromium on survival, oxygen consumption, hematological parameters and some biochemical profiles of the Indian Major Carp, *Labeo rohita*. Int. J. Environ. Res. Public Health, 2(3):456–462.

Walters L.J.Jr. and Wolrcy T.J. 1974. Transfer of heavy metal pollutants from lake Erie bottom sediments to the overlying water. Ohio State Univ., Columbus.

Ward D.R., Nickelson II R. and Finne G., 1979. J. Food Sci., 44: 920.

Watanabe T., Kiron V. and Satoh S., 1997. Trace minerals in fish nutrition. Aquaculture, 151: 185 -207.

Weigal H.J., Elmadfa F. and Jagar H F., 1984. The effect of low doses of dietary Cd oxide on the disposition of trace elements), haematological parameters and liver functions in rats. Arch. Environ. Contam. Toxcol., 13: 289-296.

Weiss H.V., Koide M. and Goldberg E.D. 1971. Scince 174: 692.

Wiener J.G., 1996. Mercury in fish and aquatic food webs. In: Abstracts for presentations given at the USGS workshop on mercury cycling in the environment. (http://toxics.usgs.gov/pubs/hg/abstracts.html).

Wo K.T., Lam P.K.S., Wu R.S.S., 1999. A comparison of growth biomarkers for assessing sublethal effects of cadmium on a marine gastropod, *Nassarius festivus*. Mar. Pollut. Bull., 39: 165-173.

Wong P.P.K., Chu L.M. and Wong C.K., 1999. Study of toxicity and bioaccumulation of copper in the silver sea bream *Sparus sarba*. Environ. Int., 25: 417-422.

Wood C.M., 2001. Toxic response of the gill. In: Target Organ Toxicity in Marine and Freshwater Teleosts, Vol. 1 (Eds: D. Schlenk and W.H. Benson), pp 1-89. London: Taylor and Francis.

Wood J.M. and Wang H.K., 1983. Microbial resistance to heavy metals. Environ. Sci. and Technol., 17: 82-90.

Wood J.M., Kennedy F.S. and Rosen C.G., 1968. Synthesis of methyl- mercury compounds by extracts of a methanogenic bacterium. Nature, 220: 173-174.

Woodal C., Maclean N., Crossley F., 1988. Responses of trout fry (*Salmo gairdneri*) and *Xenopus laevis* tadpoles to cadmium and zinc. Comp. Biochem. Physiol., 89C, 93–99.

Woodward D.F., Farag, A.M., Bergman H.L., Delonay A.J., Little E.E., Smith C.E. and Barrows F.T., 1995. Metals-contaminated benthic invertebrates in the Clark Fork River, Montana: effects on age-0 brown trout and rainbow trout. Can. J. Fish. Aquat. Sci., 52, 1994 -2004.

Wu-Yulin, Zhao-Hongru and Hou-Lanying, 1990. Effects of heavy metals on embryos and larvae of flat fish Paralichthys olivaceus. Oceanology Limnology. Sin./Haiyang-Yu-Huzhao., 21 (4): 386 - 392.

Yýlmaz Mehmet, Ali G€ul and Erhan Karak€ose., 2004. Investigation of acute toxicity and the effect of cadmium chloride ($CdCl_2$_H_2O) metal salt on behavior of the guppy (*Poecilia reticulata*). Chemosphere, 56: 375–380.

Chapter 8

Metal Toxicity Effects on Fish Behavior

Water containing toxic chemicals especially from industries such as the heavy metals (Allan and Flecker, 1995) affect the behaviors, biochemistry, and physiology of the living fauna including fish (Shilov, 1981; Radhaiah *et al.*, 1987; Vosyliene and Kazlauskiene, 1999; Gbem *et al.*, 2001). Several studies show that toxic agents may affect behavioral parameters (Sloman *et al.*, 2003; Abdul Farah *et al.*, 2004; Lefebvre *et al.*, 2004; Nakayama *et al.*, 2004; Yilmaz *et al.*, 2004). Behavior study in fact provides a unique perspective linking the physiology and ecology of an organism and its environment (Little and Brewer, 2001) since it allows the organism to adjust to external and internal stimuli in order to best meet the challenge to survive in a changing environment. Behavioral changes are further good indicators of damage to the central nervous system, as a consequence of exposure to toxic agents (Sloman *et al.*, 2003). According to Tinbergen (1963) ontogeny refers to the development of the behavior over the lifetime of an individual, as it contributes in maintaining a relationship between an organism and its environment. Eisler (1979) had suggested that changes in behavior of a chemical treated fish can be considered as an important sensitive indicator of its response to a toxicant as an animal's behavior is an integrated output of its physiological processes in response to any such affecting factor, may be environmental conditions like pH, alkalinity or the salinity or presence of any chemical toxicant. Many other workers (Little, 1990; Cooke *et al.*, 2000) have considered behavioral alterations as very sensitive indicators of a environmental constraint or stress imposed on fish. Since the behavioral changes are a direct reflection of changes in its physiological functions with the change in the surroundings, methods of monitoring and quantifying the behavioral responses have thus, become a potential alternative for assessing stress, disease, water pollution and toxic materials in water (Kane *et al.*, 2004). Some important behavioral anomalies which could be measured as an important bioindicator are given below.

1. Fish Orientation and Swimming Movements

As a fish culturist, one should become familiar with the normal behavior of a fish. If their behavior changes *e.g.*, the fish stop feeding, swim near the water surface, hanging listlessly (spiritlessly) in shallow water, gasping at the surface, or rubbing or scratching on objects, indicate that something wrong is being faced by the fish in water which need to be known. In normal raceways, a fish depending on the species, usually swim leisurely, either in mass or singly. Distribution in raceways also varies for species *e.g.*, some species prefer covered areas and the other ones may prefer uncovered areas; some concentrate toward the water inflow and others randomly distributed. Quantification of stress in fish models is often measured in terms of changes in its activity (Gerhardt, 1997). Fish on exposure to pollutants like Zn, Pb and Cd were found to show erratic, unbalanced swimming movements and tended to surface more frequently than fish in controls (Rao *et al.*, 1997). Similar observations were also made by Eaton (1974) and Abbasi and Soni (1984). Whirling movements observed in salmonids (Neil, 1957) and cichlids (Leduce, 1966) when exposed to Zn or cyanide, were attributed to stress on account of metal poisoning. The erratic swimming, jerky movement and convulsions before death were evident and the serenity varied with the concentration of the toxic chemical. The highest treatment of copper (0.12 mg Cu/l) caused lethargy and loss of equilibrium in fish. Ali *et al.* (2003) while evaluating the effect of different sublethal concentrations of copper (0, 0.15, 0.3 and 0.5 ppm) on the behavioral responses of *Oreochromis niloticus*, observed hyperactivity and reduced exploratory behavior at different levels of copper in water as compared with the control.

Gupta and Chakarbarty (1993) compared behavioral responses in freshwater teleost, *Notopterus notopterus* and *Puntius javanicus* against Zn toxicity. Many other workers also reported that toxicants can disrupt startle responses (Carlson *et al.*, 1997) and the swimming performance and activity of the fishes (Weis and Weis, 1995; Zhou and Weis, 1998). The studies exhibited more sensitivity in *P. javanicus* than the *N. notopterus*. Pb and Zn produced surfacing and whirling responses in the treated fish, and over sliming on the body and gill surface. Khunyakari *et al.* (2001) studied toxic effects of three heavy metals *viz.*, nickel, copper and zinc in *Poecilia reticulata*. Among these metals copper was found to be most active followed by zinc and nickel. Groups of the adult and healthy *Oreochromis aureus* subjected to the chronic concentration (1/10 LC_{50}) of Pb and Zn for 6 and 12 weeks (Abhas *et al.*, 2003), showed altered behavioral activities. The altered behavioral responses were also observed under investigation in the test organism exposed to Cd. Scott and Sloman (2004) also found complete elimination of the performance of behaviors that are essential to fitness and survival in natural ecosystems, frequently after exposures of lesser magnitude than those causing significant mortality. Masood *et al.* (2005) on exposing *Cyprinus carpio* to mercuric chloride, observed no sign of distress initially in groups exposed to 1.0 and 1.5 ppm, but abnormal posturing, disbalance and sluggishness became apparent after 72 hrs; all specimens of 1.5 ppm group however, had died within 124 hours, showing thus, adverse effects of mercurial toxicity even at low levels on fish behavioural responses in *C. carpio*.

Findings in Cr(VI) treated *Channa punctatus* (Mishra and Mohanty, 2008) further displayed erratic swimming and lethargic movements.

Annune *et al.* (1994) reported that zinc could cause sub-acute effects that change fish behavior. Such behaviors include lack of balance since most fins are motionless in the affected fish, agitated swimming, air gulping, periods of quiescence and death. The increase in swimming activity may be due to disruption of shoaling behavior which occurs because of the stress of the toxicant (Venkata *et al.*, 2008). Fast swimming was also observed by Yaji *et al.* (2011) in *Oreochromis niloticus* treated with Cypermethrin. Similar observation found by Ramesh and Saravanan (2008) in *Cyprinus carpio* exposed to chlorpyrifos. Vanadium was also found to induce increased movements like S jerking, threat and burst swimming in the experimental fishes. Similar observations were reported by Nimila and Nandan (2010) in *Etroplus maculates* when it is treated with lindane and by Al-Aker and Shamsi (2000) in carbaryl treated fishes. Jerking and burst swimming were also observed by Marigoudar *et al.* (2009) in *Labeo rohita* exposed to cypermethrin. The fingerlings exhibited prompt and erratic behavioral responses at extremely low concentrations of effluents compared to the adults. The adult fish also showed characteristic hyperactivities, erratic swimming, frequent surfacing, followed by sinking, loss of equilibrium and coloration and gradual onset of inactivity. The behavioral responses were increased as the metal concentration was increased. Kumari *et al.*, 2014 in *Labeo rohita* fish on exposure to Cr also recorded development of locomotory responses, frequency of swimming movements, and duration of activity significantly altered in chromium treated fishes. Chromium treated fishes also showed lethargic movement and loss of equilibrium. Likewise, Ravindran and Radhakrishnan (2015) exposed freshwater catfish, *Heteropneustes fossilis* to vanadium and found the behavioural changes including speedy movements, increased ventilator movement, abnormal coughing and jerky movements, erratic swimming, hyperactivity changes in opercular movement, copious amount of mucus secretion all over the body and loss of equilibrium as important signs to monitor the quality of water contaminated with chemicals.

Lawal and Samuel (2010) on exposure of *Poecila reticulata* to actellic was seen to result in aggressive behavior, rapid gulping of water, increased opercular movement and abnormal swimming movements. Sandahl *et al.* (2005) observed during the experiment that the fishes in chlorpyrifos treated groups were unable to take as much food in comparison to control fishes. Shahi and Singh (2010) observed various forms of abnormal behavior in *Channa punctatus* when exposed to different concentrations of rutin, taraxerol and apigenin. Susan and Sobha (2010) studied the toxic effect of fenvalerate on Indian major carps: *Labeo rohita, Catla catla* and *Cirrhinus mrigala* and observed behavioural changes such as swimming at the water surface, hyper excitation, loss of equilibrium, flaring of gills and increased mucus production. *Puntius chola* fishes exposed to the 0.2 ppm of organophosphate pesticide, chlorpyrifos (Verma and Saxena, 2013) showed a skin color change slightly from silver white due to mucus secretion and mortality was observed at 48 to 96 h. In 0.3 ppm treatment, fast and jerking movement with mortality and increased mucus production was observed. As time elapsed, fish exhibited vertical hanging and settling at the bottom. In 0.4 ppm and 0.5 ppm treatments, fishes exhibited stress symptoms, vertical hanging in the water and mortality. Skin color became brownish with a red spot in the head region, and slimy body was noticed. Fishes of all treated groups showed hyper activity and frequent surfacing to gulp air. It was also noticed that as time passed, they continued to swim near the water surface and tried to jump out from the holding tanks. Once the fishes were exhausted, they sank to the bottom of the tanks with no opercular movement and finally succumbed with their mouth opened. The nature and rapidity of onset of fish behavioural responses indicated that these compounds act on the neuromuscular system of fish. Similar changes in fish behavioural patterns were made earlier by Hülya *et al.* (2006) in *Oreochromis niloticus,* following sublethal exposure to diazinon.

Orientation and locomotor patterns were found involved in many aspects of fish behavior, such as feeding behavior, photo taxis responses, or the predator avoidance (Beitinger, 1990; Little *et al.*, 1993), and fish migration, mating *etc.* courtship which are altered under stress conditions of environmental toxicants (David, 1995; Prashanth *et al.*, 2011). Daily observation of fish behavior and feeding activity allows early detection of problems when they do occur so that diagnosis can be made before the majority of the population becomes sick. If treatment is indicated, it will be most successful when it is implemented early in the course of occurrence of the disease while the fish are still in good shape. Under these conditions the symptomatic observations play an important role. The behavioral traits more in reference and communally observed are the feeding rates, swimming movements, surfacing, schooling *etc.* Feeding and the swimming behaviors need close and continuous monitoring. The first response to any unfavorable condition is abnormal behavior; to recognize what is abnormal; you must first be familiar with what is normal. The behavioral abnormalities indicate that the fish are not feeding well or that something is irritating them. Under routine aquaculture conditions, healthy fish display a normal feeding behavior *i.e.* fish feed vigorously when food is presented or shortly thereafter. A reduced feeding activity should serve as a notice to the aqua culturist that an immediate further investigation is warranted. James *et al.* (1992) has reported reduced rates of feeding, absorption, conversion and metabolism in the presence of sub lethal concentrations of mercury in *H. fossilis*. Fish were seen refusing feed immediately after exposure, but started accepting it after about 4–5 h of their exposure.

2. Respiratory Activities and Oxygen Consumption

Measuring oxygen consumption of a fish or the ventilator activity on a continuous basis could also be a good measure of the intensity of water toxic effects. The behavioral anomalies like jerking and whirling movements while swimming, engloping air while surfing on the water or going for exodus trials in fish exposed to toxic heavy metals, virtually reflect compensation to additional oxygen requirements of the body to escape hypoxia or suffocation due to water toxicities. Frequent surfing in fish is most common feature on their exposure to toxic metals. *Colisa fasciatus* on exposure to Ni were seen to be restless, approaching water surface after every 2 to 3 minutes, along with the formation of the mucus layer on the opercula region (Chaudhary and Nath, 1985), probably to go for higher intake of air on one hand and also to protect gills from the toxicity effects. Morgan *et al.* (1988) had also used fish ventilator responses to monitor water quality in the trout streams in Tennessee. Moribund fish were found to hang lifelessly at the surface and exhibiting exaggerated opercula movements. During the hour preceding death, the fish was also found to swim erratically and some channel cat fish also hang motionless due to muscle spasms. All these observations can be considered to monitor breathing as per the quality of water contaminated with metals. Sarkar (1989) had evaluated the effect of $HgCl_2$ and $CrCl_2$ on oxygen consumption in freshwater fish, *Tilapia mossambica*. Results indicated higher O_2 consumption in fish on exposing to $CrCl_2$, as compared to that of $HgCl_2$. At 0.5 ppm of $HgCl_2$ and 0.7 ppm $CrCl_2$ onward O_2 consumption of fish progressively decreased. Aberrant breaths such as cough can also be detected (Spoor *et al.*, 1971) in toxic waters. Ghosh and Mukhopadhaya, 2000 also attributed metal induced changes in behavior of *Rita rita* fish to stress.

While integrating physiological functions with the toxic effects of a chemical most important is to follow the spectrum of animal organization, level of complexity, such as it may alter a cell function by affecting the enzyme functions or secretion of hormones or ultra structure of an organ. For example, many toxicants cause cell thickening or even their death (necrosis) in the gill epithelium, affecting its permeability to oxygen, thereby affecting animal respiration leading to failure of homeostasis. Jia (2002) determined the effect of four heavy metals (Cu, Hg, Zn and Pb) on the intensity of respiration in juvenile *Misgurnus anguillicaudatus*. It was observed that the fish were sensitive to heavy metal pollution. The respiration intensities of the different groups of fishes treated with heavy metals were higher than those of the control in which Cu showed the maximum and Pb produced the minimum effects on respiration. For example, zinc a trace element and essentially needed for some enzymatic functions in animals including fish, when present as a contaminant in water at a concentration of 1.5 mg/l, it causes a rapid drop in blood oxygen and pH over a period of less than 12 hr, and the fish die due to hypoxia, and at 4.5mg/l the fish die in less than 12 hr. Fish exposure to a lower concentration of Zn (0.8 mg/l) for 3 days, however, result in slight blood alkalosis, but no change in blood oxygen levels (Spry and Wood, 1984). At comparable toxicity levels Cr is said to be less harmful to gill tissue than Ni (Hughes *et al.*, 1979). Martínez *et al.* (2006) studied the effects of chronic hypoxic exposure on carbohydrate metabolism in the Gulf kill fish (*Fundulus grandis*). Fish were held at approximately 1.3 mg l^{-1} dissolved oxygen (3.6 kPa) for 4 weeks, after which maximal activities were measured for all glycolytic enzymes in four tissues (white skeletal muscle, liver, heart and brain), as well as for enzymes of glycogen metabolism (in muscle and liver) and gluconeogenesis (in liver). The specific activities of the glycolytic enzymes and glycogen metabolism were strongly suppressed by hypoxia in white skeletal muscle, which may reflect decreased energy demand in this tissue during chronic hypoxia. In contrast, several enzyme specific activities were higher in liver tissue after hypoxic exposure, suggesting increased capacity for carbohydrate metabolism. Hypoxic exposure affected fewer enzymes in heart and brain than in skeletal muscle and liver and the changes were smaller in magnitude, perhaps due to preferential perfusion of heart and brain during hypoxia. The specific activities of some enzymes involved in gluconeogenesis increased in liver during long-term hypoxic exposure, which may be coupled to increased protein catabolism in skeletal muscle. These results demonstrate that when intact fish are subjected to prolonged hypoxia, enzyme activities respond in a tissue-specific fashion reflecting the balance of energetic demands, metabolic role and oxygen supply of particular tissues. Furthermore, during glycolysis, the effects of hypoxia varied among enzymes, rather than being uniformly distributed among pathway enzymes.

Jezierska and Sarnowski (2002) measured the oxygen consumption of common carp and rainbow trout larvae exposed to Hg, Cd and Cu and found oxygen consumption rate as a reliable indicator of metal toxicity to fish. The metals showed interspecific variations in their effects. The level of oxygen consumption decrease indicated that the metal toxicity for the given nominal trace metal are Hg+Cu> Cu> Cd+Cu> Cd+Hg> Cd for common carp and Hg+Cu > Cu> Cd+Hg> Hg for the rainbow trout. The results showed that Cu was more toxic to both the species and the metal mixtures are more toxic than the single metal and cause greater reduction in oxygen consumption; common carp larvae being more sensitive to Cd and rainbow larvae to Hg. Banchez *et al.* (2008) also found oxygen consumption data as a sensitive indicator to reliably detect physiological alterations induced by most of the contaminants. After 2 h of exposure, they identified significant increases in oxygen consumption upon exposure to pentachlorophenol (100 and 1000 microg/L) and cadmium chloride (0.0002 and 0.002 microg/L) and atrazine (150 microg/L). In contrast, a significant decrease in oxygen flux after exposures to potassium cyanide (44 and 66 microg/L) and atrazine (1500 microg/L) was observed. No effects were detected after exposures to malathion (200 and 340 microg/L). At sub-lethal cyanide concentrations (Prashanth *et al.*, 2011) the schooling behavour of the fish was slowly changed, the ventilation rate was increased and the oxygen consumption rate was also reversed to normal. Alterations in oxygen consumption rates were correlated with respiratory distress as a consequence of impairment in oxidative metabolism. Toxicity of sodium cyanide to the freshwater fish, *Labeo rohita* studied by Prashanth *et al.* (2011) showed decreased oxygen consumption rate, increased opercular movements, increased surface behavior, loss of equilibrium, change in body color increased secretion of mucus, irregular swimming activity rapid jerk movements, partial jerk and aggressiveness at lethal concentrations. The swimming was in a cork-screw paller, rotating along horizontal axis. The opercular movements of the fish were ceased on exposure to cyanide.

Observing opercular movements (Cairns and van der Shalie, 1980) and their quantification under stress in fish is often considered important in terms of changes in its fish behavior. Various workers had reported hyperexcitability, rapid opercular movements, jerky and whirling movements, and frequent surfacing or a moribund fish (Wise *et al.*, 1987; Gill *et al.*, 1991; Dutta *et al.*, 1992; Jenkins *et al.*, 2003); loss of equilibrium, uncoordinated swimming movements and easily excitable conditions as observed in *H. fossilis*, after chronic treatment with lead nitrate (Sastry and Gupta, 1979); and irregular, erratic and darting movement, imbalance-swimming activity, surfacing and gulping of air, muscular spasms, loss of orientation and gyrating swimming movements, frequent jumping, erratic movement followed by convulsions (Shivakumar and David, 2004), spinning around their own axis (Yilmaz *et al.*, 2004), and altered positive photoactive response and swimming rate (Kannupandi *et al.*, 2001) in various chemical treated fish. Spotted snakehead fish exposed chronically to Cr (2-4 mg/l) resulted in abnormal behavior such as convulsions, loss of equilibrium, and increased opercular activity (Mishra and Mohanti, 2009). Similarly, cichlids exposed for 4 days to 3 or 6 mg/l of Cr^{III} showed a dose dependent change in a variety of behaviors like coughing, yawning, fin-flickering, jerking movements and nudge and nip behaviors (Al-Kahem, 1995). Behavioral abnormalities like increased swimming activity and breathing movements were further evidenced by Shah (2002) in fish, *Cyprinion watsoni* on exposure to 0.06 mg of Cu/l for a week. Yýlmaz *et al.* (2004) noticed some behavioral changes in fish guppy (*Poecilia reticulate*) after exposure to Cd, like imbalanced swimming, capsizing, difficulty in breathing and attaching to the surface and gathering around the ventilation filter. The increased opercular movement and corresponding increase in fish surfacing clearly indicated that fish adaptively shifts towards aerial respiration and the fish tries to avoid contact with the cyanide through gill chamber (Prashanth and Patil, 2006). The erratic swimming of the treated fish along with loss of equilibrium may be associated with the affected region in brain associated with this function (Prashanth *et al.*, 2005). Anandhan and Hemalatha (2009) in a freshwater fish, *Brachydanioreri* also observed hyperactivity, erratic swimming, changes in opercular movement and loss of equilibrium following fish exposure to aluminium. Siddiqui and Arifa (2011) in *Clarius batrachus* on exposure to copper found erratic swimming, jerky body movements, rolling the body, convulsions, mucous secretion over the body at 0.75 ppm and loss of equilibrium, rapid opercular movements, difficulty in respiration, and lethargy at 1ppm. In short the effects and the dosage were found to be directly proportional to the behavior of the experimental animal.

Assessing the acute toxicity of dimethoate on fingerlings of common carp, *C. carpio* in terms of impact on oxygen consumption in exposed fishes showed a significant decline at all concentrations (Singh *et al.*, 2009). Singh *et al.* (2009) had concluded that carps are more sensitive to dimethoate as compared to air-breathing teleosts. Decreased opercular movements in the pesticide exposed carp fingerlings probably were probably to reduce absorption of the pesticide poison through gills, as well as simultaneously resulting in reduced rate of oxygen consumption in *C. carpio*. The test fish also exhibited erratic swimming, increased surfacing, decreased rate of opercular movement, copious mucous secretion, reduced agility and inability to maintain normal posture and balance with increasing exposure time. They attributed increase in surfacing and gulping of surface waters in carps on exposure to dimethoate as the behavioral manifestations to avoid breathing in the poisoned water. Similar observation had been reported in *Anabas testudineus* after exposure to monocrotophos (Santhakumar and Balaji, 2000). Reduction in oxygen consumption in *C. carpio* has also been reported after sublethal exposure of copper (Boeck *et al.*, 1995) and antimony chloride (Chen and Yang, 2007). Coughing responses have been shown to have a direct relationship with the concentration of pollutants in water (Carlson and Drummond, 1978). Coughing response along with secretion of strands of mucus trailing from gill chambers and other parts of the body surface are frequently exhibited as fish response to metal toxicity stress. Wise *et al.* (1987) observed such behavior in *Ictalurus punctatus* and *Carassius auratus* exposed to 63 mg/l or more of nitrofurazone. Irregularity in body movement, cough rate, and ventilator rate and death rate of bluegill (*Leomis macrochirus*) were reported in fish exposed to toxic test water (Cairns and van der Shalie, 1980). Exposing fishes to chromium revealed occurrence of respiratory stress as a non-specific gill irritation response, as evident by frequent coughing, yawning and ventilation frequency (Gendusa and Beitinger, 1992). Ganeshwade *et al.* (2006) had also reported increased opercular rate and coughing in the common carps exposed to industrial pollutants in water. Radhaia *et al.* (1998) has stated that hypoxic condition of the pesticide exposed fish also contributes to increase surfacing. Hypoxic condition arises primarily due to damage of gills on account of pesticide poisoning which also hampers oxygen uptake simultaneously (Velmurugan *et al.*, 2007).

Respiration is a rhythmic neuromuscular sequence regulated by an endogenous biofeedback loop as well as by external environmental stimuli. Acute contaminant exposure can induce reflexive cough and gills purge responses to clear the opercular chamber of the irritant and can also increase rate and amplitude of the respiratory cycle as the fish adjusts the volume of water in the respiratory stream. Under toxic condition the oxygen supply becomes deficient and so the fish breathe rapidly (Susan *et al.*, 2010). The abnormal behavior in *Clarias gariepinus* fingerlings and the adults was also reported on their exposure to soap and detergent effluents by Adewoye (2010). Verma and Saxena, 2013 recorded increase in oxygen consumption as an indication of stress. Fishes in treated group were likely to have consumed oxygen more rapidly than in the control group to tolerate the stress. Oxygen consumption rate was, however, found decreased significantly with time in Zn exposed *Phascolosoma esculenta* and also in interaction with Cd concentration and time of exposer (Chen *et al.*, 2009). Kumari *et al.*, 2014 in *Labeo rohita* fish on exposure to Cr observed normal respiration altered in the fishes. Diamond *et al.* (1990) also found that the frequency and amplitude of bluegill opercular rhythms and cough responses were altered following exposure to different contaminants. *Clarias gariepinus* fingerlings exposed to lethal and sublethal concentrations of Gammalin 20 investigated for its effects on fish behavior and survival (Ezemonye and Emmanuel, 2010), revealed respiratory distress, increased physical activity, convulsions, erratic swimming, loss of equilibrium, and increased breathing activity.

3. Fish Avoidance to Metals

When an animal is exposed to a perturbation, the first line of defense is a behavioural one, most often an avoidance behavior, designed to lessen the probability of death or the metabolic costs incurred by maintaining physiological homeostasis (Schreck *et al.*, 1997). Such behavioural alterations are very sensitive indicators of a constraint level or stress imposed on fish by the environment (Scherer, 1992; Cooke *et al.*, 2000). In fishes, avoidance test is well established as a mean of showing effects of any chemical well below its lethal range, and such a test is one most sensitive sublethal response of fish to contaminants such as the heavy metals (Beitinger and Freeman, 1983). Although the behavioral effects of many heavy metals have not been evaluated, most of those that elicit a consistent avoidance response, including copper (Cu), zinc (Zn), nickel, mercury, and others, are avoided at concentrations that are much lower than lethal (Atchison *et al.*, 1987). Fish avoid metals, especially As, Cd, Cr, Co, Cu, Fe, Hg, NI, Se and Zn to varying degrees (Tierney *et al.*, 2010). Zn poisoning in fish is further said to exhibit behavioral changes like avoidance behavior with increases swimming activity and skin darkening in color at higher concentrations of Zn, along with impaired feeding and growth. Decreased swimming rate was directly related to test concentration. Of all the heavy metals Cu has been studied most often for the avoidance reaction studies in fishes. Several studies have documented that anadromous and freshwater salmonids avoid low Cu concentrations ranging from 2.3 (Sprague, 1964) to 7.3 mg/L (Giattina *et al.*, 1982). This range of values suggest that single test concentration, or steep gradient test, or mixed metals *etc.* probably influence the lowest (threshold) concentrations of Cu that are avoided. For example, Giattina *et al.*, [1982] found that rainbow trout, *Oncorhynchus mykiss* avoided 7.3 mg Cu/L in single-concentration, but 6.4 mg Cu/L in steep-gradient tests and 4.4 mg Cu/L in shallow-gradient tests. Sprague (1964) further showed that Atlantic salmon on average avoided 5.2 per cent of the incipient lethal level (ILL) of Cu and 9.2 per cent of the Zn ILL, when the fish were exposed to the single metal. The lowest concentration to cause avoidance in 50 per cent of fish tested was approximately 5 and 10 per cent of the ILLs for Cu and Zn, respectively. However, when Cu and Zn were mixed, the fish avoided 2.1 per cent of the ILL of each metal. So the fish were responding to the additive (or possibly synergistic) effects of the two metals. Hansen *et al.* (1998) observed behavioral avoidance to copper (Cu), cobalt (Co), and a Cu + Co mixture in soft water to differ greatly between rainbow trout (*Oncorhynchus mykiss*) and chinook salmon (*O. tshawytscha*). Chinook salmon avoided at least 0.7 mg Cu/L, 24 mg Co/L, and the mixture of 1.0 mg Cu/L and 0.9 mg Co/L, whereas, rainbow trout avoided at least 1.6 mg Cu/L, 180 mg Co/L, and the mixture of 2.6 mg Cu/L and 2.4 mg Co/L. Chinook salmon were also reported as more sensitive to the toxic effects of Cu in that they failed to avoid 44 mg Cu/L, whereas, rainbow trout failed to avoid 180 mg Cu/L. Furthermore, following acclimation to 2 mg Cu/L, rainbow trout avoided 4 mg Cu/L and preferred clean water, but chinook salmon failed to avoid any Cu concentrations and did not prefer clean water. The failure to avoid high concentrations of metals by both species suggests that the sensory mechanism responsible for avoidance responses was impaired. Exposure to Cu concentrations that were not avoided could result in lethality from prolonged Cu exposure or in impairment of sensory-dependent behaviors that are essential for survival and reproduction. Hansen *et al.* (1999) further concluded that copper was avoided by chinook (*O. tshawytscha*) and rainbow trout (*O. mukiss*), with chinook exhibiting higher sensitivity; fish avoided low concentrations of Cu, but not the higher concentrations. Baldwin *et al.*, 2003 had also mentioned that since Cu can impair the olfactory epithelium within minutes, conceivably a copper plume could impair neurological detection rapidly enough to prevent an olfactory-mediated behavioral response. In rainbow trout, Ni was also seen evoking concentration dependent avoidance/attraction responses like Cu; the fish was seen attracted to low (6μg/L), it however, avoided higher (24 μg/L) concentration of Ni (Giattina *et al.*, 1982). Cobalt was also found avoided at concentrations higher than that of Cu (Hansen *et al.*, 1999). Cadmium though lethal to salmonids at 1 to 1.8 mg/L; it however, is said to elicit inconsistent avoidance responses to concentrations ranging from 0.2 to 1 mg/L (Mc Nicol and Scherer, 1993).

4. Toxicity Effects on Fish Avoidance to Predators

Olfaction conveys critical environmental information to fishes, enabling activities such as locating food, habitat and mate, courtship, discriminating kins, avoiding predators and homing *etc.* A well known example is homing salmon exhibit to the odorant bouquet of their natal stream (Sholz *et al.*, 1976). A shark is another important example of showing searching behavior *i.e.* exhibiting searching behavior for a prey by following up a concentration gradient of blood (Gilbert, 1977). Likewise, male sticklebacks can discriminate between males and ovulated females (McLennan, 2004). An injured fish by releasing alarm pheromones provide tactile information may confirm the presence of pray or a predator (Brown, 2003a; Hara, 2006a). The teleost fish in general possess a well developed peripheral olfactory organ, located in bilaterally positioned olfactory chamber. Alarm substance is a chemical signal released from the fish skin epithelial cells of a prey its skin is damaged by a predator. Other prey fish detect such alarm substance by olfaction and they perform stereotyped predator-avoidance behaviors to decrease predation risk. Adaptive significance to these behaviors to prey fish in natural conditions presumably involves being inconspicuous to predators, and the behavioral response to alarm substances increase survival of the fish during encounters with predators (Chivers *et al.*, 2001). Salmonid are said to exhibit decrease in swimming and feeding activities in response to such alarm substances (Mirza and Chivers, 2001). All of these behaviors can be impaired or lost as a result exposure to toxic contaminants in surface waters. Sullivan *et al.*, 1978 found the pray fish fathead minnows more susceptible to its predator because of disrupted schooling behavior due to exposure to Cd. A temporary reduction in food consumption in the first week of exposure of rainbow trout to sublethal Cd (3 µg/l was noted by McGeer *et al.*, 2000.

Scott *et al.*, 2003 evidenced that fish exposed to water born Cd (2 µg/l) disrupted normal fish behavior to alarm substances, thereby affecting predator avoidance behavior of juvenile rainbow trout (*Oncorhynchus mykiss*). Water born exposure to 2.0 µg/l of Cd for 7 days was found to eliminate normal anti-predator behaviors exhibited in response to alarm substance. No such elimination was however, detected when fish was exposed to Cd for a shorter duration than 7-days. Interestingly, fish was seen with a normal response, when fed 3.0 µg/l of Cd for 7 days in food, in spite of the fact that it accumulated same amount of Cd as it was with water born exposure to Cd., suggesting Cd accumulation in water only interfere with olfactory system. Such a behavioral abnormality has potential impact on aquatic fish communities. Contaminant exposures have also been shown to cause reduced food odor attraction and predator scent avoidance, as well as altered alarm responses (Berejikian *et al.*, 1999; Kasumyan, 2001) Alarm responses further include many behaviors, such as dashing, freezing, and hiding (Berejikian *et al.*, 1999; Pollock *et al.*, 2003).

The fish olfactory system is very efficient in detecting alarm substances, a chemical cue released from the epithelial cells of the prêy, which elicits typical predator avoidance behavior of the conspecifics, including area avoidance, freezing, shoaling, and shelter use. Baker and Montgomery, 2001 reported that fish were unable to sense adult migratory pheromones following exposure to extremely low levels of waterborne Cd (less than 1 μg/l) for 48 hrs. Exposures of a fish to low water levels of Cd for 4-7 days, impaired the normal anti-predator behavior that a fish exhibits in response to the alarm substances and in contrast dietary Cd caused no such effect on such a behavior (Scott *et al.*, 2003). The alarm response is sometimes associated with a stress response, and so certain stress hormones may prove to be good co-relates of alarm behavior. Scott *et al.*, 2003 observed that in an unexposed rainbow trout, an alarm substance was found to increase the plasma cortisol 4 fold: whereas, in fish exposed to 2 μg/l of Cd the elevation level was only 2 fold thus, impairing alarm response. A short term elevation in plasma cortisol was also seen in response to alarm substance under control conditions, but in case of water born exposure to Cd the treated fish showed inhibition in plasma cortisol elevation also. Brown and Smith (1997) have observed similar results indicating that there exists a large degree of variability among individuals in the use of shelters for predator avoidance in juvenile rainbow trout. Cadmium is reported to act as a neuromuscular blocking agent. How Cd interrupts alarm signals in fish has been explained by some workers (Phillips, 2003; Scott *et al.*, 2003), According to Phillips (2003), when a predator strikes, it's too late for the victim, but not for the rest of the crowd if the victim can sound the alarm in time.

Metals are further well-known for their effectiveness as blockers of ion channels, such as sodium and calcium channels, inhibiting signal conduction as in case of olfactory responses. Sutterlin and Sutterlin, 1971 showed that mercuric chloride ($HgCl_2$) block the olfactory responses of Atlantic salmon to various amino acids. Copper induced olfactory toxicity has been known for more than three decades (Hara *et al.*, 1976). Cadmium has been shown to interfere with chemoreception in fish, albeit at exposure concentrations that are higher than those for copper (Thompson and Hara, 1977). They further exposed fish Arctic char (*Salvelinus aplinus*) to lake water degraded by mining effluents, consisting of Cd, Cu, Ni and Zn and found that Cd but not Ni or Zn could impair OSNs. Copper has received further considerable attention in recent years. Aluminium has also been shown to influence the electrical properties of the olfactory sensory neurons in rainbow trout (Klaprat *et al.*, 1988). Fish exposures to aluminium in low pH water caused reductions in trout olfactory responses. Hansen *et al.* (1999) also recoded interspecific differences in olfactory responses to copper, such as rainbow trout (*O. mykiss*) were more vulnerable than juvenile chinook salmon (*O. tshawytscha*). Baldwin *et al.*, 2003 observed a steady decline in the responses of OSN in coho salmons (*O. kisutch*) over 30 min exposure to copper at concentrations below 10 ug/l. The inhibitory effects of Cu on odor-evoked responsiveness of OSNs are further influenced to a certain degree by change in water chemistry, such as hardness, alkalinity, and dissolved organic carbon (Winberg *et al.*, 1992; Baldwin *et al.*, 2003).

5. Toxicity Effects on Fish Reproductive Behavior

Expression of breeding behavior in male fishes is reflexed primarily in establishing breeding territories, nests, mate attraction, courtship, spawning, parental care *etc.* and these activities in principle are controlled by the secretion of androgens (Baggerman, 1966; Borg, 1994; Oliveira and Canario, 2000). Several aspects of parental care seem to be regulated by prolactin, which in some cases may be acting together with androgens, such as in parental aggression (Liley and Stacy, 1983). Studies had further indicated a potential trade-off between androgens and parental behaviors, since androgens are implicated more in the expression of aggressive behavior, which should be reduced during parental care phase; fish species with marked parental care phase during their breeding cycles usually show a decrease in androgen levels during parental phase (Winffield *et al.*, 1990). Castration, or injecting certain anti-androgen chemicals like Cyproterone acetate or even Methyl testosterone had negative effects on aggression, nest building and courtship, whereas the luteinizing hormone (LH) or gonadotropin releasing factor (GnRH) had a positive effect on all these three behaviors. Kramer *et al.* (1969) had suggested a negative feedback mechanism of testosterone acting on reproductive behavior in *O. mossambicus*.

Most of the studies conducted on the effects of various pollutants on the reproductive performance in fishes have revealed effects on courtship included decrease or increase in frequency of displays, increased courtship duration, or performance of male-like behavior by masculinized females. Studies of parental care have found decreased nest-building activity, decreased off- spring's defense, or change in division of parental care between the sexes (Jones and Renolds, 1997). Fishes exposed to the paper/kraft mill effluents were found to exhibit signs of musculinization in females, and these females had developed a gonopodium, *i.e.* the modified anal fin used by males for internal fertilization in the family of fishes. Fish sampled from above and below the effluent outflow showed some degree of gonopodial development. There was however, no evidence of any testicular tissue, or heteromorphic sex chromosomes in the musulinized females (Jones and Renolds, 1997). Schroder and Peters (1988) contended that as a behavioral assay, courtship by guppies (*Poecilia reticulate*) was extremely sensitive to even very low concentrations of aquatic contaminants. Exposing mosquito fish (*Gambusia affinis* Holbrooki) to paper mill effluents or the paper/Kraft mill effluents revealed that exposed females became masculinized and showed more male-like courtship than control females (Howell *et al.*, 1980; Bortone *et al.*, 1989; Krotzer, 1990). The musculinized pregnant females also displayed a typical male reproductive behavior *i.e.* chasing normal and musculinized females with gonopodial swings and thrust attempts. Males exposed to the effluents matured precociously and exhibited typical but more aggressive courtship. A masculinized female was dominant over a normal male (Howell *et al.*, 1980). The two musculinized females when placed together, the smaller one behaved as males and the females showed much less aggressive behavior (Krotzer, 1990). Mosquito fish Males exposed to waterborne lead (0.5 mg/l) for 30 days while fish spawned (Weber, 1993), were found to spent significantly less time in activities directed towards ceiling of nest, including preparation of substrate for eggs, and touching ceiling with dorsal pad.

6. Toxicity Effects on Impulse Transmission

A behavior of an organism since represents an integrated result of its physiological and biochemical processes, it projects an ultimate response of an organ or body as a whole as a result of the effects of any particular environmental factor. Changes in any such behavioral act also represent a higher organizational level of biomarkers than any other considered biochemical or a physiological test (Walker *et al.*, 2003), making thus, a single behavior parameter more comprehensive in analyzing toxicity effects in fishes or any other animal. Such behaviors are prerogative to the animals as they are equipped with a neuromuscular system. The animals exhibit such movements in response to environmental influences which are perceived by them through their sensory organs and the movements are performed by their neuro-muscular system. The toxic effects of heavy metals on fish are multidirectional and manifested by numerous changes in the physiological and chemical processes of their body systems (Dimitrova *et al.*, 1994). Most important physiological effects of a toxicant associate with such a behavior include disruption of sensory, hormonal, neurological, and metabolic systems, which are likely to have profound implications on fish survival. Frequent or longer toxicant exposures often completely eliminate the performance of certain behaviors that are essential to fitness and survival in natural ecosystems. However, there has been very little toxicological research sought to integrate the behavioral effects of a toxicant with its physiological processes. Scott and Sloman (2004) also emphasized upon the need for a future integrative and multidisciplinary research in the study of behavioral toxicity to increase both the significance and usefulness of behavioral indicators in aquatic toxicology. Heavy metals can directly influence behavior by impairing mental and neurological functions, influencing neurotransmitter production and utilization and altering numerous metabolic body processes. Disruption of impulse transmission in central and peripheral nervous system in animals is one such important aspect of chemical toxicity. Anything affecting nervous system or neuron integrity, also affect behavior.

According to Singh *et al.* (2009) tremors, gradual loss of equilibrium and drowning in dimethoate exposed carps are caused by adverse effects of organophosphate on central nervous system. Exposure to N-ethylmaleimide (NEM), a reagent that binds covalently to protein sulfhydryl groups, also results in a specific reduction in sodium conductance in crayfish axons (Shrager, 1977), thus, inhibiting nerve or signal conduction. Fish though have evolved a chemical alarm system that sends a school fleeing for cover when an attack is launched on the contrary if there are traces of cadmium in the water, the alarm seems to fail, leaving the predator free to feast, which could have significant consequences for fish ecology. Chris Wood was intrigued to find that how cadmium threatens a fish's reaction to the life saving alarm. The author tested how doses of cadmium at similar concentrations to the levels fish might encounter in polluted waters affected the trout's responses. After returning the fish to freshwater they released the alarm, and after seven days of exposure to 2 µg cadmium per liter of water the trout stopped reacting to the warning, and became easy pickings for a passing predator. Scott *et al.*, 2003, knew that once cadmium got into the fish's system, it could not pass through the blood brain-barrier, either remaining in the fish's olfactory system or in its body. So if fish that had been fed cadmium lost their ability to flee, the metal was disabling their bodies' hormonal flight response. However, if fish that had been exposed to cadmium in the water lost the alarm response, the metal must have destroyed their sense of smell. Scott exposed fish to equal doses of cadmium, either in their diet or from the water, and released the skin extract into the water. Again the cadmium-bathed trout lost their ability to flee for safety. The metal had destroyed the trout's sense of smell. Scott *et al.* (2003) speculated that the cadmium disrupts the fish's reaction to the alarm signal by interfering with the calcium ion channels in the olfactory nerve.

Animal behavior is a neurotropically regulated phenomenon, mediated by neurotransmitters like acetylcholinesterase. Inhibition of acetylcholinesterase activity is a typical characteristic of many metal toxicants especially organophosphate compounds (Siang *et al.*, 2007). Acetyl cholinesterase (AChE) inhibition, altered brain neurotransmitter levels and sensory deprivation are most commonly observed links with behavioral disruption. AChE is a crucial enzyme in the nervous system of both vertebrates and invertebrates, where it is responsible for the degradation of the neurotransmitter acetylcholine in the synaptic cleft, thus, is important for the action of various neurons in signal conduction. The arresting of acetylcholine breakdown due to inhibition of acetyl cholinesterase (AChE) forms the basis of action of various toxicants. AChE activity in general is inhibited by xenobiotics like organophosphate and carbamate pesticides (Heath 1995; Key and Fulton 2002; van der Oost *et al.*, 2003; Varó *et al.*, 2003). AChE inhibition due to heavy metal exposures has also been reported to be an important attribute in affecting fish communication and behavior. The stressful breathing behavior exhibited by fish and other such abnormalities in swimming movement symptoms could also be due to the inhibition of acetylcholinesterase (AChE) activity, leading to accumulation of acetylcholine in cholinergic synapses, ensuring hyperstimulation (Kumari *et al.*, 2010). Erratic movements and the abnormal swimming movements are also said to be triggered by deficiency in nervous and muscular coordination which may be due to accumulation of acetylcholine in synaptic and neuromuscular junctions (Rao *et al.*, 2005). The results of survival tests of many other studies also indicated that metals in mixture are usually more toxic than single metal treatment and their action is synergistic (Lewis, 1978; Khangrot *et al.*, 1981; Roy and Campbell, 1995); Verma *et al.*, 1982 however, showed that some metal mixtures may behave either synergistically or antagonistically.

Hyperexcitabilty as a result of exposure to cyanide at lethal dose level may be probably due to hinderance in the functioning of the enzyme AChE in relation to nervous system *i.e.* accumulation of the enzyme is likely to cause prolonged excitatory post synaptic potential (Prashanth, 2003). This may first lead to stimulation and later cause a block in cholinergic system. Metals role is said to vary depending on the fish species, metal type and ambient conditions (Frasco *et al.*, 2005). Nemcsók *et al.* (1984) in *C. carpio* L., found no effects of Zinc chloride on AChE activity in any of the organs studied *in vivo* or *in vitro*. In contrast, paraquat competitively and copper sulphate in a mixed way, inhibited acetylcholinesterase activity. Shaw and Panigrahi (1990) reported changes in brain AChE activity in fish collected from an estuary contaminated with Hg and other metals due to inflow of the effluents of the chlor-alkali industries. Gill *et al.* (1990) in rosy barb (*Puntius conchonius*), exposed to 181 µg/l mercuric chloride for 48 h showed the AChE activity considerably lowered in their brain, gills and liver. A significant suppression in AChE activity was recorded in all the organs from both mercury and zinc intoxicated fish at all the exposure periods. Suresh *et al.* (1992) studied the effects of sub lethal concentrations of mercury (0.1 mg/l) and zinc (6 mg/l) on acetylcholinesterase activity and acetylcholine content of gill, kidney, intestine, brain, liver and muscle of the freshwater fish, *Cyprinus carpio* at 1, 15 and 30 days of exposure. Najimi *et al.*, 1997 observed Cd and Zn yielding inhibition in AChE activity. On the contrary, Heath, 1995 reported AChE stimulation found during the 24 h exposure of a fish to cadmium.

Patil and Hande (2004) conducted a study on toxic effect of zinc chloride on brain AChE activity of a marine teleost, *Arius nenga.* It was observed the that zinc exerted the inhibitory effect on cytoplasmic and membrane bound fractions of AChE. Recent evidences also suggest AChE activity inhibition by even trace metals like copper (Frasco *et al.*, 2005).They also suggested that the metal ions mercury, cadmium, copper and zinc inhibit AChE activity, which are linked with the interference of metal ions with thiol groups of thiocholine, a product of the hydrolysis of the substrate acethylcholine, and with certain buffers such as phosphate buffer (Frasco *et al.*, 2005). Another valid explanation could be that metal ion causes conformational changes that result in loss of catalytic activity (Del Ramo *et al.*, 1993). Song *et al.* (2006) have described AChE deactivation in carp brain as a result of oxidative damage caused by hexachlorobenzene exposure. Stress responses, though are basically adjustments to the physiology and behavior of fish that promotes the best chance of survival when the organism is faced with toxic substances or any other kind of threatening situation, under stress elevation in the adrenaline level also tends to occur and it is believed that it further reduces the AChE enzyme inhibition, thus, leading to faster nerve impulses under stress.

REFERENCES

Abbasi S.A. and Soni R., 1984. Toxicity of lower than permissible levels of chromium (VI) to freshwater teleost *Nuris denricus*. Environ. Pollut. Ser., A, 36: 75-82.

Abdul Farah M., Ateeq B., Niamar Ali M., Sabir R. and Ahmad, W., 2004. Studies on lethal concentrations and toxicity stress of some xenobiotics on aquatic organisms. Chemosphere, 55: 257-265.

Abhas H.H., Zaghloul K.H. and Mousa M.A., 2003. Effect of some heavy metal pollutants on some biochemical and histopathological changes in blue Tilapia, *Oreochromes aureus*. Egyptian J. Agric. Res., 80 (3): 1395-1410.

Adewoye S.O., 2010. A comparative study on the behavioral responses of *Clarias gariepinus* on exposure to soap and detergent effluents. Adv. Appl. Sci. Res., 1 (1): 89-95.

Al-Aker A.S. and Shamsi M.J.K., 2000. Comparitive study of toxicity of carbaryl and its impact on the behaviour and carbohydrate metabolism of cichlid fish, *Oreochromis niloticus* (Linnaeus,1758) and catfish *Clarias gariepinus* (Burchell, 1822) from Saudi Arabia. Egypt. J. Aquat. Biol. and Fish, 4(2): 211-227.

Ali A., Al-Ogaily S.M., Al-Asgah N.A. and Gropp J., 2003. Effect of sublethal concentrations of copper on the growth performance of *Oreochromis niloticus.* J. Appl. Ichthyol., 19(4): 183–188.

Allan J.D. and Flecker A.S., 1993. Biodiversity conservation in running waters. BioScience, 43 (1): 32–43.

Al-Kahem H.F., 1995. Behavioral responses and changes in some haematological of the adult cichlid fish, *Oreochromis niloticus* exposed to trivalent chromium. JKAU Sci., 7: 5-13.

Anandhan R. and Hemalatha S., 2009. Acute toxicity of aluminium to zebra fish, *Brachydanio rerio* (Ham.). Internet J. Veterin. Med., 7: 1-5.

Annune P.A., Lyaniwura T.T., Ebele S.O. and Olademeji A.A., 1994. Effects of sublethal concentrations of zinc on haematological parameters of water fishes, *Clarias gariepinus* (*Burchell*) and *Oreochromis niloticus* (Trewawas). J. Aquat. Sci. 9: 1-6

Atchison G.J., Henry M.G. and Sandheinrich M.B. 1987. Effects of metals on fish behavior: A review. Environ Biol Fishes 18:11–25.

Baggerman B., 1966. On the endocrine control of reproductive behavior in the male three-spinned stickleback (*Gasterosteus aculeatus* L.). Symp. Soc. Exper. Biol., 20: 427-456.

Baker C.F. and Montgomery J.C., 2001. Sensory deficits induced by cadmium in banded kokopu, Galaxias fasiatus juveniles. Environ. Biol. Fish, 62; 455-464.

Baldwin D.H., Sandahl J.F., Labenia J.S. and Scholz N.l., 2003. Sublethal effects of copper on coho salmon: impacts on nonoverlapping receptor pathways in the peripheral olfactory nervous system. Environ. Toxicol. Chem., 22: 2266-2274.

Banchez Brian C., Ochoa-Acuna H., Porterfield D. Mashall and Sepulveda Maria S., 2008. Oxygen flux as an indicator of physiological stress in Fathead Minnow (*Pimephales promelas*) embryos: a real-time biomonitoring system of water quality. Environ. Sci. Toxicol., 42(18):7010-7

Beitinger T.L. and Freeman L., 1983. Behavioral avoidance and selection responses of fishes to chemicals. Residue Rev., 90:35–55.

Beitinger T.L., 1990. Behavioral reactions for the assessment of stressing fishes. J. Great Lakes Res., 16: 495.

Berejikian B.A., Smith R.J.F., Tezak E.P., Schroder S.L. and Knudsen C.M., 1999. Chemical alarm signals and complex hatchery rearing habitats affect antipredator behavior and survival of Chinook salmon (*Oncorhynchus tshawytscha*) juveniles. Can. J. Fisher. Aquat. Sci., 56: 830-838.

Boeck G., Smeth De. and Blust R., 1995. The effect of sublethal levels of copper on oxygen consumption and ammonia excretion in the common carp, *Cyprinus carpio.* Aquat. toxicol, 32: 127-141.

Borg B., 1994. Androgens in teleost fishes. Comp. Biochem. Physiol. C., 109: 219-245.

Bortone S.A., Davis W.P. and Bundrik C.M., 1989. Morphological and behavioral characters in mosquitofish as potential bioindication of exposure to Kraft mill effluents. Bull. Environ. Contam. Toxiol., 43; 370-377.

Brown G.E., 2003. Learning about danger: chemical alarm cues and local risk assessment in prey fishes. Fish and Fisher., 4: 227-234.

Brown G. E. and Smith R. J. F., 1997. Conspecific skin extracts elicit anti-predator responses in juvenile rainbow trout. Can. J. Zool., 75: 1916-1922.

Cairns J. and van dr Shalie W.H., 1980. Biological monitoring part I: early warning systems. Wat. Res., 14: 1179-1196.

Carlson R.W., Bradbury S.P., Drummond R.A., and Hammermeister D.E., 1998. Neurological effects on startle response and escape from predation by medaka exposed to organic chemicals. *Aquat. Toxicol.*, 43(1): 51–68.

Carlson R.W. and Drummond R.A., 1978. Fish cough response–A method for evaluating quality of treated complex effluents. Water Res., 12: 1-16.

Chaudhary H.S. and Nath K., 1985. Nickel induced hyperglycemia in the freshwater fish, *Colisa fasciatus*. Water, Air and Soil Pollu., 24: 173-176.

Chen L.H. and Yang J.L., 2007. Acute toxicity of antimony chloride and its effect on oxygen consumption of common carp (*Cyprinus carpio*). Bull. Environ. Contam. Toxicol., 78(6): 459-462.

Chen Xi Xiang, Chang Yi Lu and Yong Ye, 2009. Effects of Cd and Zn on oxygen consumption and ammonia excretion in sipuncula (*Phascolosoma esculenta*). Ecotox. Environ. Safe., 72(2): 507-515.

Chivers D. P., Mirza R. S., Bryer P. J. and Kiesecker J. M., 2001. Threat-sensitive predator avoidance by slimy sculpins: understanding the importance of visual versus chemical information. Can. J. Zool. 79,867 -873

Cooke S.J., Chandroo K.P., Beddow T.A., Moccia R.D. and McKinley, R.S., 2000. Swimming activity and energetic expenditure of captive rainbow trout *Oncorhynchus mykiss* (Walbaum) estimated by electromyogram telemetry. Aquac. Res., 31: 495-505.

David M., 1995. Effects of fenvelerate on behavioral physiological and biochemical aspects of freshwater fish, *Labeo rohita* (Ham.). PhD. Thesis, S.K. Univ., Anantpur, AP, India.

Del Ramo J., Torreblanca A., Díaz-Mayans J., 1993. Toxicidad de los metales. In: Mas A., Azcue J. (Eds.). Metales en sistemas biológicos. Promociones y Publicaciones Universitarias, S.A., Barcelona, pp. 143-162.

Diamond J.M., Parson M.J. and Gruber D., 1990. Rapid detection of sub-lethal toxicity using fish ventilator behaviour. Environmental Toxicology and Chemistry., 9: 3–11.

Dimitrova M.S.T., Tsinova V. and Velcheva V., 1994. Combined effects of zinc and lead on the hepatic superoxide dismutase catalase system in carp (*Cyprinus carpio*). Compt. Biochem. Physiol. Part C, 108: 43-46.

Dutta H.M., Datta M.J.S., Roy P.K., Singh N.K. and Richmonds C.R., 1992. Variation in toxicity of malathion to air and water breathing teleosts. Bull. Environ. Contam. Toxiol., 49(2): 279-284.

Eaton J.G., 1974. Chronic toxicity to the blue-gill Lepomid macrochirus Raf. Trans. Ame. Soc., 103: 729-735.

Eisler R., 1979. Behavioral responses of marine poikilotherms to pollutants. Phil. Trans. R. Soc. (B), 286: 507-521.

Ezemonye L. and Ogbomida T.E., 2010. Histopathological effects of Gammalin 20 on African catfish (*Clarias gariepinus*). Appl. and Environ. Soil Sci., Vol. 2010, Article ID 138019: 8 pages.

Frasco M.F., Fournier D., Carvalho F., Guilhermino L., 2005. Do metals inhibit acetylcholinesterase (AChE)? Implementation of assay conditions for the use of AChE activity as a biomarker of metal toxicity. Biomarkers, 10: 360-375.

Ganeshwade R.M., Bokade P.B. and Sonwane S.R., 2006. Behavioral responses of *Cyprinus carpio* Carbofuran induced impairment in the hypothalamo exposed to industrial effluents. J. Environ. Biol., 27(1): 159-160.

Gbem T.T., Balogun J.K., Lawal F.A. and Annune P.A., 2001. Trace metal accumulation in *Clarias gariepinus* (Teugels) exposed to sublethal levels of tannery effluent. Sci. of Total Environ., 271(1–3): 1–9.

Gendusa T.C. and Beitinge T.L., 1992. Bull. Environ. Contam. Toxicol., 48 (2): 237-242.

Gerhardt A., 1998. Whole effluent toxicity with *Oncorhynchus mykiss* (Walb.): survival and behavioral responses to a dilution series of a mining effluent in South Africa. Arch. Environ. Contam. Toxicol., 35: 309-316.

Ghosh S. and Mukhopadhyay M.K. 2000. Toxicity of five industrial metals on Gangetic cat fish, *Rita rita*. Geobios., 27: 93-96.

Giattina J.D., Garton R.R. and Stevens D.G., 1982. Avoidance of copper and nickel by rainbow trout as monitored by a computer-based data acquisition system. Trans. Am. Fish Soc., 111: 491–504.

Gill T.S., Panday J. and Tewari H., 1990. Enzymes modulation by sublethal concentrations of aldicarb, phosphomidon and endosulfan in fish tissues. Pestic. Biochem. Physiol., 38: 231-244.

Gill T.S., Pandey J. and Tiwari H., 1991. Individual and combined toxicity of common pesticides on teleost, *Puntius conchonius* Hamilton. Indian J. Exp. Biol., 29: 145-148.

Gilbert P.W., 1977. Two decades of shark research; a review. BioSci., 27: 670-673.

Gupta A. Kumar and Chakrabarti Padmanabha, 1993. Toxicity of zinc to fresh-water teleosts, *Notopterus notopterus* (Pabas) and *Puntius javanicus* (Blkr). J, Freshwater Bio., 5(4): 359-363.

Hansen J.A., Marr J.C.A., Lipton J., Cacela D. and Bergman H.L., 1999. Differences in neurobehavioral responses of Chinook salmon (*Oncorhynchus tshawytscha*) and rainbow trout (*Oncorhynchus mykiss*) exposed to copper and cobalt; behavioral avoidance. Environ. Toxiol. Chem., 18: 1972-1978.

Hansen J.A., Woodward D.F., Little E.E., DeLonay A.J. and Bergman H.L., 1998. Behavioral avoidance: A possible mechanism for explaining abundance and distribution of trout species in a metals impacted river. Environ. Toxicol. Chem., 18: 313–317.

Hara T.J., 2006. Feeding behavior in some teleosts is triggered by single amino acid primarily through olfaction. J. Fish Biol., 68: 810-825.

Hara T.J. and Law Y.M.C. and Macdonald S., 1976. Effects of mercury and copper on the olfactory response in rainbow trout, *Salmo gairdneri*. J. Fisher. Res. Bd. Can., 33: 1568-1573.

Heath G.L., 1995. Alterations in cellular enzyme activity antioxidants, adenylates, and stress proteins. In water pollution of fish physiology. CRC Lewis Pub. 225pp.

Howell W.M., Black D.A. and Bortone S.A., 1980. Abnormal expression of secondary sex haracters in a population of moquitofish, *Gambusia affinis* Hollbrooki: Evidence for environmentally induced masculinization. Copeia: 676-681.

Hughes G.M., Perry S.F. and Brown V.M., 1979. A morphometric study of effects nickel, chromium and cadmium on the secondary lamellae of rainbow trout gills. Water Res., 13: 665.

Hülya D., Sevgiler Y. and Üner N., 2006. Tissue-specific antioxidative and neurotoxic responses to diazinon in *Oreochromis niloticus*. Pestic. Biochem. Phys., 84: 215-226.

James R., Sampath K., Pattu V.J. and Devakiamma G., 1992. Utilization of *Eichhornia crassipes* for the reduction of mercury toxicity on food transformation in *Heteropneustes fossilis*. J. Aquacult. Trop. 7 : 189-196.

Jenkins F., Smith J., Rajana B., Shameem U., Sandhya V. and Madhavi R., 2003. Effect of sub-lethal concentrations of endosulfan on hematological and serum biochemical parameters in the carp, *Cyprinus carpio*. Bull. Environ. Contam. Toxicol., 70: 993-997.

Jezierska B. and Sarnowski P., 2002. The effect of mercury, copper and cadmium during single and combined exposure on oxygen consumption of *Oncorhynchus mykiss Wal.* and *Cyprinus carpio* L. larvae. Arch. Pol. Fish, 10(1): 15-22.

Jia X.Y., 2002. Effect of four kinds of heavy metal on respiration intensity of juvenile *Misguxnus anguillieandatus*. J. Agricul. Life Sci, 27 (5): 556-558.

Jones Jackie C. and Renolds John D., 1997. Effects of pollution on reproductive behavior of fishes. Rev. Fish boil. and Fisherey, 7: 463-491.

Kane A.S., Salierno J.D., Gipson G.T., Molteno T.C.A. and Hunter C.A., 2004. Video based movement analysis system to quantify behavioural stress response of fish. Water Res., 38(18): 3993-4001.

Kannupandi T., Pasupathi K. and Soundarapandian P., 2001. Effect of endosulfan, HCH, copper and zinc on the phototactic and swimming rate of larvae of the mangrove carb, *Macropthalmus erato*. Ind. J. Fish, 48 (3): 313-322.

Kasumyan A.O., 2001. Effects of chemical pollutants on foraging behavior and sensitivity of fish to food stimuli. J. Ichthy. 41: 76-87.

Key P.B., Fulton M.H., 2002. Characterization of cholinesterase activity in tissues of the grass shrimp (*Palaemonetes pugio*). Pestic. Biochem. Phys., 72: 186-192.

Khangrot B.S., Derve V.S. and Rajbansi V.K., 1981. Toxicity of interactions of Zinc-Nickel, Copper-Nickel and Zin-Nickel-Copper to a freshwater teleost, *Lebistes reticulates* (Peters). Acta. Hydrochim. Hydrobiol., 5: 495-503.

Khunyakari- Rupesh P., Tare-Vrushali and Sharma R.N., 2001. Effects of some trace heavy metals on *Poecilia reticulates* (Peters). J. Environ. Biol. 22(2): 141-144.

Klaprat D.A., Brown S.B., Hara T.J., 1988. The effect of low pH and aluminium on the olfactory organ of rainbow trout, *Salmo gairdneri*. Environ. Biol. Fishes, 22: 69-78.

Kramer B., Molenda W. and Fiedler K., 1969. Behavoral effects of the anti-androgen cyproterone acetate (Schering) in *Tilapia mossambica* and *Lepomis gibbosus*. Gen. and Comp. Endo., 13: 515.

Krotzer M.J., 1990. The effect of induced masculinization on reproductive and aggressive behavior of the female mosquitofish, *Gambusia affinis affinis*. Environ. Biol. Fishes., 29: 127-134.

Kumari Kanchan, Khare Ankur, and Dange Swati, 2014. The applicability of oxidative stress biomarkers in assessing chromium induced toxicity in the Fish *Labeo rohita*. Biomed. Res. Int., 2014: 782493.

Kumari K., Ranjan N. and Sinha R.C., 2010. Effect of endosulfan on the absorption ratio of α and β chains of hemoglobin and acetylcholinesterase activity in the fish, *Labeo rohita*. *J. V.N. Karazin Kharkiv Nat. Uni.*, 12: 83–89.

Lawal, M. O. and Samuel, O. B. 2010. Investigation of acute toxicity of pirimiphos-methyl (actellic®, 25 per cent EC) on guppy (*Poecilia reticulata*, Peters, 1859). Pak. J. Biol. Sci., 13: 405-408.

Leduce G. 1966. Some physiological and behavioral responses of fish to chronic poisoning by cyanides. PhD thesis, Oregon State Univ., Corvallis.

Lefebvre, K.A., V. L.Trainer and N. L. Scholz. 2004. Morphological abnormalities and sensorimotor deficits in larval fish exposed to dissolved saxitoxin. Aquatic Toxicology, 66: 159-170.

Lewis M., 1978. Acute toxicity of copper, zinc and manganese in single and mixed salt solutions to juvenile longfin dace, *Agosia chrysogaster*. J. Fish Biol., 13: 695-700.

Liley N.R. and Stacy N.E., 1983. Hormones, pheromones and reproductive behavior in fish. In: W.S. Hoar, D.J. Randall, and E.M. Donaldson (Eds). Fish Physiology-Vol. 19: Reproduction, Part B: Behavior and fertility control, Academic Press, NY: 1-63 pp.

Little E.E. 1990. Behavoral toxicology: Stimulating challenges for a growing discipline. Environ. Toxicol. Chem. 9:1.

Little E.E. and Brewer S.K., 2001. *Target organ toxicity in marine and freshwater teleost New Perspectives: Toxicology and Environment*. Vol.2. London, UK: Taylor and Francis; Neurobehavioral toxicity in fish: 139–174.

Little E.E., Fairchild J.F., and Delonay A.J., 1993. Behavioral methods for assessing impacts of contaminants on early life stage fishes. Fisheries Soc. Symp., 14: 67-76.

Marigoudar S.R., Ahmed R.N. and David M., 2009. Impact of Cypermethrin on behavioural responses in the freshwater teleost, *Labeo rohita* (Hamilton). Wld. J. Zool., 4(1): 19-23.

Martínez Mery L., Landry Christie, Boehm Ryan, Manning Steve, Cheek Ann Oliver and Rees Bernard B. 2006. Effects of long-term hypoxia on enzymes of carbohydrate metabolism in the Gulf killifish, *Fundulus grandi*. J. Exp. Biol., 209: 3851-3861.

Masud S., Singh I.J and Ram R.N., 2005. Behavioral and hematological responses of *Cyprinus carpio* exposed to mercurial chloride. J. Environ. Biol., 26: 393-397.

McGeer J.C., Szebedinszky c., McDonald D.G. and Wood C.M., 2000. Effects of chronic exposure to waterborne cu, Cd, or Zn in rainbow trout. 1: Ionoregulatory disturbance and metaboli osts. Aquat. Toxicol., 50: 231-243.

McLennan D.A., 2004. Male brook stiklebaks' (*Culaea inonstans*) response to olfactory cues. Behav. Brain Res., 141: 1411-1422.

McNicol R.E. and Scherer E., 1993. Influence of cadmium pre-exposure on the preference-avoidance responses of lake whitefish (*Coregonus clupeaformis*) to cadmium. Arch. Environ. Contam. Toxicol., 25: 36–40.

Mirza R. S. and Chivers D. P., 2001. Are chemical alarm cues conserved within salmonid fishes? J. Chem. Ecol., 27: 1641 -1655.

Mishra A.K. and Mohanty B., 2008. Acute toxicity impacts of hexavalent chromium on behavior and histopathology of gill, kidney and liver of the freshwater fish, *Channa punctatus* (Bloch). Environ. Toxicol. and Pharmacol., 26(2):136–141.

Mishra A.K. and Mohanti B., 2009. Chronic exposure to hexavalent chromium affects organ histopathology and serum cortisol profile of a *Channa punctatus* (Bloch). Sci. Total Environ., 407: 5031-5038.

Morgan F.L., Young R.C. and Wright J.R., 1988. Developing portable computer-automated biomonitoring for a regional water quality surveillance network. In: Automated biomonitoring. Eds., D.J. Gruber and J.M. Diamond. Halsted Press, London. Ch. 9.

Najimi S., Bouhaimi A., Daubeze M., Zekhnini A. and 3 others., 1997. Use of acetylcholinesterase in *Perna perna* and *Mytilus galloprovincialis* as a biomarker of pollution in Agadir Marine Bay (South of Morocco). Bull. Environ. Contam. Toxicol., 58: 901-908.

Nakayama K., Oshima Y., Yamaguchi T., Tsuruda Y., Kang I. J., Kobayashi M., Imada N. and Honjo T. 2004. Fertilization success and sexual behavior in male medaka, *Oryzias latipes*, exposed to tributyltin. Chemosphere, 55: 1331-1337.

Neil J.H., 1957. Some effect of potassium cyanide on *Salvelinus fontinalis*. Ontario Industrial waste Conf. 4: 74-76.

Nemcsók J., Németh A., Buzás Z. and Boross L., 1984. Effects of copper, zinc and paraquat on acetylcholinesterase activity in carp (*Cyprinus carpio* L). Aquat. Toxicol., 5 (1):23-31.

Nimila P.J. and Nandan B., 2010. Effect of organochlorine pesticide lindane (dakua-HCH) on behavior of *Etroplus maculatus* (Bloch, 1795). Poll Res. 29(4), 577-582.

Oliveira Rui F. and Canario Adelino V.M. 2000, Hormones and soial behavior of cichlid fishes: a case study in the Mozambique tilapia. J. Aqua. and Aquatic, Cichlid Res., State of Art, Vol. IX: 187-208.

Patil S.M. and Hande R.S., 2004. *In vitro* studies of zinc chloride on brain acetylcholinestrases of *Arius nenga*, a marine teleost. Polln. Res., 23(4): 787-790.

Phillips K., 2003.Cadmium hits trout in the snout. J. Exp. Biol., 206: 1765-1766.

Prashanth M.S., 2003. Cypermethrin induced physiological and histological changes in freshwater fish, *Cirrhinus mrigala*. PhD Thesis, Karnatka Univ., Dharwad., India.

Prashanth M.S., David M. and Mathad S.G., 2005. Behavioral changes in freshwater fish, *Cirrhinus mrigala* (Ham.) exposed to cypermethrin. Environ. Biol., 26(13): 141-144.

Prashanth M.S and Patil Yogesh B., 2006. Behavoural surveillance of Indian major carp, *Catla catla* (Ham.) exposed to free cyanide. Curr. Sci., 9(1): 313-318.

Prashanth M.S., Sayeswara H.A. and Mahesh Anand Gougar, 2011. Effect of sodium cyanide on behavour and respiratory surveillance in freshwater fish, *Labeo rohita* (Ham.). Rec. Res. Sci. Tech., 3(2): 24-30.

Radhaiah V., Girija M. and Rao K.J., 1987. Changes in selected biochemical parameters in the kidney and blood of the fish, *Tilapia mossambica,* exposed to heptachlor. Bull. Environ. Contam. and Toxicol., 39 (6): 1006–1011.

Ramesh M. and Saravanan M. (2008) Haematological and biochemical responses in a freshwater fish *Cyprinus carpio* exposed to Chlorpyrifos. Int. J. Integrative Biol., 3(1): 80-83.

Rao L.M., Patnaik Sree and Manjula R., 1997. Acute toxicity of Zn, Pb and Cd in the freshwater catfish Mystus vittatus (Bloch). Ind. J. Fishery, 44(4): 405-408.

Ravindran Ambili and Radhakrishnan M.V., 2015. Determination of acute toxicity of vanadium to the freshwater catfish *Heteropneustes fossilis* (BLOCH.) Intl. J. Toxicol. and Appl. Pharmacol., 5(2): 14-18.

Roy R. and Campbell P.G.C., 1995. Survival time modeling of exposure of juvenile Atlanti salmon (*Salmo salar*) to mixtures of aluminium and zinc in soft water at low PPh. Aquat. Toxicol., 33: 155-176.

Sandahl J.F., Baldwin D.H., Jenkins J.J. and Scholz N.L., 2005. Comparative thresholds for acetylcholinesterase inhibition and behavioral impairment in coho salmon exposed to chlorpyrifos. Environ. Toxicol. Chem., 24(1): 136-145

Santhakumar M. and Balaji M., 2000. Acute toxicity of organophosphorus insecticide monocrotophos and its effect on behavior of an air breathing fish, *Anabas testudineus* (Bloch). J. Environ. Biol., 21(2): 121-123.

Sarkar S.K., 1989. Evaluation of two heavy metals on the oxygen consumption of *Tilapia mossambica* (Peters). Geobios., 16: 108-110.

Sastry K.V. and Gupta P.K., 1979. Enzyme alterations in the digestive system of *Heteropneustes fossilis* induced by lead nitrate. Toxicol. Lett., 3 : 145-150.

Scherer E., 1992. Behavioural responses as indicators of environmental alterations: approaches, results, developments. J. Appl. Ichthyol., 8: 122-131.

Schreck C.B., Olla B.L. and Davis M.W., 1997. Behavioral responses to stress. In Iwama, G.K.; Pickering, A.D.; Sumpter, J.P.; Schreck, C.B. (Eds), Fish stress and Health in Aquaculture. Cambridge University Press, Cambridge, pp. 145-170.

Schroder J.H. and Peters K., 1988. Differential courtship activity of competing guppy males (*Poecilla reticulate* Peters: Pisces: Poeiliidae) as an indicator for low concentrations of aquatic pollutants. Bull. Environ. Contam. Toxicol., 40: 396-404.

Scott G.R. and Sloman, K.A., 2004. The effects of environmental pollutants on complex fish behavior: integrating behavioral and physiological indicators of toxicity. Aquatic Toxicol., 68: 369-392.

Scott G.R., Sloman K.A., Rouleau C. and Wood C.M., 2003. Cadmium disrupts behavoural and physiological responses to alarm substances in juvenile rainbow trout (*Oncorhynchus mykiss*). J. Exp. Biol., 11: 1779-90.

Shah S.L., 2002. Behavioral abnormalities of *Cyprinion watsoni* on exposure to copper and zinc. Turk. J. Zool., 26: 137-140.

Shahi J. and Singh A., 2010. A comparative study on the piscicidal activity of synthetic pesticides and plant origin pesticides, to fish, *Channa punctatus*. Wld. J. Zool., 5 (I): 20-24.

Shaw B.P. and Panigrahi A.K., 1990. Brain AchE activity studies in fish species collected from a mercury contaminated estuary. Water, Air and Soil pollution, 53: 327.

Shilov I.A., 1981. Stress as an ecological phenomenon. J. Zool., 58 (6): 805–812.

Shivakumar R. and David M., 2004. "Toxicity of endosulfan to the freshwater fish, *Cyprinus carpio*. Ind. J. Ecol., 31 (1): 27–29.

Shrager P., 1977. Slow sodium inactivation in nerve after exposure to sulhydryl blocking reagents. J. Gen. Physiol., 69(2): 183-202.

Siang H.Y, Yee L.M. and Seng C.T., 2007. Acute toxicity of organochlorine insecticide endosulfan and its effect on behavior and some haematological parameters of Asian swamp eel (*Monopterus albus*, Zuiew). Pestic. Biochem. Phys., 89: 46-53.

Siddiqui Akhter Ali and Arifa Noori, 2011. Toxicity of heavy metal copper and its effect on the behavior of freshwater Indian cat fish, *Clarias batrachus* (Linn.). Current Biotica, 4: 405-412.

Singh Ram Nayan, Pandey Rakesh Kumar, Singh Narendra Nath and Das Vijai Krishna, 2009. Acute toxicity and behavioral responses of common carp *Cyprinus carpio* (Linn.) to an organophosphate (Dimethoate). Wld. J. Zool., 4 (2): 70-75.

Sloman K.A., Scott G.R., Diao Z., Rouleau C., Wood C.M. and McDonald D.G., 2003. Cadmium affects the social behaviour of rainbow trout, *Oncorhynchs mykiss*. Aquatic Toxicology, 65: 180-185.

Song S.B., Xu Y., Zhou B.S., 2006. Effects of hexachlorobenzene on antioxidant status of liver and brain of common carp (*Cyprinus carpio*). Chemosphere, 65: 699-706.

Sprague J.B., 1964. Avoidance of copper–zinc solutions by young salmon in the laboratory. J Water Pollut Control Fed., 36:990–100Sullivan J.F., Atchison G.J., Kolar D.J. and Melntosh A.W., 1978. Changes in the predatory-prey behavior of fathead minnows (*Pimephales promelas*) and largemouth bass (Miropterus sabmoides) caused by cadmium. J. Fish Res. Bd. Can., 35: 446-451.

Spoor W.A., Neitheisel T.W. and Drummond R.A., 1971. An electrode chamber for recording respiratory and other movements of free swimming animals. Trans. Am. Fish Soc., 100: 22-28.

Spry D.J. and Wood C.M., 1984. Acid-base plasma ion and blood gas changes in rainbow trout during short term toxic zinc exposure. J. Comp. Physiol., B. 154: 149.

Suresh A., Sivaramakrishna B., Victoriamma P.C. and Radhakrishnaiah K., 1992. Comparative study on the inhibition of acetylcholinesterase activity in the freshwater fish *Cyprinus carpio* by mercury and zinc. Biochem. Int., 26(2): 367-75.

Susan A.T. and Sobha K., 2010. A study on acute toxicity, oxygen consumption and behavioural changes in the three major Carps, *Labeo rohita* (Ham), *Catla catla* (Ham) and *Cirrhinus mrigala* (Ham) exposed to fenvalerate. Bio. Res. Bull., 1: 24-28.

Susan A.T., Sobha K. and Tilak K.S., 2010. A study on acute toxicity, oxygen consumption and behavioural changes in the three major carps, *Labeo rohita* (ham), *Catla catla* (ham) and *Cirrhinus mrigala* (ham) exposed to Fenvalerate. Biores. Bull., 1: 33-40.

Sutterlin A.M. and Sutterlin N., 1971. Electrical responses of the olfactory epithelium of Atlantic salmon (*Salmo salar*). J. Fisher. Res. Bd. Can., 28: 565-572.

Thompson B.E. and Hara T.J., 1977. Chemo sensory bioassay of toxicity of lake waters contaminated with heavy metals from mining effluents. Wat. Pollut. Res. Can., 12: 179-189.

Tierney Keith B., Baldwin David H., Hara T.J., and 3 others, 2010. Olfactory toxicity in fishes. Aquat. Toxicol., 96: 2-26.

Tinbergen *Niko, 1963.* On aims and methods of ethology. *Zeitschrift für Tierpsychologie (renamed Ethology in 1986) 20: 410–433.*

van der Oost R, Beyer J, Vermeulen N.P.E., 2003. Fish bioaccumulation and biomarkers in environmental risk assessment: a review. Environ. Toxicol. Pharmacol.,13: 57–149.

Varó I., Navarro J.C., Amat F., Guilhermino L., 2003. Effect of dichlorvos on cholinesterase activity of the European sea bass (*Dicentrarchus labrax*). Pestic. Biochem. Phys., 75: 61-72.

Velmurugan, B., M. Selvanayagam, EI. Cengiz and E. Unlu, 2007. The effects of monocrotophos J. to different tissues of freshwater fish *Cirrhinus mrigala*, Bull. Environ. Contam.Toxicol., 78(6): 450-454.

Venkata Rathnamma V., Vijayakumar M. and Philip G.H., 2008. Acute toxicity and behavioural changes in freshwater fish, *Labeo rohita* exposed to Deltamethrin. J. Aqua Biol. 23(2), 165-170. 40.

Verma S.R., Jain M. and Dalela R.C., 1982. A laboratory study to assess separate and in-combination effects of zinc, chromium and nickel to the fish, *Mystus vittatus*. Acta. Hydrochim. Hydrobiol., 10: 23-29.

Verma V.K. and Saxena Amita, 2013. Investigations on the acute toxicity and behavioural alterations induced by the organophosphate pesticide, chlorpyrifos on *Puntius chola* (Hamilton-Buchanan). Indian J. Fish., 60(3) : 141-145.

Vosyliene M.Z. and Kazlauskiene N., 1999. Alterations in fish health status parameters after exposure to different stressors. Acta Zool. Lituanica Hydrobiol., 9 (2): 82–95.

Walker C.H., Hopkin S.P., Sibly R.M., Peakall D.B., 2003. Principles of Ecotoxicology. Taylor and Francis, London.

Weber D.N., 1993. Exposure to sublethal levels of waterboene lead alters reproductive behavior pattern in fathead minnows (*Pinephales promelus*). Neurotoxicol., 14: 347-358.

Weis J.S. and Weis P., 1995. Swimming performance and predator avoidance by mummichog (*Fundulus heteroclitus*) larvae after embryonic or larval exposure to methylmercury. Canadian J. Fisher. and Aquat. Sci., 52(10): 2168–2173.

Winberg S., Bjerselius R., Baatrup E., Doving K.B., 1992. The effect of Cu(II) on the electro-olfactogram (EOG) of the Atlantic salmon (*Salmo salar*) in artificial freshwater of varying inorganic carbon concentrations. Ecotoxicol. Environ. Safty, 24; 167-178.

Winffield J.C., Hegner R.E., Dufty A.M. and Ball G.F., 1990. The "Challenge hypothesis": theoretical implications for patterns of testosterone secretion, mating systems, and breeding strategies. Am. Naturl., 136: 829-846.

Wise M.L., Stiebel C.L. and Grizzle J.M., 1987. Acute toxicity of nitrofurazone to channel catfish *Ictalurus punctatus*, and Goldfish, *Carassius auratus*. Bull. Environ. Contam. Toxicol., 38 : 42-46.

Yaji A.J., Auta J., Oniye S.J., Adakole J.A. and Usman J.I., 2011. Effects of Cypermethrin on behavior and biochemical indices of freshwater fish, *Oreochromis niloticus.* EJEAF Che., 10(2): 1927-1934.

Yilmaz M., Gul A. and Karakose E., 2004. Investigation of acute toxicity and the effect of cadmium chloride ($CdCl_2.H_2O$) metal salt on behavior of guppy (*Poecilla retiulata*). Chemosphere, 56: 375-380.

Zhou T. and Weis J.S., 1998. Swimming behavior and predator avoidance in three populations of Fundulus heteroclitus larvae after embryonic and/or larval exposure to methylmercury. *Aquat. Toxicol.*, 43(2-3):131–148.

Chapter 9

Metal Toxicity and the Tissue Biochemicals

1. General Adaptive Syndrome (GAS)

Like all other animals, fishes too display stress responses to unsafe environmental conditions or any other stress including exposures to toxic chemicals. In the physiological context, an acceptable working definition of stress is a nonspecific response of the body to any demand placed upon it that is an extension of a physiological state beyond the normal resting state. A physical stress means any extreme water condition like temperature, pH, hardness, oxygen, NH_3, or any toxic chemical like a pesticide/heavy metal. Any physical, chemical or any other perceived stress can evoke a non-specific response or responses in a fish, which is/are considered adaptive to enable the fish to maintain its homeostatic state and to cope with such disturbances (Barton, 2002). In physiological studies most important measure of stress is either a rate function, such as the breathing rate or changes in the oxygen consumption, or changes in the concentration of any tissue biochemical; and the below sub lethal response threshold levels are best called as 'no effect levels'. Within the sub lethal dose level however, a variety of reversible or irreversible processes may take place for any physiological adjustment to such changed conditions. The duration of exposure to a chemical further have a considerable impact on both the qualitative and quantitative characters of any such changed variable. Cannon (1929) introduced the term "homeostasis" to describe the "coordinated physiological processes which maintain a steady state with respect to a rate function in an organism." He turned his attention to the sympathetic nervous system, as an essential homeostatic system that serves to restore stress-induced disturbed homeostasis and to promote survival of the organism. Cannon was also the first to touch on the issue of specificity of stress responses. If the stressor is overly severe or long-lasting to the point that the fish is not capable of regaining homeostasis, the responses themselves may become maladaptive and threaten fish health and its well-being. Selye (1971, 1974) was known as the father of stress physiology, and he deserves much of the credit for introducing the term "stress," and for popularizing the concept of stress in the scientific and medical literature of the 20th century. He mentioned that when mammals are subjected to any stress they exhibit a generalized group of physiological responses "the nonspecific responses" of the body to any demand made upon it. Mazeaud and Mazeaud (1981) have stated that a fish respond to various stressors by a series of biochemical and physiological stress reactions, so called secondary stress responses, comparable to those of higher vertebrates. McKim *et al.* (1987) termed the group of physiological effects in response to a chemical stimulus as 'syndrome'. These changes although exhibit no definite pattern; but reflect a physiological counteraction against the toxicity stress as a result of fight and flight reactions in body tissues, and if the stressful conditions continue the adrenal cortex is stimulated to release the increased amounts of cortisol, which sustain changes caused by the adrenaline, as well as mobilization of certain amounts of plasma proteins into plasma amino acids.

As in other vertebrates, generalized stress response in fish comprises physiological responses that are common to a wide range of environmental, physical and biological stressors as follows:

i. *Primary response*: Endocrinal changes as measurable levels of the circulating catecholamine and corticosteroid levels, as a result of any physical or chemical stressor.
ii. *Secondary response*: Wide range of changes in features related to majorly the metabolism, hydro-mineral balance, cardiovascular activities, respiratory activities and the immune function; that are caused to a large extent by the stress hormones (Vijayan *et al.*, 1994) in an affected animal, as evident by certain bio-chemicals/stress proteins (hsp), enzymes or the immunological and the histo-pathological changes. These result in physiological stability in animals under changed environmental conditions, *i.e.* a homeostatic response (Barton *et al.*, 2002).
iii. *Tertiary changes*: Representing changes in the whole animal activity/ animal population, and/or survival with adaptation/or retarded growth/ or decline in activity/or the pathological state/or even the fish death, if the magnitude or duration of a stressor overwhelms. The tertiary or 'whole animal' responses to stress in fish (Wedemeyer and McLeay 1981; Wedemeyer *et al.*, 1984), such as any change in the metabolic rate or scope for activity. These changes have been suggested as a possible method for measuring stress in fish (Brett 1958; Wedemeyer and McLeay 1981).

The ability of fish to respond to stress is a part of their adaptive capacity to adjust to perturbations in their surroundings. Most stressors induce a neuroendocrine response, characterized by a rapid release of stress hormones (Catecholamine and cortisol) into the blood circulation (Gamperl *et al.*, 1994). When stressed, a fish secretes glucocorticoids and catecholamines from the adrenal tissue and elevate their adrenal and cortical levels (Randall and Perry, 1992; Balm *et al.*, 1994), thus, affecting sugar and protein metabolism (Young and Chavin 1965). These hormonal changes are the primary responses, and the physiological responses occurred are the secondary responses. Most investigations with fish have concentrated on measuring such stress indicators or changes in individual performances (Schreck,1981). These effects are termed collectively as the 'General Adaptation Syndrome (GAS)'. According to the General Adaptation Syndrome (GAS) paradigm, when a fish perceives a stressful stimulus, there is a change in biological functions (Moberg, 1985), as an organism's response to compensate for the stress.

The main endocrine stress systems of the body are the sympathetic–adrenal–medullary (SAM) axis and the hypothalamic–pituitary–adrenal (HPA) axis, which trigger the release of the catecholamine (epinephrine and nor-epinephrine) from the adrenal medulla and glucocorticoids (*e.g.*, cortisol) from the adrenal cortex, respectively. A stressor induces neuroendocrine/endocrine response by stimulating the 'Hypothalamic-pituitary-interneural axes. Most potently the release of adreno-cortico trophic hormone (ACTH) from the pituitary gland leads to a rapid release of the stress hormones: Catecholamine (CAT), and Cortisol. In teleosts, CATS are released from the chromaffin tissue and cortisol from the inter-renal tissues located in the head kidney', and also from the endings of adrenergic nerve (Randall and Perry, 1992). Cortisol is released from the inter-renal tissue, located in the head kidney in response to several pituitary hormones, but most potently to the adreno-corticotrophic hormone (ACTH; Balm *et al.*, 1994). ACTH may also stimulate the release of the catecholamine epinephrine, and chronically increased levels of cortisol may affect catecholamine storage and release in trout (Reid *et al.*, 1996). As both the chromaffin and the inter-renal tissues in fish lie in close proximity, a paracrine effect for these stress hormones may exist (Reid *et al.*, 1996). Paracrine signaling is a form of cell-to-cell communication in which a cell produces a signal to induce changes in nearby cells, altering the behavior of those cells. The endocrine (primary) and metabolic (secondary) responses of fish to stress though considered adaptive, but there is a maladaptive component to these responses, that may ultimately affect fish's survival and its general well-being. Fish respond to a stressor by eliciting a generalized physiological stress response, characterized by an increase in stress hormones and consequent changes that help maintain the animal's normal or homeostatic state (Iwama *et al.*, 1999; Barton, 2002). In some instances, the endocrine responses are directly responsible for such secondary responses, resulting in changes in concentrations of the blood constituents including metabolites and the major ions. This generalized response has been considered to be adaptive and represents the natural capacity of the fish to respond to a stress, such as an increase in plasma cortisol, catecholamine, glucose level, increase in branchial blood flow and or an increase in any muscular activity (Barton and Iwama, 1991).

Tertiary responses are the whole-animal changes in performance, such as the behavior, body growth and the body resistance, as a result of the primary and secondary responses. Such effects may range from any biochemical, physiological or behavioral responses to various physical, chemical and biological factors of the environment (Yousef, 1985), particularly those that extend their adaptive responses beyond normal range; such that chances of survival are significantly reduced. The release of stress-related hormones, support the adaptation process of the organism to changes and can protect the body in the short run. In contrast the bodily responses to stress can cause damage in the long run and can finally promote development of several stress-related diseases. The biological 'costs' of short-term adaptation to stress are described as an allostatic load, as defined by McEwen and Stellar (1993). Physiological effects, however, depend upon the nature of the receptors present in a target tissue. Based upon many such studies with salmonid and other fishes on elevations in plasma cortisol and the metabolic responses, certain generalizations can be made about stress responses depending upon:

i. Environmental factors
ii. Genetic and ontogenetic factors
iii. Fish habituate
iv. Metabolic cost
v. Intensity of the stressor
vi. Duration of the stressor
vii. Cumulative stress effects.

2. General Metabolic Effects

Biochemical, physiological and histological changes occur in fish exposed to pollutants (Sprague, 1971; Warner, 1987). Most important aspect of animal survival is maintaining an energy level required for a metabolically active function. Energy homeostasis on account of toxicity effects is thus, one major cause of chemical imbalances in tissue composition and the enzymatic functions, through secretion of adrenal-cortical hormones from the adrenal cortex. To counteract energy imbalance in tissues on their exposure to a pollutant, the living organisms need energy to metabolize the xenobiotic and its excretion through renal attributes. Sub lethal toxicity effects are majorly biochemical in origin as the most toxicants exert their effects at the basic level in a stressed organism by reacting with enzymes or metabolites and other functional components of the cell. Such effects might lead to irreversible and detrimental disturbances of integrated functions such as behavior, growth, reproduction and survival (EIFAC, 1975; Waldichuk, 1979). Toxic effect of a chemical further depends on the fish species and nature and concentration of a toxicant (Kuzumina *et al.*, 2004). It follows intuitively that as a net result of severe or prolonged stress that a fish is less capable of functioning optimally in their environment, may have reduced chances of survival. Yehuda (2007), however, in his study of the neuro-endocrinology of post-traumatic stress disorder (PTSD), has presented an interesting paradox. Initial observations of low cortisol levels in a disorder precipitated by extreme stress directly contradict the emerging and popular formulation of hormonal responses to stress- the glucocorticoid cascade hypothesis, evidencing that stress-related psycho-pathology involves hyper-cortisolism, as a cause or as a consequence of the disorder. It is now clear that insufficient glucocorticoid signaling may produce detrimental consequences as robust as those associated with the glucocorticoid toxicity.

Biological methods in risk assessment in an environment are primarily based on various qualitative and quantitative observations in the living organisms in their environment (Smolders *et al.*, 2003). The use of a living fish as a test animal compliments interpretation of both physical and chemical characteristics of the water. A toxic chemical usually does not affect a single organ or a function; rather it may tend to act mostly on certain organs or functions together due to setting up in various physiological functions or the compensatory changes, as in homeostasis. Each biological level in fact, offers unique problems and insights. Several heavy metals have been reported to stimulate inter-renal activity and plasma corticosteroid and glucose levels in fish (Pratap and Wendelaar-Bonga, 1990). Endocrine responses to stress in fish and subsequent secondary responses often vary along with it does not mean that there is necessarily a direct cause-effect relationship (Leatherland, 1985). Several studies have attempted to establish a relationship between the physiological and cellular stress responses, but there exist apparent inconsistency between these two levels of response. According to Selye (1973), following an exposure to a chemical there is a rapid elevation in the level of adrenaline and nor-adrenaline hormones, which mobilizes muscle glycogen level into blood sugar, resulting in further rise in blood sugar. Since, energy reserves are adaptive responses and these are mobilized and used to cope with the stress, an increase occurs in corticosteroid and catecholamine levels, followed by an elevation in plasma glucose and in metabolic rate.

3. Tissue Biochemicals

There have been many reports on the chronic toxicological effects of metals on biochemical composition to fishes. Umminger (1970) stressed that carbohydrates represent the principal and immediate energy precursors for fish exposed to stress conditions, while proteins are spared during chronic period of pollutant stress. Many other workers (Herpert *et al.*, 1977; Tytler and Calow, 1985) also stated that in addition to carbohydrates, lipids also represent the principal and immediate energy precursors for fishes under stress. Xenobiotic induced hypersecretion of adrenalin and cortisol triggers biochemical and physiological alterations as the metabolic effects, which includes glycogenesis, proteolysis and lypolysis resulting in hyperglycemia, depletion of tissue glycogen reserves, catabolism of muscle proteins and altered blood levels of proteins, cholesterol and free fatty acids (Thomas, 1990; Jobling, 1994; Wendelaar-Bonga, 1997). It may further result in hyperglycemia, and sometimes hypoglycemia too, if energy consumption increases the metabolic limits, leading to various physiological and pathological symptoms. A physiological response in a fish can thus, be measured by monitoring changes in any one or more of such functional components affected by any toxicant. As a result of compensation for enhanced energy requirements of the body to fight and survive under toxic conditions and to obtain physiological homeostasis in terms of maintaining optimum level of blood metabolites to counteract excessive body activities under metal or any other chemical toxicity stress, a fish gain energy by proteolysis and glycogen breakdown (glycol-genolysis), resulting in reduction in tissue carbohydrates, proteins and total tissue glycogen. This decrease might be an accommodation to counteract fight and survive under toxic conditions. Cortisol antagonizes insulin; maintain sugar level in blood by reducing intake and enhancing glycol-neogenesis; reducing protein synthesis and glucogenesis. Whereas, insulin enhances uptake of sugars in cells, enhance protein synthesis, growth stimulation and hypoglycemia. An increase in plasma glucose further indicates mobilization of all energy reserves such as tissue glycogen through glycogenolysis and it may also reflect the increased rate of metabolic activity (Umminger, 1977; Love, 1980). Moberg (1985) stated that corticosteroids are released during stress presumably to induce gluconeogenesis, to increase glucose availability for metabolism. But increased gluconeogenesis is at the expense of lipid and protein metabolism, thus, affecting fish growth. Determination of serum biochemicals, including proteins in fish further contributes to an appraisal of the fish health. Furthermore, increased corticosteroids over the time detrimentally affect immune responses and reproductive processes.

Liver and kidneys are the principle metabolic organs affected under stress. Liver being the largest store of carbohydrates in a fish serves as the primary source for blood glucose in the post food absorptive phase of an animal. The liver is also involved in detoxification of xenobiotics. In liver, there also occurs varied type of histological aberrations along with degeneration of the fatty tissue and loss of glycogen on exposure of a fish to a chemical toxicant. Toxic chemicals thus, are known to increase the hepatic dysfunctions. Alterations in carbohydrate metabolism, since is a common feature in fish exposed to toxicants, the potential effects of the xenobiotics on the liver are numerous which are often seen indirectly. The type of liver injury is often dependent upon not only on a particular toxic agent and its mechanism of action but also on the length of exposure (Jacobson-Kram and Keller, 2001). These changes are also related with depressed feeding. Further alterations in biochemical and physiological processes in a living organism are quantified by standard assessment methods. Heavy metals adversely affect various tissue bio-chemicals. Toxicity of these metals depends upon their chemical form and interaction within metals. Following are the metal toxicity effects in terms of hyper- and hypoglycemic, proteolytic and lypolytic effects *etc.* in fishes:

i. Hyper- and Hypoglycemic Fish Responses

Changes in liver glycogen concentration are almost always seen in fish exposed to a variety of chemical and physical stressors. Silbergeld (1974) mentioned blood glucose as one among most reliable and sensitive indicators of environmental stress in fish. This change might stem from the metabolic differences between the fish species and the environmental heavy metal concentrations and the duration, to which the fishes are exposed. Metals in general, produce hyperglycemia that is increased blood glucose levels in fish (Shaffi, 1979) and other animals (Chavin and Young, 1970; Clary, 1975; Mathur and Tandon, 1979). Metal toxicity in general is reported to result in hypoglycemia and hypolactemia in the affected fish. Diwan *et al.* (1979) studied the levels of blood glucose and the tissue glycogen levels in two fish *viz., H. fossilis* and *C. batrachus,* exposed to the industrial effluents. having high load of the heavy metals like Hg, Zn, Cd, Co, Cr, Pb, Cu and Fe. There was 92.2 and 95.3 per cent increase in the blood glucose level in *H. fossilis* and *C. batrachus,* respectively. In *C. batrachus* a significant decrease in glycogen level was found in liver, muscles and the gills, whereas, there was a slight increase in heart and kidney. Glycogen depletion was highest with Hg (96 per cent), followed by Ni (86 per cent) and then the Cr (75 per cent). Following acute and chronic effects of Cd in the flounder, *Pleuronectes fleusus* (Larsson, 1975), in *C. punctatus* (Bhattacharya *et al.*, 1987; Sastry and Shukla, 1990) and *H. fossilis* (Sastry and Subhadra, 1982), the studies showed a significant decrease in glycogen, lactic acid, pyruvic acid, and total liver and muscle proteins in both acute and chronic exposures. After 60 days of exposure, blood glucose and muscle glycogen levels remained below normal level, but lactic acid levels in the blood, liver and muscles and glycogen content of the liver increased at this stage, indicating drastic effects of the treatment duration. Wedemeyer *et al.* (1984) have stated that an elevation in the circulating levels of glucose (hyperglycemia), following stressful disturbances is a major metabolic response to stress and are well documented for the fishes. Presumably, the stress-induced increase in blood glucose is an adaptive response to provide an energy source for the fish during stressful condition. Several reports showed that the levels of blood glucose were significantly increased in the fishes treated with nickel, lead, copper, cadmium and mercury (Nath and Kumar, 1988; James *et al.*, 1992; Jha and Jha, 1994; Van Vuren, 1994; Al-Attar, 2005; 2006). Chaudhary (1984) stated that the muscle glycogenolysis accompanied by blood hyperlactatemia is probably the result of some direct action of nickel on the fish, *Colisa fasciatus* and the severe stress conditions caused by nickel induced pathological changes in the gills and blood may also be responsible for these metabolic changes. Hyperglycemia is accompanied with simultaneous fall in liver/muscle glycogen levels, as evidenced by Srivastava and Singh (1981) and Chaudhary and Nath (1984), as a consequence of catecholamine secretion from adrenals. A prolonged environmental stress in the fish makes the adaptation difficult and creates a weakness in the fish. This weakness is characterized by a decrease in the liver glycogen and the serum cortical levels, which subsequently create alterations in the metabolism and also shortens the lifespan of the organisms (Heath, 1995). Kumari and Kumar (1997) studied the effects of pollutants on the distribution of carbohydrates in heart, muscle, kidney, liver, brain, gills and ovary of the fish, *Channa punctatus*. These studies revealed that there was a significant decrease in the carbohydrate content suggesting the possibility of increased glycogenolysis in fish under pollution stress as compared to the control fish. A significant rise in the level of blood glucose in the experimental animal, *O. niloticus* on account of exposure to nickel was also reported by Al-Attar (2007).

A dose dependent decrease in the liver glycogen of Atlantic salmon, weighing 20.3±0.7g was observed by Soengas *et al.* (1996), following their exposure to 0.01 and 0.1mg/l Cd. A significant decrease in glycogen level in *H. fossilis* was observed only in liver and the gills. Das and Bhattacharya (2000) while studying the effect of cadmium and mercury on the blood and liver of *C. punctatus* also observed depletion in the hepatic glycogen contents, following a treatment for 15 days. Lowering of hepatic glycogen reflects reduced glycogenesis or increased glycogenolysis. Elevated serum glucose levels were reported by Benson *et al.* (1987), Partap and Wendelaar-Bonga (1990), Hontela *et al*, (1996) in fish treated with cadmium. Cattani *et al.* (1996) exposed Sea bass, *Dicentrarchus labrax* to sub lethal concentrations of 0.5 and 5 µg/l of Cd. A significant decrease in the liver glycogen and a simultaneous increase in the serum glucose were observed during the first 4 days. The fish recovered during the next 4 days also following a lower dose treatment. There was, however, a continuous decrease in the liver glycogen and a continuous increase in the serum glucose, following the higher dose treatment. Fishes exposed to sub lethal concentration of cadmium and zinc (1.12 and 4.0 ppm), respectively showed decrease in the rate of transport of glucose and fructose, which was more marked after 30 days as compared to 15 days in the two fishes. Dinodia (2001) exposed 3 freshwater fish species: *Labeo rohita, Cirrhinus mrigala* and *Cyprinus carpio* to Cd. He found decline in the carbohydrate content up to 19.2, 42.46 and 38.3 per cent, respectively in *L. rohita, C. mrigala* and *C. carpio;* whereas, decline in liver glycogen was from 11.2 to 45.2 per cent. Similar decline in liver glycogen was observed in *Garra mullya* by Sinha *et al.* (2001) after exposure to cadmium for 4 months and 45 days, respectively. A concentration dependant inverse relationship between cadmium concentration and food utilization parameters like food consumption, assimilation metabolism, assimilation efficiencies and production efficiencies were observed by Vincent *et al.* (2002). In *C. carpio* (Cicik and Engin, 2005), the glycogen levels in the liver and muscle tissues of the fish exposed to predetermined concentrations of Cd (0.05, 0.1.0.5,1 mg/l) were significantly lowered by 23.72 per cent and 29.05 per cent, respectively as compared to the control fish. The serum glucose levels of the fish were increased following all tested concentrations of Cd. Vutukuru (2005) reported that there is an appreciable decline in different biochemical constituents in various tissues in freshwater fish *L. rohita* under chromium stress. Emad *et al.*, 2005 showed high toxicity of copper to cadmium in marine fish, *Mugil seheli;* they estimated increased levels of glucose, reaching the highest values after 4 days of the treatment with Cd and Cu, while muscle glycogen was increased at first few days then dropped after 14 days even below the control level. Glucose recorded high values after exposure to Cd rather than Cu. Muscle glycogen were affected by Cu more than Cd. The Common carp, *C. carpio* exposed to sublethal concentrations of chromium (Parvathi *et al.*, 2011) for various exposure periods (8, 16, 24 and 32days) also showed serum glucose significantly elevated in the experimental fish over the control and the serum total protein was decreased significantly ($P<0.05$) in experimental fish.

Decrease in the glycogen reserves under heavy metal toxicity in the muscle and the liver tissues of the fish changes further with the fish species (Sastry and Rao, 1984; Naidu *et al*, 1984). Qayyum and Shaffi (1977) reported a decrease in the glycogen level of liver, muscles and kidney in *H. fossilis* when exposed to Hg. Short term exposure to sub lethal concentrations of Hg (36.3-60.5 µg/l) elicited significant hyperglycemia and glycogenolysis in liver and brain of juveniles and the adults of *P. conchonius* (Gill and Pant, 1981). There was hyperglycemia persisting up to four weeks, the glycogen levels in liver and brain remained low up to three weeks but an increase in myocardium glycogen was registered after 8 weeks exposure. Biochemical sequel under long term exposure however, included an increase in blood glucose over the controls in Hg (60.5 µg/l) treated fish. Hyperglycemia again subsided after 4 weeks exposure, blood glucose levels remained above control levels. In the fresh-water teleost fish, *C. punctatus* after exposure to a sublethal concentration (3µg/1) of mercuric chloride, glycogen and lactic acid contents of the liver and muscles and the blood glucose level were found decreased significantly after 60 days of exposure (Sastry and Rao, 1981), after 120 days; however (Sastry and Rao, 1984) it showed glycogen content of liver and muscles unaltered, and the muscle lactic acid level decreased significantly, and the rate of intestinal absorption of glucose was also reduced significantly. Khanee *et al.* (1991) after exposing *O. mossambicus* to sub lethal concentration of mercury (0.01 and 0.02 ppm) for four weeks, recorded significant decrease in carbohydrates in tissues of muscle, liver and gills. In the ovary of *C. batrachus* after exposure to methyl mercuric chloride for 15 days (Verma *et al.*, 2002), and the level of liver and muscle glycogen in the cichlid, *O. mossambicus* exposed to sublethal concentrations of mercuric chloride was found declined indicating increased utilization of glucose to counter act energy demand imposed by severe anaerobic stress of mercury toxicity (George *et al.*, 2012).

A trace metal like Zn is reported to cause significant depression in plasma insulin and liver glycogen in rainbow trout (Wagner and McKeown, 1982). Decline in glycogen was also reported in the liver of *L. rohita* after exposing the fish to Zn by Bengeri and Patil (1986). Likewise, Zn stress (Malik *et al.*, 1998) in Murrel, *C. punctatus* and Cd stress in the rainbow trout (Hontella *et al.*, 1996), were shown to yield marked decrease in the liver as well as the muscle glycogen, because of the decreased metabolism rate. Kumar (2002) reported decline in the carbohydrate content between 28.18 to 29.58 per cent was also reported due to Co toxicity in the freshwater fish species *L. rohita, C. mrigala* and *C. carpio*. Decline in glycogen contents has been reported in the liver of *Channa punctatus* by Srivastava *et al.* (2002) after exposure to zinc for 96 hr and 15 days; and also in *C. mrigala* (Ham.) after exposure to lead for 30 and 60 days (Kumar *et al.*, 2005). According to Clary, 1975 and Nielsen, 1977 an increase in blood glucose due to Ni, presumably results from its inhibitory effect on the insulin release or some direct effect of Ni on the pancreas. Chaudhary and Nath (1985) studied effects of Ni on blood glucose levels at a sub lethal dose of 64 ppm (0.8 of Lc_{50} value) in freshwater fish *Colisa fasciatus* for a period 96 h, and noticed the blood glucose level to be a reliable indicator of the Ni toxicity to the fish. It however, revealed that blood glucose level exhibited a steady increase due to Ni toxicity. Maximum increase of 85.08 per cent was observed at 96 hr of Ni exposure. They further suggested that the hyperglycemia in *C. fasciatus* caused by an exposure to nickel sulphate was possibly a reflection of the stress induced hormone mediated response. Additionally significant hyperglycemia were noted by Nath and Kumar (1988) in the catfish, *H. fossilis* when exposed to nickel; in the fish *L. rohita and C. gariepinus* with copper by Radhakrishaniah *et al.* (1992) and Van Vuren *et al.* (1994) and by James *et al,* (1992) in the teleost, *O. mossambicus* exposed to lead. *O. aureus* exposed chronically to Cu or Pb (Abhas *et al.*, 2003) indicated serum glucose levels increased in all the treated groups. The creatinine and the uric acid levels were also elevated. Specimens of *C. carpio*, exposed to the sub lethal concentration of Hg (0.01-0.1 mg/l), Cr (2-20 mg/l), and Ni (1.5-18 mg/l) for 7 days resulted in a significant decrease in glycogen in tissues of all carps exposed to the metals (Canli, 1996). It demonstrated a significant decrease in liver and muscle glycogen levels. A high demand of glycogen for excess energy requirements, as corroborated by excessive movements and surfacing frequency in the experimental fish. Neha-Bhatkar *et al.*, 2004 reported the impact of 30 days exposure of $CrCl_2$, $NiCl_2$ and $ZnCl_2$ on the biochemical parameters of the fish *L. rohita*. After 10 days of exposure to $CrCl_2$, $NiCl_2$ and $ZnCl_2$, blood glucose level increased by 48.11 per cent, 71.71 per cent and 68.35 per cent, respectively. However, after 20 and 30 days of treatment, significant decrease in glucose level was recorded. Liver glycogen depletion was reported as at maximum with Co and Cd, followed by Ni, but least with Zn. Virmani, 2005 have also reported similar dose dependent depletion in liver glycogen as Co=Cd>Ni>Zn. Whereas, the muscle glycogen levels were depleted in the order: Cd > Co > Ni > Zn.

Carbohydrates are the primary and immediate sources of energy. In stress conditions, carbohydrate reserve is depleted to meet energy demand. Hyperglycemic response is an indication of a disrupted carbohydrate metabolism possibly due to enhanced breakdown of liver glycogen and mediated perhaps by adreno-cortical hormones and reduced insulin secretary activity. Khanna and Gill (1975) while studying the effects of cobalt salts on the glycemia and islet histology of *Channa punctatus*, suggested the possibility of an imbalance in the pancreatic hormones involved in the carbohydrate metabolism, due to damage of pancreas tissue by the toxic substances. In *C. punctatus* administration of cobalt chloride and cobalt nitrate induced hyperglycemia along with degranulation and vacuolization of the pancreatic tissue in the initial stages while a damage of the α-cells in the later stages. Wedemeyer and McLeay (1981) reported that high levels of blood glucose are caused by disorders in carbohydrate metabolism appearing in the condition of physical and chemical stresses. Bhattacharya *et al.* (1987) studied the blood glucose and the hepatic glycogen profiles in *C. punctatus* exposed to a non lethal concentration of a pollutant. In short term test, hyperglycemia with depletion in the hepatic glycogen was recorded while in the long term exposure, hyperglycemia was registered with persistent depletion of the hepatic glycogen content. It was concluded that the hepatic glycogen is the probable source for hyperglycemia in *C. punctatus*. A maximum depletion corresponded to a dramatic increase in the blood glucose levels, thus, suggesting that some of the hepatic glycogen via the intermediate glucose 1-phosphate appeared converted to glucose which entered the circulation. This decrease in insulin producing secretion rate causes the hyperglycemia (Larsson *et al.*, 1976; Heath, 1995). Several reports have also evidenced depletion of liver glycogen energy reserves in fish as a result of severe muscle exercise (Black *et al.*, 1962; Nath and Kumar, 1988). Gill and Pant (1981) have suggested that elevated glycogen levels in heart muscles of the adults of *P. conchonius* on account of fish exposure to Hg are possibly to compensate the enhanced cardiovascular functions, as a consequence of increased blood viscosity accompanying polycythemia. Glycogen depletion may be a manifestation of an initial regulatory step which increases intermediary metabolism resulting in protection of the hepatocytes from the xenobiotics. In some situations, materials may accumulate in the liver at toxic levels and cause pathological alterations (Meyers and Hendricks, 1984; Fergusson, 1990). A prolonged environmental stress in the fish (Heath, 1995) in fact makes the adaptation difficult and creates a weakness in the fish. This weakness is characterized by a decrease in the liver glycogen and the serum cortisol levels, which subsequently create alterations in the metabolism and also shortens the lifespan of the organisms. Kralj

– Klobuear (1993) observed Cu accumulation in the liver tissue of carp treated with 200mg/L $CuSO_4$. The increased amount of Cu was accompanied with a decrease in the glycogen content.

Murugappan *et al.* (1999) found during chronic exposure to Cd, metal accumulation in the cells of Isle in pancreas of the fish, *C. carpio* and causing damage to these insulin producing cells. Depletion of glycogen under stress may be due to its direct utilization for extra energy generation due an increased demand as under hypoxia, since carbohydrates stored in the fish tissue and the organs like the muscles and the liver are to meet the energy needs during the hypoxic conditions, intensive stocking and the food shortage (Wendelaar-Bonga, 1997). Heath and Pritchard (1965) reported liver glycogen levels were depleted during acute hypoxia or physical disturbances in fish. Kumari and Kumar (1997) studied the effects of pollutants on the distribution of carbohydrates in heart, muscle, kidney, liver, brain, gills and ovary of the fish, *Channa punctatus*. The studies revealed that there was a significant decrease in the carbohydrate content suggesting the possibility of increased glycogenolysis in fish under pollution stress as compared to the control fish. The hepatic glycogen levels were reported significantly lowered than in the controls even on account of simple transportation stress in the fish, *O. niloticus* (Orji, 1998).

The elevation in the blood glucose level is in response to the increased rate of glycogenolysis or gluconeogenesis, later is the result of increased secretion of catecholamine due to stress. A variety of stressors are known to stimulate the adrenal tissues, resulting in increased level of circulating glucocorticoids (Partap and Wendelaar-Bonga, 1990; Hontela *et al.*, 1996; Gluth and Hanke, 1985; Ricard *et al*, 1998 and catecholamine's (Nakano and Tomlinson, 1967; Witters *et al*, 1991). Rainbow trout exposed for 48 h to aluminium (60 µg/l) in soft water (pH 5), showed a ten-fold increase in adrenaline and nor-adrenaline concentrations (catecholamine's (Witters *et al.*, 1991) which implies that catecholamines may play a role in aluminium exposed fish.

Ricard *et al.* (1998) reported glycogen levels decreased in the liver of *O. mykiss*, as a result of the activation of glycolytic enzymes via catecholamines under metal stress. He observed Cd could decrease the glycogen reserves in the American eel, *Anguilla rostrata* by increasing the production of catecholamines from the adreno-medula (Gill and Epple, 1993). Depletion of glycogen in liver may further be correlated with a high demand of glycogen for excess energy requirements, as noted by excessive movements and surfacing frequency in the experimental fish. Significant reductions in carbohydrates and the lipid levels in all tested tissues of *C. carpio* after Cd treatment (De Smet and Blust, 2001) were also attributed to increased catabolic processes imposed by metal intoxication. Some other investigations also showed that the heavy metals decrease the glycogen reserve in the fish (Levesque *et al*, 2002) and invertebrates (Blasco and Puppo, 1999) by affecting the activities of the enzymes that play a role in the metabolism of carbohydrates. Shukla *et al.* (2001) studied the rate of uptake of glucose, fructose and amino acid tryptophan by the intestines of the freshwater teleost fish, *C. punctatus* and *H. fossilis*, following exposer to sub lethal concentration of cadmium and zinc (1.12 and 4.0 ppm). It showed decrease in the rate of transport of glucose and fructose, which was more marked after 30 days as compared to 15 days in the two fishes. Emad *et al.*, 2005 also emphasized that high glucose levels in muscles in cadmium exposed marine fish, *Mugil seheli* were due to gluconeogenesis, *i.e.* formation of glucose and glycogen from other non-carbohydrate sources in the tissues. The same reason could be attributed to the inability of the fish to recover completely even after the toxicity of the heavy metals was over as observed by Virmani (2005).

ii. Proteolysis and Depletion in Tissue Proteins

Proteins are an important group of macromolecules, which occupy a central place in almost all structural and functional aspects of a living organism. The protein contents in a fish occur maximally in muscles, followed by in liver and gills (Sastry and Shukla, 1990; Khanee *et al.*, 1991). There are reports about significant reductions in tissue protein levels in fish following metal toxicity (Oikari *et al.*, 1982). Depressed muscle proteins were observed after exposure of the fish to low levels of pulp-mill effluents. On the contrary, Jeney *et al.*, 1996 noted that sub-chronic pulp and paper mill effluents were able to enhance total proteins in roaches (*Rutilus rutilus*). Kaur and Saxena (2001) recorded decline in body proteins in fishes inhabiting river Satluj due to contamination of inflowing water by the effluent rich Budha Nalah water in Punjab. These effluents are said to be rich in heavy metals and many other chemical toxicants (Kaur *et al.*, 2000a,b). Niak *et al.* (2004) reported that the total protein as well as total glycogen and lipid contents underwent a significant depletion in the tissue of the tannery effluent treated fish, *C. carpio*. Verma and Tonk (1983) recorded significant decline in protein content of flesh of *Notopterus notopterus* (Pallas) exposed to 0.88 and 0.44 mg/l $HgCl_2$ for 30 days. Ram and Sathyanesan (1984) reported that mercuric chloride reduced the proteins of chosen tissues of *C. punctatus*. Khanee *et al.* (1991) after exposing *O. mossambicus* to sub lethal concentration of mercury (0.01 and 0.02 ppm) for four weeks, recorded significant decrease in protein in tissues of muscle, liver and gills. The muscle content were said to gradually decreas from initial 182 mg/g to 160 mg/g after 28 days of exposure to sub lethal concentration of mercury (0.02 ppm). Similar results were obtained for liver and gill protein also. The inorganic Hg was observed first to increase then decease total proteins in *C. idella* (Shakoori *et al.*, 1994). Likewise, the biochemical components of the liver of two important penaeid prawns were found significantly reduced, following six days of exposure to 0.005 ppm and 0.01 ppm of mercuric chloride during various reproductive stages (Das *et al.*, 2001).

Many other workers have also reported decrease in tissue proteins in fishes like *C. punctatus* (Bhattacharya *et al.*, 1987: Sastry and Shukla, 1990) and *H. fossilis* (Sastry and Subhadra, 1982). Kumari (1984) reported decrease in protein content of liver and muscles by 10.3 per cent and 33.3 per cent respectively in *C. punctatus*, when exposed to nickel for 30 days, but increase of 20.1 and 7.7 per cent protein respectively, when exposed similarly to chromium (2.6 mg/l). Hilmy *et al.* (1985) after exposing *Mugil cephalus* to Cd, evidenced an increase in total protein contents in the fish liver, gill and serum; no such change was recorded in heart muscles. Exposure of *A. testudineus* (Vijayram *et al.*, 1989a) to Cd also showed a significant decline in the protein content in the body tissue. Likewise, Jana and Bandhopadhyay

(1987) exposed fish *C. punctatus* to five heavy metals *viz.*, magnesium, arsenic, lead, cadmium and chromium at sublethal concentration of 0.5, 1.0 and 2.0 ppm and found body proteins decreased significantly. Jana and Sahana (1988) studied the effects of cations such as Cu^{2+}, Cd^{2+} and Cr^{6+} on changes in proteins in a freshwater fish *Clarias batrachus*. They showed an increase of the protein content in the liver, kidney, stomach, intestine, testis and ovary, and a decrease in the muscle after Cu^{2+} and Cd^{2+} treatment as compared with control data; but the Cr^{6+} did not cause any changes of protein concentration in the kidney and testis. Dinodia (2001) also reported a decline in protein content from 21.3 to 25.5 per cent in the freshwater carps *L. rohita, C. mrigala* and *C. carpio* after exposer to Cd. De-Smet and Blust (2001) attributed reduction of proteins in liver, because of the utilization of endogenous protein for maintenance during chronic exposure to metals like Cd. Blood hematocrit, plasma glucose, plasma lactate and total tissue protein contents did not alter significantly. They have studied stress responses and the changes in protein metabolism in the common carp *C. carpio,* exposed to Cd over a period 0, 0.8, 4 and 20 m 29 days. The highest Cd concentration proved to be lethal to fish resulting in 100 per cent mortality after an exposure of 21 days. *C. punctatus* (Sastry and Shukla, 1990), also showed decrease in the protein levels in liver and muscles on exposure to Cd. Emad *et al.*, 2005 as a result of the toxicity of cadmium and copper in marine fish, *Mugil seheli,* higher values of proteins in plasma and muscle than the controls and plasma protein were affected by Cu more than Cd. Sastry and Anuradha, (2005) in the freshwater fish *C. catla* reported a significant decrease in the total protein levels in muscle of the group exposed to Cd as compared to that in the liver. The Cu exerted its maximum effect on liver.

Hota (1996) mentioned significant fall in the levels of protein in the liver, brain and intestine of *C. punctatus* on account of arsenic toxicity at 5 ppm for 75 days. Rani *et al.* (2001) investigated the sub lethal toxicity of sodium arsenite on protein metabolism in a teleost fish, *Tilalpia mossambica.* At the end of 24, 48, 72 and 96 h of exposure, it evidenced elevated protein and amino acid content. Kumar (2002) in the freshwater fish species *L. rohita, C. mrigala* and *C. carpio* found a decline in protein content at maximum up to 34.58 per cent in *C. carpio* and a minimum of 18.65 per cent in *L. rohita*. Exposure to 12 mg/l of ZnSO4 for 37 days (Shukla and Pandey, 1986) was said to cause decrease in the protein content of fish *C. punctatus*. Meenakshi *et al.* (1998) studied decrease in protein and amino acid content in different tissues of *L. rohita* fingerlings exposed to 96 h median lethal concentration of zinc sulphate. Neha-Bhatkar *et al.*, 2004 on exposing the fish *L. rohita* to $CrCl_2$, $NiCl_2$ and $ZnCl_2$, observed proteins in both liver and muscle depleted in all the experimental fish. Liver proteins recorded highest changes in contrast to lipid and carbohydrate irrespective of the species, sex, medium and exposure. Virmani (2005) clearly revealed metal concentration dependent decrease in the protein levels, while observing the effects of Ni, Co, Cd and Zn independently and as well as metals in combination in fish. Cd alone and a combination of Cd and Zn caused maximum protein reductions in the treated fish. Similar were the observations of Oikari and Nittyla (1985) and Vijayram *et al.* (1991). Kori-Siakpere *et al,* (2006) reported decreased plasma protein in *Heterobranchus bidorsalis and Clarias gariepinus* when exposed to sublethal effect of cadmium toxicity. Prabhahar *et al.* (2012) in the edible freshwater fish *C. mrigala* also recorded a significant decrease the protein levels in gills, liver and kidney of cadmium sulphate treated fishes when compared to their controls. Parvathi *et al.* (2011) reported decreased plasma proteins on exposure to chromium in *C. carpio* could be the result of the breakdown of proteins into amino acids. This could also be attributed to renal excretion or impaired protein synthesis or due to liver disorder (Kori-Siakpere, 1995).

Reduction in protein contents has been an important parameter in determining metal toxicity in a fish. Brett and Groves (1979) reported that in fishes proteins are the major source of energy. Changes in proteins are due to the increased energy cost of 'homeostatic functions', tissue repair and detoxification of the chemical stressor or due to the altered enzyme activity. According to Passow *et al.* (1961) heavy metals inactivate intracellular proteins as a result of their interaction with the proteins and the nucleic acids which get precipitated consequently, thus, interfering with the metabolism. Passow *et al.* (1961) also suggested binding of the heavy metals to enzyme proteins. According to Mustafa and Chandra (1972) the decrease in total protein contents may be due to blocking of the metabolism of amino acids by the heavy metals and such cells become incapable of synthesizing proteins. Syversen (1981) reported that, heavy metals in general interfere with protein synthesis, as concluded from decrease in protein content of all tested tissues of *O. mossambicus* after exposer to mercury. Heavy metals are also capable of displacing metals situated at the active sites of enzymes, thus, disturbing cell functions. Heavy metal ions are strongly bound by sulphydril group of proteins (Viarengo, 1985), which cause changes in the structure and enzymatic activities of proteins by forming complexes, thereby causing toxicity effects evident in the living organisms (Hodson 1988). Zalme *et al.* (1976) suggested binding of mercury ions to sulphydryl groups of proteins in the cell membranes. Black and Love (1986) showed that in cods, the major energy reserves (glycogen, lipids and protein) were mobilized or laid down in a definite sequence in different organs, so that measurements of all the three components could provide a better estimate of the nutritional status of a fish.

Decrease in the protein content of fish, *C. punctatus* on exposure to 12 mg/l of $ZnSO_4$ for 37 days (Shukla and Pandey, 1986) were also correlated with stress induced utilization of protein reserves of flesh during starvation as the fish probably depend more on protein catabolism to meet the energy demands under heavy metal stress. Severe proteolysis was also recorded in gills, kidney and intestine of freshwater fish, *C. carpio* exposed to sub lethal concentration (0.1 mg/l) of mercury (Suresh *et al.*, 1991). In *O. mossambicus* (Khanee *et al.*, 1991), muscle protein contents were found gradually decreased from 182 mg/g to 160 mg/g after 28 days of exposure to a sublethal concentration of mercury (0.02 ppm), but gradually the protein amounts increased to 172 mg/g when the treated fish were transferred to metal impoverished water, showing that theses metal effects were reversible. Similar results were also obtained for liver and gill proteins. The changes were more in gills at the initial periods of exposure (1 and 15 days), but as the period of exposure increased, these changes were more pronounced in the kidney at 30 days of exposure. There was depletion of total soluble and structural proteins, followed by increased activity of protease enzyme and progressive increase in accumulation of free amino acids. It also showed a steady enhancement in the activities of aspartate aminotransferase and alanine aminotransferase along with elevation of glutamate dehydrogenase. Such an adaptive utilization of protein to meet the energy demand under stress was also reported by Ramalingam and Ramalingam (1982) in *Sarotheroden mossambicus*. They found that reduction in liver and muscle proteins was due to the maintenance of energy supply irrespective of the nature of the toxicant. Heath (1987) reported that reduction in the availability of carbohydrates for energy was partially compensated by increasing the activity of glutamate dehydrogenase and amino oxidase enzymes; which are the enzymes to control utilization of amino acids for energy, as also demonstrated by the decline in protein value in plasma and muscle (exhausted) during first 7 days from the exposure to different concentrations of Cd and Cu, due to supplying fish with energy to do vital activities. Significant decline in the protein content in the body tissue on exposure of *A. testudineus* (Vijayram *et al.*, 1989a) to Cd also indicates that during chronic periods of heavy stress due to metals, fish utilized the endogenous proteins. Sastry and Shukla (1990) also stated that decrease in the protein levels in liver and muscles of *C. punctatus* on exposure to Cd was probably due to utilization of endogenous proteins by fish during chronic periods under stress. Along with reduced body glycogen levels, the majority of the toxicity treatments in the fish further evidence reduced proteins and the lipid amounts, indicating stress induced increase in the utility of body metabolites to counteract physiological stress (Heath, 1995). Hota (1996) mentioned a significant fall in the level of cholesterol and ascorbic acid along with proteins in the liver, brain and intestine of *C. punctatus* on account of As toxicity at 5ppm exposed for 75 days. They also attributed loss of proteins to arsenic toxicity in *C. punctatus* resulting in increased proteolysis and utilization of amino acids in various catabolic reactions.

The proteolytic enzyme activity was also reported declined maximally (32.6 per cent) in *L. rohita* with 5.0 ppm Cd toxicity (Dinodia, 2001). Following sub lethal toxicity of sodium arsenite in a teleost fish, *Tilalpia mossambica* (Rani *et al.*, 2001), total protein content and free amino acid content in liver, gill, brain and muscle were found to exhibit significant alterations throughout the investigation in relation to that of control. It was suggested that the fish responded to the stressful situations by gearing up its metabolic activity, as revealed by the elevated protein and amino acid contents. Desai *et al.*, 2002, mentioned a gradual decrease in the level of liver and brain protein due to proteolysis was the result of stress of nickel (up to 40ppm for 40 days) toxicity in freshwater fish *C. punctatus*. Virk and Sharma (2003) stated that reduction in proteins in gills of *C. mrigala* was due to the increased energy cost of homeostasis, tissue repair and detoxification during stress or it could be due to altered enzyme activities under effect of acute toxicity of nickel and chromium. Protein and carbohydrates were reduced significantly in both treatments and reduction was dose dependent in the case of nickel, but not chromium. Pathan *et al.* (2009) concluded that impairment in fish liver on account of paper mill effluents was created due to metabolic crisis as a result of the effects of paper mill effluents. The fish is known to stop feeding in a toxicant rich medium, as a result it causes breakdown of the tissue proteins (proteolysis) in fish under stress and starvation, to provide energy, causing thus, a reduction in its tissue proteins. Another reason for the depletion in its tissue protein content may be the lesser protein synthesis or may also be due to utilization of amino acids in other catabolic reactions.

iii. Changes in Cholesterol and other Lipid Levels

Lipids another important source of energy in fish, are primarily present in mesenteries, liver cells and the muscles. Major circulatory lipids of fish are free fatty acids and tri-acyl glycerols (TAG) and most of these are rapidly shuttled in between tissues as free fatty acids and are slowly delivered as TAG or phospholipids (Weber and Zwingelstein, 1995). About 30 per cent of the circulating free fatty acids are the unsaturated fatty acids of 20-22 carbon chain. Most of the longer chain unsaturated fatty acids are utilized in the synthesis of cholesterol, membrane phospholipids and eicosanoid compounds such as prostaglandins, prostacyclins, thromboxanes, and leukotrienes. Cholesterol and the phospholipids play an important role in fish physiology, cholesterol as the precursor of steroid hormones including sex and adrenal hormones and the phospholipids mainly for the membrane integrity. Cholesterol as the precursor of steroids constitutes a major part of cell membranes also. Cholesterol levels in a fish, however, vary with the stage of its life cycle. In a catfish serum, it is said to vary with the spawning season, as these were recorded to be the highest (643 mg/100 ml) in the month of November and a lowest (310 mg) in August at the end of the spawning period (Tandon and Chandra, 1976). Larsson and Fang, 1977 analyzed serum cholesterol in 22 species of marine fishes (Cyclostomi, Holocephali, Elasmobranchi and Teleostii). They found it to range in between 190-921 mg/100 ml except in Elasmobranchi in which it varied between 86-214 mg/100 ml (Larsson and Fange, 1977).

Shorter chain fatty acids are used primarily for energy metabolism (Weber and Zwingelsein, 1995). Later also help in the formation of lipoproteins and enhance fat mobilization as well as in absorption of fat- soluble vitamins in the intestine. Triglycerides are also used as substrate for gluconeogenesis during high stocking density of brook charr, *Salvelinus fontinalis* (Vijayan *et al.*, 1990). The eicosanoids are the local hormones, which produce specific effects on target cells close to the site of their formation. These are rapidly degraded, thus, are not transported to distal sites within the body. But in addition to participating in intercellular signaling, there is evidence of their involvement in intracellular signal cascades. These eicosanoids have various roles in inflammation, fever, regulation of blood pressure, blood clotting, and immune system modulation, control of reproductive processes and tissue growth, and regulation of the sleep/wake cycle. Lipids catabolism as in case of protein is, however, not very important during acute stress such as hypoxia, as the amount of oxygen required to metabolize lipids is said to be lacking.

Changes in cholesterol levels in the fish blood inhabiting contaminated waters are in general stated to be less sensitive to water pollutants, although these appear changed under stress. Cholesterol, however, being an essential structural component in tissue membranes as steroid cholesterol, liver failure may cause release of cholesterol into the blood. Heavy metals are known to produce hazardous effects on cell structure, especially on the membranes, possibly because of their effects on depletion in cholesterol levels. In fact a relationship appears to exist in between serum cholesterol concentration, serum protein levels, and anemia as observed in fish living in lead-contaminated water. There are reports indicating significant decrease in the blood cholesterol levels (Larsson, 1975; Johansson and Larsson1978), as a result of long term exposure of different fish species to Cd toxicity in fishes. Rafia *et al.*, 2008 showed appreciable decline in total lipids with decline in total glycogen, and total protein contents of the fish in presence of toxins, resulting in decreased productivity of fish population. Some investigators (Aly *et al.*, 1993; Billy *et al.*, 1995; Jehan and Motlag, 1995) have proposed that heavy metals inhibit cholesterol synthesis, which leaves insufficient cholesterol for the maintenance of cell membranes. This causes hemolysis and the release of hemoglobin and membrane associated proteins into the serum. Goodman and Noble (1968) found that the liver, plasma, and erythrocytes comprise a pool of cholesterol that is rapidly and continuously turning over (in terms of hours to days). Consequently, changes in serum cholesterol (or absence of changes) reflect changes in the cholesterol content of erythrocyte membranes and the liver. Hemolytic anemia has been observed in a number of organisms exposed to lead. Some workers have considered changes in the blood cholesterol levels together with RBC count, Hb content and the Hct per cent in the fishes exposed to the different dose levels of different heavy metals as indicative of metal toxicity. Sahoo and Mukherjee (1999) in the fingerlings of rohu fish (*L. rohita*) weighing between 29-38 g, recorded a significant decrease in the blood cholesterol levels and various other cell entities like RBC count and hematocrit per cent as a result of heavy metal toxicity in fishes. Fishes further exhibit seasonal changes in blood cholesterol levels. Tandon and Chandra (1976) in a catfish found serum cholesterol level to be the highest (643 mg/100 ml) in November month and a lowest (310 mg) in the month of August at the end of the spawning period.

Dutta and Haghighi (1986) suggested that heavy metals, such as methyl-mercury inhibit cholesterol synthesis in fish, *Lepomis macrochirus*. As a result, there is insufficient cholesterol for the maintenance of cell membranes. This causes hemoglobin and membrane-associated proteins to be released into the serum. Along with depression in serum cholesterol concentration, Dutta and Haghighi (1986) also found a significant depression in packed cell volume of lead-exposed fish. The inorganic form of the Hg is further said to be more effective than methyl-Hg in lowering lipid levels in fish, which in fact is reverse of what usually occurs with this form of Hg. Kirunbagaran and Joy (1992) also observed greater reduction effects due to inorganic Hg. There was reduction in free and esterified cholesterol in *Clarius* on exposure to both the forms of Hg for 180 days. Reduction in lipids along with cholesterol was attributed to a general decline of energy stores in the Hg treated fish. Mukherjee *et al.* (1994), In case of chronic exposure of *C. carpio* to a lethal dose of mercuric chloride and cadmium chloride on gonadal function, investigated that hypothalamic extract or the Gonadotropin Releasing Hormone (Gn RH) induced both transport of cholesterol to ovary and *in vitro* release of pituitary gonadotropin (GTH).

Tewari *et al.* (1987) found that the serum cholesterol concentration of the fish *Barbus conchonius* exposed to 47.4 mg Pb2+/1 for thirty days were depressed by 30.3 per cent. In addition to high decrease in cholesterol in blood, Tewari *et al.*, 1987 also noted significant decrease in cholesterol in liver, ovary and testis of rainbow trout on chronic exposure to lead for 30-60 days. Cholesterol concentrations in other organs of the same fish were depressed by 38.9 per cent (liver) and 60.0 per cent (testes). They also reported a 31.9 per cent decrease in packed cell volume and a 14.9 per cent decrease in serum hemoglobin concentration of lead-exposed fish. Ruparelia *et al.* (1989) found that the serum cholesterol concentration of the fish, *O. mossambicus* exposed to 18 mg Pb^{2+}/L for 21 days were depressed by more than 50 per cent. They further found decreased serum glucose and increased serum protein levels in the lead treated fish. Ferrando and Andreu-Moliner (1991ab) also recorded a significant decline in the blood cholesterol levels in fish on exposure to heavy metals, whether individual or in the combination. Virk and Sharma (1999) studied decrease in biochemical components (lipids, cholesterol and phospholipids as well as the proteins) of the liver of *C. carpio,* following 60 days of exposure of safe and sublethal concentration of nickel (7.5, 15.0 mg/l) and chromium (10.0, 20.0 µg/l). According to them these biochemicals were significantly reduced in their various reproductive phases like pre-spawning, spawning and the post spawning periods as compared to the control fish following an exposure of safe and sub lethal concentration of Ni (7.5, 15.0 mg/l) and Cr (10.0, 20.0 µg/l) for a period of 60 days. It is presumed that reduction in circulating cholesterol level in the blood is either due to its more utilization during corticosteroidogenesis *i.e.* synthesis of corticiosteroids, as it is precursor for steroid hormones, or its depressed *de novo* synthesis (Dutta and Haghigh, 1986; Tewari *et al.,* 1987). Likewise, alterations lipid metabolism along with carbohydrates of freshwater fishes were reported by Kaur and Saxena (2001) in the fish species inhabiting the grossly polluted waters of Satluj in Punjab by the Budha Nallah (BN) and by Amudha *et al.* (2002) in *O. mossambicus* due to diary effluents. Dhanapakiam and Ramaswamy, (2001) recorded cholesterol, serum glucose and serum proteins decreased in *C. carpio* at sublethal levels of heavy metals over a period of 30 days, but the levels of serum chloride and Ca concentration were increased. A significant reduction in liver cholesterol content has been reported after Cd treatment of *Garra mullya* for 4 months (Sinha *et al.,* 2001). Sindhe *et al.* (2002) reported a decline in cholesterol of both the liver and ovary of *N. notopterus* after exposure to Cd and Hg for 2 months. Sindhe *et al.* (2002) also evidenced a gradual decrease in the lipid content both in the liver and ovary. Liver being the metabolic seat of the body attempts to compensate for biochemical depletions in the ovary. The reduction in total lipid content may be a result of disturbed vitellogenesis, sterioidogenesis and/or reduced enzyme activity (Sindhe *et al.,* 2002). They further suggested that the reduction in hepatic and ovarian cholesterol content may result in altered vitellogenisis and steroidogenesis. Observations made by Verma and Neera (2007) too indicated similar influences on the cholesterol level in murrel, *C. punctatus* on chronic exposures to Hg and Cd. Lowering of cholesterol may be attributed to an increase in lipid utilization for meeting additional energy requirements under stress conditions, as suggested by Srivastava *et al.* (2002). Exposure of *C. punctatus* to Ni for 30 days (Desai *et al.,* 2002) and Zn for 15 days (Srivastava *et al.,* 2002) also results in a similar trend in the liver. Decline in the lipid content of liver was also observed in female *H. fossilis* exposed to nickel chrome electroplating effluents for 120 days (Gupta, 1991) and *C. carpio* exposed to zinc electroplating effluent for 90 days (Kaur, 1992). Mercury is reported to reduce gonadotrophin release which causes impairment in yolk formation by the oocytes in *C. punctatus* (Ram and Sathyanesan, 1984). Zinc, being a heavy metal, may also have a similar influence on the hypothalamic center in *C. punctatus* resulting in yolk depletion, a few other investigators (Singh and Reddy, 1990; Canli, 1995; Yang and Chen, 2003) evidenced an increase in serum cholesterol concentrations in the metal-exposed fishes. Likewise, copper (Singh and Reddy, 1990) was also found to cause a steady increase in cholesterol reaching almost twice the levels in controls in Indian catfish, *H. fossilis* after 30 days of continuous exposure. Desai *et al.*, 2002, suggested high liver cholesterol level in freshwater fish, *C. punctatus* due to nickel toxicity (up to 40ppm for 40 days), was the result of hepatic dysfunction and its accumulation in brain with a gradual decrease in the levels of liver and brain proteins.

iv. Metal Effects in Combination

Most of the toxic metals since occur in combination with other metals, salts or pesticides in water particularly in industrial zones, study of their combined effects in fish further forms an important aspect. Some metals in combination causes relatively further higher toxicity effects in fishes. Interactions in between toxic chemicals towards animal in an aquatic system may be synergistic, antagonistic or additive, and may result in increase or decrease, acceleration or inhibition in biochemical constituents in tissues. The interaction between zinc and other metals in fish have been investigated by few workers (Brown and Dalton, 1970; Spehar, 1976; Broderius and Smith, 1979). Results from these experiments ranged from antagonistic to synergistic responses. James *et al.* (1991a) have reported toxic effects of copper, zinc and cadmium in *O. mossombicus*. A significant decrease in protein, carbohydrate and lipid were observed in muscles, liver, gills and the whole body of *O. mossambicus* exposed to metals individually and in combinations except when exposed to copper alone. Combinations of Cu+Cd, Cu+Zn and Zn+Cd markedly reduced the protein, carbohydrate and lipid in all the tissues tested. Copper was however, reported as most toxic, followed by zinc and cadmium; Zn in combination with Cd was also less toxic than its combination with Cu or a combination of Cd and Cu. Cd, Cu+Zn and Cu+Cd caused a reduction in the liver carbohydrate content in fishes to 32, 54 and 63 per cent, respectively as compared to controls. Acute toxicities of copper, cadmium and mercury to the freshwater fish, *Varicorhinus barbatus* and *Zacco barbata*, studied by Shyong and Chen (2000) revealed the order of metal toxicity to *V. barbatus* as: Hg>Cu>Cd, while to *Z. barbata* it appeared as: Cu>Hg>Cd.

Shukla *et al.* (2002), showed toxic effects of cadmium individually and in combination with other metals on the nutritive value of freshwater fish *C. punctatus*. The Cd treated fish showed a significant increase in the total liver lipids for 15 days, but the effects were reversed after 60 days when it showed a significant decrease in liver and muscle fat content. Sindhe *et al.* (2002) in the fish, *Notopterus notopterus* on exposer to metals, revealed decrease in cholesterol contents along with in proteins in both ovary and liver in the order of $HgCl_2$>$HgCl_2$+$CdCl_2$>$CdCl_2$>control, indicating

$HgCl_2$ as more toxic, its toxicity, however, was reduced on combination with Cd salts. Jain and Sharma (2003) in some freshwater fishes further indicated enhanced effects of Hg alone and in combination with Co. They reported liver carbohydrates, protein and glycogen levels were reduced by 8.61, 7.09 and 8.10 per cent, respectively due to cobalt toxicity; 12.84, 13.02 and 13.57 per cent, respectively due to Hg and 35.73, 25.52 and 27.27 per cent, respectively due to Hg+Co treatment. A combination of Hg with endosulfan enhanced further the rate of reduction in these components up to 40.15, 30.13 and 45.25 per cent, respectively. Sastry and Anuradha (2005) reported greater reducing effects of Cd and Cu in combination on proteins, glycogen and lactic acid in liver as well as in muscles of the freshwater fish, *C. catla*. Mittal (2004) in *C. mrigala* observed maximum reducing effects of As+Hg treatment, reducing carbohydrates (33.5 per cent), proteins (29.5 per cent) and glycogen (40.0 per cent) contents in liver. Sarita, 2006 also showed maximum reduction in tissue biochemicals with Hg+As in *C. carpio*. She evidenced protein contents reduced maximally in kidneys (41.7 per cent), followed by liver (36.8 per cent), muscles (28.3 per cent), blood (23.6 per cent) and brain (16.9 per cent) with this combination of metals at 0.5 ppm concentration. She, however, observed chromium to induce an increment in protein quantity in muscles and liver of the fish species studied. Virmani (2005) while observing the effects of certain metals like Ni, Co, Cd and Zn independently and in combination, clearly revealed a dose dependent decrease in protein contents in various tissues as compared to control. She observed maximum reduction in the treatment of Cd alone and with the combination of Cd and Zn. Garg *et al*, (2009) in Indian major carps *L. rohita*, *C. mrigala* and *C. catla* to metals also recorded *a* significant reduction in carbohydrate and lipid contents in muscles and gills, and the metal effects were at maximum with Cd+As+Zn, followed by a combination of Cd+As; with minimum effects of Zn alone. Garg *et al*, 2009 while assessing the effects of sub lethal exposure of heavy metals cadmium, arsenic and zinc on Indian major carps: *L. rohita, C. mrigala* and *C. catla,* showed significant reductions in both carbohydrate and lipid contents in muscles and gills in all the three fish species.

4. Conclusion

In conclusion, carbohydrates though being the primary and immediate source of energy are depleted in fish tissues under stress conditions to meet the immediate energy demand. Glycogen is localized in sarcoplasm under the sarcolemma, in sarcoplasmic reticulum and throughout the sarcoplasmic core. Increase in blood sugar levels (hyperglycemia) following metal toxicity is the result of glycogenolysis. Once blood sugars are utilized it reflects hypoglycemic conditions in affected fish tissues. Depletion in tissue proteins and lipids levels further points towards exhaustion of cell energy to meet high demand of fish in stressful condition. During stress condition a fish need more energy to detoxify the toxicants and overcome stress. Since fish have a very little amount of carbohydrate stored in tissues, the next alternative source of energy in a fish are the proteins to meet its increased energy demand. The observed depletion of protein fraction may have been due to their degradation and possible utilization of degraded products for metabolic purpose. Decline in serum proteins like the albumin and globulin as well as cholesterol levels are important characteristics of aflatoxin toxicity in fish. Zacharia *et al.*,

2003, recorded dose dependent changes in the serum protein profile of the fish, *O. mossambicus* treated with aflatoxins beta-1. Histochemical studies (Pathan *et al.*, 2009) on protein, lipid and glycogen contents of liver in *Rasbora daniconius* fish, exposed to sub lethal concentrations (1.9 per cent and 0.95 per cent) of the paper mill effluents showed a progressive decrease in staining intensity to Mercury bromophenol blue (Hg-BPB), Sudan black B and Bets's Carmine. The magnitudes of these changes were dose dependent, and it was concluded that paper mill effluents create metabolic crisis and impairment in fish liver. The depletion in level of protein, lipid and glycogen points towards exhaustion of cell- energy to meet high demand of fish in stressful condition. The protein amounts showed an initial increase up to day-21, but declined thereafter, by the day-30. These changes are attributed to DNA damage, which in turn leads to changes in protein profile of the affected animal.

Emad *et al.*, 2005 recorded higher values of total lipids in plasma and muscle than the controls and the muscle lipids were said to be affected more by Cu than Cd in marine fish, *Mugil seheli.* No significant differences were observed in the whole body crude protein content of fish, *O. niloticus* (Ali *et al.*, 2003) exposed to different concentrations of copper as compared with the control. Liver moisture and ash contents increased whereas the crude protein, fat, nitrogen free extract and gross energy contents decreased. The liver glycogen level also decreased significantly with the copper concentration increase in water. Rafia *et al.*, 2008 also reported lipid content in fish muscles inversely related to the level of As, Hg and Pb. The freshwater fish, *Pomadasy agyrew* (22.86) was said to contain highest total lipid concentration as compared to the marine fish, *Liza subviridus* (14.04), indicating that marine water fish use more lipids contents due to heavy load of metals, as also suggested by other workers (Michael *et al.*, 1986; Ozmen *et al.*, 2006). hypothalamic pertaining to the hypothalamus.

Depletion in these biochemicals in ovary was also referred as metabolic compensations to the increased energy requirements for various reproductive functions under toxicity stress (Verma and Srivastava, 2007). Verma and Srivastava (2007) in *C. punctatus* evidenced decline in glycogen, lipid and protein levels in liver and ovary on exposure to Hg and Cd. Decline in glycogen in the liver and ovary was also shown to be dose and duration dependent decrease. These changes in glycogen levels were attributed to its utilization for maturation and development of oocytes in ovary as well as a result of higher production of energy molecules required for metabolic activities of the ovary. The bio-chemical changes in ovary (Verma and Srivastava, 2007) in *C. punctatus* further reflect a decline in reproductive capacity and breeding of fish. The decline in protein contents as noted above may be due to blocking of protein synthesis, as suggested by Vijayram *et al.* (1989). Hepatic proteins may be depleted due to proteolysis (Dhanpakian and Ramasamy 2001, Desai *et al.*, 2002) by increased activity of lysosomal enzyme or they may be mobilized for yolk formation in the developing oocytes. Short term (7 days) exposure of the yearlings of *Cyprinus carpio* to safe concentration of $HgCl_2$ (0.5 ppm) and study of changes 7 days after withdrawal of the treatment (Masud *et al.*, 2009) further revealed noticeable degenerative histophysiological changes in both ovary and liver after exposure, which were more prominent in the group with abnormal behaviour. After withdrawal of the $HgCl_2$ treatment the recovery was apparent in both organs but was more appreciable in liver. These observations indicated that even safe concentration of $HgCl_2$ might not be fully devoid of deleterious influence on reproductive functions in *C. carpio*. Srivastava and Verma, 2009 on observing similar decline in various biochemical parameters studied in *C. punctatus* reported an adverse influence of Zn on metabolism as well as reproductive activity of the fish. Exposing *C. punctatus* to sub-lethal concentrations of zinc (10, 15 and 25 Zn mg l-1) for15 days was found to result in a significant decline of hepatic and ovarian glycogen, cholesterol, total proteins and total lipids. They also found DNA and RNA contents in the ovary simultaneously declined with an increase in dose and duration of the experiment. Exposure of *C. carpio* and *C. mrigala* to zinc for 60 days (Dhawan and Kaur, 1997) also results in a decline of total proteins and lipids in the ovary.

Al-Attar (2007) in the freshwater fish, *O. niloticus* exposed to Ni showed significantly elevated levels of serum glucose, cholesterol, total protein, albumin, amylase, lipase. Palaniappan *et al.* (2008) showed that the liver tissues are vulnerable to arsenic intoxication and that arsenic intoxication induces significant alteration in the major biochemical constituents such as lipids, proteins and nucleic acids, which can be easily evidenced by Fourier transform infrared (FTIR) spectroscopy. He focused on the arsenic induced biochemical changes in the liver tissues of freshwater fish, *L. rohita*, using FTIR spectroscopy. The spectrum of liver tissue is quite complex and contains several bands arising from the contribution of different functional groups belonging to proteins, lipids and others. The FTIR spectra revealed significant differences in absorbance intensities and areas between control and arsenic intoxicated liver tissues; this variation shows the alteration in biochemical contents like proteins and lipids in the liver tissues due to arsenic intoxication. The decreased band areas of symmetric and asymmetric CH_2 stretching modes observed in the arsenic intoxicated tissues suggests the decreased composition of the acyl chains in the arsenic intoxicated liver tissues of *L. rohita*. The decrease in the band areas and intensities of amide bands in the arsenic intoxicated tissues indicate a decrease in the protein quantity of the system and the destructive effect of the arsenic in the liver tissues. The decrease in the band areas of symmetric and asymmetric stretching modes of phosphor-diester groups suggest a decrease in the relative content of the nucleic acids in the arsenic intoxicated liver tissues.

Consciously, increasing water contamination appear lethal to fish in inducing physiological/genotoxic effects, like the induction of chromosomal aberrations, micronuclei formation or the DNA modulations, as reported by Manna *et al.*, 1985, with aldrin in the erythrocytes of *O. mossambicus*. The toxic chemicals further, cause a variety of indirect rather subtle ill-effects (Al-Sobti, 1985). Chaudhary (2004) reported decrease in liver and muscle protein and total DNA and RNA on exposure of *C. gariepinus* to sublethal concentration of cadmium and the decrease was fully in proportion to the length of the exposure period. Similar decline in RNA and DNA levels were also observed in liver tissue of *C. punctatus* when exposed to lead acetate (Ramalingam *et al.*, 2000). Deposition of mercuric chloride and significant decrease in RNA/DNA ratio after 9 days of exposure was also noticed in pre-spawning ovary of *L. rohita* as well as after mercury treatment of *L. rohita* for 30 days by Aditya *et al.*

(2002). This decline may be a result of chromosomal aberration and fragmentation as reported by Gupta *et al.* (2006), in a freshwater teleost, *H. fossillis* (Bloch) treated with zinc. Lowering of RNA/DNA ratio and DNA and RNA content also indicates an increased catabolism of proteins which corroborates well with the decline in proteins. Muley *et al.* (2000), found significant alterations in the DNA and RNA contents in gills, liver and brain of the common carp, *C. carpio* exposed to 96-h LC_0 (98 and 504 ppm) and LC_{50} (121.8 and 594 ppm) concentrations of cadmium chloride and lead acetate, respectively. Both heavy metals decreased DNA content in all tissues. Cd and Pb toxicity decreased RNA content in liver and brain and increased it in gills.

REFERENCES

Abhas H.H., Zaghloul K.H. and Mousa M.A., 2003. Effect of some heavy metal pollutants on some biochemical and histopathological changes in blue Tilapia, *Oreochromes aureus*. Egyptian J. Agric. Res., 80 (3): 1395-1410.

Aditya A.K., Chattopadhyay S. and Mitra S., 2002. Effect of mercury and methyl parathion on the ovaries of *Labeo rohita* (Ham). J. Environ. Biol., 23: 61-64.

Al-Attar A.M., 2005. Biochemical effects of short-term cadmium exposure on the freshwater fish, *Oreochromis niloticus*. J. Biological Sci., 5: 260-265.

Al-Attar A.M., 2006. The physiological response of the fish, *Cyprinion mhalensis* to mercury intoxication. J. Egypt. Ger. Soc. Zool., 51: 123-137.

Al-Attar Atef M., 2007. The influences of nickel exposure on selected physiological parameters and gill structure in the teleost fish, *Oreochromis niloticus*. J. Biol. Sci., 7: 77-85.

Ali K.S., Dorgai L., Gazdag A., Abraham M. and Hermesz E., 2003. Identification and induction of Hsp70 gene by heat shock and cadmium exposure in carp. Acta Biol. Hung., 54 (3-4): 323-334.

Al-Sobti K., 1985. Carcinogenic- mutagenic chemicals induced chromosomal aberrations in the kidney cells of three cyprinids. Comp. Biochem Physiol., 82: 489-493.

Aly Mil Kim H.C., Renner S.W., Boyarsky A., Kosmin M. and Paglia D.E., 1993. Hemolytic anemia associated with lead poisoning from shotgun pellets and the response to succeeding treatment. Am J Hematol., 44: 280- 3.

Amudha P., Sangeetha G. and Mahalingam S., 2002. Dairy effluent induces alterations in the protein, carbohydrate and lipid metabolism of a freshwater teleost fish, *Oreochromis mossambicus*. Pollut. Res., 21(1): 51-53.

Ayas Z., Ekmerkci G., Yerli S.V. and Ozmen M., 2007. Heavy metal accumulation in water, sediments and fishes of Nallihan Bird Paradise, Turkey J. Environ. Biol., 28, 545-549.

Balm P.H.M., Pepels P., Helfrich S., Hovens M.L.M. and Wendelaar Bonga S.E. 1994. Adrenocorticotropic hormone in relation to interregnal function during stress in tilapia (*Oreochromis mossambicus*). Gen. Comp. Endocrinol., 96: 347-360.

Barton B.A., 2002. Stress in fishes: a diversity of responses with particular reference to changes in circulating corticosteroids. Integ. Comp. Biol., 42: 517-525.

Barton B.A. and Iwama G.K., 1991. Physiological changes in fish from stress in aquaculture with emphasis on the response and effects of corticosteroids. Ann. Rev. Fish Dis., 1: 3-26.

Barton B.A., Morgan J.D. and Vijayan M.M., 2002. Physiological and condition-related indicators of environmental stress in fish. In: Biological Indicators of Aquatic Ecosystem Stress (ed. S. M. Adams), pp. 111-148. Bethesda, USA: American Fisheries Society.

Bengeri K.V. and Patil H.S., 1986. Respiration, liver glycogen and bioaccumulation in *Labeo rohita* exposed to zinc. Indian J. Comp. Anim. Physiol., 4: 79–84.

Benson N.U., Essein J.P., Willium A.B. and Bassey D.E., 2007. Mercury accumulation in fishes from tropical aquatic ecosystems in the Niger Delta, Nigeria. Curr. Sci., 92: 781-785.

Bhattacharya R.K., Francis A.R. and Shetty T.K., 1987. Modifying role of dietary factors on the mutagenecity of alfatoxin B_1: *in vivo* effects of vitamins. Mutat. Res., 188 : 121-128.

Billy G.G., Miller C.A., Pallone M.N., Donachy J.H. and Pierce W.S., 1995. Hemolytic differences among artificial cardiac valves used in a ventricular assist pump. Artif. Org., 19: 339-43.

Black S. and Love R.M., 1986. The sequential mobilization and restoration of energy reserves in tissues of Atlantic cods during starvation and feeding. J. Comp. Physiol. B: Biochem. Syst. Environ. Physiol., 156: 469-479.

Black E.C., Conner A.R., Lam K. and Chin W., 1962. Changes in glycogen, pyruvate and lactate in rainbow trout (*Salmo gairdneri*) during and following muscular activity. J. Fish Res. Bd. Can., 19: 409.

Blasco J. and Pupo J., 1999. Effects of heavy metals (Cd, Cu, Pb) on aspartate and alanine aminotransferase on *Ruditapes philippinarum* (Mollusca:Bivalva). Comp. Biochem. Physiol. C., 122: 253-263.

Brett J.R., 1958. Implications and assessment of environmental stress. pp. 69-83. In: The investigation of fish-power problems, P.A. Larkin (ed.). H.R. MacMillan, Lectures in Fisheries, Univ. British Columbia, Vancouver, B.C.

Brett J.R. and Groves T.D.D., 1979. Physiological energetic, Ch., 6. In: Fish Physiology. Vol. III; Hoar W.S. and Randal D.J. (Eds). Academic Press, New York.

Broderius S.J. and Smith L.L. Jr., 1979. Lethal and sublethal effects of some mixtures of cyanide and hexavalent chromium, zinc or copper to the fathead minnow (*Pimephales promelas*) and rainbow trout (*Salmo gairdnairi*). J. Res. Board. Can., 36: 164-172.

Brown V.M. and Dalton R.A., 1970. The acute lethal toxicity to rainbow trout mixtures of copper, phenol, zinc and nickel. J. Fish Biol., 2: 211-216.

Canli M., 1995. Effect of mercury, chromium, and nickel on some blood parameters in the carp, *Cyprinus carpio*. Turk. J. Zool., 19: 305-311.

Canli M., 1996. Effects of Hg, Cr and Ni on glycogen reserves and protein levels in tissues of *Cyprinus carpio*. Turk. J. Zool., 20(2): 161-168.

Cannon W.B., 1929. Organization for physiological homeostasis. Physiol. Rev., 9: 399–431.

Cattani O., Serra R., Isani G., Raggi G., Cortesi P. and Carpene E., 1996. Correlation between metallothionein and energy metaboliosm in sea bass, *Dicentrarchus labrax*. Comp. Biochem. Physiol., 113 (2): 193-199.

Chaudhary S., 2004. Study of ecobiology and physiology of African catfish *Clarius gariepinus* with special reference to effect of cadmium on reproduction. Ph.D. Thesis, MD University, Rohtak.

Chaudhary H.S. and Nath K., 1985. Nickel induced hyperglycemia in the freshwater fish, *Colisa fasciatus*. Water, Air and Soil Pollut., 24: 173-176.

Chaudhary H.S., 1984. Nickel toxicity on carbohydrates metabolism of a freshwater fish, *Colisa fasciatus*. Toxicol. Lett., 20: 115-121.

Chavin W. and Young J E., 1970. Factors in the determination of normal serum glucose levels of gold fish, *Carassius auratus* L. Comp. BIochem. Physiol., 33: 629-653.

Cicik B. and Engin K., 2005. The effects of Cd on levels of glucose in serum and glycogen reserves in the liver and muscle tissues of *Cyprinus carpio* (L.). Turk. J. Anim. Sci., 29: 113-117.

Clary J.J., 1975. Nickel chloride-induced metabolic changes in the rat and guinea pig. Toxicol. Appl. Pharmacol., 31: 55-65.

Das S. and Bhattacharya T., 2000. Effect of cadmium and mercury on the blood and liver of *Channa punctatus*. The Fifth Ind. Fisheries Forum (17-20 Jan) CIFA, Bhubaneshwar, India. 51-123pp.

Das S., Patro S.K. and Sahu B.K., 2001. Biochemical changes induced by mercury in the liver of penaeid prawns *Penaeus indicus* and *P. monodon* (Crustacea: Penaeidae) from Rishikulya estuary, East coast of India. Indian J Mar. Sci., 30(4): 246-252.

Das K.K. and Dasgupta S., 2002. Effect of nickel sulfate on testicular steroidogenesis in rats during protein restriction. Environ. Health Perspect., 110: 923-926.

Dawson A.B., 1935. The hemopoietic response in the catfish *Ameriurus nebulosus* to chronic lead poisoning. Biol. Bull., 3: 235-346.

De-Smet-Hans and Blust-Ronny, 2001. Stress responses and changes in protein metabolism in carp, Cyprinus *carpio* during cadmium exposure. Ecotoxicol. Environ. Saf., 48(3): 255-262.

Desai H.S., Nanda B. and Panigrahi J., 2002. Toxicological effects on some biochemical parameters of freshwater fish, *Channa punctatus* (Bloch.) under the stress of nickel. J. Environ. Biol., 23 (3): 275-277.

Dhanapakiam P. and Ramaswamy V.K., 2001. Toxic effects of copper and zinc mixtures on some haematological and biochemical parameters in common carp, *Cyprinus carpio* (L.). J. Environ. Biol., 22 (2): 105-111.

Dhawan A. and Kaur K., 1997. Effect of zinc on maturation and breeding potential of *Cyprinus carpio* and *Cirrhina mrigala*. Intern. J. Environ. Stu., 53: 265-274.

Dinodia G.S., 2001. Studies on the effects of cadmium toxicity in some freshwater fishes. M.Sc. Thesis, Depart. Zool. and Aquacul., CCS HAU, Hisar.

Diwan A.D., Hingorani H.G. and Naidu N.C., 1979. Levels of blood glucose and tissue glycogen in two live fish exposed to industrial effluents. Bull. Environ. Contam. Toxicol., 21: 269-272.

Dutta H.M. and Haghighi A.Z., 1986. Methylmercuric chloride and serum cholesterol level in bluegill (*Lepomis machrochirus*). Bull. Environ. Contam. Toxicol., 36: 181-185.

EIFAC. 1975. Report on fish toxicity testing procedures. Prepared by European Inland Fisheries. Technical paper, 24.

Emad H., Abou El-Naga, Khalid M., El-Moselhy and Mohamed A. Hamed., 2005. Toxicity of cadmium and copper and their effect on some biochemical parameters of marine fish, Mugil sehel. Egyptian J. Aqua. Res., 31: 60-71.

Fergusson J.E., 1990. The heavy elements: Chemistry, environmental impact and health. Pergamon Press, Oxford. p 614.

Ferrando M.D. and Andreu-Moliner E., 1991a. Changes in selected parameters in the brain of the fish, *Anguilla anguilla*, exposed to lindane. Bull. Environ. Contam. Toxicol., 47: 459-464.

Ferrando M.D. and Andreu-Moliner E., 1991b. Effect of lindane on the blood of a freshwater fish. Bull. Environ. Contam. Toxicol., 47: 465-470.

Gamperl A.K., Vijayan M.M. and Boutilier R.G., 1994. Experimental control of stress hormone levels in fishes: techniques and applications. Rev. Fish Biol. Fish., 4: 215-255.

Garg S., Gupta R.K., Jain K.L., 2009, Sublethal effects of heavy metals on biochemical composition and their recovery in Indian major carps. J. Hazard Mater. 163(2-3):1369-84.

George K. Roy, Malini N.A. and Sandhya Rani G.O., 2012. Biochemical changes in liver and muscle of the cichlid, *Oreohromis mossambicus* (Peters) exposed to sublethal concentrations of mercuric chloride. Ind. J. Fish, 59(2): 147-152.

Gill T.S. and Epple A., 1993. Stress related changes in the hematological profile of the American eel (*Anguilla rostrata*). Ecotoxicol. Environ. Saf., 25: 127-135.

Gill T.S. and Pant J.C., 1985. Erythrocytic and leucocytic responses to cadmium poisoning in freshwater fish, *Puntius conchinius*. Environ. Res., 36: 327-337.

Gluth G. and Hanke W., 1985. A comparison of physiological changes in carp *Cyprinus carpio*, induced by several pollutants at sub lethal concentrations. Ecotox. Environ. Saf., 9: 179-188.

Goodman D.S. and Noble R., 1968. Turnover of plasma cholesterol in man. J. Clin. Invest., 47: 231-41.

Gupta N., Gupta D.K., Verma V.K. and Krishna G., 2006. Zinc-induced changes on chromosomes of a freshwater teleost, *Heteropneustes fossilis* (Bloch). Asian J. Exp. Sci., 20: 281-288.

Gupta P., 1991. Effect of electroplating effluent on certain tissues of freshwater teleost teleost fish of the class Osteichthyes, having the skeleton completely ossified., *Mystus bleekeri*, using bioassay statistical and histopathological methods. Arch. Hydrobial., 4: 427-434.

Heath A.G., 1987. Water pollution and fish physiology. CRC Press, Inc., Florida. 245 pp.

Heath A.G., 1995. Water pollution and Fish physiology. Second Ed. Lewis Publishers. 359 pp.

Heath A.G. and Pritchard A.W., 1965. Effects of severe hypoxia on carbohydrate energy stores and metabolism in two species of freshwater fish. Physiol. Zool., 38: 325.

Harpert A., Rodwell V.W. and Mayer A., 1977. A review of physiologial chemistry. 16th Edn. Lang Medial Publ., California.

Hilmy A.M., Shabana M.B. and Daabees A.Y.,1985. Effects of cadmium toxicity upon the *in vivo* and *in vitro* activity of proteins and five enzymes in blood serum and tissue homogenates of *Mugil cephalus*. Comp. Biochem. Physiol., 81C: 145-153.

Hodson P.V., 1988. The effect of metal metabolism on uptake, disposition and toxicity in fish. Aquat Toxicol., 11: 3-18.

Hontela A., Daniel C., and Ricard A.C., 1996. Effects of acute and subacute exposures to Cd on the inter-renal and thyroid function in rainbow trout, *Onchorhynchus mykiss*. Aquat. Toxicol., 35:171-182.

Hota S., 1996. Arsenic toxicity to the brain, liver and intestine of a freshwater fish, *Channa punctatus*. Geobios., 23: 154-156.

Iwama G.K., Vijayan M.M., Forsyth R.B. and Ackerman P.A., 1999. Heat shock proteins and physiological stress in fish. Am. Zool., 39: 901-909.

Jacobson-Kram D, KA Keller. 2001. Toxicology testing handbook. New York: Marcel Dekker.

Jagadeesan G. and Sankarsami Pillai S., 2007. Hepato-protective effect of taurine against mercury induced toxicity in rat J. Environ. Biol., 28, 753-756.

Jain K.L., Sharma M., 2003. Toxic effects of mercury and cobalt on biochemical composition of freshwater fish, *Cyprinus mrigala*. Proc. 3rd Interact. Workshop (December 17-18): pp 203-207.

James R, Sampath K. and others, 1992. Effect of lead on respiratory enzymes activity, glycogen and blood sugar levels of the teleost, *Oreochromis mossambicus* (Peters) during accumulation and depuration. Asian Fish Sci. 9: 87-100.

James R.V., Jancy Pattu, Devakiamma G. and Sampath K., 1991. Impact of sublethal levels of mercury on glycogen and selected respiratory enzymes in *Heteropneustes fossilis* and role of water hydracinth in reduction of Hg toxicity. Ind. J. Fish, 30: 249-252.

Jana S.R. and Bandhopadhyay N., 1987. Effects of heavy metals on some biochemical parameters in freshwater fish, *Channa punctatus*. Environ. Ecol., 5: 488-493.

Jana S. and Sahana S.S., 1988. Effects of copper, cadmium and chromium cations on the freshwater fish, *Clarias batrachus* L. Physiol. Bohemoslov., 37(1): 79-82.

Jehan Z.S. and Motlag D.B., 1995. Metal induced changes in the erythrocyte membrane of rats. Toxicol. Lett., 78:127-133.

Jeney Z., Valtonen E.T., Jeney G., Jokinen E.I.,1996: Effects of pulp and paper mill effluent (BKME) on physiology and biochemistry of the roach (*Rutilus* rutilus L.). Arch. Environm. Contam. Toxicol., 30: 523-529.

Jha B.K. and Jha B.S., 1994. Ovarian histopathology in urea and ammonium sulphate intoxicated freshwater teleost, *Heteropneustes fossilis*. J. Environ. Pollut., 1: 145-148.

Jobling M., 1994. Fish Bioenergetics. Vol. 13, Chapman and Hall, London, UK.

Johansson-Sjobeck M.L. and Larsson A., 1978. The effect of Cd on the haematology and on the activity of delta-amino levulinic acid dehydratase (ALA-D) in blood and haemopoetic tissues of the flounder, *Pleuronectes americanus*. Environ. Res., 17: 191-204.

Joshi P.K., Bose M. and Harish D., 2002. Haematological changes in the blood of *Clarias batrachus* exposed to mercuric chloride. J. Ecotoxicol. Environ. Monit., 12: 119-122.

Kaur K., 1992. Effect of industrial effluents on reproductive and developmental biology of *Cyprinus carpio* (Linn.) Ph.D. Thesis. Punjab Agril. Univ. The Punjab Agricultural University (PAU), Ludhiana, Punjab is one of the State Agricultural Universities in India. It was established in 1962 and is the oldest agricultural university in India after GBPUA&T, Pantnagar. , Ludhiana. 168 pp.

Kaur T. and Saxena P.K., 2001. Impact of pollution on the flesh of some fishes inhabiting river Satluj waters – a biochemical study. Ind. J. Environ. Hlth., 44(1): 58-64.

Kaur Tejinder, Saxena P.K. and Jior R.S., 2000. Impact of pollution in Budha Nalla brook on the occurrence of fishes in river Satluj. Ind. J. Fish, 47(1): 49-53.

Khanee K.M.A., Nazeemul James R., Sivakumar K. and Muthuramalakshmi M., 1991. Sublethal effects of mercury on growth and biochemical composition and their recovery in *Oreochromis mossambicus*. Ind. J. Fish, 38(2): 111-118.

Khanna S.S. and Gill T.S., 1975. Effect of cobalt salts on the glycemia and islet histology of *Channa punctatus* (Bloch). Acta. Anat., 92: 194.

Kirunbagaran P. and Joy K. P., 1992. J. Fish Biol., 41: 305-313.

Kori-Siakpere O., Ake J.E.G., and Others, 2006. Sublethal effects of Cadmium on some selected hematological parameters of *Heteroclarias*. Inter. J. Zool. Res., 2 (1): 77-83.

Kori-Siakpere O., 1995. Some alterations in haematological parametersin *Clarias isheriensis* (Sydenham) exposed to sublethal concentration of water-born lead. Biosci. Res. Commun., 8: 93–98.

Kralj–Klobuear Nada, 1993. Copper Sulphate and relation to glycogen content in liver parenchymal cells. Proceedings: Trace elements in man and Animals. Place-Dresdan, Njemaeka, 16-21 May. pp 386 – 389.

Kumar A. (Ed.), 2002. Ecology of polluted waters Vol. 1. APH Publ. Corp., New Delhi.

Kumar A.K. and Achyuthan H., 2007. Heavy metal accumulation in certain marine animals along the East Cost of Chennai, Tamil Nadu, India. J. Environ. Biol., 28: 637-643.

Kumar P., Gupta Y.G., Tiwari V., Tiwari S. and Singh A., 2005. Hematological and biochemical abnormalities in *Cirrhinus mrigala* (Ham.) induced by lead. J. Ecophysiol. Occupl. Hlth., 5: 213-216.

Kumari Anitha S. and Kumar S.R.N., 1997. Effect of Polluted water on histochemical localization of carbohydrates in a freshwater teleost, *Channa punctatus* (Bloch) from Hussain Sagar Lake, Hydrabad, Andhra Pradesh. Pollut. Res., 16 (3): 197 -200.

Kumari Kumari S., 1984. Studies on toxicity of heavy metals to *Channa punctatus* with special reference to nickel and chromium. Ph.D. Thesis, Merrut Univesity, Meerut.

Kuzumina V.V., Golovanova I.L., Shishin M.M. and Smirnova E.S., 2004. Influence of heavy metals on feeding behaviour and digestive process. The sixth Syrian, Egyptian conference on the chemical and petroleum engineering. Al Baath University Homes, Syria.

Larsson A and Fang R., 1977. Cholesterol and free fatty acids in the blood of marine fish. Compar. Biochem. Physiol., 57: 191-196.

Larsson A, Johansson-Sjobeck M L. and Fange R., 1976. Comparative study of sime haematological and biochemical blood parameters in fishes from the Skagerrak. J. Fish Biol., 9: 425-440.

Larsson A., 1975. Some biochemical effects of Cd on fish: In: Sublethal effects of toxic chemicals on aquatic animals¸ Koeman J.H. and Jitwa Strik (eds). Elsevier Scient. Pub. Co. Amsterdam, 3-13pp.

Leatherland J.F., 1985. Studies of the correlation between stress-response, osmoregulation and thyroid physiology in rainbow trout, *Salmo mirdnerii* (Richard.). Comp. Biochem. Physiol., 8OA: 523-531.

Levesque H.M., Moon T.W., Campbell P.G.C and Hontela A., 2002. Seasonal variation in carbohydrate and lipid metabolism of yellow perch (*Perca flavescens*) chronically exposed to metals in the field. Aquatic toxicol., 60: 257-267.

Love R.M., 1980. The chemical biology of fishes, vol. 2: Advances 1968-1977. Academic Press, London, U.K.

Malik D.S., Sastry K.V. and Hamilton D.V., 1998. Effects of zinc toxicity on biochemical composition of muscle and liver in Murrel (*C. punctatus*). Environ. Internat., 24(4): 433-438.

Manna G.K., Benerrgee G. and Gupta S., 1985. Micronucleus test in the peripheral erythrocytes of the exotic fish, *Orechromis mossambicus*. Curr. Sci., 28: 176-179.

Masud Sahar, Singh I.J. and Ram R.N., 2009. Histophysiological responses in ovary and liver of *Cyprinus carpio* after short term exposure to safe concentration of mercuric chloride and recovery pattern. J. Environ. Biol., 30(3) 399-403.

Mathur A.K. and Tandon S.K., 1979. Chemosphere, 8: 893.

Mazeaud M.M., and Mazeaud F., 1981. Adrenergic responses to stress in fish. In: Stress and fish, A.D. Pickering (ed.). Academic Press, London, U.K. pp 49-75.

McEwen B.S. and Stellar E., 1993. Stress and the individual: Mechanisms leading to disease. Arch. Intern. Med., 153(18): 2093-101.

McKim J.M., Bradbury S.P. and Niemi G.L., 1987. Fish acute toxicity syndromes and their use in the QSAR approach to hazard assessment. Environ. Hlth. Prospect., 71: 171.

Meenakashi V., Indira N. and Pugazhendhi K., 1998. Influence of zinc sulphate on the freshwater edible fish, *Labeo rohita* fingerlings. U.P. J. Zool., 18(3): 181-183.

Meyers T.R. and Hendricks J.D., 1985. Histopathology. In: Fundamentals of Aquatic Toxicology. (Eds. Rand, G.M. and Petrocelli, S.R.), Hemisphere, Washington, D.C., 283-331pp.

Michael L.F., Pennypacker K.R., Drummond K.A. and Blem C.R., 1986. Concentration and location of metabolic substrates in fast toadfish sonic muscle. Copeia, 4: 910-915.

Mittal V., 2004. Studies on the toxic effects of some heavy metals in freshwater fish, *Cirrhinus mrigala*. PhD. Thesis Chaudhary Charan Singh Haryana Agricultural University, Hisar., pp145.

Moberg G.P., 1985. Biological response to stress: key to assessment of wellbeing? In: Animal stress, G.P. Moberg, ed. Am. Physiol. Soc., Bethesda, MD., pp. 27-49.

Mukherjee D., Kumar V., Chakraborti P., 1994. Effect of mercuric chloride and Cd chloride on gonadal function and its regulation in sexually mature common carp, *Cyprinus carpio*. Biomed. Environ. Sci., 7: 13–24.

Muley D.V., Kamble G.B. and Bhilave M.P., 2000. Effect of heavy metals on nucleic acids in *Cyprinus carpio*. J. Environ. Biol., 21: 367–370.

Murugappan R.M., Muthusamy M. and Gunasundari M., 1999. Genotoxicity and bioaccumulation of BHC and cadmium on the freshwater fish, *Cyprinus carpio*. J. Environ.Biol., 20: 313-316.

Mustafa S. J. and Chandra S., 1972. Adenosine deaminase and protein pattern in serum and cerobro-spinal fluid in experimental manganese encephalopathy. Arch. Toxicol., 28: 279-285.

Naidu K.A., Abhinender K., Ramamurthi R., 1984. Acute effects of mercury toxicity on some enzymes in liver of teleost, *Sarotherodon mossambicus*. Ecotoxicol. Environ. Safety., 8: 215 – 218.

Nakano T. and Tomlinson N., 1967. Catecholamines and carbohydrate concentration in rainbow trout, *Salmo gairdneri*, in relation to physical disturbance. J.Fish.Res. Bd.Can., 24:1701-1715.

Nath K. and Kumar N., 1988. Hyperglycemic response of *Heteropneustes fossilis* exposed to nickel. Acta Hydrochem- Hydrobiol., 16: 333-336.

Neha Bhatkar, Vankhede G.N. and Dhande P.R., 2004. Heavy metal induced biochemical alterations in the freshwater fish, *Labeo rohita*. J. Ecotoxiol. Environ. Monit., 14(1): 1-7.

Naik S. J. K., Devi V. V and Piska S., 2004. Toxicity of tannery effluent on carbohydrate, protein, lipid constituents in the selected tissue of *Cyprinus carpio* (L.). J. Aqua. Biol., 19(1): 177-181

Nielsen F.H., 1977. Advances in modern technology, Vol-2; Toxicology of trace elements. Hemishere Publ. Corp., Washington, 129pp.

Oikari A. and Nittyla J., 1985. Subacute physiological effects of bleached Kraft-pulp mill effluent (BKME) on the liver of rainbow trout, *Salmo gairdneri*. Ecotoxicol. Environ. Saf., 10: 159-172.

Oikari A., Holmbom B. and Bister H., 1982. Uptake of resin acids into tissues of trout (*Salmo gairdneri* Richardson). Ann. Zool. Fennici, 19: 61-64.

Orji R.C.A., 1998. Effect of transportation stress on hepatic glycogen of *Oreochromis niloticus* (L.). Naga: 20-22.

Ozmen M., Gungordu A., Kucukbay F.Z. and Gnler R.E., 2006. Monitoring the effects of water pollution on *Cyprinus carpio* in Karakaya Dam Lake Turkey. Ecotoxicol., 15: 157-169.

Palaniappan P.L.R.M. and Vijayasundaram V. 2008. FTIR study of arsenic induced biochemical changes on the liver tissues of freshwater fingerlings (*Labeo rohita*). Romanian J. Biophys., 18(2):135–144.

Partap H.P. and Wendelaar-Bonga S.E., 1990. Effect of water- borne cadmium on plasma cortisol and glucose in the cichlid fish, *Oreochromis mossambicus*. Comp. Biochem.Physiol., 95: 313-317.

Parvathi Kumar, Palanivel Sivakumar, Mathan Ramesh and Sarasu. 2011. Sublethal effects of chromium on some biochemical profiles of the freshwater teleost, *Cyprinus carpio*. J. Appl. Biol. and Pharm. Tech. 2: 294-300.

Passow H., Rothstein A. and Clarkson T.W., 1961. The general pharmacology of heavy metals. Pharmacol. Rev., 13: 185-224.

Pathan T.S., Thete P.B., Shinde S.E., Sonawane D.L. and Khillare Y.K., 2009. Histochemical changes in the liver of freshwater fish, *Rasbora daniconius*, exposed to Paper-Mill effluents. Emir. J. Food Agric., 21 (2): 71-78.

Prabhahar C., Saleshrani K., Tharmaraj K. and Vellaiyan M. 2012. Studies on the Effect of Cadmium Compound on the Biochemical Parameters of Freshwater Fish in Cirrhinus mrigala. International Journal of Pharmaceutical and Biological Archives 2012; 3(1):69-73.

Qayyum M.A. and Shaffi S.A., 1977. Changes in tissue glycogen of a freshwater catfish, *Heteropenustes fossilis* (Bloch) due to mercury intoxication. Curr. Sci., 46: 652.

Radhakrishaniah K, Venkataramana P and others, 1992. Effect of lethal and sublethal concentrations of copper on glycolysis in the liver of the freshwater teleost, *Labeo rohita* (Ham). J.Environ.Biol., 13: 63 68.

Rafia Azmat, Farha Aziz and Madiha Yousfi, 2008. Monitoring the effect of water pollution on four bioindicators of aquatic resources of Sindh Pakistan. Res. J. Environ. Sci., 2 (6): 465-473.

Ram R.N. and Sathyanesan A.G., 1984. Mercuric chloride induced changes in the protein, lipid and cholesterol levels of the liver and ovary of the fish, *Channa punctatus*. Environ. Ecol., 4: 23-29.

Ramalingam K. and Ramalingam K., 1982. Effects of sublethal levels of DDT, malathion and mercury on tissue proteins of *S. mossambicus*. Proc. Ind. Acad. Sci., 91: 501-505.

Ramalingam V., Vimaldevi V., Narmadaraj R. and Prabhakaran P. 2000. Effect of lead on haematological and biochemical changes in freshwater fish, *Cirrhinus mrigala*. Pollut. Res., 19 (1): 81-84.

Randall D.J. and Perry S.F., 1992. Catecholamine. In Fish Physiology, vol. XII B (ed. W. S. Hoar, D. J. Randall and T. P. Farrell), New York, London: Academic Press, pp. 255-300.

Rani A Shobha, Sudharsan R., Reddy T.N., Reddy P.U. and Raju T.N., 2001. Effect of arsenite on certain aspects of protein metabolism in freshwater teleost, *Tilapia mossambica* (Peters). J Environ Biol., 22 (2):101-4.

Reid S.G., Vijayan M.M. and Perry S.F., 1996. Modulation of catecholamine storage and release by the pituitary–interrenal axis in the rainbow trout (*Oncorhynchus mykiss*). J. Comp. Physiol., B 165: 665-676.

Ricard A.C., Daniel C., Anderson P. and Hontella A., 1998. Effects of subchronic exposure to cadmium chloride on endocrine and metabolic functions in rainbow trout, *Oncorhynchus mykiss*. Arch. Environ. Contam. Toxicol., 34(4): 377-381.

Ruparelia S.G., Verma Y., Mehta N.S. and Salyed S.R., 1989. Lead-induced biochemical change in freshwater fish, *Oreochromis mossambicus*. Bull. Environ. Contam. Toxicol., 43: 310- 314.

Sahoo P.K. and Mukherjee S.C., 1999. Normal ranges for diagnostically important haematological parameters of laboratory-reared rohu (*Labeo rohita*) fingerlings. Geobios., 26: 31-36.

Sanders B.M., Nguyen J., Martin L.S., Howe S.R. and Coventry S., 1995. Induction and subcellular localization of two major stress proteins in response to copper in the fathead minnow, *Pimephales promelas*. Comp. Biochem. Physiol., C 112: 335-343.

Sarita, 2005. Effects of heavy metals on enzyme activity in some freshwater fishes. PhD thesis, Deptt. Zoology., CCS HAU., Hisar, Haryana: 195 + xxii.

Sarkar B., Chatterjee A., Adhikari S. and Ayyappan S., 2005. Carbofuran- and cypermethrin-induced histopathological alterations in the liver of *Labeo rohita* (Hamilton) and its recovery. J. Appl. Ichthyol., 21, 131.

Sastry K.V and Anuradha, 2005. Effect of cadmium on some biochemical and enzymological parameters in the freshwater fish, *Catla catla*. Nat. Symp. Challenges and Opportunities in Animal Science Edu. and Res. (17-19 March, 2005), CCS Univ., Meerut.

Sastry K.V. and Rao D.R., 1981. Enzymological and biochemical changes produced by mercuric chloride in a teleost fish, *Channa punctatus*. Toxicol. Lett., 9 (4): 321-328.

Sastry K.V. and Rao D.R., 1984. Effects of mercuric chloride on some biochemical and physiological parameters of the freshwater murrel, *Channa punctatus*. Environ. Res., 34: 343 – 350.

Sastry K.V. and Shukla V., 1990. Toxic effects of cadmium on some biochemical and physiological parameters in the teleost fish, *Channa punctatus*. Biojournal, 2: 325-332.

Sastry K.V. and Subhadra K., 1982. Effect of cadmium on some aspects of carbohydrate metabolism in a freshwater catfish, *Heteropneustes fossilis*. Toxicol. Lett., 14(1-2): 45-55.

Schreck C.B., 1981. Stress and compensation in teleost fishes: Response to social and physical factors. In A. D. Pickering (ed.), Stress and fish, Academic Press, New York, pp 295–321.

Selye H., 1973. The evolution of the stress concept. Am. Sci., 61692-699.

Selye H., 1974. Stress without distress. New York American Library, New York.

Shaffi S.A., 1979. Lead toxicity: Biochemical and physiological imbalance in nine freshwater teleosts. Toxicol. Lett. 4: 155-161.

Shakoori A.R., Javed Iqbal M., Latif Mughal A. and Syed Shahid A., 1994. Biochemical changes induced by inorganic merury on the blood, liver and muscles of freshwater Chinese grasscarp, *Ctenopharyngodon idella*. J. Ecotoxicol. Environ. Monit., 4: 81-92.

Shukla J.P. and Pandey, K. 1986. Effects of sublethal concentration of zinc sulphate on the growth rate of fingerlings of *Channa punctatus*, a fish water murrel. Acta Hydrochem. Hydrobiol. 14: 195-197.

Shukla V., Rathi P. and Sastry K.V., 2001. Effect of toxicants on the intestine transport in fishes. Himalayan J. Environ. Zool., 15(2): 129-136.

Shukla V., Rathi P. and Shastry K.V., 2002. Effect of cadmium individually and in combination with other metals on the nutritive value of freshwater fish, *Channa punctatus*. J. Environ. Biol., 23(2): 105-110.

Shyong W.J. and Chen H.C., 2000. Acute toxicities of copper, cadmium and mercury to the freshwater fish, *Varicorhinus barbatus* and *Zacco barbata*. Acta Zoologica Taiwanica, 11(1): 33-45.

Silbergeld E K., 1974. Blood glucose a sensitive indicator of environmental stress in fish. Bull. Environ Contam. Toxicol., 11: 20-25.

Sindhe V.R., Veeresh M.V. and Kulkarni R.S., 2002. Ovarian changes in response to heavy metal exposure to the fish, *Notopterus notopterus*. J. Environ. Biol., 23(2): 137-141.

Singh H.S. and Reddy T.V., 1990. Effect of copper sulfate on hematology, blood chemistry and hepato-somatic index of an Indian catfish, *Heteropneustes fossilis* (Bloch) and its recovery. Ecotoxicol. Environ. Saf., 20: 30-35.

Sinha M.P., Kumar M., Pilan S.T. and Srivastava S., 2001. Calorific calorific generating heat measurable in calories. changes in liver, ovary and muscle of hill stream *Garra mullya* (Sykes) due to cadmium toxicity. J. Ecophysiol. Occup. Hlth., 1: 375-384.

Smolders R., Bervoets L., Wepener V. and Blust R., 2003. A conceptual framework for using mussels as Biomonitors in Whole Effluent Toxicity. Human and Ecol. Risk Assessm., 9: 1-20.

Soengas J.L., Agra L., Carballo B., Andres M.D. and Viera J.A.R., 1996. Effects of an acute exposure to sublethal concentrations of cadmium on liver carbohydrate metabolism of Atlantic Salmon. Bull. Environ. Contam. Toxicol., 57(4): 625-631.

Spehar R.L., 1976. Cadmium and Zinc toxicity to flagfish, Jordanella floridae. J. Fish Res. 28: 1107-1112.

Sprague J.B. 1971.Measurement of pollutant toxicity to fish. III. Sublethal effects and 'safe' concentrations. Water Res., 5:245–266.

Srivastava A. K. and Singh N. N., 1981. Acta Pharmacol. Et toxicol. 48: 26.

Srivastava A.K., Srivastava S.K., Mishra D., Srivastava, S. and Srivastava S.K., 2002: Ultimobranchial gland of freshwater teleost, Heteropneustes fossilis in response to deltamethrin treatment. Bull. Environ. Contam. Toxicol., 68 (4): 584-591.

Srivastava Rama and Neera Srivastava, 2008. Changes in nutritive value of fish, Channa punctatus after chronic exposure to zinc. J. Environ. Biol., 29: 299-302.

Srivastava Neera and Verma Hemlata, 2009. Alterations in biochemical profile of liver and ovary in zinc-exposed freshwater murrel, *Channa punctatus* (Bloch). J. Environ. Biol., 30(3) 413-416.

Suresh A., Sivaramakrishna B., Victoriamma P.C. and Radhakrishnaiah K., 1991. Shifts in protein metabolism in some organs of freshwater fish, *Cyprinus carpio* under mercury stress. Biochem. Int., 24(2): 379-389.

Syversen T.L.M., 1981. Effects of repeated dosing of methyl mercury on *in vivo* protein synthesis in isolated neurons. Acta Pharmcol. Toxicol., 50: 391-397.

Tandon R.S. and Chandra S., 1976. Cyclic changes in serum cholesterols of freshwater catfish, *Clarias batrachus*. Zeitschrift – fur – tierphysiologie – Tierernahrung – und – filtermittelkunde, 36(4): 176-183.

Tewari H., Gill T.S., Pant J., 1987. Impact of chronic lead poisoning on the hematological and biochemical profiles of a fish, *Barbus conchonius*. Ham. Bull. Environ. Contain. Toxicol., 40:198-203.

Thomas P., 1990. Molecular and biochemical responses of fish to stressors and their potential use in environmental monitoring. In: Biological Indicators of Stress in Fish, Adams, S.M. (Ed.). American Fisheries Society, Bethesda, Maryland.

Tytler P. and Calow P., 1985. Fish Energetics- New Perspectives. Croom-Helm Publications.London and Sydney. 369pp.

Umminger B.L., 1970. Physiological studies on super coolcd fish *Fundulus heteroclitus*. Carbohydrate metabolism and survival at sub zero temperature. J. Expt. Zool., 17: 159-174.

Umminger B.L., 1977. Relation of whole blood sugar concentrations in vertebrates to standard metabolic rate. Coup. Biochem. Physiol., 56A: 457-460.

Van Vuren J.H., van der Merwe M. and du Preeze H.H., 1994. The effect of copper on the blood chemistry of *Clarias gariepinus* (Clariidae). Ecotoxicol. Environ. Safety, 29: 187-199.

Verma Hemlata and Srivastava Neera, 2007. Biochemical changes in the liver and ovary of freshwater murrel, *Channa punctatus* after chronic exposure to zinc. Bull. Pure and Appl. Sci. (Zoology). (www.thefreelibrary.com/Biochemical changes)

Verma B.P., Sinha A.P. and Nomani M.M.R., 2002. Effect of methyl mercuric chloride (MMC) on ovarian cycle of *Clarias batrachus*. Aquacult., 3: 17-27.

Verma S.R. and Tonk I.P., 1983. Effect of sublethal concentration of mercury on the composition of liver, muscle and ovary of *Notopterus notopterus*. Water Air Soil Pollut., 20: 287-292.

Viarengo J.S., 1985. Biochemical effects of trace metals. Mar. Pollut. Bull. 16(4): 153-158.

Vijayan M.M., Ballantyne J.S. and Leatherland J.F., 1990. High stocking dnsity alters the energy metabolism of brook charr, *Salvelinus fontinalinus*. Aquacul., 88; 371-381.

Vijayan M.M., Reddy P.K., Leatherland J.F. and Moon T.W., 1994. The effects of cortisol on hepatocyte metabolism in rainbow trout: a study using the steroid analogue RU486. Gen. Comp. Endocrinol., 96: 75-84.

Vijayram K., Geraldine P., Varadarajan T.S., George J. and Loganathan P., 1989. Cadmium induced changes in the biochemistry of an air-breathing fish, *Anabas testudineus*. J. Ecobiol., 1 (4): 245-251.

Vijayaram K., Loganathan P. and Janarthanan S., 1991. Liver dysfunction in fish, *Anabas testudineus* exposed to sublethal levels of Pulp and Paper mill effluent. Environ. Ecol., 9 (1): 272 – 275.

Vincent S., Ambrose T. and Selvanayagam M., 2002. Impact of cadmium on food utilization of the Indian major carp, *Catla catla* (Ham.). J. Environ. Biol., 23(2): 209-212.

Virk S. and Sharma R.C., 1999. Biochemical changes induced by nickel and chromium in the liver of *Cyprinus carpio*. Pollutn. Res. 18(3): 217-222.

Virk S. and Sharma A., 2003. Changes in the biochemical constituents of gills of *Cirrhinus mrigala* following exposure to metals. Ind. J. Fish, 50: 113-117.

Virmani Shelly, 2005. Effects of heavy metals on histopathological and biochemical changes in *Cyprinus carpio* (L) and *Cirrhinus mrigal (Am). PhD thesis, Deptt. Zoology., CCS HAU., Hisar, Haryana*: 195+xxxiii

Vutukuru S.S. 2005. Acute effect of hexavalent chromium on survival, oxygen consumption, hematological parameters and some biochemical profiles of the Indian major carps, *Labeo rohita*. Int. J.Environ. Res. Pub. Helth., 2(3): 456-462.

Wagner G.F. and McKeown B.A., 1982. Changes in plasma insulin and carbohydrate metabolism of zinc-stressed rainbow trout, *Salmo gairdneri*. Canadian J. Zool., 60(9): 2079-2084.

Waldichuk M., 1979. Review of the problems. In: The assessment of the sublethal effect of pollutant in the sea (Ed. H. A. Cole), pp399-424.

Warner B., 1987. MNRAS, 227: 23

Weber J-M. and Zwingelstein G., 1995. Circulatory substrate fluxes and their regulation. In: Metabolic Biochemistry, vol. 4 (ed. P. W. Hochachka and T. P. Mommsen):15–32.

Wedemeyer G.A. and McLeay D.J., 1981. Methods for determining the tolerance of fishes to environmental stressors. In: Stress and fish, A.D. Pickering (Ed.). Academic Press, London, U.K., pp. 247-275

Wedemeyer G.A., McLeay D.J., and Goodyear C.P., 1984. Assessing the tolerance of fish and fish populations to environmental stress: the problem and methods of monitoring. In: Contaminant effects on fisheries, V.W. Cairns, P.V. Hodson, and J.O. Nriagu (eds). John Wiley and Sons, New York, N.Y., pp 163-195.

Wendelaar-Bonga S.E., 1997. The Stress response in fish. Physiol. Rev., 77: 591-625.

Writters H.E., Van Puymbroeck S. and Vanderborght O.L.J., 1991. Adrenergic response to physiological disturbances in rainbow trout, *Onchorhynchus mykiss*, exposed to aluminium at acid pH. Can. J. Fisheries and Aquat. Sci., 48: 414-20.

Woerr J.A., Huff W.E. and Hamilton P.B., 1987. Increased sensitivity to staphylococcal betahaemolysin of erythrocytes from chicken during alfatoxicosis. Poul. Sci., 66: 1929-1933.

Yang J. and Chen H.C., 2003. Effects of gallium on common carp, *Cyprinus carpio:* Acute test, serum biochemistry and erythrocyte morphology. Chemospher., 53: 877-882.

Yehuda R., 2007. HPA alterations in PTSD. Encyclopedia of Stress: 359-364.

Young J.E. and Chavin W. 1965. Effects of glucose, epinephrine or glucagon upon serum glucose levels in the goldfish, *Carassius auratus* L. Am. Zool., 5: 688-689.

Yousef M.K., 1985. Stress physiology: definition and terminology. In: Stress physiology in livestock, vol. I, Basic Principles, M.K. Yousef (Ed). CRC Press, Boca Raton, FL., pp. 3-7

Zacharia S., Thomas P.C. and Paulton M.P., 2003. Changes in serum proteins and immune response in *Oreochromis mossambicus* induced by aflatoxin B_1. Ind. J. Fish, 50 (4): 421-426.

Zalme R.C., McDowell E.M., Nagle R.B., McNeiel J.S., Flamenbaum W. and Tramp B.F., 1976. Virchows Arch. Abt. Zellpath., 22: 197.

Chapter 10

Hematological Aberrations

Blood together with lymph is a homeostatically controlled internal liquid environment and the chemical transporting medium. Fish blood is known to have two main components: red blood cells (RBCs) and plasma. Unlike mammals, there is no bone marrow or the lymph nodes in fishes. The blood cells in a teleost fish are produced in the hematopoietic tissue (hemocytoblasts), located primarily in the spleen and kidney (Satchell, 1991). Most important changes studied in fish blood of clinical value are the decline in blood volume, glucose and protein levels, plasma enzymes, cell counts, eosinophyll sedimentation rate (ESR), blood haemoglobin concentration and the hematocrit. The procedures are well standardized (Blaxhall and Daisley, 1973; Houston *et al.*, 1993). Other characteristic feature of the fish blood is that the erythrocytes are nucleated and are similar in size to the leucocytes, in contrast to the much larger cells in mammals; and the fish haemoglobin (Hb) further posses about 50 to 100 fold high oxygen handling capacity depending upon its concentration in the blood. Fish blood instead of the mammalian platelet cells, contain thrombocytes. Initial changes due to toxic chemicals are well manifested in this transporting medium that is the blood. It is well known that industrial wastes in an aquatic environment produce considerable toxic effects on organisms, especially on fish (Sjöbeck *et al.*, 1984). The stress caused by environmental pollution is also known to change the structure of red and white blood cells (Larsson *et al.*, 1984). As an indicator of pollution, blood parameters are used in order to diagnose and describe the general health condition of some fish. Blood parameters, such as RBC counts and estimating the hemoglobin (Hb), hematocrit (Hct), mean cellular volume (MCV), mean cellular hemoglobin (MCH), thrombocytes (PLT) are the most common criteria used in the toxicity studies in fish. hematocrit (Hct) also known as packed cell volume (PCV) or erythrocyte volume fraction (EVF), is the volume percentage (per cent) of red blood cells in blood; more the red blood cells are there in the blood, it carries more the oxygen. So, hematocrit (Hct) is a representation of the oxygen carrying capacity of the blood. Plasma is mostly water, along with some plasma proteins, glucose and a variety of other organic and inorganic chemicals and the enzymes, which in fact reflect functional status of the vital body organs in an animal, such as the liver, muscles, kidney, brain and the lungs. Serum is the left out plasma after the blood is allowed to clot naturally. Hematology is thus, stated as the study of chemical composition of the plasma and the quantification of blood cell components. Blood clotting is another such an important process which helps in checking the blood flow out of the injured tissues. It involves some of the plasma proteins and some filtered out blood cells, leaving behind the yellowish-red fluid called serum. Presence of aflatoxins are also known to prolong the blood clotting time (Woerr *et al.*, 1987), thus, resulting in severe pathological disorder. Changes in plasma lactate, hydro-mineral balance, and the circulating lymphocytes *etc.* are some other important secondary responses to stress (Wedemeyer *et al.*, 1984).

Hematology like mammals has been in frequent use as an index of health and pathological status in a number of fish species (Blaxhall, 1972; Heath 1987), and the hematological techniques thereby, are frequently applied as the most common method to determine especially sub-lethal effects following different types of stress conditions, like exposure to pollutants, diseases, hypoxia *etc.* (Andrews *et al.*, 1966; Christensen *et al.*, 1972; Duthie and Tort, 1985; Larsson *et al.*, 1985). Jantrarotai *et al.*, 1990 also evidenced severe anemia together with change in hematocrit (Hct), haemoglobin concentration, and erythrocyte and leukocyte counts; as the signs of acute alfatoxin toxicity. Sahoo and Mukherjee (1999) made a study for two months in the fingerlings of rohu (*L. rohita*) weighing between 29-38 g, so as to establish normal ranges for several hematological parameters and the serum protein levels. Larsson and Fang, 1977 studied 22 species of marine fishes (Cyclostomi, Holocephali, Elasmobranchi and Teleostii) and found Hb to range from 86-214 mg/100 ml in Elasmobranchi to 190-921 mg/100 ml in other fish species. Like the bio-chemical responses, as discussed earlier in chapter-9, most of the hematological responses are also known to be mediated by the cortical steroid hormones.

1. Blood Cells and Hemoglobin

Correct estimates of the blood Hb, Hct and the cell counts in blood are the most challenging aspects. Hematocrit (Hct) also termed as 'packed cell volume' as it determines the relative volume of the packed cells per volume of plasma; is a most commonly used assay for monitoring hemolysis (Tewari *et al.*, 1987; Redondo *et al.*, 1995; Billy *et al.*, 1995). Stated below is the schematic model explaining the metal toxicity and its impact on fish hematology:

Metal Toxicity	
Gill damage	Inhibit Hb synthesis
Reduced O_2 intake into Blood.	Delay maturity of RBCs, Erythropoiesis or anemia

Metal toxicity in general may induce either erythropoiesis or anemia. Erythropoiesis is triggered as a typical stress response in fishes. Hematological changes have been detected in a variety of fish species following different types of stress conditions, like exposure to pollutants, diseases, hypoxia *etc.* (Andrews *et al.*,

1966; Christensen *et al.*, 1972; Duthie and Tort, 1985). Elevations in hematological parameters such as Hct, Hb, and RBC have been reported in *C carpio* exposed acutely to copper (Svobodova, 1982). An elevation in RBC numbers resulting in increased Hct and Hb could be consequences of blood cell reserve release. Evaluating hematological and immunological effects of some heavy metals during short term exposure in common carp, *Cyprinus carpio* for 3 hours, such as 5- 20 mg/l of lead, copper, cadmium or zinc and their subsequently transfer to clean water showed the hematocrite value and erythropoietic rate increased (Witeska (2005). Rajalakshmi *et al.*, 2014 exposed fingerlings of (7-9 cm) estuarine fish, *Lates calcarifer* to acute concentration of cadmium plus nickel (3.0 ppm) for 96 hours. They observed the haemoglobin level decreased throughout the exposure period, showing a minimum per cent of 11.94 and 51.88 at the end of 24 and 48 hrs, respectively, whereas the maximum percentage decrease of 91.66 and 95.67 was observed at the end of 72 and 96 hrs, respectively.

Hg tends to concentrate in the kidney of teleosts (Penreath, 1976), where it probably inhibits a heme biosynthetic enzyme, uroporphyrinogen-1 synthatase, thus, affecting the RBC profile. Brouwer and Brower-Hoexum, 1985 also reported Hg to induce the deformability of the erythrocytes which could contribute to their early destruction in affected fish. Exposure of red blood cells of the carp to an increased concentration (0.05-0.3 mmol/l) of Cu and Hg ions also decreased membrane fluidity and induced conformational alterations in internal peptides and proteins, and changed internal viscosity of the red blood cells. This was probably due to metal–protein interactions inside the cells (Gwozdzinski, 1992). Gill and Pant (1981) demonstrated simultaneous changes in the blood *i.e.* erythropenia in juveniles and the adults of *Puntatus conchonius* when exposed to Hg, 36.3 and 60.5 micro-gm of $HgCl_2$/l in soft and hard waters. A fall in the haemoglobin after exposure of 1 –3 week and polycythemia after an exposure of 6-8 weeks was recorded. Increased number of RBC in Hg treated fish in fact was to counter the damage and the loss of gill surface due to metal poisoning, as well as its subsequent impact on gas exchange capacity of the gills. Adverse effects due to acute toxicity of $HgCl_2$ on the fish, *A. testudineus* during 24 to 96 hr exposure (Sinha and Kumar, 1992); in *C batrachus* (Joshi *et al.*, 2002) and in *C. carpio* (Masud *et al.*, 2005) were also reported.

Study of the effects of inorganic mercury on hematological parameters and hepatic oxidative stress enzyme activity in olive flounder, *Paralichthys olivaceus* (Kim *et al.*, 2012) showed major hematological parameters were significant decreased, such as the red blood cell count, hematocrit value, and hemoglobin level at 8 mg Hg/kg BW. They also found remarkably low levels of calcium and chloride, with reduced osmolality. Recently, in *Labeo rohita*, the LC_{50} dose of mercury (57.074 mg/L) for 96 h was also reported to result in significant changes in the physico-chemical properties such as hemoglobin, osmotic fragility and blood cell counts (Chitra and Jayaprakas, 2013). Methyl-mercury (MethylHg), though highly toxic in general than the inorganic Hg, has been said to cause less effects in producing anemia in these fish (Lock *et al.*, 1981). Niimi and Lowe-Jinde, 1984 recorded no discernible hematological effects in rainbow trout on their exposure to 15 µg/L of waterborne methylHg for 75-119 days, although the fish accumulated whole body Hg levels of 3-12 mg/kg. Effects of methylHg in food on hematocrit values, however, were equivocal (Rogers and Beamish, 1982; Bidwell and Heath, 1983). In *Clarias gariepinus*, Hounkpatin *et al.*, 2012 demonstrated that cadmium and particularly mercury and combination of cadmium and mercury are high stressors threatening the aquaculture. Exposer of fish to these metals revealed significant decline in blood parameters at all periods with mercury; the highest decrease in blood parameters occurred after three weeks with high concentration of combined cadmium and mercury.

Spleen in fact in some teleost fish species has been reported to serve as an erythropoitic and a potent blood storage organ, sequestering blood cells under resting conditions and releasing them to circulation under stress (Yamamoto 1988). Spleen contraction after stress has been detected in fish (Abrahamsson and Nilsson, 1975). Cells released from spleen would then have a lower MCV value. Witeska, 2005 evaluated whether short-term exposures to high concentrations of heavy metals may induce stress symptoms in fish blood in common carp. Exposing fish to 10, 5, 10 or 20 mg/l of lead, copper, cadmium or zinc, respectively, and subsequently transferred to clean water, showed an increase in hematocrit value without a substantial changes in red blood cell count (which indicates swelling of the cells), and an increase in erythropoietic rate (indicated by an increase in percentage of immature cells in circulation).Fish exposed to high concentration of cadmium quickly develop calcium deficiency and low blood hemoglobin levels (Bryan, 1976). The Hb and the Hct decreased in eels (*A. anguilla*) and perch (*P. flexus*) during short-term as well as a long term exposures to Cd in sea water (Larsson, 1975). Nine weeks exposure to sublethal levels of Cd in brackish water reduced Hb, Hct and RBC counts in flounder, *P. flexus* (Johansson-Sjobeck and Larsson, 1978). The prolonged exposure for 90 days to sublethal dose level of Cd in soft water caused a reduction in Hb, Hct and RBC counts in fish, *P. conchonius* (Gill and Pant, 1985). These reports revealed that the high level of Cd in the test water causes hematological changes much earlier as compared to the lower levels of Cd. However, in contrast to the above observations, Smith *et al.* (1976) and Calabrese *et al.* (1975) did not observe any statistically significant changes in the these three hematological variables in catfish (*I. punctatus*) and flounder (*Pseudopleuronectes americanus*), respectively, when exposed to Cd in tap water and sea water. The present study differs a little from the above observations and indicates the reduction in Hb, Hct and RBC. Certain ice fish which even lack Hb, are reported to have compensatory adaptation, such as a large cardiac out-put. Gill and Pant (1985) observed similar changes on account of the stress of Cd on the fish, *Punctatus conchonius*. They found significant decrease in Hb, packed cell volume (PCV) and the total erythrocyte count (TEC).

Cd treatments in fish was said to decrease the levels of hepatic glycogen (Singhal and Murali, 1979), but significantly elevating the concentration of blood glucose and urea. On the contrary, De Smet and Blust (2001) after exposing common carp, *C. carpio* to Cd for a period 29 days found no significant alterations in blood hematocrit, plasma glucose, plasma lactate and total tissue protein contents. Johansson-Sjobeck and Larsson (1978) reported a decrease in Hb, Hct and RBC count in Flounder, *Pleuronectes flexus* on exposing the fish to Cd for 9 weeks. The anemia because of cadmium in flounder, *Pleuronectes flexus* recovered in non-contaminated water in one year (Larsson *et al.*, 1985). Hematocrit values of fish examined after 10 weeks exposure, showed a general decline with the increasing cadmium concentration. Blood parameters of *C. carpio* exposed to sublethal concentrations of cadmium nitrate and mercuric chloride (0.30 ppm) for 90 hours were studied by Beena and Viswaranjan (1987). The results showed that there was significant decrease in erythrocyte count, hematocrit and hemoglobin content. However, the leucocyte count, thrombocyte count and blood clothing time of the fish did not change significantly. Vijayram *et al*, 1989b showed decreased Hb and RBC counts after 3 weeks in *Anabas testudineus* on its exposure to Cd, but continued exposure resulted in an increased RBC count in this fish. Vijayram *et al.* (1989a) studied the stress of Cd on *A. testudineus*. RBC count and Hb concentration decreased along with leucocytosis upto 3 weeks. Kumar *et al.* (1999) reported that TLC (total leucocyte count) declined after an initial increase, when the fish *A. testudineus* was exposed to the different concentration of mercuric chloride whereas TLC, Hb content and the Hct registered an increase in a dose dependent fashion. Continued exposure to CD, however, resulted in an increase in the RBC count. Studies by Dinodia (2001) on the effects of Cd toxicity showed a similar trend of reduction in the Hb content, RBC count and per cent PCV, may be due to dysfunctions of several physiological biochemical processes. Maximum effect was evident in case of RBC count where a reduction of 33.3 per cent was induced by Cd toxicity (0.05ppm). Significant changes, thus, caused rendered the fishes incapable of normal growth. Kumar (2002) also recorded a trend of reduction in the Hb content, RBC count and per cent PCV was also revealed.

Some parameters which were unchanged after 24 h showed a significant variation after 96 h (plasma glucose and lactate levels). They stated that decrease in hematocrit as a result of exposure to cadmium in the dogfish, *Scyliorhinus canicula*, was indicative of blood dilution. The RBC increase in number could be a consequence of blood cell release combined with cell shrinkage, probably due to osmotic alterations of blood by the action of metal. Tort and Hermandez-Pascual (1990) however, found no significant change in any blood parameter with Cd in dogfish. Das and Bhattacharya, 2000 after administrating $HgCl_2$ and $CdCl_2$ in *C. punctatus* observed Hb, total RBC and WBC cells decreased along with depletion of hepatic glycogen after 15 days. Sankaraperumal *et al.* (1990) reported the synergistic effects of Cd and Zn on the erythrocytes and opercular activities of the fishes, *Lepidocephalichthyes thermalis* and *Amblypharyngodon melettinus.* Acute toxicities of copper, cadmium and mercury to the freshwater fish *Varicorhinus barbatus and Zacco barbata* were studied by Shyong and Chen (2000). The order of metal toxicity to *V. barbatus* is reported as: Hg>Cu>Cd, while to *Z. barbata as*: Cu>Hg>Cd. Results from these experiments ranged from antagonistic to synergistic response.

2. Anemia and Inhibition of the Enzyme ALAD

The term anemia is applied to any condition where the Hb concentration in the blood is abnormally low, whether due to reduced number of RBC or to an inadequate concentration of Hb in the RBC cells. Jantrarotai *et al.*, 1990 evidenced severe anemia together with change in hematocrit (Hct), haemoglobin concentration, and erythrocyte and leukocyte counts; as the signs of acute alfatoxin toxicity.

Presence of aflatoxins are also known to prolong the blood clotting time (Woerr *et al.*, 1987), thus, resulting in severe pathological disorder. In addition to the metals like Hg (Fletcher and White, 1986) and cadmium (Gill and Pant, 1985; Houston *et al.*, 1993, many other chemicals are also capable of causing varying degrees of anemia in fish, such as the organochlorinre pesticides (Venkateshwarlu *et al.*, 1990), the fungicide chlorothalonil (Davies, 1987). The winter flounder and striped bass after exposer to Hg (inorganic) for 60 days, exhibited a distinct anemia (Dawson, 1982). The magnitude of the change was smaller in the flounder even though a higher Hg concentration was used with this species. Anemia has also been reported in the plaice (*Pleuronectes platessa*) with higher and shorter duration exposures (Fletcher and White, 1986). Heavy metals present in the paper mill waste in Assam were said to result in deformation of the surface of the RBCs, leading to anemia in the fish exposed to contaminated waters (Roy and Gupta, 2003).

According to Hernberg (1976), lead or a mixture of copper caused anemia by inhibiting hemoglobin synthesis. Copper though an essential micronutrient, is said to cause haemolytic anemia in mammals by inhibiting glycolysis in the RBCs, denaturing the Hb and oxidizing glutathione (Fairbanks, 1967). In the fish, however, RBC since are said to possess greater intake capacity of oxygen than mammals, Cu does not appear to affect a fish sufficiently (Heath, 1987). At higher concentrations Cu is stated to be highly toxic. McKim *et al.*, 1970, have reported an increase in the Hct, Hb and RBC counts in trout fish when exposed for 6 days to a concentration far below the LC_{50}. By 21 days of continuous exposure to Cu, however, returned the Hct value to normal; and by 11th month Hb and blood counts had also returned to normal, indicating possibly adaptations to lower Cu doses in the fish. McKim *et al.* (1970), however, stated an increase in plasma water level during chronic exposure to Cu in trout. Christensen *et al.* (1972) reported some changes in the blood of the brown bullhead (*Iclalurus nebulosus*) on exposure to Cu at 3.4-10.4 µg/l for 6, 30 and 600 days, and they recorded significantly increased quantities of blood glucose and hematocrit in the treated fish as compared to the controls. Hematocrit and the proteins, however, were found decreased after 30 days and Hb and glucose were again found increased after 600 days. Waiwood (1980) also made similar records in trout exposed to Cu and he attributed these changes to shift of water from the plasma to muscles cells, thereby producing increasing Hb concentration. Same were the observations of Wilson and Tayler, (1993) in trout exposed to lethal dose of Cu. Cyriac *et al.* (1989) used hematological studies as an index of fish health. They studied the hemoglobin and hematocrit values in fish, *Oreochromis mossambicus* (Peters) after short term exposure to copper and mercury. The studies revealed that hemoglobin content increased significantly in Cu and Hg dosed fishes at 120 and 168 hours. The Hb values were significantly low in Cu treated fish than that of Hg treated one. There are further many reports indicating hematological and physiological changes in the blood of juveniles of the freshwater fish, *Prochilodus scrofa* as determined after acute exposure to 20, 25, and 29 mg Cu L^{-1} (Mazon *et al.*, 2002). It caused significant increase in the hematocrit and red blood cell values; Hb and leukocytes increased following copper exposure and were significantly higher in fish exposed to 29 mg Cu L^{-1}. Copper exposures were further reported to impair iono-regulations, along with chloride cell hypertrophy induction. The changes in red blood cells suggest a compensatory response to respiratory surface reduction of gills (tissue damage and cell proliferation) in order to maintain oxygen transference from water to the tissues. *O. aureus* exposed chronically to Cu (Abhas *et al.*, 2003) showed hemoglobin contents increased in the fish, but decreased in fish intoxicated by Pb. Fish exposure to Cu was further reported to interfere with branchial ion transport and affecting various blood parameters such as plasma ion concentrations, hematological parameters, and the enzyme activities in blood and liver (Stagg and Shuttleworth, 1982).

Lead causes anemia in mammals due to specific inhibition of the enzyme delta-aminolevulinic acid dehydratase (ALAD), required in the early stages of synthesis of Hb in the hemopoitic tissue, and shortening of the life-span of circulating erythrocytes (Hemberg *et al.*, 1976). Trout fish exposed to waterborne lead at concentrations varying between 10-300 µg/L also showed anemia only at higher concentrations, since it was found to cause a dose dependent inhibition (21-86 per cent) in the enzyme ALAD in the erythrocytes and spleen (Johansson-Sjobeck and Larsson, 1979). At lower concentrations of lead, it was suggested that there is a possibility of the presence of compensatory amount of this enzyme in the erythrocytes of the treated fish. Recovery of ALAD activity after exposure to lead is reported to be further very slow in the affected fish. Johansson-Sjobeck and Larsson, 1979 recorded only a slight recovery in this enzyme after 7 weeks in metal free water, showing the possibility of continuous presence of lead in the erythropoietic tissue of the treated fish. Lead is said to be retained in the kidney for about a month at least following its up take from the medium (Reichert *et al.*, 1979).

Dawson (1935) observed decreased erythrocyte and eosinophil counts in catfish (*Ameriurus nebulosus*) exposed to water containing 1 per cent lead acetate, and he described an anemic condition that became more severe with increased lead exposure. Chronic exposure of the cyprinid *Barbus sp.* to lead for up to 60 days at 47µg/l was reported to result in severe reductions (12-31 per cent) in RBC count, Hb concentration, Hct and MCV with a severe anemic condition (Tewari *et al.*, 1987). MCV is the mean corpuscle volume. The test animals after exposure to lead toxicity when transferred to Pb free water improved the status of hematological parameters and normal state was found to be restored in 45 – 65 days, depending upon the Pb dose level. Gupta *et al.* (2004) also found the RBC count, Hb content and the packed cell volume (hematocrit) of *L. rohita*, *C. mrigala* and *C. carpio* decreased when the fishes were exposed to the different lead treatments (2.5 and 5.0 ppm). The reduction in the RBC count at 5.0 ppm was maximum in the fish, *C. carpio* (41.47 per cent) and minimum in *C. mrigala* (26.51 per cent) as compared to control fishes. The Hb content showed almost similar reduction in all the three fish species exposed. Packed cell volume at 5.0ppm declined maximum in *C. carpio* (42.64 per cent) and minimum in *C. mrigala* (32.37 per cent), showing thus, marked species dependent differences in fish responces to lead dose levels. This reduction may be due to increased destructions of RBC's leading to an anemic state. Likewise in *Colisa fasciatus*, acute exposure to sub lethal dose of lead produced hemolytic anemia due to the lysis of erythrocytes with concomitant decrease in hematocrit, Hb and RBC number (Srivastava and Mishra, 1979b).

A similar pattern was reported by Koyama and Ozaki, 1984 in *C. carpio* after cadmium exposure, where the authors also observed hematocrit decreased along with maintenance of RBC cells and the increase of circulating reticulocytes; indicating a premature release of these cells into the blood. Cd caused an abnormally large number of malformed erythrocytes (Newman and McLean, 1974; Houston *et al.*, 1993) which implies a lesion in the machinery for forming blood cells. Acute toxicities of cadmium on blood of tilapia, *Oreochromis mossambicus* (Peters) were investigated by Ruparella *et al.* (1990). Hemoglobin, hematocrit and RBC counts were reported decreased significantly. Cd has been reported to cause anemia in different fish species at a lower as well as higher concentrations (*e.g.*, Newman and McLean, 1974; Larsson *et al.*, 1976; Larsson *et al.*, 1985; Gill and Pant, 1985; Houston *et al.*, 1993). Houston *et al.*, 1993 suggested slowing of maturity of RBC, and their breakdown as the only possibilities of causing anemia in Cd treated flounders.

Test for anemia in the affected fish in polluted waters, however, does not appear to be good physiologically as there is some evidence that many fish species in general at low ambient temperatures (<12°C) are able to tolerate mild anemia. Measurements of the enzyme delta-aminolevulinic acid dehydratase (ALA-D) in the kidney have shown a compensatory increase as a result of exposure of Cd (Johansson-Sjobeck and Larsson, 1978). This indicates that the initial steps of the synthesis of Hb are not blocked by Cd, rather Hb synthesis still goes down in spite of the fact that enzyme activity increased. Larsson *et al.*, 1985 further found that Cd induced anemia in flounders is a reversible process. They found a year's recovery in non-contaminated water abolished anemia, but the carbohydrate metabolism remained still in disorder possibly due some permanent damage or lesions in the insulin secreting pancreatic cells. Cd is said to have no effect on this enzyme in mammals too. According to Heath (1987) Cd has been reported to cause anemia in a variety of fish species at a wide range of concentrations. It was said to produce a compensatory increase in the enzyme, ALAD in the kidney of the fish as a result of Cd exposure, thus, showing an increased Hb synthesis. The net effect is a slowing down of erythrocyte maturation (Houston *et al.*, 1993). In addition to the reduction in the capacity to produce normal blood cells, some workers have also noted an increased rate of erythrocyte breakdown (Houston *et al.*, 1993). Palace *et al.* (1993) reported that the erythrocytes from rainbow trout exposed to Cd for 181 days had a tendency for lysis. This increased cell fragility was attributed to lipid peroxidation of the erythrocyte membranes which affects membrane fluidity. The possibility of reduction in absorption of iron from the gut was ruled out by Larson *et al.*, 1976. They exposed flounders to Cd at 1 ppm for 15 days and observed anemia in the treated fish. Since, the fish was not fed during the treatment period there were no chances of any iron absorption in the gut. Studies on the red carp, *C. carpio* showed that the haem-oxygenase activity was greatly enhanced in the hepato-pancrease, but it appeared reduced in the kidney after treatment with Cd (Ariyoshi *et al.*, 1990).

Cobalt is also known to inhibit the enzyme 5-aminolevulinate synthatase, one of the enzymes involved in heme synthesis (Besur *et al.*, 2014). The final step of heme biosynthesis, catalyzed by the enzyme, ferrochelatase (FECH) also known as heme synthase, is the insertion of ferrous iron into protoporphyrin (PP) to produce heme. Under conditions of iron deficiency, lead poisoning or excessive formation of PP, zinc protoporphyrin (ZnPP) is formed as a by-product in the heme biosynthesis pathway, resulting from insertion of Zn^{2+} rather than Fe^{2+} into protoporphyrin by FECH. FECH is also capable of inserting cobalt into PP with the formation of cobalt protoporphyrin (CoPP), a metalloporphyrin with physiologic effects that are of interest. A marked decline in the RBC count as well as the Hb content was caused due to Co toxicity independently or in combination with Zn in both the fish species in the present studies. Co is somewhat interesting in that it inhibits the enzyme 5-aminolevulinate synthetase, one of the enzymes involved in heme synthesis. When *Sarotherodon mossambicus*, an Indian freshwater teleost, was exposed for 15 days to Co, a marked increase in erythrocyte count within 5 days was reported. The total Hb content of the blood was, however, declining. The fish were over compensating for the inhibited Hb synthesis (Venkateshwarlu *et al.*, 1990). Lowe-Jinde and Niimi (1996) observed a reduction in the number of erythroblasts-proerythrocytes in goldfish, *Carassius auratus* and rainbow trout exposed to cadmium. Besides, from the observation of heavy metals accumulation in kidney, spleen and liver by the same authors, it is conceivable that heavy metals might have suppressed the activity of this hematopoietic tissue. It also was supported by Gill and Epple (1993) who attributed anaemia to impaired erythropoiesis caused by a direct effect of metals on haematopoietic centers (kidney/spleen). Galhoom *et al.*, 2000 compared data for the effects of heavy metals on some blood parameters in mugil fish (*Mugil cephaleus*). Hematology revealed non-significant decrease in Hb and MCV, but a significant decrease in MCH in metal affected fish as compared to the fish inhabiting normal metal free waters.

3. Changes in Leucocyte Counts

Relative numbers of lymphocytes, granulocytes and thrombocytes are considered as very sensitive parameter to pollutant stresses. Like all other large animals, overall reduction in leucocyte (Leucocytopenia) in response to stress though a non-specific response, is also a characteristic feature in fishes. Changes in leucocyte counts are further said to be sensitive to metal toxicity (Larsson *et al.*, 1985) and possibly other pollutants as well, as these affect body tissues. McLeay and Gordon (1977) proposed the term 'leukocrit' to the estimated per cent volume of leucocytes in the circulating blood. A very few reports however, are available in the fish on the study of differential white blood cell counts which are affected upon exposure to the chemical pollutants (Johansson-Sjobeck and Larsson, 1978). Significant differences were observed in the total number of leucocytes and the number of lymphocytes between the fish that received arsenic and cadmium and the healthy fish. Cd has been shown to change the leukocyte differential count. Newman and McLean (1974) reported a dose-dependent increase (three fold) in neutrophils and a dose-dependent decrease (also nearly three-fold) in lymphocytes in the Cunner, *Tautogolabrus adspersus*. This contrasts with the "markedly enhanced lymphocyte count" as observed by Larsson *et al.* (1985) in perch from a river polluted with Cd. It was suggested that this contradiction in lymphocyte counts was reported possibly due to a different pathological organisms present in water in these two different studies. Papoutsoglou and Abel (1988) investigated sublethal toxicity effects and accumulation of cadmium in *Tilapia aurea*. Thrombocytosis was reported in the fish *Colisa fasciatus* when exposed to Cd (Srivastava and Mishra, 1979a). The cytological shift observed in this study was said to be similar to the shift observed between small and large lymphocytes in Cd treated rats and mice (Oshawa and Kawai, 1981). The responses of different white blood cells in this study might be secondary responses and were partly expected to be mediated through an increase in pituitary inter-renal activity, as reported earlier by Donaldson (1981) during stress condition of the fish. Lowe-Jinde and Niimi (1983) in rainbow trout, *Salmo gairdneri* in response to cadmium exposures also found a decline in erythrocyte counts, leucocyte and lymphocytes. The WBC count also significantly decreased in all the treated fishes (Dhanapakiam and Ramaswamy, 2001). On the contrary, Bhoopathy *et al.* (2000) reported that the WBC count increased while the RBC count decreased from the control levels in *O. mossambicus*, when the fish was exposed to mercuric chloride for 24 days. Hb and Hct however, increased when the fish was exposed to $HgCl_2$ for 30 days. Gupta *et al.* (2001) in *Labeo rohita* and *Catla catla* exposed to cadmium, found decreased mean leukocyte and per cent lymphocyte counts, whereas neutrophilia was found to occur in both the species after their exposer to cadmium. The study shows that cadmium causes sufficient stress to fish and ultimately affect its growth and performance.

Kotsanis and his coworkers (2000) reported some changes in hematological parameters at early stages of the rainbow trout, *Oncorhynchus mykiss*, subjected to metal toxicants: arsenic, cadmium and mercury. A decrease in total leukocyte count was observed. Danabas *et al.* (2010) while evaluating hematological and biochemical responses of *Capoeta trutta*, captured in Munzur river, Tunceli, Turkey concluded that the most rapid detection of changes in fish after the exposure to xenobiotics can be detected and the hematological and biochemical parameters could be ranked as possible biomarkers of pollution. During the study of the effects of safe concentration of zinc and copper on the haematology of a freshwater fish, *Labeo rohita* exposed for 60 days (Summarwar, 2012), hematological parameters *viz.*: Hb, RBC and PCV (Hct) were noted to decline more during initial phase (1st to 15th day) of exposure compared to later phase. However, in case of WBC the decrease was relatively less during initial phase compared to later phase. Çelik *et al.*, 2013 recorded significant changes in the hematological and innate immune parameters *etc.* in tilapia (*Oreochromis mossambicus*), exposed to several zinc concentrations *in vivo*. In all groups exposed to zinc, a decrease in the erythrocyte count (RBC), WBC counts and lymphocyte percentage and an increase in hemoglobin (Hb). A decrease in white blood cell (WBC) count occurred with medium and high concentrations. As per hematocrit (Hct) values, a decrease with high concentrations and an increase with low and medium concentrations were found. This decrease presumably resulted from the toxic effect of zinc on WBC or the stress caused on the cell production activity of the spleen (Yamamoto, 1988; Fýrat, 2007). In addition, the decrease in WBC count can be associated to the increase in the secretion of corticosteroid and cortisol hormones, since these hormones play an important role in the prevention and healing of inflammation. Suppressive effect of cortisol hormone on lymphocyte production was also suggested by Rink and Kirchner, 2000 and Nussey *et al.*, 2002). In addition, it was found that neutrophil percentage significantly increased in all zinc applied concentrations and monocyte percentage was found higher in high zinc applications compared to those in the control group on the 14th day. This effect may be associated to the damage of metals on the defense cells of the fish (Witeska, 2005). It has been further reported that zinc decreased WBC count, neutrophiles and lymphocytes in carp fish (Witeska, 2005), in *O. mossambicus* (Buthelezi *et al.*, 2000) and in *C. gariepinus* (Ololade and Ogini, 2009). On the other hand, it has been found that zinc increased WBC count, lymphocytes, basophyles, and neutrophiles in Mozambique tilapia (Sampath *et al.*, 1998), and WBC count in *Heteroclarias* sp (Kori-Siakpere and Ubogu, 2008). This increase in WBC count may be as a result of the prevention of damage caused by zinc in the gill, kidney, and liver tissues (Buthelezi, 2000; Nussey *et al.*, 2002). This may be explained by a reaction of the defense mechanism of the fish by leucocytosis under pathological conditions and against foreign bodies (John, 2007).

Other metals that stimulate blood cells/Hb production in fish are Ni (Ghazaly, 1992), Zn (Mishra and Srivastava, 1979; Hilmy *et al.*, 1987) and the hexavalent chromium (van der Putte *et al.*, 1982). According to Mishra and Srivastva (1979) total red cell counts, haematocrit, total white cell counts and the number of small lymphocytes/1000 cells decreased significantly in freshwater teleost, *Colisa fasciatus* on its exposure to zinc. Kori-Siakpere and Ubogu, 2008 evaluated some hematological changes resulting from the exposure of a freshwater fish, *Heteroclarias* sp. to sublethal concentrations (5.0 and 10.0 mg L^{-1}) of zinc in water for a period of fifteen days. These treatments were said to cause a dose dependent decrease in hemoglobin values, coupled with a decrease in hematocrit values and red blood cell counts are obvious indication of anemia of the norm chronic type. The total white blood cell counts and the differential white blood cell counts were decreased except for the lymphocytes in which there was a slight increase. Plasma level of protein and glucose were also lower in the exposed fish when compared to the control. Likewise, Shahi *et al*, 2013 in observed similar changes in *Channa punctatus* exposed to sub-lethal concentrations of some pesticides. Pesticides caused a dose dependent decrease in haemoglobin values coupled with a decrease in haematocrit values and red blood cell counts. The total white blood cell counts and the differential white blood cell counts were decreased except for the lymphocytes. Alterations in the hematological parameters were dose and duration dependent and can thus, serve as a useful physiological index. Study of the effects of waste water containing heavy metals in the order: Fe > Mn > Zn > Co > Ni > Cu = Cr in *C.punctatus* (Javed and Usmani, 2013) revealed significant decrease in red blood cell count (RBC)count and hemoglobin (Hb) contentrespectively, but white blood cell (WBC) counts showed significant ($P<0.01$) increase when compared to the control.

Kotsanis *et al.* (2000) investigated changes in hematocrit, red cells/mm^3, white cells/mm^3, red/white cell ratio and white blood cell differential counts at early stages of rainbow trout subjected to metal toxicants-arsenic, cadmium and mercury. Recently, Tripathy and Qureshi, 2014 also noted decrease in total red blood cell count, white blood cell count and heamocrite value as well as an increase in number of granulocytes excluding neutrophils and decrease in number of agranulocytes and neutrophils in fish, *Channa gachua* on its exposures to heavy metals like Cr,

Cu and Co. Mazon *et al.*, 2002 also recorded significant increase in the hematocrit and red blood cell values; Hb and leukocytes following copper exposure and these values were significantly higher in fish exposed to 29 mg Cu L^{-1}, and the differential leukocyte percentage was found to display a significant reduction in lymphocytes and an increase in neutrophils in fish exposed to 25 and 29 mg Cu L^{-1}. Increase and decrease in hematological indices means that fish exposed to effluents containing heavy metals was under stress. Study of the effects of inorganic mercury on hematological parameters and hepatic oxidative stress enzyme activity in olive flounder, *Paralichthys olivaceus* (Kim *et al.*, 2012) showed major hematological parameters were significant decreased, such as the red blood cell count, hematocrit value, and hemoglobin level at 8 mg Hg/kg BW. Zn was also found to produce haemolytic anemia due to the lysis of eruthrocytes with concommitant decrease in hematocrite, Hb and RBC number in *H. fossilis* when exposed to Zn (Goel and Kalpana, 1985). Interaction between Zn and other metals has been investigated in the fish by a few workers (Brown and Dalton, 1970; Spehar, 1976; Broderius and Smith, 1979). The zinc concentration caused a significant decline in hemoglobin, hematorcit and RBC count, whereas, MCHC remained unaffected. Plasma glucose declined significantly.

4. Erythrocyte Morphological Anomalies

The RBCs morphological anomalies in the MCH (mean corpuscular haemoglobin), MCHC (mean corpuscular haemoglobin content) and MCV (mean corpuscular volume are usually measured as a measure of chemical toxicity effects in animals including mammals. Cadmium has been further said to cause development of abnormally large number of malformed RBCs in blood of cunner fish (*Tautogolabrus adspersus*) (Newman and McLean, 1974), which implies a lesion in the machinery forming the blood cells. Red blood cell indices i.e MCV, MCH and MCHC exhibited change during the prolonged exposure to Cd. The increase in MCV and MCH was 133 per cent and 247 per cent of the control values, while MCHC did not show much change. Increase MCV and MCH were observed in rainbow trout (*S. gairdneri*) and in *P. conchonius* during 30 days exposure to sublethal dose of Pb and Cd respectively (Johanson Sjobeck and Larsson, 1978; Gill and Pant, 1985). MCV and MCHC were also reduced significantly, whereas rise in MCHC was significant at 3 days at 1.0 ppm and 10.00 ppm. Kumari and Banerjee (1986) reported effects of sublethal concentration of zinc chloride (300 mg/l) and mercuric chloride (0.2 mg/l) on the blood parameters in *A. testudineus* after 15, 30, 45, 60 and 75 days of exposure. They reported a permanent decrease in the erythrocyte count, Hb content and the erythrocyte sedimentation rate. The MCV, however, was found increased even after an exposure for 75 days. Massaro, 1974 reported that methylHg binds reversibly with the Hb in erythrocytes, thus, may affect binding of O_2 molecules with Hb, in presence of methylHg in the fish blood. Kumar *et al.*, 1999 reported that when the fish, *Anabas testidinus* was exposed to $HgCl_2$ at different concentrations, total erythrocytes recorded a decline after an initial increase, whereas Hb, total leucocytes and Hct registered a dose dependent increase.

In fish, *C. carpio* at sublethal levels of heavy metals over a period of 30 days there occurred a significant decrease of the total RBC count, haemoglobin content and hematocrit (Hct) per cent (Dhanapakiam and Ramaswamy, 2001). The MCH, MCHC and MCV increased depending upon the exposure period. A significant decrease was observed in the fish treated with 0.1 ppm of Hg for 24h on Hct and the RBC counts; whereas, change in MCH and MCHC were non-significant. Interestingly the changes in RBC, Hct, and Hb were also significant in fish treated with Hg at 1ppm for 48h, but the increases in MCH (15.79 per cent), MCV (12.06 per cent) and MCHC (3.11 per cent) at the same treatment in tench were statistically non significant. Dogfish subjected to exposure of sublethal dose of Cd (25 mg/l) for 24 and 96 h by Tort and Torres (1988) also showed a decrease in Hct, MCV and increased Hb concentration, RBC count and MCHC after a 24 h exposure. Sahoo and Mukherjee (1999) established that the morphology of the blood cells depicted a wide variation in Hct and MCV of the individual healthy fish.

Sub-lethal concentrations of zinc administered to freshwater teleost, *Channa punctatus* for a period of 135 days brought about significant hematological alterations (Tyagi and Srivastava, 2005). Sublethal concentrations (5.0 and 10.0 mg L^{-1}) of zinc was said to cause a dose dependent decrease in hemoglobin values, coupled with a decrease in hematocrit values and red blood cell counts, the hematological indices MCHC, MCH and MCV were also lowered (Kori-Siakpere and Ubogu, 2008). Çelik *et al.*, 2013 recorded significant changes in tilapia (*Oreochromis mossambicus*), exposed to several zinc concentrations *in vivo*. In all groups exposed to zinc, along with a decrease in the erythrocyte count (RBC) and lymphocyte percentage and an increase in mean corpuscular volume (MCV), mean corpuscular hemoglobin (MCH) values and neutrophile percentage occurred. The RBC count, haemoglobin and haematocrit content progressively decreased while WBC count, MCV, MCH and MCHC increased. *In vitro* studies by Akahori *et al.* (1999) on the effect of Zn exposure on some properties of erythrocyte membrane of the carp also showed reduced fluidity of the lipid bi-layer, both in the middle and near the aqueous interface. Zn however, was said to have no significant influence on Na-K ATPase activity in the erythrocyte membrane. Zn led to the marked decrease of hemolytic resistance of the cells. It was concluded that the Zn may be toxic to carp erythrocytes at higher concentrations causing the changes in the membrane fluidity and hemolytic resistance. Changes in hematological parameters and plasma glucose in fish, *Cyprinion watsoni*, in response to zinc and copper treatments was also reported by Shah *et al.* (1995).

Fish erythrocyte morphology is more sensitive to various environmental agents than basic red blood parameters (Zeni *et al.*, 2002; Witeska, 2004), and cellular anomalies are sometimes observed without distinct decrease in their values (Hoffer *et al.*, 1992; Witeska *et al.*, 2010). Gill and Pant (1986) observed erythrocyte swelling, poikilocytosis, vacuolation, amitosis, deformation, and deterioration of cell membranes in *Barbus conchonius* exposed to chromium, and nuclear aberrations such as chromatin condensation, nuclear puffs, and chromatin leakage in the same species subjected to cadmium intoxication (Gill and Pant, 1987). Erythrocyte anomalies may result from various physiological disturbances. Eiras *et al.* (1996) reported cytoplasmic inclusions and erythrocyte degeneration: irregular outlines, nuclear displacement, condensation and leakage of nuclear material, and cytoplasmic vacuolation, as a result of infection with the erythrocyte necrosis virus.

Various workers have reported erythrocyte anomalies such as presence of Heinz bodies, poikilocytosis, clumped chromatin, ragged cell membranes, altered staining properties, and hemolysis in erythrocytes of *Oncorhynchus kisutch* from contaminated water (Buckley, 1996) and erythrocyte echinocytosis (abnormal RBC cells with spikes/echines at its membrane outer surface) in *Ictalurus melas* sublethally (Zeni *et al.*, 2002). Witeska (2004) observed toxicity of metals to carp erythrocytes ranged (according to the frequency of erythrocyte anomalies): Pbe"Zn>Cd>Cu, and the changes induced by various metals were similar: nuclear malformation, chromatin condensation, cell swelling, and malformation. Shah and Altindag (2004) reported blood cell deformities in tench (*Tinca tinca*) on their exposure to Hg treatments.

Karuppasamy *et al.* (2005) reported increased fragility, rupture of erythrocyte membrane, and hemolysis in *Channa punctatus*, sub-lethally exposed to cadmium. According to Sharma *et al.* (2007), fish erythrocytes were more sensitive to water pollution compared to the other biological endpoints and thus, should be included in the routine fish bioassay. Frequencies of nuclear anomalies such as irregular nucleus shape, vacuolation, binuclei and micronuclei that indicate genotoxic effects often increase in fish exposed to water pollution (Guha and Khuda-Bukhsh, 2002; Cavas *et al.*, 2005). Witeska *et al.* (2011) evaluated the effect of short-term (6.5 mg/l for 3h) and long-term (4 weeks) *in vivo* and *in vitro* exposure to cadmium (0.65 mg/l for 4 week) on the morphology of common carp erythrocytes. They recorded anomalies such as chromatin condensation at the nucleus border, nuclear malformation, cell body malformation, cytoplasm vacuolation, and swelling and hemolysis. In fish subjected to long-term cadmium exposure (and in some degree in fish after short-term exposure) an increase in erythroblast frequency occurred, which indicates hematological compensation for erythrocyte damage. Cadmium nitrate induced more anomalies in carp erythrocytes (particularly nucleus malformations and cytoplasm vacuolation) than chloride and sulfate.

5. Metal Effects on Oxygen Consumption

Factors like blood oxygenation, plasma skimming, seasonal variations, fish nutrition and the environmental stress affect both the blood composition and the blood cell status. For example, blood coming out of the gills and flowing in the dorsal aorta since, contain more of oxygen than in blood that flows in the ventral aorta (taking blood to the gills), is said to possess higher Hct value (Soivio *et al.*, 1981), due to skimming of plasma in the gills. Same is to be true about the blood, taken out for any analysis from the caudal peduncle of a fish, which also contain more oxygen than the blood taken directly from the heart. It has long been accepted that the respiratory properties of blood, notably the oxygen-carrying capacity and the shape and position of the oxygen equilibrium curve have been responsive to evolutionary selection pressure. Hypoxia *i.e.* deficiency in the amount of oxygen reaching the tissues, is said to cause swelling of the blood cells which increase their volume in a unit blood, thereby increasing the Hct value. Higher Hct value indicates increased blood viscosity *i.e.* blood may get so thick that heart may find it difficult to pump it, which may even lead to a heart attack.

Heavy metals are known to affect oxygen carrying capacity (OC), rate of oxygen consumption and ventilation in fish due to their damage to gill lamellae as well as the metabolites. Reduction in OC, appear to be a function of reduced RBC counts and the Hb contents. The toxicity of lead, in addition to other blood parameters such as RBC counts, Hb contents and Hct also influenced the oxygen carrying capacity. Lead free water, however, was found to show recovery signs in treated fish. James *et al.* (1993) exposed *O. mossambicus* to sublethal concentration of Pb and observed a dose dependent decrease in RBC count and Hb, thereby leading to anemia. It may be due to the inhibition of erythropoesis and the haem-synthesis and increase in the destruction of RBC, thus, negatively influencing oxygen carrying capacity of blood to 6.5 and 5.9 ml O_2/g at the end of exposure to 17.5 and 35.0 mg/l of lead, as against 11.0 ml in the controls, leading respectively to 67 and 81 per cent decline in OC. The test animals after exposure to lead toxicity when transferred to Pb free water improved the status of hematological parameters and normal state was found to be restored in 45 – 65 days, depending upon the Pb dose level. Even tilapia fish on its exposer to Hg at 0.02 ppm level were seen with reverse effects to some extent. Skidmore, 1970 also recorded reduction in oxygen consumption in rainbow trouts, possibly due to gill damage to Zn sulphate. Kumari and Shahi, 2014 while exposing the fish, *Cirrhinus mrigala* (Ham.) to lethal and sublethal concentrations of cobalt chloride, recorded increased O_2 uptake rate initially, followed by a decline up to 54.47 per cent at 240hr of exposure to lethal concentration (92.00 mg/l) and up to 28.80 and 10.65 per cent in sublethal concentration at 960hr of exposure. Cadmium induced gill necrosis, decreased rate of oxygen consumption and increased ventilation frequency of fish (Voyer *et al.*, 1975); thus, amounting to decreased oxygen consumption and increase in ventilation frequency in chronically exposed fish. Durve *et al.* (1980) reported decrease in oxygen consumption in fish exposed to copper and it accompanied heavy deposition of mucus on the gills and skin. On evaluating the effects of $HgCl_2$ and $CrCl_2$ on the oxygen consumption of a freshwater fish, *T. mossambica,* Sarkar (1989) observed that the oxygen consumption of the fish exposed to $CrCl_2$ was higher than exposed to $HgCl_2$. At 0.5 ppm of $HgCl_2$ and 0.7 ppm $CrCl_2$ onwards oxygen consumption progressively decreased in the fish. Varghese *et al.* (1992) measured the oxygen consumption in *Barytelphusa cunicularis* (Westwood), a freshwater carp, in order to study the stress effect caused by the heavy metals. Normal oxygen consumption was found affected when a fish was exposed to mercuric chloride, copper sulphate and zinc sulphate from 0 to 12 h. Similarly, crabs exposed to sublethal concentrations of the same pollutants for 24, 48, 72 and 96 h showed an elevation in the blood sugar level with a maximum increase at 48 h. The results indicate a switch towards glycolysis in order to overcome the anaerobic stress caused by the heavy metals. Chromium generally induces a hypoxic condition in the fish (van der Putte *et al.*, 1982), and an increase in oxygen capacity of the blood was viewed merely as an adaptation to the altered respiratory homeostasis, caused by metal stress and these changes do not involve the blood forming tissues. Similar effects also occur during acclimation to environmental hypoxia, as acute stress cause adrenergic stimulation of the spleen to contract and release stored RBCs into blood circulation (Nilson and Grove, 1974). Sarkar (1999) further showed that fish, *Cyprinus carpio* exposed to mixture of $CuSO_4$ and $CdSO_4$ at different ratios (1:1, 2:2, 3:3, 4:4, and

5:5), exhibited less oxygen consumption than when exposed to individual metals. At 0.320 and 0.317 mg/l of copper sulphate and cadmium sulphate, the oxygen consumption of fish decreased by 9 and 12 per cent of the control, respectively. Sankaraperumal *et al.* (1990) reported the synergistic effects of Cd and Zn on the erythrocytes and opercular activities of the fishes, *Lepidocephalichthyes thermalis* and *Amblypharyngodon melettinus.* At 0.320 and 0.317 mg/l of copper sulphate and cadmium sulphate, the oxygen consumption of fish decreased by 9 and 12 per cent of the control, respectively. Espina *et al.* (2000) studied the effects of 96 h exposure to sublethal cadmium concentrations on oxygen consumption rate and oxygen extraction efficiency in juvenile *Ctenopharyngodon idella.* The rate of oxygen consumption was found increased by Kale Monica and Kulkarni (2003) in the initial period of analysis up to 24 h. there after it started decreasing, attaining almost normalcy after 60 to 72 h, indicating high tolerance of this fish for cadmium probably because of an efficient mechanism of metal depuration and detoxification. Samuel and Samuel and Thateyus (2004) investigated the toxicity of cadmium in *Oreochromis mossambicus* using different physiological parameters. Increase in exposure period and concentration of cadmium resulted in the increase of oxygen consumption rate. Bhamre *et al.* (2004) studied effects of heavy metals *viz.*, $HgCl_2$, $CuSO_4$ and $CdCl_2$ on oxygen consumption of freshwater mussel, *Parreysia favidens* after 24, 48, 72 and 96 h exposures. Metabolic rate and oxygen consumption decreased up to 96 hours. Van Aardt and Booysen (2004) has reported effects of Cd on oxygen consumption, plasma chloride and bio accumulation in *Tilapia sparrmanii.* Effects of cadmium compounds on common carp, *Cyprinus carpio* administered via an intra peritoneal injection have also been assessed by Brucka-Jastrzebska and Protasowicki (2004). Samuel and Thateyus (2004) investigated the toxicity of cadmium in *Oreochromis mossambicus* using different physiological parameters. Increase in exposure period and concentration of cadmium resulted in the increase of oxygen consumption rate.

Fishes regulate their metabolic rate over a range of dissolved oxygen concentrations; however, at some point a further reduction in oxygen tension will produce a shift from a metabolic rate that is independent of oxygen concentration to one that is dependent on oxygen level. The point is referred to as the 'critical oxygen tension' (Ultsch *et al.*, 1978). In response to a low concentration of dissolved oxygen in the water, the fish can respond in two ways: the blood flow can be increased by opening up secondary lamellae further to increase the effective respiratory area. It, however, may be difficult to increase significantly the blood flow rate through the capillaries itself, but the concentration of red blood corpuscles can be increased to raise the oxygen carrying capacity of the blood per unit volume simultaneously. The latter can be achieved by reducing the blood plasma volume (*e.g.* by increasing the urine flow rate) in the short term, and by releasing extra blood corpuscles from the spleen in the longer term (Svobodova *et al.*, 1993). At the same time, the ventilation rate is increased to bring more water into contact with the gills within a unit of time. There are, however, limits to the increased flow attainable; the space between the secondary lamellae is narrow (in trout it is about 20 µm) and water will tend to be forced past the tips of the primary lamellae when the respiratory water flow is high, thus, by-passing the respiratory surfaces. These reactions are quite adequate to compensate for the normal fluctuations of energy demands of the fish and of dissolved oxygen concentrations in the water. One of the consequences, however, of an increased ventilation rate will be that there will be an increase in the amount of toxic substances in the water reaching the gill surface, where these can be absorbed (Boyd and Tucker 1998).

Oxygen carrying capacity test under metal toxicity stress is one of the most important parameters in fishes. Fish tolerance for low oxygen may relate in part to the metabolic rate of a fish (Verheyen and Decleir, 1994). O_2 threshold level for salmonid is reported as 120mm Hg and for non-salmonids as 95 mm Hg. The cyprids are unique as these can convert lactate into ethanol as the end-product of anaerobic respiration, which is easily excreted from the fish. This adaptation make the cyprinids to enhance their adaptation to hypoxia, thus, especially resistant to O_2 stress than that of *C. carpio* and it can survive in cold water even under anoxia (Holopainen and Hyvärinen, 1985), and common gold fish is able to live at total anoxia even at 20°C for 22 hrs. Lowest level of oxygen tolerance in cyprinid is reported to be 5ppm, which is lower than in salmonid (7ppm) and the trout. Gold fish can survive anoxia for 24 hr at 20°C. Early fish stages are further said to be more resistant to oxygen stress due to anaerobic mode of respiration. Hypoxia induces hyperventilation with reduced resistant to blood flow, enhanced O_2 supply by gills, increased O_2 carrying capacity of Hb and reduced activity of AChEase enzyme. Hypoxia is also said to inhibit AChEase enzyme activity in perch and trout. Catecholamines cause vasodilatation, decreases plasma volume and urine output, and cause lamellar fusion in gills. AChEase induces reverse effects. *In vivo*, copper sulphate produces iono-regulatory or respiratory disturbances that imply an increase in energy consumption to restore homeostasis (Carvalho and Fernandes, 2006). In rainbow trout (*O. mykiss*), exposures of only 4 h to 1.65 μM of waterborne copper caused hypoxia in the gill epithelium, the activation of hypoxia-inducible 'factor-1α' as well as an increase in the metallothionein mRNA levels (van Heerden *et al.*, 2004).

An increase in the Hct value has also been reported in the fish subjected to a non specific stress (such as a physical stress on account of fish being lifted out of the water), and this increase occurs within minutes of the onset of stress (Casillas and Smith, 1977). A part of this change is due to the swelling of the erythrocytes which occurs when the fish blood cells are exposed to a hypoxic environment (Soivio and Nikinmaa, 1981). It apparently does not matter whether the blood cells are exposed to the hypoxia *in vitro* or *in vivo* (Soivio and Nikinmaa, 1981; Swift and Lloyd, 1974). Thus, any dose of the pollutant that results in any damage to the fish gills, and a subsequent internal hypoxia can be expected to cause an increase in the Hct also. Environment stresses as these affect fish osmoregulatory functions also, may also decrease or increase the blood volume due to water absorption or its elimination under stress, thus, affecting the Hct values. Chemicals *viz.*, ozone (Wedemeyer *et al.*, 1979), Ni (Ghazaly, 1992), Zn (Mishra and Srivastava, 1979; Hilmy *et al.*, 1987), and the hexavalent Cr (Van der Putte *et al.*, 1982) that stimulate the blood cell/Hb production, generally induce a hypoxic condition in the fish. Since RBC consumes higher amounts of oxygen, delay in oxygen intake may lead to hypoxia.

The ability to make short-term adaptive changes in the O_2 delivery system in response to hypoxic exposure may be typical for vertebrates in general, rather than a feature seen only in those organisms which encounter environmental hypoxia on a regular basis as evidenced by Wells *et al.* (1989), who had also examined the effects of hypoxic exposure on whole-blood oxygen-affinity in an Antarctic fish, *Pagothenia borchgrevinki.* In general, the actions of toxicants are aggravated during hypoxia, through a variety of mechanisms. Some species are much more tolerant of hypoxia than others, leading to differential survival during extended periods of hypoxia (Poon *et al.*, 2002). The major pathological-anatomic changes include a very pale skin color, congestion of the cyanotic blood in the gills, adherence of the gill lamellae, and small hemorrhages in the front of the ocular cavity and in the skin of the gill covers. In the majority of predatory fish the mouth opens spasmodically and the operculum over the gills remains loosely open (Svobodova *et al.*, 1993). More than that, fish reduces food intake, leading to a reduction in growth. Reproduction is inhibited, and both fertilization success and larval survival are compromised. Oxygen carrying capacity (OC) of fish blood is further badly hampered after exposure to pollutants on account of damaged blood cell profile.

This Antarctic fish on exposure to low oxygen levels resulted in a significant (66 per cent) rise in hemoglobin concentration, and erythrocyte [ATP levels] decreased by approximately 27 per cent. There was no evidence for erythrocyte swelling. Aberrant gill morphology showed unexpectedly high erythrocyte ATP levels. Oxygen-carrying capacity increased by approximately 40 per cent in hypoxic fish and was correlated with a 34 per cent decrease in spleen mass. Under hypoxia, energy utilization is decreased and it is associated with a shift from aerobic to anaerobic metabolism. To reduce energy expenditure under this situation, fish move to water at lower temperature, and reduce activity, reproduction, feeding, and protein synthesis. Transcription mediated by increased levels of hypoxia-inducing factor 1 (HIF-1) is reduced, which also up-regulates genes involved in erythropoiesis, capillary growth and glucose transport. All these responses are directed at maintaining cellular oxygen homeostasis and reducing energy expenditure, thereby augmenting survival of the animal during hypoxia. Nikinmaa and Jensen (1986) made observations on Rainbow trout under stress, obtained from a medium at a temperature below 5°C. The fish showed, as a result of the stress a marked increase in extracellular acidosis, which had both respiratory and metabolic components and the dorsal aortic red cell pH also decreased from 7.33 to 7.12. This also caused a further drop in the arterial oxygen saturation, although arterial oxygen tension remained unchanged. Additionally, no red cell swelling was observed. The carbon dioxide and red cell chloride data also suggest that bicarbonate influx into the red cells and, consequently carbon dioxide excretion increased under stress; since the ventral aortic oxygen tension remained unchanged, the oxygen saturation of venous blood was also not affected by stress.

REFERENCES

Abhas H.H., Zaghloul K.H. and Mousa M.A., 2003. Effect of some heavy metal pollutants on some biochemical and histopathological changes in blue tilapia, *Oreochromes aureus*. Egyptian J. Agric. Res., 80 (3): 1395-1410.

Abrahamsson T. and Nilsson S., 1975. Effects of nerve sectioning and drugs on the catecholamine content in the spleen of the cod, *Gadus morhua*. Comp. Biochem. Physiol., C 51: 231-233.

Akahori A., Jozuviak Zofia, Gabryelah Teresar and Gondko Roman., 1999. Effect of Zn on carp (*Cyprinus carpio* L.) erythrocytes. Biochem. Physiol., 123 (3): 209-215.

Andrews A.K., Van Vallin B.J. and Stabbings B.E., 1966. Some affects of heptachlor on bluegills, *Lepomis macrochirus*. Trans. Amer. Fish Soc., 95: 297.

Ariyoshi T., Shiiba S., Hasegawa H. and Arizono K., 1990. Profile of metal binding proteins and heme-oxygenase in Red carp treated with heavy metals, pesticides and surfactants. Bull. Environ. Contam. Toxicol., 44: 643-649.

Beena S. and Viswaranjan S., 1987. Effect of cadmium and mercury on the Hematological parameters of the fish, *Cyprinus carpio*. Env. And Ecol., 5: 726-732.

Besur Siddesh, Hou Weihong, Schmeltzer Paul, and Bonkovsky Herbert L., 2014. Clinically important features of porphyrin and heme metabolism and the porphyrias. metabolites. 2014 Dec; 4(4): 977–1006.

Bhamre P.R., Desai A.E. and Deoray B.M., 2004. Effect of heavy metals on oxygen consumption of freshwater mussel, *Parreysia favidens*. Pollut. Res., 23 (3): 459-460.

Bhoopathy S., Gunasegar N. and Mandella M.V., 2000. Mercuric Chloride induced dysfunction in *Oreochromis mossambicus* at sublethal dose level. Environ. Issues Manag., 6: 209-215.

Bidwell J.R. and Heath A.G., 1993. An in situ study of rock bass (*Ambloplites rupestris*) physiology: effect of season and mercury contamination. Hydrobiologia, 264(3): 137-152.

Billy G.G., Miller C.A., Pallone M.N., Donachy J.H. and Pierce W.S., 1995. Hemolytic differences among artificial cardiac valves used in a ventricular assist pump. Artif. Org., 19: 339-43.

Blaxhall P.C, 1972. The Hematological assessment of the health of freshwater fish, a review of selected literature. J. Fish Biol., 4: 593-604.

Blaxhall P.C. and Daisley K.W., 1973. Routine hematological methods for use with fish blood. J. Fish Biol., 5: 771.

Boyd C.E. and Tucker C.S., 1998. Pond Aquaculture Water Quality Management. Kluwer Academic Publishers, Boston, MA: 700 pp.

Broderius S.J. and Smith L.L. Jr., 1979. Lethal and sublethal effects of some mixtures of cyanide and hexavalent chromium, zinc or copper to the fathead minnow (*Pimephales promelas*) and rainbow trout (*Salmo gairdnairi*). J. Res. Bd. Can., 36: 164-172.

Brouwer M. and Brouwer-Hoexum T., 1985. Mechanism of Cu(II) and Hg(II) induced loss of red blood cell deformability. Fed. Proc., 44: 859.

Brown V.M. and Dalton R.A., 1970. The acute lethal toxicity to rainbow trout mixtures of copper, phenol, zinc and nickel. J. Fish Biol., 2: 211-216.

Brucka-Jastrzebska E. and Protasowicki M., 2004. Elimination dynamics of cadmium, administered by a single intraperitoneal injection in common carp, *Cyprinus carpio* L. Acta. Ichthyol. Piscat., 84: 167-180.

Bryan G.W., 1976. Some aspects of heavy metal tolerance in aquatic organisms. In: Effects of pollutants on Aquatic organisms, A.P.M. Lockwood (Ed). Cambridge Univ. Press, NY: 7

Buckley J.A., 1996. Heinz body hemolytic anemia in *Coho salmon* (*Oncorhynchus kisutch*) exposed to chlorinated wastewater. J. Fish. Res. Bd. Can., 34: 224.

Buthelezi P.P., Wepener V. and Cyrus D.P., 2000. The sublethal effects of zinc at different water temperatures on selected Hematological variables in *Oreochromis mossambicus*. Afr. J. Aquat. Sci., 25:146- 151.

Calabrese A., Thurberg F.P., Dawson A. and Wenzloff D.R., 1975. Sublethal physiological stress induced by cadmium and mercury in the winter flounder (*Pseudopleuronectes americanus*). In: Sublethal effects of toxic chemicals on aquatic animals, Koemann J.H. and Strik Jjtwa (Eds.). Elsvier Scientific Publishing Company, Amsterdam: 15-21.

Carvalho C. S. and Fernandes M. N., 2006. "Effect of temperature on copper toxicity and hematological responses in the neotropical fish, *Prochilodus scrofa* at low and high pH. Aquaculture251(1): 109-117.

Casillas E. and Smith L.S., 1977. Effects of stress on blood coagulation and hematology in rainbow trout (*Salmo gairdneri*). J. Fish Biol. 10: 481.

Cavas T., Garanko N. N., Arkhipchuk V. V., 2005. Induction of micronuclei and bi-nuclei in blood, gill and liver cells of fishes chronically exposed to cadmium chloride and copper sulphate. Food and Chem. Toxicol., 43: 569.

Çelik E.S., Kaya H., Yilmaz S., Akbulut M., Tulgar A., 2013. Effects of zinc exposure onthe accumulation, haematology and immunology of Mozambique tilapia, *Oreochromis mossambicus.* Afr. J. Biotechnol, 12(7):744–753.

Chitra S. and Jayaprakash K., 2013. Effect of Mercury on blood components of freshwater edible fish, *Labeo rohita*. J. Acad. Indus. Res. Vol. 1(12): 774-777.

Christensen G.M, NcKim J.M., Brungs W.A. and Hunt E.P., 1972. Changes in the blood of the brown bullhead, *Ictalurus nebulosus* following short and long term exposure to copper (II). Toxico. Appl. Pharm., 23: 417-427.

Crandall C.A. and Goodnight C.J., 1963. The effects of sublethal concentrations of several toxicants to the common guppy, *Lebistes reticulates*. Trans. Am. Microsc., 82:59-73.

Cyriac P.J., Antony A. and Nambison P.K.N., 1989. Haemoglobin and haematocrit values in the fish, *Oreochromis mossambicus* (Peters) after short term exposure to copper and lead. Bull. Environ. Contam. Toxicol., 43: 315-320.

Danabas D., Yildirim N.C., Gulec A.K., Yildirim N. and Kaplan O., 2010. An Investigation on some Hematological and biochemical parameters in *Capoeta trutta* (Heckel 1843) from Munzur river (Tunceli,Turkey). J. Ani. Veter. Adv., 9: 2578-2582.

Das S. and Bhattacharya T., 2000. Effect of cadmium and mercury on the blood and liver of *Channa punctatus*. The Fifth Ind. Fisheries Forum (17-20 Jan) CIFA, Bhubaneshwar, India: 51-123.

Davies P.E., 1987. Physiological, anatomic and behavioral changes in the respiratory system of *Salmo gairdneri* Rich. on acute and chronic exposure to chlorothalonil. Comp. Biochem.Physiol., 88C: 113-119.

Dawson A.B., 1935. The hemopoietic response in the catfish, *Ameriurus nebulosus* to chronic lead poisoning. Biol. Bull., 3: 235-346.

Dawson M.A., 1982. Effects of long – term mercury exposure on hematology of striped bass, *Morone saxatilis*. Fish. Bull., 80: 389.

De-Smet-Hans and Blust-Ronny, 2001. Stress responses and changes in protein metabolism in carp, *Cyprinus carpio* during cadmium exposure. Ecotoxicol. Environ. Saf., 48(3): 255-262.

Dhanapakiam P. and Ramaswamy V.K., 2001. Toxic effects of copper and zinc mixtures on some Hematological and biochemical parameters in common carp, *Cyprinus carpio* (L.). J. Environ. Biol., 22 (2): 105-111.

Dinodia G.S., 2001. Studies on the effects of cadmium toxicity in some freshwater fishes. M.Sc. Thesis, Depart. Zool. and Aquacul., CCS HAU, Hisar.

Donaldson E.M., 1981. The pituitary-interrenal axis as an indicator of stress in fish. In: Stress and Fish, Pickering, A.D. (ed.) Academic Press, New York: 11-47.

Durve V.S., Gupta P.K. and Khangarat B.S., 1980. Toxicity of copper to the freshwater teleost, *Rasbora daniconius neilgerieriensis* (Ham.). Nat. Acad. Sci. Lett., 3: 221-223.

Duthie G.C. and Tort L., 1985. Effects of dorsal artic carrulation on the respiratory and haematology of the Mediterranean dog-fish, *Sagliorhinus canicula*. Comp. Biochem. Physiol., 81A: 879-883.

Eiras J.C., Costa G., Biscoito M, Davies A.J., 1996. Suspected viral erythrocytic necrosis (VEN) in the intertidal fish, *Mauligobius maderensis* from Madeira, Portugal. J. mar. biol. Ass. U.K., 76: 545.

Espina S., Salibian A. and Diaz F., 2000. Influence of cadmium on the respiratory function of the grass carp, *Ctenopharyngodon idella*. Water Air and Soil Pollut., 119: 1-10.

Fairbanks U.F., 1967. Copper sulphate-induced haemolytic anaemia. Arch. Intern. Med., 120: 428-436.

Fýrat Ö., 2007. Effects of metal (Zn, Cd)and metal mixtures (Zn+Cd) on physiological and biochemical parameters in blood tissues of *Oreochromis niloticus*. PhD Thesis, Çukurova University, Turkey.

Fletcher T. and White A., 1986. Nephrotoxic and Hematological effects of mercuric chloride in the plaice (*Pleuronectes platessa*). Aquat. Toxicol., 88: 77.

Galhoom K.L., Laila G., Rizk G. and El-ezzaway M.H., 2000. Some biochemical and hematological parameters in Mugil fish (*Mugil caphalcus*) reared in Bahr El-Bakar drain. Egypt. J. Agril. Res., 78: 1-13.

Ghazaly, K., 1992. Sublethal effects of nickel on carbohydrate metabolism, blood and mineral contents of *Tilapia nilotica*. Water Air Soil Pollut., 64: 525.

Gill T.S. and Pant J.C., 1981. Effects of sublethal concentration of mercury in a teleost, *Puntius conchonius*: Biochemical and Hematological responses. Indian J. Expt. Biol., 19: 571-573.

Gill T.S. and Epple A., 1993. Stress-related changes in hematological profile of the American eel (*Anguilla rostrata*). Ecotoxicol. Environ. Saf., 25: 227-233.

Gill T.S. and Pant J.C., 1985. Erythrocytic and leucocytic responses to cadmium poisoning in freshwater fish, *Puntius conchinius*. Environ. Res., 36: 327-337.

Gill T.S. and Pant J.C., 1986. Chromatin condensation in the erythrocytes of fish following exposure to cadmium. Bull. Environ. Contam. Toxicol., 36: 199.

Gill T.S., Pant J.C., 1987. Hematological and pathological effects of chromium toxicosis in the freshwater fish, *Barbus conchonius* Ham. Water, Air and Soil Pollut., 35: 241.

Goel K.A. and Kalpana G.,1985. Hematological characteristics of *Heteropneustes fossilis* under the stress of zinc. Ind. J. Fisher., 32: 256-59.

Guha B., Khuda-Bukhsh A.R., 2002. Efficacy of vitamin-C (L-ascorbic acid) in reducing genotoxicity in fish (*Oreochromis mossambicus*) induced by ethyl methane sulphonate. Chemosphere, 47: 49.

Gupta N., Gopal-Krishna, Gupta D.K. and Shalaby S.I., 2001. Immunological responses in carps (*Labeo rohita* and *Catla catla*) exposed to cadmium. Egyptian J. Veter. Sci., 35: 105-112.

Gupta R.K., Jain K.L., Yadava N.K. and Dharmendra Kumar, 2004. Lead contamination in aquatic medium and its impact on body tissue composition in Indian major carps. Proc. Nat.Workshop Rational Use Water Res. Aqua. (CCS, HAU Hisar, March 18-19, 2004): 208-210.

Gwozdzinski, 1992. Structural changes of proteins in fish red blood cells after copper and mercury treatment. Arch. Environ. Contam. Toxicol., 23 (4): 426- 430.

Heath A.G., 1987. Water pollution and fish physiology. CRC Press, Inc., Florida: 245 pp.

Hernberg S., Nikkanen J., Mellin G. and Lilius H., 1976. Delta-aminolevulinic acid dehydratase as a measure of lead exposure. Arch. Environ. Hlth., 21: 140-149.

Hilmy A.M., Shabana M.B. and Said M.M., 1980. Hematological responses to mercury toxicity in the marine teleost, *Aphanius dispar* (Rüpp). Comp. Biochem. Physiol., C 67: 147-158.

Hoffer R., Weyrer S., Kock G., Pittracher H., 1992. Heavy metal intoxication of arctic charr (*Salvelinus alpines*) in a remote acid Alpine lake. FAO/EIFAC/XVII/92/ Symp. E31, Lugano, Switzerland.

Holopainen, I.J., Hyvärinen, H., 1985. Ecology and physiology of crucian carp (*Carassius carassius* L.) in small Finnish ponds with anoxic conditions in winter. Verh. Int. Ver. Theor. Angew. Limnol., 22: 2566–2570

Hounkpatin Armelle S Y, Ogunkanmi Adebayo Liasu, Alimba C. G. and 5 others, 2012. Hematological study of *Clarias gariepinus* exposed to chronic and subchronic doses of cadmium, mercury and combined cadmium and mercury. Science and Nature, 4 (2): 1-19.

Houston A., Blahut S., Murad A. and Amikrtharaj P., 1993. Changes in erythron organization during prolonged cadmium exposure: an indicator of heavy metal stress. Can. J. Fish. Aquat. Sci., 50: 217.

James R.V., Alagurathinam S. and Sampath K., 1993. Hematological changes and oxygen consumption in *O. mossambicus*, exposed to sublethal concentration of lead. Ind. J. Fisher., 38: 49-54.

Jantrarotai I.V., Lovell R.T. and Grizzle J.M., 1990. Acute toxicity of AFB1to channel catfish. J. Aqua. Anim. Helth., 2: 237-247.

Javed Mehjbeen and Usmani Nazura, 2013. Hematological indices of *Channa punctatus* as an indicator of heavy metal pollution in waste water aquaculture pond, Panethi, India. African J. Biotech., 12(5): 520-525.

Johansson-Sjobeck M.L. and Larsson A., 1978. The effect of Cd on the haematology and on the activity of delta-amino levulinic acid dehydratase (ALA-D) in blood and haemopoetic tissues of the flounder, *Pleuronectes americanus*. Environ. Res., 17: 191-204.

Johansson-Sjobeck M.L. and Larsson A., 1979. Effects of inorganic lead on aminolevulinic acid and hydratase activity and hematological variables in the rainbow trout, *Salmo gairdneri*. Arch. Environ. Contam. Toxicol., 8: 419-431.

John P.J., 2007. Alteration of certain blood parameters of freshwater teleost, *Mystus vittatus* after chronic exposure to Metasystox and Sevin. Fish Physiol. Biochem., 33: 15-20.

Joshi P.K., Bose M. and Harish D., 2002. Hematological changes in the blood of *Clarias batrachus* exposed to mercuric chloride. J. Ecotoxicol. Environ. Monit., 12: 119-122.

Kale Monica K. Kulkarni G.D., 2003. Corelative changes in the cadmium bioaccumulation and oxygen consumption in a freshwater fish, *Rasbora daniconius*. J. Aquat. Bio., 18: 97-102.

Karuppasamy R., Subathra S., Puvaneswari S., 2005. Hematological responses to exposure to sublethal concentration of cadmium in air breathing fish, *Channa punctatus*. J. Environ. Biol., 26: 123.

Kim Jun-Han, Lee Jung-Sick and Kang Ju-Chan, 2012. Effect of inorganic mercury on hematological and antioxidant parameters on olive flounder, *Paralichthys olivaceus*. Fish Aquat. Sci., 15(3): 215-220.

Kori-Siakpere O. and Ubogu E.O., 2008. Sublethal Hematological effects of zinc on the freshwater fish, *Heteroclarias* sp. (Osteichthyes: Clariidae). African J. Biotech., 7(12): 2068-2073.

Kotsanis N., Iliopoulou-Georgudaki J. and Kapata-Zoumbas K., 2000. Changes in selected hematological parameters at early stage of the rainbow trout, *Oncorhynchus mykiss*, subjected to metal toxicants: arsenic, cadmium and mercury. J. Appl. Ichthyol., 16(6): 276-278.

Koyama J. and Ozaki H., 1984. Hematological changes of fish exposed to low concentration of cadmium in the water. Bull. Jap. Soc. Sci. Fish, 50: 199-203.

Kumar Arvind (ed.), 2002. Ecology of polluted waters, Vol. 1. APH Publ. Corp. Darya Ganj, New Delhi.

Kumar Suresh, Lata Swarn and Gopal Krishna, 1999. Acute toxicity of deltamethrin to the freshwater teleosts, *Heteropneustes fossilis* and *Channa punctatus*. Proc. Acad. Environ. Biol., 8(1): 83-85.

Kumarasamy A. and Chandran M.R., 2001. Effect of ascorbic acid on the immune response of the catfish, Mystus gulio (Hamilton), to different bacterins of *Aeromonas hydrophila*. Fish and Shellfish Immunol., 11(4): 347-55.

Kumari K. and Banerjee V., 1986. Effect of sublethal toxicity of Zn, Hg and Cd on peripheral haemogram in *Anabas testudineus* (Bloch.). Uttar P.J. Zool. 16 (2): 241-250.

Kumari Rachana and Shahi R.N., 2014. Effect of cobalt chloride on the oxygen consumption and ventilation rate of a freshwater fish, *Cirrhinus mrigala* (HAM). Intl. J. Res. Eng. and Technol., 3(2): 279-282.

Larsson A and Fang R., 1977. Cholesterol and free fatty acids in the blood of marine fish. Compar. Biochem. Physiol., 57: 191-196.

Larsson A., 1975. Some biochemical effects of Cd on fish: In: Sublethal effects of toxic chemicals on aquatic animals¸ Koeman J.H. and Jitwa Strik (eds). Elsevier Scient. Pub. Co. Amsterdam: 3-13.

Larsson A., Bengtsson B.E. and Svanberg O., 1976. Effects of pollutants on aquatic organisms, Lockwood A.P.M. (Ed.) Soc. For Exp. Biol. Seminar Series. Vol.2, Cambridge Univ. Press: 35-43.

Larsson A., Haux C. and Sjobeck M., 1985. Fish physiology and metal pollution: Results and experiences from laboratory and field studies. Ecotoxicol. Environ. Safety, 9: 250.

Larsson A., Haux C., Sjöbeck M.L.and Lithner G., 1984. Physiological effects of an additional stressor on fish exposed to a simulated heavy-metal-containing effluent from Sulfice Ore Smeltery. Ecotoxicol. Environ Safety, 9: 118-128.

Lock R.A.C., Cruijsen P.M.J.M. and van Overbeeke A.P., 1981. Effects of mercuric chloride and methylmercuric chloride on the osmoregulatory function of the gills in rainbow trout, *Salmo gairdneri* Rich-ardson. Comp. Biochem. Physiol. C Comp. Pharmacol., 68: 151-159

Lowe-Jinde J., Niimi D., 1986. Hematological characteristics of rainbow trout *Salmo gairdneri* (Richardson), in response to cadmium exposure. Bull. Environ. Contam. Toxicol., 37: 375-381.

Lowe–Jinde L. and Niimi A.J., 1983. Influence of sampling on the interpretation of Hematological measurements of rainbow trout (*Salmo gairdneri*). Can. J. Zool., 61: 396.

Massaro E.J., 1974. Pharmacokinetics of toxic elements in rainbow trout, USEPA. Ecol. Res. Ser. No. EPA-660/3-74-027, Washington, DC. 30p

Masud S., Singh I.J. and Ram R.N., 2005. Behavioural and hematological responses of *Cyprinus carpio* exposed to mercurial chloride. J. Environ. Biol., 26: 393-397.

Mazon A.F., Monteiro E.A.S., Pinheiro G.H.D. and Fernadez M.N., 2002. Hematological and physiological changes induced by short-term exposure to copper in the freshwater fish, *Prochilodus scrofa*. Braz. J. Biol., 62(4): 10pp

McKim J.M., Christensen G.M. and Hunt E.P., 1970. Changes in the blood of brook trout (*Salvelinus fontinalis*) after short-term and long-term exposure to copper. J. Fish. Res. Bd. Canada, 27: 1883-1889.

McLeay D.D. and Gordon M.R., 1977. Leucokrit: a simple hematological technique for measuring acute stress in salmonid fish, including stressful concentrations of pulp mill effluents. J. Fish Res. Bd. Can., 34: 2164.

Mishra J. and Srivastava A.K., 1979. Malathion induced hematological and biochemical changes in the Indian catfish, *Heteropneustes fossilis*. Environ. Res., 30: 393.

Niimi A.J. and Love-Jinde L., 1984. Differential blood cell ratios of rainbow trout (*Salmo gairdneri*) exposed co mechyl-mercury and chlorobenzenes. Arch. Environ. Concam. Toxicol., 13: 303.

Nikinmaa M. and Jensen F.B., 1986. Blood oxygen transport and acid-base status of stressed trout (*Salmo gairdnerii*): Pre- and post-branchial values in winter fish. Comp. Biochem. Physiol., 84 A: 391-396.

Nilsson S. and Grove D. J., 1974. Adrenergic and cholinergic innervation of the spleen of the cod, *Gadus morhua*. Eur. J. Pharm., 28: 135–143.

Nussey G., van Vuren J.H.J. and du Preez H.H., 2002. The effect of copper and zinc at neutral and acidic pH on the general haematology and osmoregulation of *Oreochromis mossambicus*. Afr. J. Aquat. Sci. 27:61-84

Ololade I.A. and Ogini O., 2009. Behavioural and hematological effects of zinc on African Catfish, *Clarias gariepinus*. Int. J. Fish. Aquat., 1: 22- 27.

Oshawa M. and Kawai K., 1981. Cytological shift in lymphocytes induced by cadmium in mice and rats. Environ. Res., 24: 192-200.

Palace V.P., Majewski H.S. and Klaverkamp J.F., 1993. Interactions among antitoxicant defences in liver of rainbow trout (*Oncorrhyncus mykiss*) exposed to cadmium. Canad. J. Fish Aq. Sci., 50: 156.

Papoutsoglou S.E. and Abel P.D., 1988. Sublethal toxicity and accumulation of cadmium in *Tilapia aurea*. Bull. Environ. Contam. Toxicol., 41: 404.

Pentreath R.J., 1976. The accumulation of mercury from food by the plaice, *Pleuronectes platessa* L. J. Exp. Mar Biol. Ecol., 25: 51-65.

Poon W.L., Hung C.Y. and Randall D.J., 2002. The effect of aquatic hypoxia on fish. Depart. Biol. and Chem., City Univ. Hong Kong, Kowloon, Hong Kong, SAR, China. EPA/600/R-02/097.

Redondo P.A., Alvarez A.I., Diez C., Fernandez-Rojo F., Prieto J.G., 1995. Physiological response to experimentally induced anemia in rats: a comparative study. Lab. Anim. Sci., 45: 578-83.

Reichert W.L., David A.F. and Malins C.D., 1979. Uptake and metabolism of lead and cadmium in coho salmon, *Onchorhynchus kisutch*. Comp. Biochem. Physiol. Pharmacol., 63: 229-234.

Rink L. and Kirchner H., 2000. Zinc-altered immune function and cytokine production. J. Nutr., 130: 1407-1411.

Rogers W.D. and Beamish F.W.H., 1982. Dynamics of dietary methylmercury in rainbow trout. *Salmo gairdneri*. Aquat. Toxicol., 2: 271-290.

Roy I. and Gupta M.N., 2003. Selectivity in affinity chromatography. In: Isolation and purification of proteins, Mattiasson B. and Kaul-Hatti R. (Eds.) Marcel Dekker Inc., New York: 57-94.

Ruparella S.G., Verma Y., Salyed S.R. and Rawal U.M., 1990. Effect of cadmium on blood of tilapia, *Oreochromis mossambicus*, during prolonged exposure. Bull. Environ. Contam. Toxicol., 45: 305-312.

Sahoo P.K. and Mukherjee S.C., 1999. Normal ranges for diagnostically important Hematological parameters of laboratory-reared rohu (*Labeo rohita*) fingerlings. Geobios., 26: 31-36.

Sampath K., James R. and Ali K.M.A., 1998. Effects of copper and zinc on blood parameters and prediction of their recovery in *Oreochromis mossambicus* (Pisces : Cichlidae). Ind. J. Fish., 45: 129-139.

Samuel Y. and Thateyus A.J., 2004. Effect of cadmium on the respiration of *Oreochromis mossambicus*. J. Exp. Zool., 7(1): 113-116.

Sankaraperumal, G., Rajan, M.K. and Mohandoss, A. 1990. Synergistic effect of Cadmium and Zinc on the erythrocyte and opercular activity of the fishes (*Lepidocephalichthyes thermalis*) and (*Amblypharyngodon melettinus*). Environ. Ecol., 8 (4): 1213 – 1216.

Sarkar S.K., 1989. Evaluation of two heavy metals on the oxygen consumption of *Tilapia mossambica* (Peters). Geobios., 16: 108-110.

Sarkar S.K., 1999. Effect of two heavy metals (copper sulphate and cadmium sulphate) on the oxygen consumption of the fish, *Cyprinus carpio* (Linn.). U. P. J. Zool., 19(1): 13-16.

Shah S.L. and Altindag A., 2004. Hematological parameters of tench (*Tinca tinca*) after acute and sub chronic exposure to lethal and sublethal Hg treatments. Bull. Environ. Contam. Toxicol., 73: 911-918.

Shah S.L., Hafeez M.A., and Shaikh S.A., 1995. Changes in Hematological parameters and plasma glucose in the fish, *Cyprinon watsoni*, on exposure to zinc and copper treatment. Pak. J. Zool., 24: 50-54.

Shahi J., Chauhan S. and Singh A., 2013. Comparative study on the Hematological effect of synthetic and plant origin pesticides on fish, *Channa punctatus*. Indian J. Nat. Products and Resources, 4(1): 48-53.

Sharma K.P., Sharma S., Sharma S., Singh P.K., Kumar S., Grover R., Sharma P.K., 2007. A comparative study on characterization of textile waste waters (untreated and treated) toxicity by chemical and biological tests. Chemosphere, 69: 48.

Shyong W.J. and Chen H.C., 2000. Acute toxicities of copper, cadmium and mercury to the freshwater fish, *Varicorhinus barbatus* and *Zacco barbata*. Acta. Zoologica Taiwanica, 11(1): 33-45.

Singhal R.L. and Merali Z., 1979. Cadmium toxicity, Marcel Dekker New York.

Sinha T.K.P. and Kumar K., 1992. Acute toxicity of mercuric chloride to *Anabas testudineus* (Bloch). Environ. Ecol., 10: 720-722.

Sjobeck M.L., Haux C., Larsson A., Lithner G., 1984. Biochemical and hematological studies on perch, *Perca fluviatilis*, from the cadmium-contaminated river Eman. Ecotoxicol. Environ. Saf., 8: 303.

Skidmore J. F. 1970: Respiration and Osmoregulation in rainbow trout with gills damaged by zinc sulphate. J. Exp. Biol., 52: 481-494.

Smith B.P. Hejtmanick E. and Camp B.J., 1976. Acute effects of cadmium on catfish (*Ictalutas punctatus*). Bull. Environ. Contam. Toxicol., 15: 271-277.

Soivio A. and Nikinmaa M., 1981. The swelling of erythrocytes in relation to the oxygen affinity of the blood of the rainbow trout, *Salmo gairdneri* Richard. In: Stress and Fish (Pickering A.D., Ed.), London: Academic Press: 103-119 pp.

Soivio A., Nyholm K. and Westman K., 1981. The role of gills in the responses of *Salmo gairdneri* during moderate hypoxia. Comp. Biochem. Physiol., 70A: 133.

Spehar R.L., 1976. Cadmium and zinc toxicity to *Jordanella floridae*. Ecol. Res. Series, 3: 76-96.

Srivastava A.K. and Mishra G., 1979a. Blood dyscrasia in a teleost fish (*Colisa faciatus*) associated with cadmium poisoning. J. Comp. Pathol., 89: 609-613.

Srivastava A.K. and Mishra Shashikala, 1979b. Blood dyscrasia in a teleost, Colisa fasdatus after acute exposure to sublethal concentrations of lead. J. Fish Biol., 14: 199-203.

Stagg R.M. and Shuttleworth T.J., 1982. The accumulation of copper in *Platichthyes nesus* L. and its effects on plasma electrolyte concentrations. J. Fish Biol., 20: 491-500.

Summarwar Sudha, 2012. Hematological investigation on *Labeo rohita* following chronic exposure to zn and cu. International Journal of Advanced Life Sciences, 4: 43-48.

Svobodova Z., Richard L., Jana M. and Blanka V., 1993 Water quality and fish health. EIFAC Technical paper 54 Tanzania Fisheries Division. Aquaculture in Tanzania, Ministry of Natural Resources and Tourism Fisheries Division.

Svobodova Z., 1982. Changes in some Hematological parameters of the carp after intoxication with $CuSO_4$. Bulgarian Vyak Usta Ryo. Hydrobiol., 18: 26-29.

Swift D.J. and Lloyd E., 1974. Changes in urine flow rate and haematocrit value of rainbow trout (*Salmo gairdneri*) exposed to hypoxia. J. Fish Biol. 6: 379.

Tewari H., Gill T.S., Pant J., 1987. Impact of chronic lead poisoning on the hematological and biochemical profiles of a fish, *Barbus conchonius*. Ham. Bull. Environ. Contain. Toxicol., 40:198-203.

Tort L. and Hernandez-Pascual M.D., 1990. Hematological effects in dogfish, *Scyliorhinus conicula* after short term sublethal concentration of lead. J. Fish. Biol., 14: 199-203.

Tort L. and Torres P., 1988. The effect of sublethal concentrations of cadmium on Hematological parameters in the dogfish, *Scyliorhinus canicula*. J. Fish Biol., 32: 277-282.

Tyagi A. and Srivastava N., 2005. Hematological response of fish, *Channa punctatus* (Bloch) to chronic zinc exposure. J. Environ. Biol. Acad., 26(2 Suppl): 429-432.

Ultsch G.R., Boschung H. and Ross M.J., 1978. Metabolism, critical oxygen tension, and habitat selection in darters (Etheostoma). Ecology, 59: 99-107.

Van Aardt W.J. and Booysen J., 2004. Water hardness and the effects of Cd on oxygen consumption, plasma chlorides and bioaccumulation in *Tilapia sperrmanii*. Water SA., 30: 57-64.

Van der Putte L., Laurier M.B.H.M. and Van Eijk G.J.M., 1982. Respiration and osmoregulation in rainbow trout (*Salmo gairdneri*) exposed to hexavalent chromium at different pH values. Aquat. Toxicol., 2: 99.

Van Heerden D., Vosloo A., Nikinmaa M., 2004. Effects of short-term copper exposure on gill structure, metallothionein and hypoxia-inducible factor-1 alpha (HIF-1 alpha) levels in rainbow trout (*Oncorhynchus mykiss*). Aquat. Toxicol., 69: 271-280.

Varghese G., Naik P.S. and Katdare M., 1992. Respiratory responses and blood sugar level of the crab, *Barytelphura cunicularis* (West wood), exposed to mercury, copper and zinc. Ind. J. Exp. Biol., 30: 308-312.

Venkateshwarlu P., Rani V., Janaiah C. and Prasad M., 1990. Effect of endosulfan and kelthane on haematology and serum biochemical parameter of the teleost, *Clarias batrachus*. Ind. J. Comp. Anim. Physiol., 8: 8.

Verheyen E.R. Bluse and Decleir W., 1994. Metabolic rate, hypoxia tolerance and aquatic surface respiration of some lacustrine and riverine African cichlid fishes (Pisces: Cichlidae). Comp. Biochem. Physiol., 107A: 403-411.

Vijayram K., Geraldone K. and Longasnathan P., 1989b. Hematological responses of an air breathing fish, *Anabas testudineus* to cadmium. J. Ecobiol., 1: 15-19.

Vijayram K., Geraldone K., Varadarajan T.S., George J. and Longasnathan P., 1989a. Cadmium induced changes in the biochemistry of an air breathing fish, *Anabas testudineus*. J. Ecobiol., 1: 245-251.

Voyer R.A., Yevich P.P. and Barszec C.A., 1975. Histological and toxicological response of the mummichog, *Fundulus heteroclitus* to combinations of low levels of cadmium and dissolved oxygen in freshwater. Water Res., 9: 1069-1074.

Waiwood K.G., 1980. Changes in hematocrit of rainbow trout exposed to various combinations of water hardness, pH, and copper. Trans. Ame. Fisher. Soc., 109(4): 461.

Wedemeyer G.A., McLeay D.J., and Goodyear C.P., 1984. Assessing the tolerance of fish and fish populations to environmental stress: the problem and methods of monitoring. In: Contaminant effects on fisheries, Cairns V.W., Hodson P.V., and Nriagu J.O. (eds). John Wiley and Sons, New York, N.Y., pp 163-195.

Wedemeyer G.A., Nelson N.C. and Yasuake W.T., 1979. Physiological and biochemical aspects of ozone toxicity to rainbow trout (*Salmo gairdneri*). J. Fish. Res. Bd. Can., 36: 605.

Wells R.M.G., Grigg G.C., Beard L.A. and Summers G., 1989. Hypoxic responses in a fish from a stable environment: blood oxygen transport in the antarctic fish, *Pagothenia borchgrevinki*. J. Exp. Biol., 141: 97-111.

Wilson R.W. and Taylor E.W., 1993. The physiological responses of freshwater rainbow trout, *Oncorhynchus mykiss*, during acutely lethal copper exposure. J. Comp. Physiol., B 163: 38-47.

Witeska M., 2004. The effect of toxic chemicals on blood cell morphology in fish. Fresenius Environ. Bull., 12A:1.

Witeska M., 2005. Stress in fish - Hematological and immunological effects of heavy metals. Electronic J. Ichthyol., 1: 35-41.

Witeska M., Kondera E. and Szczygiels K., 2011. The effects of cadmium on common carp erythrocyte morphology. Polish J. of Environ. Stud., 20(3): 783-788.

Witeska M., Kondera E., Szymañska M., Ostrysz M., 2010. Hematological changes in common carp (*Cyprinus carpio* L.) after a short-term lead (Pb) exposure. Polish J. Environ. Stud., 19: 825.

Woerr J.A., Huff W.E. and Hamilton P.B., 1987. Increased sensitivity to staphylococcal betahaemolysin of erythrocytes from chicken during alfatoxicosis. Poul. Sci., 66: 1929-1933.

Yamamoto K.I., 1988. Contraction of spleen in exercised freshwater teleost. Comp. Biochem. Physiol., 89: 65-66.

Zeni C., Bovolenta M.R., Stagni A., 2002. Occurrence of echinocytosis in circulating RBC of black bullhead, *Ictalurus melas* (Rafinesque), following exposure to an anionic detergent at sublethal concentrations. Aquat. Toxicol. 57: 217.

Chapter 11

Immunological Disorders

1. Immune Functions

The potential of using immune functions in monitoring contaminants or the diseases is probably very large (Weeks *et al.*, 1992), however, these have not been used much so far possibly due to the shortage of easily applied procedures that would have allowed routine monitoring of the fish in polluted waters. The Immune system's primary role is to protect the body against damage from any source that proves dangerous including viruses, bacteria, fungi, and other microorganisms as well as internal threats such as cancer. Like the fish physiology in general, fish immunity is strongly influenced by the abiotic and the biotic factors. Water temperature is generally considered as one of the strongest abiotic factors affecting fish physiology and the immune functions. Aquatic life is currently being exposed to chemical contamination by increasing variety of anthropogenic activities that can induce many different mechanisms of toxicity, each contributing to varying degrees of deleterious effects (Correia *et al.*, 2003). Immunocompetence is vital in maintaining the overall health of an organism and is extremely sensitive to toxins, such as heavy metals (Institoris *et al.*, 2001). Immunotoxicity is simply defined as the study of the interaction of xenobiotics with the immune system (Kimber, 1991) which incorporates the ability of chemicals both to compromise immune function and to produce specific tissue damage or allergy. Immunological aspects in a fish thus, need some review of toxicological impact on fish's self protection system against diseases, in a polluted environment. It is clear from a number of studies that the immune system is exquisitely sensitive for assessing toxic effects of chemicals of environmental concern (Luster and Rosenthal, 1993). Natural and artificial environmental stress factors appear to suppress immune functions. Of the numerous environmental stress factors considered, pollutants, handling/confinement and low temperature are probably the best studied forms in fish. All three forms of stress factors have been shown to suppress components of both the innate (non-specific) and adaptive arms of the immune system (Bly *et al.*, 1997). A better understanding of the mechanism(s) resulting in immune-suppression should facilitate future *in vivo* manipulations to reduce susceptibility to disease in aquaculture situations.

These immunological assays, originally developed in rodents, have been adapted for use in a variety of animal species (Weeks *et al.*, 1992) including fish. The impairment of innate immunity may be more significant in fish than in mammals, as mounting an adaptive or acquired immune response takes longer in fish (Alexander and Ingram, 1992). The first line of body defense in a fish *i.e.* the first physical barriers, are the scales and the skin; the mucus containing the bacteriolytic lysosomes is the first chemical barrier (Lie *et al.*, 1989). A wide variety of pollutants, low pH and the physical stresses due to handling are said to stimulate increased secretion of mucous by the mucous glands located in gills and skin. The lysosomes help in sloughing off the harmful organisms. Stomach acidity is another line of defense to kill bacteria. Tissue macrophages, blood monocytes and neutrophils are the internal nonspecific response organelles and they act as second line of defense in an animal, and these act as phagocytes against all kind of invading harmful microbes in a fish body. Macrophages are the inter-digitating dendritic cells and these scavenge materials and other so-called antigen-presenting cells (APCs), from the blood and tissues and digest them. Neutrophils are constantly produced in the marrow and released into the blood to search for invading pathogens.

An immune system comes into play when both the primary and secondary modes of defense are unable to check the entry of various pathogens into the fish body. The animal's third and perhaps the ultimate line of defense is its immune system. The majority of immunological studies have suggested an immune-suppression effect associated with a decrease in water temperature (Bly and Clem, 1992; Hutchinson and Manning, 1996b), which may make a fish more susceptible to diseases (Goede and Barton, 1990). Teleost fish though possess an immune system mechanism similar to mammals *i.e.* the fish possess both the non-specific (innate or natural) and the specific mechanisms (acquired or adaptive) (Pasquier, 1991). However, substantial differences exist between the immune systems of poikilo- and homoiothermic organisms. According to Ainsworth *et al.* (1991), the specific system of immunity is more sensitive than the non-specific defense at lower water temperature, and it was assumed to be more important for poikilothermic than for homoeothermic vertebrates. The specific system of immunity is distinguished from the non-specific defenses by four features: specificity, diversity, self/non-self recognition, and memory. Specificity: means the ability of the immune system to recognize and eliminate particular microorganisms and the foreign molecules called antigens; Diversity is the ability of the immune system to respond to millions of kinds of invaders, each recognized by a unique antibody producing lymphocyte; Self/Non-Self recognition, *i.e.* the immune system is "trained" during its early developments to recognize differences in between self and non-self tissues or particles. That is why the immune system doesn't attack animal's own cells. Thus, there are no antigen receptors for an animal's own molecules; and memory refers to the immune system's ability to remember antigens, it has encountered first and to react quickly and effectively again when these enter a body second time. This is also known as acquired immunity and it may be active- conferred by recovery from a particular infectious disease, or passive *i.e.* transferred from one individual to another or obtained through vaccination.

Major functions of the immune system are: 1. How does the immune system identify and respond to pathogens and, 2. In what way the fishes response to such Immuno-toxicological effects to metals. The B-lymphocytes are most important specific immune response blood cells, whose response is induced by individual antigens. Antibodies inactivate antigens and mark them for destruction by the macrophages. Like mammals, fish do not possess bone marrow to produce B-lymphocytes. In fish, B-lymphocytes are produced in the head kidney, gut-associated tissues and the spleen. Some of these B-lymphocytes termed as the plasma cells are activated by the specific antigens to multiply. These start secreting antibodies which are specific to the antigens responsible for their secretion. Some of the B-lymphocytes are differentiated to act as the memory cells that serve to remember the kind of antigen responsible for their activation, so that upon their subsequent exposure to the same antigen, the secondary antibody responses are much more rapid and larger. Following under mentioned are the steps for various antigen immunity specific responses of the B-lymphocytes:

1. The first step involves preparation of a molecule of foreign protein, referred to as an antigen, for identification. The antigens are the specific molecules mostly the proteins present on the cell surface of an invading microbe. Once inside a phagocyte cell small pieces of the antigens (peptides) enter a special processing center, a vacuole called the compartment for peptide loading (CPL), where they are bound to a special class of membrane proteins called major histo-compatibility complex (MHC) molecules. The CPL somehow attaches the antigen to a class II MHC molecule and embeds it into the plasma membrane for all passing immune cells to respond. It is this antigenic flag around which various lymphocytes rally.
2. MHC molecules are categorized as: Class I molecules, found on all nucleated cells; Class II, found on B-cells and macrophages; and class III are the various soluble proteins that make up the complement. T-cells have T-cell receptors, which fit like lock and key to the Class I or Class II MHC molecules on normal healthy cells. T-cell receptors are actually capable of recognizing a specific antigen which is attached to the MHC molecule. T-cells can only be activated when the antigen-MHC site is occupied (first signal), and a second less understood site is also involved as the second signal. Conversely, anytime a T-cell gets the first signal but not the second it will be inactivated-perhaps by apoptosis *i.e.* programmed cell death.
3. When a lymphocyte first encounters an antigen it recognizes it and is stimulated to divide; it initiates the Primary Immune Response. The process involves clone selection *i.e.* the rapid development of a single line or lines of T- and B- lymphocytes from a vast and diverse army of lymphocytes.

Satchell, 1991 has described 5 other major arms of the specific immune system, out of which except the first one all others *i.e.* the last 4 types (2-5) are poorly understood so far in fishes: T-lymphocytes, responsible for a cell mediated response are produced in the thymus in the young stage of a fish; the sub-populations of these cells aid in the production of antibodies and other such chemicals that suppress antigens. NK- cells, also termed as the natural-killer cells destroy stressed and infected cells and tumors. Interferon, are cytokines which interferes with replication of viruses and bacteria. Transferrins, bind with the free iron and deprive microorganisms of iron. An assortment of all such molecules agglutinate pathogens and aid phagocytosis.

The hormones, cortisol and catecholamine have a marked effect on numerous immunological functions. The general adaptive response in a fish to any stress is characterized by the elevated levels of cortisol and catecholamine, which may also be a primary mechanism for immune system suppression in fish exposed to a pollutant. These hormones as discussed in Chapter 9 are involved primarily in the maintenance of homeostasis in animals in the face of stress. Many workers (Thomas and Lewis, 1987) have now evidenced that cortisol suppresses several of the immune functions in fish, and these eliminate the confounding effects of a chemical contaminant or any other stressor. Tripp *et al.*, 1987 carried out *in vitro* studies, observing cortisol suppressing the mitogenic response of Coho salmon lymphocytes. A mitogen is a chemical substance that encourages a cell to commence cell division, mitosis. Suppression of immune function by exogenous cortisol has also been shown to make trout more susceptible in a dose dependent manner to the common bacterial and fungal diseases (Pickering and Pottinger, 1989). Schreck and Bradford (1990) have evidenced that lymphocytes exhibit a negative feedback and these suppresses secretion of cortisol in fishes, which is incidentally reverse of what occurs in mammals (Sapolsky *et al.*, 1987). Weeks *et al.*, 1992 reported that catecholamines in mammals are elevated during stress and these further mobilize blood granulocytes and macrophages, thereby causing their number to rise in the blood. Fish exposed to sediments contaminated with petro-chemicals such as diesel are reported to cause an increase in leucocyte number with low doses, but suppression with high doses (Tahir *et al.*, 1993).

Fish are an especially important animal group from the perspective of innate immunity and eco-toxicology (Ingram, 1980). Alexander and Ingram (1992) reported that the innate immune system is a fundamental defense mechanism of fish. Tort *et al.* (2003) has defined the fish immunity system, consisted of both the cellular and humoral immune responses and head kidney as the important central endocrine organ, homologous to mammalian adrenal glands, releasing corticosteroids and other hormones. Like the mammals, most of the generative and secondary lymphoid organs are also found in fish, except for the lymphatic nodules and the bone marrow (Press and Evensen, 1999). The head kidney assumes hemopoietic functions, and is the principal immune organ responsible for phagocytosis (Danneving *et al.*, 1994), antigen processing (Kaattari and Irwin, 1985) and formation of IgM and immune memory through melanomacrophagic centers (Tsujii and Seno, 1990). The thymus, another lymphoid organ situated near the opercular cavity in teleosts, produces T-Lymphocytes, involved in allograft rejection, stimulation of phagocytosis and antibody production by B-cells (Powell, 2000). The involution of thymus in fish is more dependent on hormonal cycles and seasonal variations than on the age (Press and Evensen, 1999). The thymus in fish involutes at sexual maturity in some species, but retain in its normal size in others. These centers may retain antigens as immune complexes for long periods. Fish immune cells show the same main features like that of other vertebrates, and lymphoid and myeloid cell families are also determined. The monocyte/macrophage cell lineage is the most studied in fish although no exact cell specific markers are available. Therefore, the term macrophage is often used for phagocytic cells independent of the differentiation status and the anatomical location. The majority of cytokines identified to date and of data concerning cytokine regulation has been obtained in monocyte/macrophage cell cultures (Secombes *et al.*, 2001). As in other vertebrates the most exposed tissues are the gut, respiratory organs and the skin, and the lymphoid tissue associated to tegument is distributed around the skin, gills and intestine, thus, complementing physical and chemical protection provided by these structures. Among the epidermal secretions, antimicrobial proteins- lysozyme, phosphatases and trypsin are often found, whose amount and activity are species specific (Fast *et al.*, 2002). In addition to lymphocytes, the gill lamellae have pillar cells belonging to the reticulo-endothelial system (Dalmo *et al.*, 1997). Mucus is further a most important barrier in fish, firstly to act as substrate for the antibacterial mechanisms; secondly, the mucus covers most of the fish external surfaces *i.e.* the skin and the gills. Although this is a general trend, it is very clear that most freshwater species have a higher production of mucus compared with marine species. In addition, production of mucus is significantly increased when subjected to stressing situations, such as chemical aggressions (Demers and Bayne, 1997).

It is assumed that the response of fish to immune challenge is mainly based on the innate immune response. Lysozyme activity is one of the most studied innate responses in fish. Lysozyme can act on the peptide-glycan layer of the bacterial cell walls, resulting in the lysis of the bacterial cell. Lysozyme has been found in mucus and ova (Ellis, 1999). Serum lysozyme, probably coming from peritoneal macrophages and blood neutrophils has been used as an indicator of non-specific immune response (Tort *et al.*, 1998). The lysozyme response has been found to be variable in its potency depending on the species and the tissue location but is present in all species studied (Grinde *et al.*, 1988). It appears that the lysozyme response in fish may be induced very rapidly and not only related to bacterial presence but also to other alarm situations such as after stress (Rotllant and Tort, 1997). Thus, lysozyme in fish would be involved in the overall alarm response, acting as an acute-phase protein. Fish is the first group that shows antibody activity and therefore a simple profile of one Ig appears in this group. Cytokines as modulators of the immune response have been little studied in fish. A significant number of cytokines are functionally active in teleosts (Secombes, 1996), and haematopoiesis, apart from affecting cortisol secretion. Another potentially important cytokine, Tumour Necrosis Factor alpha (TNF-α), has also been cloned in various fish species (Bobe and Goetz 2001; Holland *et al.*, 2002). Furthermore, TNF-like protein activity has been shown to induce apoptosis, and to enhance neutrophil migration and macrophage respiratory burst activity (Qin, 2001).

2. Immunological Observations

Like mammals, metals in fish are known to cause immunological inhibition and stimulation, primarily depending upon the type of the function measured. The impairment of innate immunity may be more significant in fish than in mammals, as mounting an adaptive or acquired immune response takes longer in fish (Alexander and Ingram, 1992). Lysozyme is one of the important bactericidal enzymes of innate immunity (Demers and Bayne, 1997). Lysozyme activity in fish blood is sensitive to environmental contaminants (Bols *et al.*, 2001). Copper sulphate is commonly used as a protective measure against a number of external diseases like columnaris infections in fish ponds, and to improve their resistant to the disease causing bacteria (MacFarlane *et al.*, 1986). Columnaris disease is caused by the bacterium, *Flexibacter columnaris,* it may result in acute or chronic infections in both coldwater and warm water fish. Striped bass (Hetrick *et al.*, 1982) when exposed to Cu at 7 and 10 µg/l for 96 h, prior to challenge with two naturally occurring bacterial species (*Vibrio anguillarum and Pasteurella piscicida*), was observed to increase susceptibility of the fish to the disease organisms. So while Cu may be protective for some external diseases, defense against internal infections can be compromised by prolonged exposure to this metal. An air-breathing catfish (*S. fossilis*) exposed to Cu at its different concentrations for 28 days, also exhibited a dose-dependent depression of antibody production, phagocyte activity of spleen and kidney microphages and a prolongation of eye allograft rejection time (Khangarot and Tripathi, 1991). The later response indicates suppression of the T cell activity. Anderson *et al.* (1989) also reported copper to cause immuno-suppression of antibody producing cells in rainbow trout when tested *in vitro*. Copper in comparison to Zn is further said to be more potent with regards to suppression of lymphocytes number. A dose-dependent suppression of kidney lymphocyte number and natural cytotoxic cells was observed in zebrafish exposed for 7 days at different concentrations of Cu or Zn (Rougier *et al.*, 1994). The effect of Cu on macrophage activity was however, found to be more complex. The authors reported Cu to cause a marked decrease in macrophage activity both *in vitro* and *in vivo*, but Zn caused a modest increase under the same conditions. No explanation was however, provided about the stimulatory behavior of Zn, since the same response was observed both *in vitro* and *in vivo* conditions, this implies to the fact that the response was not an artifact.

Exposure of brown head catfish to lead up to 183 days, was reported to produce a reduction in spleen size, but an increase in leukocyte number and an especially large increase in the number of thrombocytes (Dawson, 1935). Later, Crandall and Goodnight (1963) made similar observations in guppies, exposed for at least 60 days to lead and Zn. They further recorded reduced lymphoid tissue in the head kidney. O'Neil (1981) exposed trout and carps to Ni, Zn and Cr, and found these metals suppressing to a variable extent the primary humoral responses to an Intra-peritoneal inoculation of MS-2 bacteriophages. He further stated that metals may block the active site of antibody molecule and disturb the metabolism, ionic balance and cellular division of immune-competent cells. MethylHg is also said to severely inhibit at least some components of the immune system (Roales and Perimutter, 1977). Blue gaurami (*Trichogaster* sp.) a tropical freshwater species, on exposer to

MethylHg for 4-5 days, and the testing for viral neutralizing showed some inhibition of the neutralization occurred. No bacterial agglutination was detectable in these fish exposed to MethylHg. Witeska, 2005 obtained data about metal induced stress symptoms in fish blood in common carp exposed to lead, and showed that a short-term exposures to high levels heavy metals induced stress reactions in fish. In the red blood cell system adaptive changes prevailed, preparing the organism to an increased energy expense, while the changes observed in the white blood cell system indicate a considerable immunosuppressive effect of stress. In the white blood cell system – a decrease in total leukocyte count took place caused by a considerable drop in lymphocyte count. In some cases (particularly in Pb and Cd-intoxicated fish) percentage of neutrophils increased. Therefore even a short-term exposure to heavy metals induces a persistent stress in fish which may render them more susceptible to diseases.

Spazier *et al.*, 1992, following a chemical spill (including chlorinated hydrocarbons and metals) found eels to have a considerable histo-pathological damage in the spleen. In addition melanomacrophage centers were reported lacking in eels suffering from the aftermath of the spill. Wester *et al.*, 1994 suggested melanomacrophage centers as the useful biomarkers for immunotoxicology in part because they are considered as the primitive analog of the mammalian lymph follicles. Melanomacrophage centers are fairly easy to assay using standard histological methods. Their density may decrease in fish from contaminated waters or along a pollution gradient. Wester *et al.*, 1994 suggested that increased accumulation of cytotoxic waste however, may increase the density of these centers. Morra, 1993 reported suppression of both lymphocyte proliferation and antibody production in channel catfish exposed to acidified waters (pH 2) for 2 weeks, along with suppression of the secondary antibody response, may be due to the suppression of the ability of B cells to process antigen. Suppression of lysosome activity under pollutant stress could be another important factor in affecting phagocytosis in fishes. The enzyme lysozyme is present in the lysosomes of the neutrophils and the macrophages which appear secreted into the blood by these cells. Mock and Peters (1990) found that fairly high doses of ammonia can cause a depression in its secretion in the blood thus, affecting the death of the antigens. Secombes *et al.* (1992) also found that the sewage sludge caused a decrease in leucocyte bactericidal activity and in antibody-secreting cells.

Likewise, the effects of Cd were also found to vary with the type of immune cells observed. Cd is eported to have a marked effect on differential leukocyte counts in fish. In cunners (a marine teleosts), Newman and MacLean (1974) reported a dose-dependent three-fold increase in neutrophils, contrary to this lymphocytes showed almost a three-fold decrease (again dose-dependent). Decrease in lymphocytes was attributed to their reduced production, as the blast cells which are lymphoid progenitors were also found decreased. Stomberg *et al.*, 1983 also reported lesions in the haematopoitic areas, following Cd exposure in fathead minnows. Murad and Houston, 1988 recorded similar results in goldfish after 3 week exposure to Cd, except that the effects were neither dose-dependent nor as large, possibly due to differences in its sensitivity to Cd. Increased number of neutrophils may be conceived as to compensate the loss of lymphocytes. The observations of MacFarlane *et al.*, 1986 on exposed juvenile striped bass to Cd at 12-30 µg/L for 5 days provided some evidence to this assumption. They challenged the treated fish with a bacterium which causes columnaris disease. The Cd-exposed fish were found slightly less susceptible to the disease than the controls. A variable that could have a considerable effect on the modulation of fish immune response to Cd could be lack of cortisol production response in fish to Cd. The fish in general, in response to a stressor exhibit an elevated level of plasma cortisol; Cd, however, was said to be as an exception (Thomas and Neff, 1984). Thuvander (1989) in rainbow trout exposed for 12 weeks to a very low concentration of Cd (0.7-3.6 µg/l) observed suppression of T lymphocyte function, whereas, the B cell antibody response to a bacterial challenge was found enhanced. Hutchinson and Manning (1996a), have reported significant reductions in the respiratory burst of phagocytes for all cadmium exposed marine fish, dab (*Limanda limanda*), compared with control. Gupta *et al.* (2001) studied immunological responses in *Labeo rohita* and *Catla catla* exposed to cadmium. Decreased mean leukocyte and per cent lymphocyte counts were observed, whereas neutrophilia was observed in both species exposed to cadmium. The study showed that cadmium caused sufficient stress to fish and ultimately affected its growth and performance. Witeska (2005) evaluated hematological and immunological effects of heavy metals during short term exposure in common carp, *Cyprinus carpio.* Common carp were subjected for 3 hours to 10, 5, 10 or 20 mg/l of lead, copper, cadmium or zinc respectively and subsequently transferred to clean water. Hematocrite value and erythropoietic rate increased, with a decrease in total leukocyte count. Likewise, Rajalakshmi *et al.*, 2014 on exposing fingerlings of (7-9 cm) an estuarine fish, *Lates calcarifer* to acute concentration of cadmium plus nickel (3.0 ppm) for 96 hours recorded the lysozyme activity declined up to 48 hours (17.85) and thereafter it recovered up to 96 hours (30.15). The results showed that a short term exposure to low levels heavy metals induced stress reaction in fish. While the changes observed in the white blood cell system indicates a considerable immunosuppressive effect of metal stress. Therefore, even a short- term exposure to heavy metals induces a persistent stress in fish which may render them more susceptible to disease. The significant alteration in the non specific immune response levels serve as a biomarker of pollutant exposure and effects.

Fishes are also largely used for the assessment of aquatic environment quality and are accepted as bioindicator of environmental pollution. Occurrence of aquatic pollutants (such as heavy metals) has been correlated to alterations in the fish immune system and the incidence of infectious diseases. Even very low sub lethal doses of certain heavy metals can have profound effects upon the structure and/ or functions of the immune system that could be almost as harmful as direct toxic doses. *In vitro* studies applying antibody detection/phagocyte activity techniques for testing various toxicants, are highly useful as these require not only much less time (few hrs only) and fewer animals, it also does not require expensive facilities (Anderson *et al.*, 1989)). Key players are the cytotoxic leucocytes often called N-K cells. Faisal *et al.*, 1991 reported a depression in N-K activity in *Fundulus* sampled from a river in Virginia heavily contaminated with a variety of chemicals including poly-nuclear aromatic hydrocarbons and creosote. Weeks *et al.* (1986) recorded

30-40 per cent decrease in macrophage chemo-tactic activity in those fish from the contaminated river, the activity, however, returned to normal after the fish was brought back for 3 weeks to normal clean water. Stimulation of macrophages chemiluminescent in fish fundulus was also recorded from the same contaminated river by Kelly-Reay and Weeks-Perkins, 1994. The mechanism involved, however, still remain obscure, it was suggested that it may involve increased lymphokinines or a direct effect of some chemical constituent in the water. Ellsaesser and Clem, 1986 used this technique to test some metals and found Cu causing a strong inhibition of the phagocyte activity, aluminium somewhat less; whereas, Cd caused initially a stimulatory effect followed by a variable decrease. Manganese, an essential metal, is known to have low toxicity in fish (Jones, 1964), as well as it stimulate N-K cell activity in carp both *in vitro* and *in vivo* (Ghanmi *et al.*, 1990). It has been also reported to have a strong *in vitro* enhancing effect on phagocytosis by carp macrophages (Cossarini-Dunier, 1987). There are further enough evidences to support the cytological damage in animals including fish also, due to higher concentration of heavy metals in aquatic environment (O'Neill, 1981). Tripathy and Qureshi, 2014 In their study on the morphological, histological and cytological impacts of heavy metals on lymphoid organ of *Channa gachua*, recorded red spots, loosening of scales and discoloration of skin along with tissue destruction in ground matrix and the lymphoid follicles, depicting various levels of damage in lymphoid organs due to exposure of selected heavy metals (Cr, Cu and Co); the maximum damage was observed in case of cobalt.

The maladaptive component of the overall stress response that can be considered as the "cost of doing business", which is also manifested by a reduction in available energy within the scope for activity, a decline in glycogen reserves, and a decrease in circulating lymphocytes *etc.* If the stressor is persistent, these changes could lead to reduced growth as a consequence of continued gluconeogenesis and reduced metabolic scope, metabolic exhaustion as depleted energy stores and increased disease incidences due to reduced immuno-competence. There is an increasing evidence to show that the functional capacity of the immune system of fish is reduced by stress (Ellsasser and Clem, 1986; Fries, 1986) and that this effect may be mediated by cortisol (Thomas and Lewis, 1987). Zinc has been reported to cause immune suppressive effect on fish (Dunier and Siwicki, 1993; Sanchez-Dardon *et al.*, 1999; Witeska, 2005). Zinc can activate natural killer cells (NK), macrophages, antigen-specific cytotoxic, T-lymphocytes and cytokines cells (Chandra and Au, 1980; Bhaskaram, 2002), and in high concentrations it decreased the natural killer cells (Ferry and Donner, 1984). In addition, zinc has been show to stimulate neutrophiles, monocytes, and macrophages, if used in optimum ratios, and triggered the production of oxygen which plays an important role in killing pathogens (Shankar and Prasad, 1998). However, zinc in high doses inhibits macrophage activation, mobility, phagocytosis, and decreases the oxygen production in humans and animals (Shankar and Prasad, 1998). Çelik *et al.* (2013) in tilapia (*Oreochromis mossambicus*), exposed to several zinc concentrations *in vivo* showed a decrease in phagocytic activity and an increase in lysozyme and myeloperoxidase activities with medium and low zinc concentrations. A decrease was found in nitroblue tetrazolium (NBT) activity with medium and high concentrations; in the lysozyme and myeloperoxidase activities was found with high concentrations.

Biller-Takahashi *et al.* (2015) has recently assessed the relationship between antioxidant status and immunity, and had evaluated the effect of dietary supplementation with organic selenium, given at various levels for 10 days, on the antioxidant and immune system of the pacu fish (*Piaractus mesopotamicus*). Fish fed a diet containing 0.6 mg Se-yeast kg^{-1} showed significant improvement in antioxidant status, as well as in hematological and immunological profiles. Specifically, they had the highest counts for catalase (CAT), glutathione peroxidase (GPx), glutathione S-transferase (GST), red blood cells, and thrombocytes; the highest leukocyte count (particularly for monocytes); and the highest serum lysozyme activity. There was also a positive correlation between GPx and lysozyme in this group of fish. These findings indicate that short-term supplementation with 0.6 mg Se-yeast kg^{-1} reestablished the antioxidative status, allowing the production of innate components which can boost immunity without the risk of oxidative stress. Ascorbic acid (Vitamin C), one of the essential vitamins for normal physiological activities of any organism. has also been demonstrates to possess an immunostimulatory effect on the humoral and cell mediated immunity of the bagrid catfish, *Mystus gulio*. The vitamin supplemented lipopolysaccharide (LPS) vaccinated group exhibited greater immune (both humoral and cell mediated) responses than its formalin killed (FK) and heat killed (HK) bacterin vaccinated counterparts.

REFERENCES

Ainsworth A.J., Dexiang C., Waterstrat P.R. and Greenway T., 1991. Effect of temperature on the immune system of channel catfish (*Ictalurus punctatus*). I. Leucocyte distribution and phagocyte function in the anterior kidney at 10°C. Comp. Biochem. Phys., A100: 907-912.

Alexander J.B. and Ingram G.A., 1992. Noncellular nonspecific defense mechanisms in fish. Ann. Rev Fish Dis., 2: 249 - 279.

Anderson D.P., Dixon O.W., Bodammer J.E. and Lizzio E.F., 1989 Suppression of antibody producing cells in rainbow trout spleen sections exposed to copper *in vitro*. J. Aquat. Anim. Hlth., 1: 57.

Bhaskaram P., 2002. Micronutrients malnutrition, infection and immunity: an overview. Nutr. Rev., 60: 40-45.

Biller-Takahashi Jaqueline D., Takahashi Leonardo S., Mingatto Fábio E. and Urbinati Elisabeth C., 2015. The immune system is limited by oxidative stress: Dietary selenium promotes optimal antioxidative status and greatest immune defense in pacu, *Piaractus mesopotamicus*. Fish and Shellfish Immunol., 47 (1): 360–36.

Bly J.E. and Clem L.W., 1992. Temperature and teleost immune functions. Fish Shellfish Immun., 2: 159-171.

Bly J.E., Quiniou S.M., Clem L.W., 1997. Environmental effects on fish immune mechanisms. Dev. Biol. Stand., 90: 33-43.

Bobe J. and Goetz F.W., 2001. Molecular cloning and expression of a TNF receptor and two TNF ligands in the fish ovary. Comp. Biochem. Physiol. B Biochem. Mol. Biol., 129: 475-481.

Bols N.C., Brubacher J.L., Ganassin R.C. and Lee L.E.J., 2001. Ecotoxicology and innate immunity in fish. Dev. Comp. Immunol., 25: 853 - 873.

Çelik E.S., Kaya Hasan, Yilmaz Sevdan, Akbulut Mehmet and Tulgar Arýnç, 2013 Effects of zinc exposure on the accumulation, haematology and immunology of Mozambique tilapia, *Oreochromis mossambicus*, African J. Biotechnol., 12(7): 744-753.

Chandra R.K. and Au B., 1980. Single nutrient deficiency and cell-mediated immune responses. I. Zinc. Am. J. Clin. Nutr., 33: 736-738.

Correia, A.D., Costa M.H., Luis O.J. and Livingstone D.R., 2003. Age-related changes in the antioxidant enzyme activities, fatty acid composition and lipid peroxidation in whole body *Gammarus locusta* (Crustacea: Amphipoda). J. Exp. Mar. Biol. Ecol., 289: 83 - 101.

Cossarini-Dunier M., 1987. Effects of the pesticides atrazine and lindane and of manganese ions on cellular immunity of carp, *Cyprinus carpio*. J. Fish Biol., 31: 67-73.

Crandall C.A. and Goodnight C.J., 1963. The effects of sublethal concentrations of several toxicants to the common guppy, *Lebistes reticulates*. Trans. Am. Microsc., 82:59-73.

Dalmo R.A., Ingebritgsen K. and Bogwald J., 1997. Non-specific defence mechanisms in fish, with particular reference to the reticuloendothelial system (RES). J. Fish Diseases, 20: 241-273.

Danneving B.H., Lauve A., Press C. and Landsverk T., 1994. Receptor-mediated endocytosis and phagocytosis by rainbow trout head kidney sinusoidal cells. Fish Shellfish Immunol., 4: 3-18.

Dawson A.B., 1935. The hemopoietic response in the catfish *Ameriurus nebulosus* to chronic lead poisoning. Biol. Bull., 3: 235-346.

Demers N.E., and Bayne C.J., 1997. The immediate effects of stress on hormones and plasma lysozyme in rainbow trout. Dev. Comp. Immunol., 21: 363 - 373.

Dunier M. and Siwicki K., 1993. Effects of environmental contaminants and chemotherapeutics on fish defense mechanisms. In: Siwicki A.K., Anderson D.P. and Waluga J. (Eds.) Fish Diseases Diagnosis and Preventions Methods. IRS, Olsztyn, Poland: 100-108.

Ellis A.E., 1999. Immunity to bacteria in fish. Fish Shellfish Immunol., 9: 291-308.

Ellsaesser C.F., and Clem L.W., 1986. Hematological and immunological changes in channel catfish stressed by handling and transport. J. Fish Biol., 28: 511-521.

Faisal M., Weeks B.A., Vogelbein W.K. and Huggett R.J., 1991. Evidence of aberration of the natural cytotoxic-cell activity in *Fundulus heteroclitus* (Pisces, Cyprinodontidae) from the Elizabeth river, Virginia. Vet. Immunol. and Immunopathol., 29(3-4):339–351.

Fast M.D., Sims D.E., Burka J.F., Mustafa A. and Ross N.W., 2002 Skin morphology and humoral non-specific defense parameters of mucus and plasma in rainbow trout, coho and atlantic salmon. Comp. Biochem. Physiol., 132: 645-657.

Ferry E. and Donner M., 1984. *In vitro* modulation of murine natural killer cytotoxicity by zinc. Scand. J. Immunol., 19:435-445.

Fries C.R., 1986. Effects of environmental stressors and immunosuppressants on immunity in *Fundulus heteroclitus*. Ain. Zool. 26: 271-282.

Ghanmi Z., Bouabhaia M., Alifuddin M., Troutaud D. and Deschaux P., 1990. Modulatory effect of metal ions on the immune response of fish: *in vivo* and *in vitro* influence of $MnCl_2$ on N-K activity of carp pronephros cells. Ecotoxicol. Environ. Safety, 20: 241.

Goede R.W. and Barton B.A., 1990. Organismic indices and an autopsy-based assessment as indicator of health and condition of fish. Am. Fish Soc. Symp., 8: 93.

Grinde B., Jolles J. and Jolles P., Purification and characterization of two lysozymes from rainbow trout (*Salmo gairdneri*). Eur. J. Biochem., 173: 269-73.

Gupta N., Gopal-Krishna, Gupta D.K. and Shalaby S.I., 2001. Immunological responses in carps (*Labeo rohita* and *Catla catla*) exposed to cadmium. Egyptian J. Veter. Sci., 35: 105-112.

Hetrick F.M., Robertson B.S. and Tsai C.F., 1982. Effect of heavy metals on the susceptibility and immune response of striped bass to bacterial pathogens. Nat. Ocean. Atm. Adm. Publ., 82112603: 32.

Holland J.W., Pottinger T.G. and Secombes C.J., 2002. Recombinant interleukin- 1 beta activates the hypothalamic-pituitary-interrenal axis in rainbow trout, *Oncorhynchus mykiss*. J. Endocrinol., 175: 261- 267.

Hutchinson T.H., Manning M.J., 1996a. Effect of *in vivo* cadmium exposure on the respiratory burst of marine fish (*Limanda limanda* L.) phagocytes. Mar. Environ. Res., 41: 327–342.

Hutchinson T.H. and Manning M.J., 1996b. Seasonal trends in serum lysozyme activity and total protein concentration in dab (*Limanda limanda* L.) sampled from Lyme Bay, UK. Fish Shellfish Immun., 6: 473-482.

Ingram G.A., 1980. Substances involved in natural resistance of fish to infection: A review. J. Fish Biol., 16: 23 – 60.

Institoris L., Siroki O., Undeger U., Basaran N., Banerjee B.D. and Desi I., 2001. Detection of the effects of repeated dose combined propoxur and heavy metal exposure by measurement of certain toxicological, hematological and immune function parameters in rats. Toxicol. 163: 185 - 193.

Jones J.R.E., 1964. Fish and River Pollution. Pub., Butterworth's, London: 203 pp.

Kaattari S. and Irwin M.J., 1985 Salmonid spleen and anterior kidney harbor populations of lymphocytes with different B cell repertories. Dev. Comp. Immunol. 9: 433-444.

Kelly-Reay K. and Weeks-Perkins B.A., 1994. Determination of the macrophage chemiluminescent response in *Fundulu heteroclinus* as a function of pollution stress. Fish and Shellfish Immunol., 4: 95-105.

Khangarot B.S., Takroo R., Singh R.R. and Srivastava S.P., 1991. Toxicity and bioconcentration of hexachlorocyclohexane (HCH) in an air-breathing catfish, *Saccobranhus fossilis* (Bloch). Bull. Environ. Contam. Toxicol., 47: 904-911.

Kimber, I. 1991. Toxicology and the immune system: a perspective.Hum. Exp. Toxicol., 10, 445 - 449.

Lie O., Evensen O., Sorenson. and Froysadal, 1989. Study on lysozyme activity in some fish species. Dis. Aqut. Org., 6: 1-5.

Luster M.I. and Rosenthal G.J., 1993. Chemical agents and the immune response. Environ. Hlth. Perspect., 100: 219 - 226.

MacFarlane G.T., Cummings J.H. and Allison C., 1986. Protein degradation by human intestinal bacteria. J. Gen. Microbiol., 132: 1647-56.

Mock A. and Peters G., 1990. Lysosome activity in rainbow trout, *Oncorhynchus mykiss* stressed by handling, transport and water pollution. J. Fish Biol., 37: 873.

Morra D.S., 1993. Effects of acidic water on immune responses in the channel catfish, *Ictalurus punctatus*. J. Immunol., 150 (2): 16A.

Murad A. and Houston A.H., 1988. Leucocytes and leucopoetic capacity in goldfish, *Carasius auratus*, exposed to sublethal levels of cadmium. Aquat. Toxicol., 13: 141-154.

Newman M.W. and MacLean S.A., 1974. Physiological response of the cunner, *Tautogolabrus adspersus*, to cadmium. VI. Histopathology, Nat. Ocean. Atomos. Adm. Tech. Rep., NMES SSRF-681: 27.

O'Neill J.G., 1981. Heavy metals and the humoral immune response of freshwater teleosts. In: Stress and Fish. Pickering A.D. (Ed.), Academic Press, New York: 328.

Pasquier Du L., 1993. Evolution of the Immune System. New York: Raven Press. Pickering A.D. and Pottinger T.G., 1989. Stress responses and disease resistance in salmonid fish: Effects of chronic elevations of plasma cortisol. Fish Physiol. Hem., 7: 253-258

Powell D.S., 2000. Immune system. In Ostrander GK (ed.). The laboratory fish. Academic Press : 441-449.

Press C.Mc.L. and Evensen O., The morphology of the immune system in teleost fishes. Fish Shellfish Immunol. 9: 309-318.

Qin Q.W., Ototake M., Noguchi K., Soma G.I., Yokomizo Y. and Nakanishi T. 2001. Tumor necrosis factor alpha (TNFalpha)-like factor produced by macrophages in rainbow trout, *Oncorhynchus mykiss*. Fish Shellfish Immunol.,11: 245-256.

Rajalakshmi K., Ezhilmathy R. and Chezhian A., 2014. Immunological and hematological changes induced by short term exposure to mixed heavy metals (Cd+Ni) in *Lates calcarifer* (bloch, 1790). Internat. J. R. Biol. Sci., 4(3): 69-74.

Roales R.R., and Perlmutter A., 1977. The effect of sub lethal doses of methylmercury and copper applied singly and jointly on the immune response of the blue gourami to viral and bacterial antigen, Arch. Environ. Contam. Toxicol., 38:325-331.

Rotllant J. and Tort L., 1997 Cortisol and glucose response after acute stress in the sparid red porgy (*Pagrus pagrus*) previously subjected to crowding stress. J. Fish Biol., 51: 21-28.

Rougier F., Troutaud D., Ndoye A. and Deschaux P., 1994. Non-specific immune response of zebrafish, *Branchyodanio rerio* (Hamilton-Buchanan) following copper and zinc exposure. Fish Shellfish Immunol., 4: 115.

Sanchez-Dardon J., Voccia I., Hontela A., Chilmonczyk S., and 4 others, 1999. Immuno-modulation by heavy metals tested individually or in mixtures in rainbow trout (*Oncorhynchus mykiss*) exposed *in vivo*. Environ. Toxicol. Chem., 18:1492-1497.

Sapolsky R., Rivier C., Yamamoto G., Plotsky P. and Vale W., 1987. Interleukin-1 stimulates the secretion of hypothalamic corticotrophin- releasing factor. Sci., 238: 522–524.

Satchell G.H., 1991. Physiology and form of fish circulation. Cambridge: *Cambridge University Press.*

Schreck C.B. and Bradford C.S., 1990. Inter-renal corticosteroid production: potential regulation by the immune system in the salmonid. Prog. Clin. Biol. Res., 342: 480-6.

Secombes C.J., 1996. Cytokines in fish: an update. Fish Shellfish Immunol., 6: 291-304.

Secombes C.J., Fletcher T.C., White A. and 3 others., 1992. Effects of sewage sludge on immune responses in the dab, *Limunda mimanda*. Aquat. Toxicol., 23: 217.

Secombes C.J., Wang T., Hong S., Peddie S., Crampe M., Laing K.J., *et al.*, 2001. Cytokines and innate immunity of fish. Dev. Comp. Immunol. 25: 713-723.

Shankar A.H. and Prasad A.S., 1998. Zinc and immune function: the biological basis of altered resistance to infection. Am. J. Clin. Nutr., 68: 447-463.

Spazier E., Storch V. and Braunbeck T., 1992. Cytopathology of spleen in eel *Anguilla anguilla* exposed to a chemical spill in the Rhine River. Diseases Aqua. Organis., 14: 1–22.

Stromberg P.C., Ferlante J.G. and Carter S., 1983. Pathology of lethal and sublethal exposure of fat head minnow, *Pimephalus promelasto* cadmium. A model for aquatic toxicity assessment. J. Toxicol. Environ. Hlth., 11: 247.

Tahir A., Fletcher T., Houlihan D. and Secombes C.J., 1993. Effect of short exposure to oil-contaminanted sediments on the immune response of fish, *Limunda mimanda*. Aquat. Toxicol., 27: 71.

Thomas P. and Lewis D.H., 1987. Effects of cortisol on immunity in red drum, *Sciaenops ocellatus*. J. Fish Biol., 31, Suppl. A: 123-127.

Thomas P. and Neff J.M., 1984. Effects of a pollutant and other environmental variables on the ascorbic acid content of fish tissues. Mar. Environ. Res., 14: 489-491.

Thuvander A., 1989. Cadmium exposure of rainbow trout, *Salmo gairdneri* Richardson: Effects on immune functions. J. Fish Biol., 35(4): 521–529.

Tort L., Balasch J.C. and Mackenzie S., 2003. Fish immune system. A cross road between innate and adaptive responses, 22 (3): 277-286.

Tort L., Rotllant J. and Rovira L., 1998. Immunological suppression in gilthead sea bream, *Sparus aurata* of north-west Mediterranean at low temperatures. Comp. Biochem. Physiol., 120A: 175-179.

Tripathi Alok and Qureshi T. A., 2014. An experimental studies on the effect of heavy metals on lymphoid organ of *Channa gachua*. Intl. J. Eng. Sci. Invention, 3 (6): 44-52.

Tripp R.A., Maule A.G., Schreck C.B. and Kaa, 1987. Cortisol mediated suppression of *Salmon* lymph *in vitro*. Dev. Comp. Immunol., 11: 565-576.

Tsujii T. and Seno S., 1990 Melano-macrophage centers in the aglomerular kidney of the sea horse (Teleost): morphologic studies on its formation and possible function. Anat. Rec., 226: 460-470.

Weeks B.A., Anderson D.P., Dufour A.F., Fairbrother A., and 3 others. 1992. Immunological biomarkers to assess environmental stress. In: Biomarkers: Biochemical, Physiological and Histological Markers of Anthropogenic Stress (Huggett R.J., Kimerle R.A., Mehrle P.M. Jr. and Bergman H.L., Eds.), Lewis Publishers, Boca Raton, F.L. Ch. 5.

Weeks B.A., Anderson D.P., DuFour A.P., Fairbrother A., Goven A.J., Lahvis G.P. and Peters G., 1986. Immunological biomarkers to assess environmental stress. In Biomarkers: Biochemical, Physiological and Histological Markers of Anthropogenic stress, Huggett RJ. (Ed).

Wester P.W., Vethaak A.D. and van Muiswinkel W.B., 1994. Fish as biomarkers in immunotoxicology. Toxicol., 86: 213.

Witeska M., 2005. Stress in fish - Hematological and immunological effects of heavy metals. Electronic J. Ichthyol. 1: 35-41.

Chapter 12

Metal Toxicity and the Tissue Enzyme Aberrations

Environmental stressors and the associated risks have always been an inherent part of a society culture (Adeyemo *et al.*, 2008). Pollutants, either individually or in combination may cause sub lethal or the latent effects at the cellular, organ and individual organism level, posing a serious threat to the survival of the animals (Scott *et al.*, 2003); and fish being an aquatic animal is most vulnerable to water contamination especially the metals (Alinnor, 2005). The toxic effects of metals in fish are multidirectional and manifested by numerous changes in various physiological and biochemical processes of their body organs (Dimitrova *et al.*, 1994). Enzymes are biocatalyst, and are produced by the living cells. These catalyze broad spectrum of reactions in tissues and are exceedingly efficient and specific in catalyzing a reaction and utilization of a substrate. Their synthesis and concentration in a cell are, however, under genetic control and are greatly influenced even by very small amounts of a substrate. These accelerate and regulate the rate of spontaneous biochemical reactions inside as well as outside a cell, such as those involved in metabolism. Enzymes are thus, immensely vital for various cell activities as these are highly adaptive to the environmental changes. Metals such as lead *e.g.* waterborne lead such as derived from various human activities from the use of utensils for cooking in past (Ramesh *et al.*, 2009). It has been known to cause toxicity effects more pronounced on various enzymatic and neurological functions (Hodson *et al.*, 1984). Changes in certain enzymatic activities have been found further in extensive use as a tool to diagnose any stress in mammal and the factors loke temperature, substrate concentration and pH is known to influence various enzymatic activities (Hoar and Randall, 1978; David, 1985; Heath, 1987a, 1995). Enzymes being the organic chemicals, and as reaction specific, thus, have been of much concern to the biologists.

Alterations in metabolic rates by the action of an enzyme at the molecular level since form basis for self-regulation in an organism, measuring enzyme activity thus, is a classical means by which the health of a fish population in a medium could be paradoxically assessed (Landis and Yu, 1995). Any alteration in enzyme activity can occur due to any xenobiotic, such as a metal or a pesticide and can be instrumental in affecting the health of an organism. Nevertheless, the direction of change (*i.e.* increase or decrease in cell enzyme activity) and the relative assessment of effects of the environmental change make it further a valuable reference to the fish health. Following category of enzymes such as the digestive enzymes which participate in converting complex food molecules into simple forms absorbable across the cell membrane, and the metabolic enzymes involved in food assimilation, energy production, molecular transformations and hormonal actions *etc.* inside cells/ tissues, play an important role in dealing with changes in fish functions following their exposures to the pollutants.

1. Digestive Enzymes

Fish species in general differ in their feeding habits, mode of feeding, structure of their alimentary canal and the location of digestive glands, as well as composition of their food; the basic pattern of secretion and functions of various digestive enzymes, appears almost same in all the fish species (Gratzek, 1993). Digestive enzyme activities, however, are said to be species specific (Hidalgo *et al.*, 1999). Hidalgo *et al.* (1999) conducted a comparative study on the digestive and amylolytic activities in six species of fish with different nutrition habits. Among all the species studied, trout and carp fishes were found to show the highest digestive proteolytic activity, whereas, eels had the lowest proteolytic activity, and the ratio of total amylase to total proteolytic activity was found to be higher in omnivorous fish species. Major digestive enzymes as influenced by the metal toxicants are reported to be the amylase, maltase and lactase (the sugar digesting enzymes); pepsin, trypsin and peptidases (protein digesting) and lipase, the lipid digesting enzyme (Heath, 1995). Brown, 1976 claimed that although chronic toxicity affects at a sub lethal concentration of a metal, it may not kill an animal but it changes the ability of an organism to respond to the environment, thus, likely to shorten its life. The digestive enzyme acivities in general appear inhibited following fish exposures to lower metal dose levels in a digestive organ. The amylase activity in *H fossilis* was reported significantly inhibited due to metal toxicity in all the three fish organs– stomach, intestine and liver (Sastry and Gupta, 1979ab), while maltase and lactase activities were reported inhibited only in liver. Golovanova *et al.* (1999) studied *in vitro* effects of cadmium (Cd, 0.5-50 mg/l) on the total amylolytic, sucrase and proteinase activities of intestinal mucosa in eleven freshwater teleosts. Total amylolytic activity in burbot (*Lota lota*), crucian carp (*Carassius auratus*) and common carp (*C. carpio*); sucrase activity in blue bream were significantly decreased by cadmium at 50 mg/l. Nevalennyi and Bednyakov (2004) reported influence of Cd ions at 0.25 mg/l on the alpha-amylase and maltase activities of carp *C. carpio*. Maltase activity was found decreased by 45 per cent against the control by day 10, and then it increased again on day 30 followed by again a decrease to 30 per cent over controls at the end of the experiment. Alpha-amylase activity was at 43 per cent of the controls on day 40 and it increased to 92 per cent of the controls on day 40 to day 50.

Effects of mercury (Hg) were studied in *C. punctatus* by Sastry and Gupta, 1978a and of Pb in *H. fossilis* (Sastry and Gupta, 1979a). These studies showed contrasting effects in these fishes especially with reference to their treatment dose level and the time of exposure. *C. punctatus* (Sastry and Gupta, 1978a) treated with 1.8 mg/liter of mercuric chloride showed in general no significant effects on the digestive enzymes within the experimental period of 96 hours, longer exposures, however, were shown to slightly increase the activities of the enzymes: amylase, pepsin and trypsin in both the intestine and the pyloric caeca. There were no significant alterations in the activities of the peptidases. On the contrary, in the teleost fish, *H. fossilis* (Sastry and Gupta, 1979a), following chronic treatment with 2.8 mg/l lead nitrate, activities of all the digestive enzymes were reported inhibited in the intestine. Enzymatic inhibition due to lead was shown to be reversed by treatments with chelating agents like EDTA (Sastry and Gupta 1979a), thus, indicating that such changes are reversible. Stomach enzymes were not affected in general by the lead treatment. Mercuric chloride (0.3 mg/l) in *Channa punctatus* was also reported to inhibit the activities of all the enzymes examined in intestine and pyloric caeca and digestion of proteins and lipids may be more affected by mercury than the digestion of some carbohydrates (Sastry and Gupta, 1980). It significantly inhibited glucose6phosphatase activity in the intestine and pyloric caeca. No marked alterations were observed in the activities of maltase and lactase except for elevation in maltase activity and inhibition in lactase activity in the intestine and pyloric caeca after 15 days of treatment. Three peptidases (aminotripeptidase, glycylglycine dipeptidase and glycyl1leucine dipeptidase) showed decreased activities in all the parts of the digestive canal. A decrease was also observed in the activity of lipase except for the stomach where inhibition after 15 days was insignificant.

Chronic treatments with heavy metals appeared more effective in inhibiting the activities of peptidases, such as inhibition by chronic mercury and lead treatments in *C. punctatus* (Sastry and Gupta, 1978 b,c) and *H fossilis* (Sastry and Gupta 1979b). Sastry and Gupta (1979b) further showed an elevation in pepsin activity in the stomach of *H. fossilis* with Cd, it, however, showed inhibition of trypsin in the intestine and also in the activities of aminotripeptidases and glycylglycine dipeptidase. They also evidenced cadmium to produce structural damage in the digestive organs, liver and intestine and a decrease in the digestive efficiency by inhibiting the activity of a number of enzymes like trypsin in *H. fossilis*. Mukhopadhyay and Hajra (1986) also mentioned that an increase in proteolytic activity in the alimentary canal was responsible for breakdown of food proteins, and it also advantageously influenced growth and food utilization by fishes. Increase in protein content of diet was to increase proteolytic activity of the intestine. Some heavy metals at low concentration treatments may behave like the trace metals in producing beneficial effects, but at higher concentrations these may cause inhibition. Yoshitomi *et al.* (1998) observed that carps fed with food pellets containing a range of Cd concentrations, showed a significant increase in trypsin activity in the intestine with an increase in Cd concentration in the diet. Golovanova *et al.* (1999) while studying *in vitro* effects of cadmium (0.5-50 mg/l) on the proteinase activities of intestinal mucosa in eleven freshwater teleosts, found total proteolytic activity in burbot and pipe significantly decreased by cadmium at 50 mg/l. Kotorman (2000) reported that presence of Cd^{2+} did not influence the activities of carboxypeptidase-A and trypsin in carps; it, however, caused 10-20 per cent inhibition in alpha-chymotrypsin and lipase enzymes. With the exception of trypsin, Pb^{2+} also inhibited the activities of all other investigated enzymes.

All enzyme activities were decreased at higher Zn^{2+} concentrations. Inhibitory effects of heavy metal ions on the activity of enzymes are said to occur due to the their action to displace metals situated at the active site of an enzyme, or binding of the metal with the enzyme protein (Passow *et al.*, 1961; Hodson, 1988). Eichhorn (1975) has also pointed out that heavy metals can also bind with sites other than the active sites of an enzyme molecule and can produce both beneficial and adverse effects depending upon the concentration of a heavy metal. In case of Cd vs Zn interactions in humans, Calabrese and Moore (1981) has proposed that persons exposed to elevated levels of cadmium via occupation, diet, or cigarette smoking will exhibit a diminished capability to metabolize ethanol via the action of the zinc dependent enzyme alcohol dehydrogenase (ADH), the way cadmium can inactivate ADH by replacing its zinc in liver. Increased activity of trypsin was also reported by Brafield and Koodie (1994) in the intestinal tissue of *C. carpio* after its exposure to dietary Zn. Carps fed with pallets containing a range of zinc concentration were also shown with a significant increase in trypsin activity in the intestine with the increasing zinc dose in diet (Kotorman, 2000). Cu^{2+} was seen to produce only slightly influence on the trypsin and lipase activities, but its effects on the activity of alpha-chymotrypsin distinctly visible. Dionodia (2001) after Cd treatments in commonly available freshwater fish species; *L. rohita, C. mrigala* and *C. carpio* also reported marked reductions the proteolytic activity. He, however, found inter-specific variations in fish responses to Cd, evidencing a decline up to 55.6 per cent in the enzyme activity in *C. mrigala* 32.25 per cent in *C. carpio* and 46.04 per cent in *L. rohita* at 5 ppm Cd as compared to controls. Dinodia *et al.* (2003) reported an inverse relationship between cadmium concentration and the proteolytic enzyme activity in *C. mrigala* and *L. rohita*. Proteolytic enzyme activity declined by 20.0 per cent more at 2.5 ppm Cd and by 55.6 per cent more at 5 ppm Cd in *C. mrigala* over the control. In *L. rohita* it showed 32.6 per cent higher decline at 5 ppm Cd, as compared over the controls. Hans and Ronny (2001), however, reported that the concentrations of free amino acids and the activities of proteases increased at day 4 in gills, liver, and kidney of a carp exposed to 4 and 20 mM Cd, and in gills and kidney at day 29 in carp exposed to 4 mM Cd. They suggested that the observed proteolysis is intended to increase the role of proteins in energy production under stress.

Kuzumina *et al.* (2004) studied effects of trace metals Cu and Zn in fishes (carp, bream and roach) and showed metal dependent, dose dependent and species specific toxic effects on fish species; small concentrations appeared vitally important for the normal development of the fish and in large doses these appeared toxic. Copper ($CuSO_4$) and zinc ($ZnSO_4$) concentrations of 0.01-100 mg/l were reported to influence carbohydrase activity of a fish intestinal mucosa *in vitro*.

The inhibition was more pronounced in alpha-chymotrypsin emzyme (more than 50 per cent), and Cu produced more toxic effects in comparison to zinc in all the fish species studied. 25 mg/l of copper reduced amylolytic activity in roach by 38 per cent, but zinc practically produced no change. Lower concentrations (0.1 - 1.0 mg/l) of these metals caused an increase in enzyme activity, especially in carp fry. The increasing concentrations of copper and zinc gradually reduced the intestinal mucosa proteolytic activity in all the fish species. Enzyme activity reduction at higher concentrations of copper (10 and 25 mg/l) for different species also showed different effects *i.e.* in carp and bream it showed reduction by 35-40 per cent, in perch by 55-60 per cent and in pike between 75-80 per cent, as compared to the control. These results testify that even allowable concentrations of copper and zinc in food are considerably capable to reduce activity of enzyme. They mentioned that the Cu and Zn contents in waters, usually does not exceed 0.8 and 0.07 mg/l, respectively. In natural foods however, due to biological magnification, their concentrations can reach up to 24 and 160mg/kg and in artificial foods (food synthesized by microorganisms) up to 300 and 480 mg/kg, respectively. Sarita (2005) after treating *C. carpio* and *C. mrigala* fishes with Hg and other metals *viz.*, As, Ni and Cr as single metal and in combination with As evidenced a considerable decrease in the activities of enzymes: protease, lipase and amylase. Arsenic at 0.05ppm registered a maximum reduction in activity of all these enzymes in both the species. As in combination with Cr showed further increase in reduction in protease and lipase enzyme activity, while As in combination with Hg affected the activity of amylase enzyme by reducing to maximum.

Mehrotra (1996) reported a decrease in absorption rate due to damaged villi of intestine. According to Kuzumina (1996) in the freshwater teleosts, digestive enzyme activity appeared affected by feeding behavior, biochemical composition and ambient stress. Various histochemical studies (Khillar and Wagh, 1988; Khare, 1993) have suggested that rupture of cells and deformation of tissues affect the functional activity of the digestive enzyme and this may interfere with digestion. Decline in proteolytic enzyme activity in freshwater fishes following Cd treatments was also attributed to ruptured intestinal cells and damaged intestinal villi by Dionodia (2001). Metal stressed fish was also found to show negative effects on its feeding behavior as the metal treated fish did not feed well (Dinodia *et al.*, 2003).

2. Metabolic Enzymes

Fish metabolism has a key role in polluted waters as fish have to spend more energy to maintain homeostasis and to overcome the toxicity stress, thus, affecting both liver and kidney as the primary target organs of chemical toxicity (Friberg *et al.*, 1985). The treatment of the fish with a chemical toxicant may lead to a marked increase in the activities of any such metabolic enzymes like dehydrogenases (LDH, SDH, MDH), transaminases (GOT and GPT) or phosphatases (acid and alkaline) in liver/serum; the magnitude of the effect however, being dependent on the type, concentration and duration of exposure of a chemical. The increase and decrease in the activity of these metabolic enzymes is a clear cut evidence of changes in the metabolic environment in a fish, thus, influencing further dietary intake, growth and survival of a fish in the contaminated medium. Most important aspect of alterations in these enzymes has been their excessive release in the blood; as the chemicals are said to damage and rupture cells of the liver and gills and the muscle cells. Quantitative bioassay of activity of any such enzyme thus, may be especially a good indicator for assessing damage to fish organs like the liver (Dixon *et al.*, 1987) on account of a pollutant.

i. Dehydrogenases

Major dehydrogenases being frequently analyzed are: Lactate dehydrogenase (LDH), Malate dehydrogenase (MDH), Glutamate dehydrogenage (GDH) and Succinate dehydrogenage (SDH). These enzymes are the components of aerobic and anaerobic oxidative cycles in animals involved in the production of energy rich ATP molecules. Pyruvate dehydrogenase (PDH) is another important dehydrogenase which catalyzes oxidative decarboxylation of pyruvate to form acetyl-CoA. GDH is a mitochondrial enzyme and is involved in the oxidative deamination of glutamate. GDH play an important role in osmoregulation by regulating synthesis of proline and alanine (Willett and Burton, 2003); as through synthesis of glutamate it contributes to the regulation of intracellular amino acids during salinity changes of the external medium. It was also suggested that GDH would favor amino acid storage into proteins, especially haemocyanin when the fish is transferred to a hypo-osmotic medium. SDH (succinate dehydrogenase) ia an enzyme in the Kreb's cycle acting on succinate; it binds to the inner mitohondrial membrane unlike the other enzymes of the cycle which are found in the matrix. Its suppression in an organ reflect an oxidative stress and animal's reliance on anaerobic glycolytic pathway (James *et al.*, 1996). SDH is thus, known to be specific indicator of liver cell damage and parenchymal hepatic diseases. SDH activity is further said to rise rapidly in case of liver damage, it however, decreases very shortly after peaking.

Sastry and Rao, (1981) in a fresh-water teleost fish, *C. punctatus* on its exposure to a sub lethal concentration (3 µg/1) of mercuric chloride for 15, 30 and 60 days revealed that the activities of all the three dehydrogenases, lactic-DH, pyruvic-DH and succinic-DH were inhibited significantly after 60 days of treatment except for LDH in brain. Though gills and muscles showed weak activity of these dehydrogenases, their percentage inhibition was more marked in these tissues than in others. Sastry and Rao (1984) in the same fish also found the activities of LDH, PDH, SDH, and MDH decreased significantly except an insignificant decrease in LDH activity in brain and an insignificant increase in MDH activity in kidney. On the contrary, Gill *et al.* (1990) exposed rosy barb (*Puntius conchonius*) to 181 µg/l mercuric chloride for 48 h and recorded increased LDH activity in the cardiac and skeletal muscles. Shaffi (1993) reported mercury to cause greatest inhibition in LDH, MDH and SDH activities in liver in three freshwater fish species, with progressively lesser inhibition in muscles, brain, kidney and gill. Similar observations were reported by Ozretic and Krajnovic -Ozretic (1993) in a marine fish grey mullet (*Mugil auratus* Risso). Radhakrishnaiah *et al.* (1993) exposed freshwater fish *C. carpio* to the sublethal concentration of mercury (0.1 mg/l) and zinc (6.0 mg/l). They found distinct changes in the energy metabolism of gill, liver and muscle after 1, 15 and 30 days. The rate of oxygen consumption and SDH activity decreased in the organs in a mercury exposed fish at all the exposure periods in the order of: 1 > 15 < 30 days. On the contrary, in a zinc exposed fish an increase was observed in these parameters in the order of 1 > 15 > 30 days. The activity of LDH and the levels of pyruvate and lactate were also reported increased in all the three organs of the fish at the three exposure periods studied in both the metal media. This increase was also in the order 1 > 15 < 30 days and 1 > 15 > 30 days in the organs of the fish exposed respectively to mercury and zinc, respectively.

LDH activity in fish occur at higher order in muscle tissue than in other tissues, making thus, total serum LDH levels a potential biomarker of muscle damage (Huggett *et al.,* 1992). An increase in LDH activity with subsequent decrease in glucose concentration indicates an enhanced rate of glycolysis due to chemical stress. Furthermore significant decrease in the activity of fish liver succinate dehydrogenase suggests that anaerobic metabolism is being favored over aerobic oxidation of glucose through Kreb's cycle in order to mitigate the energy crisis for survival. Regnault and Batrel, 1987 in shrimps, evidenced that GDH activity in general is positively body weight dependent, but with regards to metabolic functions alone it appeared negatively correlated with shrimp weight. Its measurable increase in serum levels is indicative of hepato-cellular necrosis. Its activity is reported to be almost at one third level in kidney than in liver. Likewise, SDH, GDH activity is also reported (Ozretic and Krajnovic -Ozretic, 1993) to increase in mullets following an injection of phenol and $CClO_4$ in a fish. GDH and AO (L-amino acid oxidase) activities were elevated in most of the tissues after mercury treatments in this fish in all tissues except GDH and AO in gills and muscles where a decrease was observed. XO (Xanthine **Oxidase**) activity in brain, gills, and kidneys was significantly elevated, but no marked alteration was noted in other tissues.

Erythrocytic Glu-6-phosphate dehydrogenase (G-6-PD) is another important metabolic enzyme, considered as a predisposing factor in the causation of haemolytic disease, leading to oxidative damage to red blood cell membrane and haemolysis. Inherited deficiency of G-6-PD is a major genetic and public health problem in malaria endemic areas of India (Balgir, 2007). G-6-PD deficiency has been reported in certain tribes of Orissa in India, which may result in morbidity, mortality and neonatal wastage. Metals like Hg also alter G-6-PD levels in affected fishes. Sastry and Gupta, 1980 on exposing *Channa punctatus* to mercuric chloride (0.3 mg/l) also found significantly inhibited glucose-6-phosphatase activity in the intestine and pyloric caeca along with inhibition of the activities of all the enzymes examined in intestine and pyloric caeca and digestion of proteins and lipids. Sastry and Rao (1984) has reported decreased glucose-6-phosphatase and hexokinase activities in *C. punctatus* after exposure to sub lethal concentration of $HgCl_2$ for 120 days in all the tissues, but hexokinase activity deceased only in intestine, kidney, and liver following fish exposure to sub lethal concentration of mercuric salts for 120 days. Both acute and chronic exposures of a freshwater fish, *Ophiocephalus* (*Channa punctatus*) to mercuric chloride (1.8 mg/L; Sastry and Sharma. 1980) has also revealed the activities of glucose-6-phosphatase, succinic dehydrogenase, lactic dehydrogenase, pyruvic dehydrogenase, and cholinesterase inhibited. Maximum inhibitions occurred after four days with cholinesterase, and 30 days with glucose-6-phosphatase, Na^+, K^+

adenosine triphosphatase, and the three dehydrogenases. Asztalos and Nemcsok (1985) correlated the magnitude of increase in LDH activity in carp tissues to the tissue damage and organ necrosis. In case of brown trout (*Salmo trutta fario*) exposed to river and sewage plant effluents, increased LDH activity in blood had been used as sensitive biomarker (Escher *et al.*, 1999). Similar supporting examples are also available with other aquatic animals. Earnshow *et al.*, 1986, observed slowed down motility in spermatozoa in marine mussel *Mytilus edulis* due to the inhibition in respiratory enzymes by the heavy metal ions. Cd was reported to decrease the glycogen reserves in *H. fossilis* by stimulating the glycolytic enzymes like LDH, PDH and GDH (Sastry and Subhadra, 1982). Larsson *et al.*, 1985 were of the opinion that sub lethal concentration of Cd reduced the aerobic metabolism in *Salmo gairdneri*. Hilmy *et al.* (1985) reported that the sensitivity of the assayed enzymes to Cd in fish, *M. cephalus* varied in different tissues. Christensen *et al.* (1982) stated metals like Cd, Cu and Zn are the known inhibitors of LDH activity. Rani and Ramamurthi (1987) also reported elevation in LDH in tissues of metal-intoxicated *T. mossambica*. Gupta and Rajbanshi, 1988 recorded accumulation of highest concentration of Cd in gills of *C. punctatus*, because of the inhibition in GDH and SDH activity during early days after treatment of Cd in *C punctatus* could be attributed to the damage of gill surface on account of their direct exposure to contaminated water. In *Donax trunculus* (Mizrahi and Achituv, 1989) reported Cd and Hg as high inhibitors of LDH and C-oxidase at 1 and 10 ppm concentrations, and Pb and Cu even at higher concentrations showed little inhibition in these enzymes. Sastry and Shukla (1990 a, b) also evidenced decreased activities of LDH, SDH and MDH along with glucose-6-phosphatase, and hexokinase in liver and muscles of *C. punctatus* after acute and chronic exposures to cadmium (1.12 mg/l) for 120 days. Sastry and Sunita (1983) in *C. punctatus* after exposure to Cd, Pb, Hg and Ni, reported alternations in the malate dehydrogenase (MDH) in the liver and muscles as well as disturbances in intestinal absorption of xylose. On the contrary, Sastry and Shukla (1994) reported an increase in the activities of hexokinase, LDH and SDH for 15 and 30 days both in liver and muscles of *C. punctatus* chronically exposed to cadmium. The enzyme activities, however, were decreased after 60 days of exposure. Sastry and Shukla (1993) also reported higher activity of these enzymes in Cd treated fish. Increased activity of SDH in Cu treated fish liver is also an indication of enhanced glycolysis *i.e.* higher anaerobic metabolism. Metal toxicity indicates degeneration of gills, thus, reducing oxygen availability and shifting to anaerobic pathway of energy metabolism. Sastry *et al.*, 1997a after observing elevation in the activity of enzymes of glycolysis and Kreb's cycle in a number of tissues, suggested that the organism adjust to mitigate the toxicity effects by increasing their metabolic rate on exposer to Cd, as it was higher in treated fish than control fish. Inhibition in the activity of PDH and SDH in the liver after 15-day exposure, suggest their interference in oxidative metabolism, which are further supported by the decreased levels of lactic acid in blood and liver and increase in pyruvic acid in Cu affected fish. A dose dependent increase in the glycogen phosphorylase and a dose dependent decrease in glycogen synthatase were also observed by Hontella *et al.* (1996), thereby suggesting an increased glycogenolytic potential in the liver of salmon exposed to Cd. An increase in the phosphofructo-1-kinase activity was also observed after 8-hr exposure to Cd that resulted in a stress leading to increased glycogenolysis. Almeida *et al.*, 2002, also indicated a decreased glycolytic activity following Cd treatment to fish. They observed decreased glycogen contents and decreased glucose uptake in white muscle in Nile tilapia, *O. niloticus* exposed to *in vivo* cadmium contamination, along with simultaneous decreased LDH and creatine phosphokinase (CK) activities, without any alterations in protein levels, suggesting a metabolic shift of carbohydrate metabolism to maintain muscle protein levels in white muscles. In red muscles however, in increased glucose uptake along with increased LDH and CK activities was observed with Cd exposure. Creatine phosphokinase or creatine kinase (CK), also known as phospho-creatine kinase, is an enzyme that consumes adenosine triphosphate (ATP) to catalyze conversion of creatine to create phosphocreatine (PCr) and adenosine diphosphate (Wallimann *et al.*, 1992). In tissues and cells that consume ATP rapidly, especially skeletal muscle, and brain, PCr serves as an energy reservoir for the rapid buffering and regeneration of ATP *in situ*, as well as for intracellular energy transport by the PCr shuttle or the circuit (Wallimann and Hemmer, 1994). Thus, creatine kinase is an important enzyme in such tissues.

There have been further various studies on the effects of sub lethal doses of metals singly and in mixture on the activities of key respiratory enzymes SDH (Succinate) and GDH in the freshwater fishes. James *et al.* (1991) reported suppression of SDH in *H. fossilis* due to Hg (0.01 and 0.03 ppm). Cu, Zn and Cd in various combinations were also said to either enhanced or lowered as compared to individual metal. James *et al.* (1992) studied the effects of sublethal doses of Cu, Cd, Zn, singly and in mixture on the activities of key respiratory enzymes SDH (Succinate) and GDH in the freshwater fish, *O. mossambicus* and their recovery following withdrawal of the metal treatments. On the basis of 96 hr LC_{50} Cu was found to be highly toxic, followed by Zn and Cd, and the tri-metal combination (Cu + Zn + Cd) was extremely toxic than any other combination; Zn + Cd was least toxic. A significant gradual decrease in SDH with a concomitant increase in GDH activity was observed in liver, brain, muscle and gill of of the fish exposed to these metals; thus, suggesting a metabolic shift from aerobiosis to anaerobiosis due to metal action. Exposed individuals when transferred to metal impoverished water showed an improvement in SDH activity only; continuous decline in GDH activity suggested further slow reversal to aerobic metabolism in metal free water and the need for more time for complete recovery of the fish. Visvikio and Rachlin (1994) observed that metals may act synergistically where the toxic response is greater than the sum of their individual toxicities. Sastry *et al.* (1997a) analyzed effects of sub lethal concentration of Cd (1.62 ppm) and Cu (0.3 ppm) individually and in combination in *C. punctatus*. The activities of LDH, MDH, and GDH were found increased in muscles, kidney and gills. In gills their activities however, were inhibited in first 15 days after metal treatment. In case of combined metal effects, the activity of all the enzymes was further elevated in all tissues.

LDH in fact is a key enzyme as it acts as a center for a delicately balanced equilibrium between the catabolism and anabolism of carbohydrates for ATP production, by promoting breakdown of glucose to lactate under aerobic conditions, as well as in anaerobic glycolysis. The decrease in SDH activity and elevation in

LDH activity leads to the accumulation of pyruvate and lactate in the tissues of the fish which may cause tissue acidosis (Gargiulo *et al.*, 1996; Valarmathi and Azariah, 2003). Thus, glycogenolysis is the initial response of a fry on exposure to copper (Emad *et al.*, 2005). Increase in the rate of oxygen consumption and the activities of ICDH (Isocitrate dehydrogenase) and SDH on all the days of exposure, suggests an elevation in oxidative metabolism, thereby the animal could derive more energy through it to counter act the stress caused by the metal as well as the alkaline pH. At this pH the metal permeability through the cell membranes might be less, leading to accumulation of metal in the cells within tolerable limits. Gradual decrease in the degree of elevation in the levels of energetics with the increase in exposure period indicates the adaptation of the fry to the toxic stress on prolonged exposure; thereby all the metabolic activities come to a normal level. These changes thus, indicate that the weak alkaline pH acts as an inhibitor to the toxicity of copper and might have prevented its binding with the active sites of the enzymes (Ghillebaert *et al.*, 1995; Wu *et al.*, 2006). Reddy *et al.* (2008) suggested that an increase in the level of lactate and LDH activity is an indication of shift to less efficient glycolysis *i.e.* anaerobic glycolysis, and an increase in pyruvate level, rate of oxygen consumption, and ICDH and SDH activities suggest an elevation in oxidative metabolism. They further concluded that the carp fry can tolerate and develop resistance to sub lethal concentrations of copper in alkaline habitats as the fish could derive the necessary energy through the elevation of oxidative metabolism. LDH catalyze the inter-conversion of lactate and pyruvate, in the presence of NAD/NADH. It also plays an important role for the synthesis of glycogen (Glucogenesis) in tissues from lactate, and in generating DPNH and ATP in organs like heart from lactate in the citric acid cycle. Variability in enzyme activity in different organs could be attributed to differences in responses of these organs to pollutants in terms of their absorption and accumulation.

There are further reports of enzymatic alterations due to chronic lead exposures in the freshwater catfish, *H. fossils* (George *et al.*, 2011). The activities of all the four enzymes in gills and pyruvate and succinate dehydrogenase activities in muscle decreased. Feeding tilapia (*O. niloticus*) on diet containing lead @ 100, 400, and 800 µg/g for 60 days (Dai *et al.*, 2009), showed LDH activity in liver and kidney affected by dietary Pb in a dissimilar way, *i.e.* it showed Pb concentration-dependent decrease in LDH in kidneys, whereas in liver it appeared enhanced in a Pb concentration-dependent manner. Ramalingam and Indra (2002) in snails on their exposure to Cu also observed decrease in glycogen with simultaneous increase in lactate, suggesting a shift to anaerobic state under stress. Likewise in crab, *Sesarma quadratum* after exposing to two sub lethal concentrations of copper chloride for 21 days (Santhana and Azariah, 2003) disturbances in tissue enzyme activities as of LDH and SDH were noticed. LDH activity was significantly elevated in muscle and hepato-pancreas tissues, whereas, SDH activity was suppressed in the crab tissues of the muscle, gills and hepato-pancreas for 21 days. Inhibition in SDH activity was said to be due to impairment of oxidative metabolic cycle and dependence on the anaerobic glycolytic pathway to meet the increased energy demands.

Sorbitol dehydrogenase is another such important metabolic enzyme, also referred in general as SDH by various workers, but it differs significantly from the well known succinic dehydrogenase. Sorbitol dehydrogenase is known to be involved in the conversion of D-fructose to D-sorbitol primarily in the cytoplasm and mitochondria of the liver, kidney and seminal vesicles of humans and other animals. It has also been observed to exhibit changes in its activity following metal toxicity in fishes; and measuring its activity is said to be a specific indicator of any damage to liver cells and any parenchymal hepatic disease. Most of the workers have emphasized upon serum sorbitol dehydrogenase (So-DH) activity as the best measure to assess liver damage on account of chemical poisoning in fish, even more sensitive than the histopathology. Its activity in serum is usually low but increases during acute episodes of liver damage (Dooley *et al.*, 1979). Dixon *et al.*, 1987 in rainbow trout, observed copper not only exerting a dose dependent rise in So-DH activity in liver, but the order of magnitude of its activity was much higher than all other enzymes. The plasma activity peaked after around 48 h of exposure to the contaminant, even before any histopathological damage was detected in liver.

ii. Transaminases

Transaminase enzymes play a vital role in carbohydrate-protein metabolism in fish and other organisms' tissues (Eze, 1983). These enzymes catalyze the transfer of an amino group from one molecule to a keto acid, forming a new amino-keto acid. The keto acids are important for the functioning of Krebs cycle for energy molecule production. Most important aminotransferases are glutamic-oxaloacetic transaminase (GOT) and glutamic-pyruvic transaminase (GPT). These enzymes are also termed as aspartate aminotransferase (AST) and alanine aminotransferase (ALT), respectively. Few workers (Hans and Ronny, 2001; Rani *et al.*, 2001) have mentioned these as AAT and ALAT, respectively. ALT and AST are the most sensitive tests for the diagnosis of liver diseases and the extent of damage (Kew, 2000). Shakoori *et al.* (1990), reported that the increase in blood enzymatic activity is either due to: i. leakage of these enzymes from hepatic cells; or ii. Increase in their synthesis; and/or iii. enzyme induction. Tietz (1987) and Campbell (1984) reported that these enzymes are liberated into the blood stream when the hepatic parenchyma cells are damaged.

Copper is a strong inhibitor of fish plasma GOT (AST) when the blood is exposed to the metal *in vitro* (Christensen, 1971). McKim *et al.* (1970) observed increase (+60 per cent) in plasma GOT during first 21 days in brook trout exposed to 32-39 µg/l of Cu, after 11 month of fish exposure to Cu, however, lowered GOT levels almost 30 per cent less than controls. Initial increase was attributed to direct heart damage by the Cu, but during a long time exposure; as accumulation of Cu increased in liver, it started declining plasma GOT. Cu accumulates in liver much readily than in heart. Likewise, elevations in both plasma GOT and GPT were reported in pike on exposure to phenol (McKim *et al.*, 1970; Kristofferoson *et al.*, 1974) and sewage in the water (Wieser and Hinterleitner, 1980). Durve *et al.*, 1980 reported Cu poisoning to decrease oxygen consumption in *Rasbora daniconius* along with copious deposition of mucous on the gills and skin, possibly due to degeneration of gill lamellae; as also observed by Gupta and Rajbanshi (1986) who in *L. rohita* after its exposure to 0.365 mg/l of Cu had also observed hyperplasia, hypertrophy and fusion of secondary gill lamellae.

Following schematic model illustrate the impact of copper toxicity in a fish gill tissue:

i. Cu intake in a gill, results in hyperplasia/hypertrophy,

ii. Damaged gill lamellae induce mucous secretion and reduce O_2 consumption; followed by an increase in glycolysis and increase in LDH/MDH activity;

iii. There occur an increase in anaerobic metabolism, and a simultaneous increase in the activity of aminotransferases (GPT and GOT) along with decease in proteins (proteolysis) and deamination.

Proteins are highly sensitive to heavy metals (Jacobs *et al.*, 1977). Metal effects in disturbing protein metabolism by altering protein metabolism and causing increase in activity of aminotransferases and increased proteolysis are reported in fishes. David *et al.* (2004) showed decrement of total structural and soluble proteins; whereas, free amino acids and activities of protease, aspartate aminotransferase and alanine amino-transferase significantly increased in cypermethrin exposed *C. carpio*. Ammonia content decreased but urea and glutamine increased. It was also observed that alterations steadily increased with the period of exposure and exhibited tissue specificity. Variations indicated its toxic effect on the cellular metabolism thereby leading to impaired protein synthetic machinery. Depletion of protein content has been observed at all exposure periods in the muscles, intestine and brain of the fish as a result of mercury chloride (Martin *et al.*, 2008) toxicity. Cell necrosis results in the release of GPT and GOT in serum and Cd is well known to cause histological changes in fish liver, kidney and gills (Tafanelli and Summerfelt, 1975). Data compared with the fish inhabiting normal metal free waters, revealed that liver and kidney functions showed no significant decrease in proteins in serum as well as urea and creatine. Histopathological observations of these organs showed slight hemorrhage, inter-sinusoidal hemorrhage, leucocytic cell aggregation and vacuolar degeneration of the liver. Kidneys revealed slight focal and inter-tubular hemorrhage. Oxidation of amino acids is also mobilized by aminotransferases. Elevated levels of any of these enzymes on account of any pathological or physiological disorder in a tissue, amount to proteolysis in the affected tissues. Increased activities of these enzymes firther indicate liver damage. Karan *et al.* (1998) investigated the consequent effects of widely used algaecide copper sulphate for the control of phytoplankton in lakes, reservoirs and ponds and after a 14-day period of exposure at five concentrations ranging in between 0.25-4.0 mg/l $CuSO_4$ (values ranging from approximately 5 to 7 per cent of the 96 h LC_{50}) and a recovery period of the same duration. They reported that there was an increase in the activity of AST and ALT as well as of alkaline phosphatase (AP) in the blood serum and gills of *C. carpio*. After a recovery period, compared with the end of treatment a decrease in enzyme activities of each enzyme was evident indicating eventual recovery from Cu-induced stress (the only exception being the ALT activity in gills at the highest $CuSO_4$ concentration). Chronic exposure to mercury has also been reported to cause an increase in plasma

GOT and GPT within 4-8 weeks in a marine teleost fish, *Aphanius dispar* (Hilmy *et al.*, 1981). Increase in GOT was almost more than double than that of GPT. Sastry and Sharma (1980) also stated about increase in the level of both GOT and GPT in the serum of *C. punctatus* after acute exposure to mercury.

There have been some studies indicating inhibition in the activity of transaminases in certain fish organs following their exposures to metals. Christensen (1971) also attributed reduction in AST and ALT enzyme activity to inhibitory effects of a metal. Exposure of the rosy barb (*P. conchonius*) to 181µg/l mercuric chloride for 48 h (Gill *et al.*, 1990b), showed both AST and ALT enzymes inhibited in different organs. Reberts *et al.* (1979) evidenced that cadmium could cause an induction or inhibition of variety of cellular enzymes in marine fish rainbow and red trout when exposed to sub lethal concentration for 2 or 3 months. Sastry and Sunita (1983) reported transaminase enzyme activities increased in the blood of *C. punctatus* after an exposure to Cd, Pb, Hg and Ni. Brafield and Koodie (1994) while analyzing the effect of high dietary zinc in carp, *C. carpio*, attributed rise in amino acids levels to increase in proteolysis, which normally leads to increased transamination. Sastry *et al.*, 1997 analyzed effects of sub lethal concentration of Cd (1.62ppm) and Cu (0.3ppm) individually and in combination in *C. punctatus* after 30 days on enzyme activities in liver. Several other reporters had also showed that these blood enzymes were highly increased in a fish treated with metals like cadmium, zinc and copper (Benson *et al.*, 1987; Hilmy *et al.*, 1987; Singh and Reddy, 1990; Karan *et al.*, 1998; Atif, 2005). Increased transaminase activity in kidney and brain of *N. notopterus* was also reported after exposure to Cd (Yamawaki *et al.*, 1986). Reduced levels of both AST and ALT enzymes, within 2 days of exposure of marine fish, *Mugil seheli* to different concentrations of Cd (2.0 and 0.5 ppm) and Cu (0.5 and 0.2 ppm) were also reported in both plasma and muscle tissue by Emad *et al.*, 2005. Both these protein-carbohydrate metabolism enzymes (AST and ALT) showed low levels than control group due to its sharing in transforming proteins to glycogen. De La Torre *et al.* (1998) reported that the activity of ATPase was decreased while AST and ALT in muscle of *C. carpio* was not affected. Galhoom *et al.*, 2000 compared data for the presence of heavy metals in water and their effects on some physiological parameters in *Mugil* fish (*Mugil cephaleus*).

Cyriac (1990) observed a differential response in the liver and kidney in the GOT activity of the metal treated fish; in the liver GOT activity was increased during metal intoxication, whereas, the enzyme activity was much lower in the kidney. The activity of GPT was, however, low in both the liver and kidney. The increase in GOT in the liver was attributed to the possibility of shift from alanine transmission to aspartate transmission, resulting in an increased synthesis of GOT and decreased GPT. Increased activity of GPT due to Cd is further an indication of formation and excretion of ammonia. Differential behaviour of Cd metal in different tissues was attributed to the fact that these accumulate Cd differently as per their path of intake as well as their metabolic status. Gupta and Rajbanshi (1988) reported highest Cd accumulation in kidneys. In the fish *C. carpio* exposed to 0, 0.8, 4 and 20mµM Cd, a decrease in the crude protein content and increased activity of AST and ALT (Hans and Ronny, 2001) were attributed to proteolysis, intended to increase the role of proteins in the energy production during Cd stress. De Smet and Blust (2001) reported that there is an increase in the activities of GOT and GPT in *C. carpio* exposed to cadmium. Such changes in biochemical levels under the effect of chromium toxicity might results in impairment of energy requiring vital processes, and hence give an idea about the health status of the fish population. Antonella and Landriscina (1999) also suggested that cadmium alters hepatocyte cell membrane structure and concomitantly induces changes in mitochondrial membranes resulting in elevation of these enzymes. Exposure of a teleost fish, *T. mossambica* (Rani *et al.*, 2001) to sublethal toxicity of sodium arsenite was also reported to elevate the activities of the AST and ALT enzymes in liver, gill, brain and muscle in relation to control. It was suggested that the fish responded to the stressful situations by gearing up its metabolic activity, as revealed by the elevated activities of AAT and ALAT.

Similar observations and suggestions were stated by many experimental investigations on animals treated with heavy metals (Singh and Reddy, 1990; Karan *et al.*, 1998; Bersenyi, 2004; Kim and Kang, 2004; Al-Attar, 2005). Neha-Bhatkar *et al.*, 2004 reported the impact of 30 days exposure of $CrCl_2$, $NiCl_2$ and $ZnCl_2$ on some enzymatic changes of the fish *L. rohita*. They observed increased serum proteins and serum glutamic oxaloacetic transminase (SGOT) and serum glutamic pyruvate transminase (SGPT) contents in all the test metals for 30 days. An increase in SGOT and SGPT has also been reported earlier in *C. carpio* (Mathan, 2006), *Sparus auratain* (Antonella and Landriscina,1999), *C. auratus gibelio* (Zikic *et al.*, 2001), *C. carpio* (De la Torre *et al.*, 2000) and *C. gariepienus* (Babu *et al.*, 2006) after exposures to various metals. Mathan (2006) opined that an increase these enzymes may have resulted from tissue damage and increased synthesis of the enzymes to defend against stress.

Similarly, De la Torre *et al.* (2000) and Zikic *et al.* (2001) attributed increased SGOT and SGPT activity to liver cell damage with concomitant liberation of transaminases into the circulation. De la Torre *et al.*, 2000 and Babu *et al.*, 2007 recorded an increase in transaminases and attributed it to alterations in amino acid catabolism after Cd treatment of *C. carpio* and *C. gariepienus* respectively. Similarly, Adham (2002) recorded an increase in SGOT and SGPT of *C. batrachus* in Lake Maryut, and these changes may be a sign of damaged liver which in turn leads to leakage of these cellular enzymes into the blood. Al-Attar and Atef (2007) in the freshwater fish, *O. niloticus* exposed to Ni, revealed a significant increase in the activity of ALT and AST after three weeks of nickel exposure, indicating considerable hepato-cellular damage. In these nickel-exposed fish, *Oreochromis niloticus,* while the levels of serum glucose, cholesterol, total protein, albumin, amylase, lipase, alanine aminotransferase and aspartate aminotransferase were significantly elevated (Atef and Al-Attar, 2007). Feeding tilapia (*O. niloticus*) on diet containing lead @ 100, 400, and 800 μg/g for 60 days (Dai *et al.*, 2009), showed ALT and AST activities along with LDH in liver and kidney affected by dietary Pb but in a in a dissimilar way; *i.e.* Pb concentration-dependent decreases in ALT, AST and LDH activities were observed in kidney, while these enzyme activities in liver appeared enhanced in a Pb concentration-dependent manner. SGPT and SGOT enzymes also showed a significant increase with an increase in dose and duration of the exposure in a freshwater murrel, *C. punctatus* (Bloch), after exposure to 10 mg/l and 15 mg/l and 25 mg/l zinc for 8,10 and 15 days (Verma and Srivastava, 2010). According to Verma and Srivastava (2010), increased transaminase (SGOT and SGPT) activity in *C. punctatus* exposed to zinc, may either be due to leakage of enzymes across damaged plasma membranes and/or an increased synthesis of enzymes by the liver as a defense against stress. The data of Parvathi *et al.* (2011) showed that the exposure of chromium caused significant elevations in the activities of blood serum GOT and GPT levels.

To corroborate this view ruptured and totally damaged hepatocytes have been observed in an earlier study conducted under similar conditions (Verma and Srivastava, 2008b). Accumulation of metals by the liver has been reported to be an important factor influencing enzyme activity (Dubale and Shah, 1981 and Humtsoe *et al.*, 2007); high accumulation of zinc in the liver has been recorded earlier during a similar investigation (Verma and Srivastava, 2008b); this could be instrumental in causing cell damage leading to an increase in transminases. Tilak *et al.* (2005) stated that elevation of SGOT activity could also provide oxaloacetate required for gluconeogenic pathway to meet additional supply of glucose required for production of energy during oxidative metabolism. Thus, elevation in the levels of SGOT and SGPT in liver of *C. punctatus* or other fishes can be considered a response to stress, induced by metals like zinc, for contributing to gluconeogenesis and/or energy production necessary to meet excess energy demands. This appeared corroborated further by excessive movements as noted in experimental fish. Verma and Srivastava, 2010 confirmed the earlier reports that these enzymes are a very sensitive index for hepatic and ovarian changes occurring due to adverse condition.

iv. Phosphatases

Phosphatases are another important group of hydrolytic (Lysosomal) enzymes, capable of removing inorganic phosphate (Dephosphorylation) from certain organic phosphate esters; such as hexose phosphates (glucose-6-phosphate), glycerophosphates and nucleotides derived from the diet and after the digestion of nucleic acids by nucleases (Mayes, 1985). These in fact catalyse the hydrolytic cleavage of phosphoric acid ester. They are designated as either alkaline (ALP) or acid phosphatase (ACP), according to their pH optima. ALP catalyses the reactions at alkaline medium (pH 8-9) and ACP at acidic (pH <7). ALP has also become a useful tool in molecular biology laboratories and as a label for enzyme immunoassays. Removal of the phosphate groups allows radio labeling (replacement by radioactive phosphate groups) in order to measure the presence of the labeled DNA through further steps in an experiment and in humans elevated levels of this enzymes is a condition associated with certain medical conditions (Li-Fern and Rajasoorya, 1999) or syndromes, *e.g.* hyperphosphatasia with mental retardation syndrome, HPMRS). The estimations of both the ALP and ACP enzymes have received much attention in many tissues (McComb *et al.*, 1979). The interests in these enzymes, has been more in vascular smooth muscles, because of the early histological and biochemical findings of alterations in lysosomes (Watanabe *et al.*, 1981) and plasma membranes (Kwan *et al.*, 1979; Leake *et al.*, 1982) associated with hypertensive disease. Hypertensive vascular disease may be associated with a general de-arrangement of cell membrane phosphor-hydrolase function which includes acid phosphatase, as elevation in ACP enzyme was associated with hypertensive disease (Watanabe *et al.*, 1981). Kwan and Ito, 1987 showed both these enzymes associated with non-vascular smooth muscles, but these were different in response to Mg^{2+} and EDTA.

Alkaline phosphatase (ALP), a brush border enzyme is involved in transphosphorylation reactions. It has been implicated in the absorption of nutrients across the intestinal wall. ALP occurs practically in all animal tissue at alkaline pH and occurs in wide variety of animal species. Liver and intestinal ALPs in fish like rainbow trout, are relatively much sensitive to thermal variations, but less sensitive to inhibition by beryllium, EDTA and amino acids, are almost insensitive to 2-mercaptoethanol. ALP is highly concentrated in the intestinal mucosa and kidney of most animals. In humans it is now well established that ALP exist in three different forms of isoenzymes, one for liver, bone and kidney; second for intestinal tissue and third for placenta (Stigbrand and Fishman, 1984). According to Pearse (1961) high ALP activity indicates increased phosphate transfer from one alcohol to other rather than hydrolysis of phosphate ester. This has been implicated in the absorption of nutrients across the intestinal wall. Mayes (1985) reported involvement of this enzyme in glycogen metabolism and capable of removing inorganic phosphate from certain organic phosphate esters such as hexose phosphate (like glucose-6-phosphate), glycerophosphates and nucleotide derived from the diet and the digestion of nucleic acids by nucleases. Norman *et al.* (1975) reported that the increase in alkaline phosphatase activity was correlated with an increase in calcium transport activity. These iso-enzymes differ in their amino acid sequence. Animals other than humans and other higher primates lack the placental ALP. These co-enzymes differ in their thermal stability and sensitivity to various inhibitors like EDTA, Zn, beryllium, cyanide, 2-mercaptoethanol, or certain amino acid derivatives such as L-homoarginine and L-phenylalanine (Yora and Sakagami, 1986).

ALP extracted from liver was found weakly inhibited by Hg when the exposure was *in vitro* (Jackim *et al.*, 1970). In other words, Hg is inhibitory of ALP, but seemingly induces synthesis of the molecule in the intact liver, which compensates more than for any inhibition. The results of Sastry and Gupta (1978a) following the treatment with 1.8 mg/liter (LC_{50}) of mercuric chloride in *C. punctatus* showed that mercury inhibits the activities of phosphatases in the liver. Acute exposure of a freshwater fish *Ophiocephalus* (*Channa*) *punctatus* to mercuric chloride (1.8 mg/L), inhibited alkaline phosphatase activity while chronic treatment (0.3 mg/L) increased its activity (Sastry and Sharma. 1980), and maximum inhibitions occurred after four days with alkaline phosphatase and cholinesterase. Hilmy *et al.*, 1981 recorded 50 per cent decrease in ALP activity in 8 days in the marine fish on exposure to mercury and they attributed this decline to Hg displacing cofactor Mg from the enzyme. Gill *et al.* (1990) in rosy barb (*P. conchonius*) exposed to 181 µg/l mercuric chloride for 48 h, also recorded increased ALP activity in all the organs examined except in kidneys. Kothari *et al.*, 2003 on exposure of a freshwater fish *H. fossilis* for 3-30 days to $HgCl_2$ (0.1 mg/l) observed significant alteration in protein and alkaline phosphate and alanine transminase in intestine.

ALP inhibition in fish organs has been used as an indicator of environmental pollution (Linhardt and Walter, 1965). Its inhibition in liver indicates that the transphosphorylation reactions are adversely affected in this organ. Heavy metals showed tissue-specific effects on ALP activity (Heath, 1995). Decrease in ALP activity further indicates an altered transport of phosphate (Engstrom, 1964) and an inhibitory effect on cell growth and proliferation (Goldfischer *et al.*, 1964). Decrease in ALP activity in intestinal mucosa is also said to reduce Ca absorption (Sugawara and Sugawara, 1975). Both the adenosine triphosphatase and alkaline phosphatase are brush-border enzymes which mediate membrane transport (Goldfisher *et al.*, 1964). ALPs are shown to contain Zn (Mathis, 1958), and possess Mg bonding sites. Zinc inhibits the ALP activity by competing with magnesium-binding sites (Cathala *et al.*, 1975). Exposure of a the murrel fish, *C. punctatus* (Bloch) to 10 mg/l and 15 mg/l and 25 mg/l zinc for 8,10 and 15 days (Verma and Srivastava, 2010) was reported to inhibited ALP in the liver but it increased its activity progressively in the ovary. Decline in ALP in liver was attributed to impairment of phosphoprylation or changes in permeability and disruption of lysosomes and mitochondria, as suggested by Sarasus and Andal (2005) and Humtsoe *et al.* (2007). An increase in ALP in Zn exposed ovary in *C. punctatus* (Verma and Srivastava, 2010) also supports earlier similar reports on intestine, liver and gills of the catfish, *H. fossilis* (Bloch) by Kothari and Soni, (2004a,b) and Jat and Kothari, (2006). Kothari and Soni (2004) reported sublethal exposure of $ZnSO_4$ (150 mg/l) to freshwater cat fish *H. fossillis* for 30 days, caused significant alteration in gill function as revealed by elevated ALP over control. An increase in ALP suggests increased glucose and phosphate transfer (Jat and Kothari, 2006); it is also known to be an indicator of histological damage (Yang and Chen, 2003 and Atli and Canli, 2007). Histological damage in the ovary of this fish has earlier been reported after exposure to same Zn concentrations (Verma and Srivastava, 2008a). Sub-lethal concentration of lead nitrate (20 ppm) in freshwater murrel, *Channa punctatus* was found to show an increased oxidative stress in the treated fish; with an increase in the activity of ALP and a decline in ACP; with simultaneous inhibition in SDH (succinate dehydrogenase); increase in ALP was attribute to adverse effects of lead nitrate on the glucose re-absorption and trans-phosphorylation; whereas, decline in ACP was attributed to tissue damage (George *et al.*, 2011).

Kumari (1984) in *Channa punctatus* found ALP activity significantly higher in brain and liver and low in gills, intestine and kidney after exposing the fish to Cr. Goel and Sastry, 1973 studied distribution of ALP in the digestive system of *C batrachus, O punctatus* and some other fishes. Intense activity of ALP was reported in the brush border of mucosa and lamina propria and the hepatic cells in the liver of these fish. Sastry and Subhadra (1985) reported *in vivo* effects of cadmium in the freshwater catfish, *H. fossilis* in decreasing the activity of ALP in the liver, kidney, and intestine; but elevation was recorded in ovary and muscles. Similar inhibition in ALP activity was reported due to Cd by Sastry and Gupta (1979ab), lead (White, 1977), mercury (Abdel Tawwab *et al.*, 2004; Hussein *et al.*, 2012) and other aquatic pollutants (Hinton *et al.*, 1973). A decrease in ALP activity in the brush border of intestine of *C. carpio* and also observed by Sastry and Gupta, 1979 in *H fossilis* due to Cd and suggested that heavy metals presumably decreases the absorption of calcium from fish intestine due to metals. Cd was found to affects alkaline phosphatase activity in *C. carpio* differently as per the exposure period (Nevalennyi and Bednyakov, 2004), *i.e.* it remained unchanged after 10 days then increased to peak on day 30 and decreased to 40-50 per cent of control at 50-60 days. Cd treated freshwater rosy barb also showed ALP activity unchanged in liver and gills; but stimulated in kidney and ovary and inhibited in gut (Gill *et al.*, 1991). Cd in sub-lethal doses ranging in between 2-10 mg/l, however, was found to increase the activity of alkaline phosphate in liver, glutamic oxaloacetic transminase in liver and kidney and glutamic pyruvic transminase in kidney and serum in *C carpio* (Neskovic *et al.*, 1996). Simultaneous feeding of poly-unsaturated fatty acids to fish maintained these parameters near normal values. Sarita (2005) also recorded ALP enzyme activity reduced in all the tissues of *C. mrigala* and *C. carpio* except in muscles, and intestine of *C. mrigala*, where it showed an increment over control. Dai *et al.* (2009) demonstrated that the inhibitory effects of dietary Pb on alkaline phosphatase, Na, K- ATPase, Ca, and Mg-ATPase activities in both liver and kidney were Pb concentration-dependent.

Acid phosphatase (ACP), also known as a lysosomal enzyme, helps in the autolysis of a cell after its death and thus, is a good indicator of stress conditions in a biological system (Verma *et al.*, 1980). ACPs are heterogenous glycoproteins whose multiplicity is based at least in part on variations in the carbohydrate residues. ACPs are reported to be distributed more in the spleen, prostate, liver and red blood cells, and these are more active at pH 5.0. At least three types of ACPs were differentiated (Goodlad and Mills, 1957) on the basis of their pH optima. Janska *et al.*, 1986 isolated two molecular forms of acid phosphatase (ACP phase-1 and ACP phase-II in catfish liver). They evidenced these merely as two different glycosylated forms with similar protein units, ACP phase-II form being with 25 times more glucose, it possesses a lower thermal stability. Both forms however, differ in their sensitivity to sulphydryl-blocking reagents like NEM (N-ethylmaleimide). Sulfhydryl groups play a role in the catalytic properties of this enzyme, and NEM inhibits ACP phase-I by 7 per cent and ACP phase-11 by only 35 per cent. The importance of free –SH groups for ACPase activity was also said to be consistent with the activating effect of reducing agent like ascorbic acid. Activity of ACPase-II increased initially and it rapidly declined after the addition of ascorbate, whereas ACPase-I was without any initial effect and it was found enhanced later. As a result when one enzyme was activated, the other one was either unaffected or inhibited.

Impairment of protein synthesis through RNA depletion by increased acid phosphatase levels was reported in fish tissue under pesticide toxicity (Nateson, 1985). Baby Shaikila *et al.* (1993) attributed the changes (either activation or inhibition) in acid and alkaline phosphatases in liver tissue of sevin exposed *Sarotherodon mossambicus* to the altered enzyme activity caused by the pesticide through interaction with regulators and cofactors, such an interaction caused by the heavy metal treatments. Acid phosphate is considered an important biomarker enzyme to assess the lysosomal changes *in vivo* because it is localized almost exclusively in the particles and is realeased alongwith lysosomal hydralases. Increase in lysosomal activity in the injured cells occurs as a part of the prenecrotic changes

(Novikoff, 1961 and Moneserrate *et al.,* 1972). Lysosomes further accommodate a variety of metals and organic compounds (Moore, 1985). Weiss *et al.* (1986) found large secondary and tertiary lysosomes in fish following copper toxicity and hepatic copper tends to be sedimented in this fraction of homogenate. Gill *et al.* (1992) speculated that since, ACP enzyme is associated with the lysosome activity its elevation reflects proliferation of lysosomes in an attempt to sequester the toxic copper ions. Cadmium is said to unstabalize lysosomal membranes. Kohler, 1991 suggested that this organelle on exposure to Cd, first show an adaptive change *i.e.* an increase in its number and size due to metal accumulation, and if the metal load increases the membrane is destabilized and result in release of its enzymes. The inhibition of the ACP activity in gills may be due to disintegration of cells affected by excessive stress of heavy metal treatment. Sastry and Gupta (1979 ab) observed increased activity level of ACP in some organs of *H. fossilis*, and correlated with low level of tissue proteins due to breakdown of proteins by phosphatase enzyme, as indicated by Dubale and Awasthi, 1982. These elevated ACP levels would probably result in an altered protein synthesis, resulting in impairment of synthetic pathway of metabolic enzyme of metabolically active tissues.

According to Zonek *et al.* (1966) an increase in acid phosphatase activity is due to increased pinocytosis. Biocides, such as metals are also known to produce cytotoxic action and changes in membrane fragility (Vijayendra and Vasudev, 1984). Significant elevation of ACP level was reported in liver and intestine and an insignificant increase in stomach of *H. fossilis* after chronic treatment with lead nitrate (Sastry and Gupta, 1979a) and cadmium chloride (Sastry and Gupta, 1979b). Acute exposure of a freshwater fish *Ophiocephalus* (*Channa*) *punctatus* to mercuric chloride (1.8 mg/L), (Sastry and Sharma. 1980) showed acid phosphatase and Na^+, K^+ adenosine triphosphatase activities elevated after four days, but chronic exposure inhibited their activities. Naidu *et al.* (1984) also reported a significant decline in hepatic ACP of *Sarotherodon mossambicus* after an acute exposure (96 hrs.) to mercury. According to them impaired oxidative and transphosphorylative activities and utilization of carbohydrates during mercury toxicosis causes this depletion. Supporting the above view, Bhatnagar and Bana (1993) reported a decrease in hepatic acid phosphatase in *C. gachua* after exposure to thiodan and rogar for 30 days; they suggested that the decline may be due to uncoupling of phosphorylation.

Acid phosphates (ACP) and N-acetyl-B-D-glucosaminidase (NAG), both being the lysosomal enzymes are released in the plasma (Versteeg and Giesy, 1985). They however, found an increase in both these ACP and NAG enzymes in the plasma after chronic exposure of the bluegills to Cd for up to 32 days; the enzymes AST and ALT (transminases) however, remained unaffected. No histological change was detectable in the fish liver, but decrease in growth rate of the fish was seen. Maximum inhibition occurred after 15 days with acid phosphatase. *H. fossilis* treated with 0.26 mg/liter of cadmium for 60 days (Sastry and Subhadra, 1985) showed ACP activity inhibited in liver, ovary and gills, but increased in kidney and intestine. They attributed elevations in ACP activity to the increase in lysosomal activity inside injured cells, occurring as a part of pre-necrotic changes or enzyme induction as pointed out earlier by Christensen (1975). Likewise, Gill *et al.*(1990)

recorded decreased ACP activity in the liver, gills of Cd treated freshwater rosy barb, observed decreased levels in ACP of gills could be taken as uncoupling of oxidative phosphorylation in fish blood under heavy metal toxicity as opined by Versteeg and Giesy 1985; Dalela *et al.* (1982) and Verma *et al.* (1980). Kumari (1984) studied the effect of chromium on acid phosphatase in *Channa punctatus*. An increase in ACP was also seen after copper exposure in some tissues (Srivastava and Pandey, 1982; Gill *et al.*, 1992). Rema and Babu (1999) studied the effect of zinc and mercury on the activities of the enzymes, acid phosphatase, along with aminotransferase, AST and a dehydrogenase, GDH (glutamate dehydrogenase) in *O.mossambicus* and reported variability in activities of these enzymes depending upon the duration of exposure to a toxicant. Decrease in hepatic ACP was also reported in *Hypophthalmichthys molitrix* and *Catla catla* after exposure to distillery effluent for 30 days (Sarasus and Andal, 2005)) and in *L. rohita* to Arsenic (Humtsoe *et al.*, 2007). Among the different tissues examined, increase in acid phosphatase activity was recorded in brain, gills and kidney. In the rest of the tissues significant inhibition was observed. The increase was highest in kidney (237.5 per cent) and decrease was the maximum in liver (65.3 per cent). ACP was also found inhibited both in the liver and ovary in a freshwater murrel, *C. punctatus* (Bloch), after exposure to 10 mg/l and 15 mg/l and 25 mg/l zinc for 8,10 and 15 days (Verma and Srivastava, 2010).

The estimated changes in the levels of both the ACP and ALP enzymes in various tissues may also be due to disruption of lysosomal membranes under heavy metal toxicity. Decrease in ACP activity in a fish tissue on account of exposure to zinc could also be directly correlated with metal uptake, resulting in increased concentrations of Zn in organs affected. Significant accumulation of zinc was reported both in the ovary and liver of fish in an earlier study in which histological damage was observed in the ovary of the fish after exposure to Zn concentrations (Verma and Srivastava, 2008a, b). Humtsoe *et al.* (2007) also subscribed to this view. Jeelani and Shaffi (1989) studied the chronic effect of sub lethal dose of mercuric nitrate in *Channa punctatus* in relation to ACP and ALPs. Middle region of kidney registered maximum rise and fall respectively, of ACP and ALPs. Right lobe of liver showed more rise and fall of ACP and ALP activity, respectively. Similarly, left gill showed more pronounced effects than the right one. The observed enzymatic variations were related to the biochemical constituents and physiological co-ordination rendered by these tissues. Effect of exposure to sub lethal concentration of mercuric chloride (0.2ppm) for 24, 48, 72 and 96 h had also been investigated on liver phosphatase activity of an air breathing fish, *Channa punctatus* (Hota and Pradhan, 1994). A significant increase in the phosphatase activity (both acid and alkaline) was observed. ACP measurements in liver macrophage aggregates (MA-ACP) of different fish species are used as a marker for a pollution-induced modulation of the digestive capacity of phagocytes, since these non-specific immune responses play a central role in the maintenance of animal's health. MA-ACP activity was reported dependent on temperature and season but, nevertheless, distinctions between differently polluted areas were visible in all sampling campaigns with lowest MA-ACP activity in fish from the polluted areas of the German Bight and the Israeli coast of the Mediterranean Sea. For mercury and copper, a significant correlation could be observed between residue concentrations in fish tissues and

MA-ACP activity. In all cases, except mercury which showed a positive correlation, ACP activity was suppressed in animals with a high contaminant burden. MA-ACP activity turned out to give reliable and consistent results for a quantification of immune-modulation in both fish species. Sarita, 2005 recorded alteration in ACP and ALP activity of various tissues *viz.*, muscles, intestine, gills and kidney was observed in *C. carpio* and *C. mrigala* under the impact of sub lethal concentrations of various heavy metals alone and in combination. Muscle and intestine in both the fish species and kidney in *C. carpio* registered an increase in ACP activity, while gills in both the species showed a decrease over the controls. Increase in ACP level was recorded highest in kidney, followed by muscles and intestine. The ALP activity was decreased in kidney, gill and intestine of *C. carpio* and gills of *C. mrigala* and elevation were recorded in muscles of both the species.

REFERENCES

Abdel Tawwab M., Shalaby A. M.E., Ahmad M. H. and Khattab Y. A.E., 2004. Effect of supplementary dietary L-ascorbic acid (Vitamin C) on mercury detoxification, physiological aspects and growth perfomance of Nile Tilapia (*Oreochromis niloticus* L.) Proc. 6th Internat. Symp. Tilapia in Aqua. (12-16 Sept.2004). Internl. Conv. Center Roxas Boulevard Manilla, Phillipines: 159-171.

Adeyemo O.K., Ajani K., Adedeji O.B. and Ajiboue O.O., 2008. Acute toxicity and blood profile of adult *Clarius garipinus* exposed to lead nitrate. Internl. J. Hematol., 4: 2.

Adham K.G., 2002. Sublethal effects of aquatic pollution in Lake Maryut on the African sharp tooth catfish, *Clarias gariepinus* (Burchell, 1822). J. Appl. Ichthyol., 18: 87-94.

Al-Attar A.M., 2005. Biochemical effects of short-term cadmium exposure on the freshwater fish, *Oreochromis niloticus*. J. Biological Sci., 5: 260-265.

Al-Attar and Atef M., 2007. The influences of nickel exposure on selected physiological parameters and gill structure in the teleost fish, *Oreochromis niloticus*. J. Biol. Sci., 7: 77-85.

Almeida J.A., Diniz Y.S., Marques S.F., Faine L.A., Ribar B.O., Burneiko R.C. and Novelli E.L. 2002. The use of the oxidative stress responses as biomarkers in Nile tilapia (*Oreochromis niloticus*) exposed to *in vivo* cadmium contamination. Environ. Int. 27 (8): 673-679.

Antonella V. and Landriscina C., 1999. Changes in liver enzyme activity in the teleost *Sparus auratain* responses to cadmium intoxication. Ecotoxicol. Environ. Safety, 43: 111-116.

Asztalos B. and Nemcsok J., 1985. Effect of pesticides on the LDH activity and isozyme pattern of carp (*Cyprinus carpio* L.) sera. Comp. Biochem. Physiol., C. 82(1): 217-219.

Atef M. Al-Attar, 2007. The Influences of nickel exposure on selected physiological parameters and gill structure in the teleost fish, *Oreochromis niloticus*. J. Biol. Sci., 7: 77-85.

Atif F., Pervez S., Pandet S., Ali M. and Kaur M., 2005. Modulatory effect of cadmium exposure on deltamethrin-induced oxidative stress in *Channa punctatus*. Arch. Environ. Contam. Toxicol., 49; 371-377.

Atli G. and Canli M., 2007. Enzymatic responses to metal exposures in a freshwater fish, *Oreochromis niloticus*. Comp. Biochem. Physiol., 145: 282-287.

Babu Velmurugan, Selvanayagam Mariadoss, Cengiz Elif I. and Unlu Erhan, 2007. The effects of fenvalerate on different tissues of freshwater fish *Cirrhinus mrigala*. J. Environ. Sci. Hlth. Part-B, 42, 157-163

Baby Shaikila, I., Thangavel, P. and Ramaswamy, M., 1993. Adaptive trends in tissue acid and alkaline phosphatases of *Sarotherodon mossambicus* (Peters) under sevin toxicity. Indian J. Environ. Health, 35 (1): 36-39.

Baldwin D.H., Sandahl J.F., Labenia J.S. and Scholz N.I., 2003. Sublethal effects of copper on coho salmon: impacts on nonoverlapping receptor pathways in the peripheral olfactory nervous system. Environ. Toxicol. Chem., 22: 2266-2274.

Balgir R.S., 2007. Genetic burden of red cell enzyme glu-6-phosphate dehydrogenase in two major SC tribe of Northwestern Orissa, India. Curr. Sci., 92: 768-774.

Benson W.H., Baer K.N. and others, 1987. Influence of cadmium exposure on selected hematological parameters in freshwater teleost, *Notemigous chrysoleucas*. Ecotoxicol. Environ. Saf., 13: 92-96.

Bersenyi A., Gy Fekete S., Szilagyi M., Berta E., Zoldag L. and Glavits R., 2004. Effects of nickel supply on the fattening performance and several biochemical parameters of broiler chickens and rabbits. Acta. Vet. Hungarica, 52: 185-197.

Bhatnagar M.C. and Bana A.K., 1993. Pesticides induced histophysiological alterations in liver of *Channa gachua*. Proc. Acad. Environ. Biol., 2: 115-118.

Brafields A.E. and Koodie A.V., 1994. The effect of high dietary zinc on trypsin activity in carp, *Cyprinus carpio*. J. Fish Biol., 45(1): 169-172.

Brambila E.M., Achanzar W.E., Qu W., Webber M.M. and Waalkes M.P., 2002. Chronic arsenic-exposed human prostate epithelial cells exhibit stable arsenic tolerance: mechanistic implications of altered cellular glutathione and glutathione S-transferase. Toxicol. Appl. Pharmacol., 183: 99–107.

Brown G.W., 1976. Biochemical aspects of detoxification in the marine environment. In: Malline D C. and Sargent J R., Eds. Biochemical and physiological; perspectives in marine biology, vol 3. Acad. Press. NY., pp 319-406.

Calabrese Edward J. and Moore Gary S., 1981. Does cadmium exposure reduce the metabolism of ethanol? Medical Hypotheses, 7: 703-706.

Campbell E.J., Dickinson C.J. and others.1984. Clinical physiology. Bulter and Tanner Ltd. London.

Christensen G.M., 1971. Effects of metal cations and other chemicals upon the *in vitro* activity of two enzymes in the blood plasma of the white sucker, *Catostomus commersoni*. Chem. Biol. Interact., 4: 351-359.

Christensen G.M., 1975. Biochemical effects of methyl mercuric chloride, cadmium chloride and lead nitrate of embryos and elevins of the brook trout, *Salvelinus fontinalis*. Toxicol. Appl. Pharmacol., 32: 191-197.

Christensen G.M., Oslon D. and Reidel B., 1982. Chemical effects on the activity of eight enzymes: a review and discussion relevant to environmental monitoring. Environ. Res., 29: 247-255.

Cohen S.M., Arnold L.L., Uzvolgyi E., Cano M., St. John M., Yamamoto S., Lu X., Le X.C., 2002. Possible role of dimethylarsinous acid in dimethylarsinic acid-induced urothelial toxicity and regeneration in the rat. Chem. Res. Toxicol., 15: 1150- 1157.

Cyriac P.J., 1990. Toxiological studies on *Oreochromis mossambicus*. PhD thesis, Div. Marine Biol., Microbiol. and Biohem., School Mar. Sci., Cochin Univ. Sci. Technol. Cochin.

Dai Wei, Fu Linglin, Du Huahua, Jin Chengguan and Xu Zirong, 2009. Changes in growth performance, metabolic enzyme activities, and content of Fe, Cu, and Zn in liver and kidney of tilapia (*Oreochromis niloticus*) exposed to dietary Pb. Biol. Trace Elem. Res., 128(2):176-183.

Dalela R.C., Saroj R. and Verma S.R., 1982. Acid phosphatase activity in tissue of *Notopterus notopterus* chronically exposed to phenolic compounds. Proc. Indian Acad. Sci., 91(1): 7-12.

David M., Mushigeri S.B., Shivakumar R. and Philip G.H., 2004. Response of *Cyprinus carpio* (Linn.) to sublethal concentration of cypermethrin: alterations in protein metabolic profiles. Chemosphere, 56 (4): 347-352.

David W.M. Jr., 1985. The chemistry of respiration: Harper's review of biochemistry, 20th Edn. Lange Medical Publications, Maruzen Co. Ltd., California, p 618.

De La Torre F.R., Salivian A. and Ferrari L., 1998. Enzyme activities as biomarkers in freshwater pollution: response of fish branchial (Na-K) ATPase and liver transminases. Environ. Toxicol., 14: 313- 319.

De la Torre F.R., Salibian A. and Ferrari L., 2000. Biomarkers assessment in juvenile *Cyprinus carpio* exposed to waterborne cadmium. Environ. Poll., 109: 277-282.

De-Smet-Hans and Blust-Ronny, 2001. Stress responses and changes in protein metabolism in carp, *Cyprinus carpio* during cadmium exposure. Ecotoxicol. Environ. Saf., 48(3): 255-262.

Dimitrova M.S., Tishinova T. and Velcheva V., 1994. Combined effects of zinc and lead on the hepatic superoxide dismutase-catalase system in carp. Comp. Biochem. Physiol., 108C: 43-46.

Dinodia G.S., Gupta R.K., Jain K.L. and Yadava N.K., 2003. Cadmium toxicity and its effects on the proteolytic enzyme activity in some freshwater carps. Proc. 3rd Interact. Worksh. Fish Prod. Using Brackish Water in Arid Ecosystem (Eds. S.K. Garg and A.T. Arasu): 195-199.

Dinodia G.S., 2001. Studies on the effects of cadmium toxicity in some freshwater fishes. M.Sc. Thesis, Depart. Zool. and Aqua., CCS HAU, Hisar.

Dixon D.G., Hodson P.V. and Kaiser K.L., 1987. Serum sorbitol dehydrogenase activity as an indicator of chemically induced liver damage in rainbow trout. Environ. Toxicol. Chem., 6: 685.

Dooley J.F., Turnquist L.J., and Racich L., 1979. Kinetic determination of serum sorbitol dehydrogenase activity with a centrifugal analyser. Jhingeran Clin. Chem., 25 (12): 2026-2029.

Dubale M.S. and Awasthi M., 1982. Biochemical changes in liver and kidney of a catfish, *Heteroneustes fossilis* exposed to dimethoate. Comp. Physiol. Ecol., 7 (2): 111-114.

Dubale M.S. and Shah P., 1981. Biochemical alteration induced by cadmium in the liver of *Channa punctatus* (Bloch). Environ. Res., 26: 110-118.

Durve V.S., Gupta P.K. and Khangarat B.S., 1980. Toxicity of copper to the freshwater teleost, *Rasbora daniconius neilgerieriensis* (Ham.). Nat. Acad. Sci. Lett., 3: 221-223.

Earnshow M J., Wilson S., Akaberly H B., Butler R D. and Marriott K.R.M., 1986. The action of heavy metals on the gametes of the marine mussel, *Mytilus edulis*. III- The effect of applied copper and zinc on sperm motility in relation to ultrastructure damage and intracellular localization. Mar. Environ Res., 20: 261-278.

Eichhorn G.L., 1975. Active sites of biological macromolecules and other interactions with heavy metals. In "Ecological Toxicol Research: Effect of heavy metals and organo-halogen compounds" (A.D. McIntyre and C.F. Mills, Eds.), Vol. V. Plenum, New York.

Emad H., Abou El-Naga, Khalid M., El-Moselhy and Mohamed A. Hamed., 2005. Toxicity of cadmium and copper and their effect on some biochemical parameters of marine fish, *Mugil sehel*. Egyptian J. Aqua. Res., 31: 60-71.

Engstrom L., 1964. Studies on bovine liver alkaline phosphatase phosphate incorporation. Biochem. Biophys. Acta., 92 : 71.

Escher M., Wahli T., Buttner S., Meier W. and Burkhardt-Holm P., 1999. The effect of sewage plant effluent on brown trout (*Salmo trutta fario*): a cage experiment. Aquat. Sci., 61: 93-110.

Eze L.C., 1983. Isonlorid inhibition of liver glutamate oxaloacetic transaminase from goat (*Carpa hercus*). Internl. J. Biochem., 15: 13-16.

Rossilkh A.R., Capula M., Crosetti D., Campton D.E. and Sola, L., 1998. Genetic divergence and phylogenetic inferences in five species of Mugilidae (Pisces: Perciformes). Marine Biology, 131: 213 – 218.

Friberg L., Kjellstorm T., Elinder C G. and Norberg G. F., 1985. Cadmium and health: A toxicological and epidemiological appraisal, Vol I (CRC prass, Boca Raton, Florida), pp 248.

Galhoom K.L., Laila G., Rizk G. and El-ezzaway M.H., 2000. Some biochemical and hematological parameters in Mugil fish (*Mugil caphalcus*) reared in Bahr El-Bakar drain. Egypt. J. Agril. Res., 78: 1-13.

Gargiulo G., P. De Girolamo, L. Ferrara, D. Soppelsa, A. Andereozzig and P. Battaglini., 1996. Action of cadmium on the gills of *Carassius auratus* (L.) in the presence of catabolic NH3. Arch. Environ. Contam. Toxicol., 30: 235-240

George K.Roy, Malini N.A., Rajan Archana and Deepa R., 2011. Enzymatic changes in the kidney and brain of freshwater murrel, *Channa striatus* (Bloch) on short term exposure to sub-lethal concentrations of lead nitrate. Ind. J. Fish., 58(4): 91-94.

Gill T.S., Tewari H. and Pande J., 1990. Use of the fish enzyme system in monitoring water quality: effects of mercury on tissue enzymes. Comp. Biochem. Physiol.C., 97(2): 287-92.

Gill T.S., Tewari H. and Pande J., 1991. *In vivo* and *in vitro* effect of cadmium on selected enzymes in different organs of the fish, *Barbus conchorinus* Ham. (Rosy barb). Comp. Biochem. Physiol. 100C: 501.

Gill T.S., Tewari H. and Pande J., 1992. Short and long-term effects of copper on the rosy barb (Puntius *conchonius*). Ecotoxicol. Environ. Safty., 23: 294.

Goel K.A. and Sastry K.V., 1973. Distribution of alkaline phosphatase in the digestive system of a few teleost fishes. Acta. Histochem. Bd., 47: 8-14.

Goldfischer S., Essner E. and Novikoff A.B., 1964. The localization of phosphatase activities at level of ultrastructure. J. Histochem. Cytochem., 12 : 72-95.

Golovanova I.L., Kuzmina V.V., Gobzhelian T.E., Davlov D.F. and Chiko G.M., 1999. In votro effects of cadmium and DDVP dichlorovos on intestinal carbohydrase and protease activities in freshwater teleosts. Comp. Biochem. Physiol., 122(1): 21-25.

Gradecka D., Palus J., and Wasowicz W., 2001. Selected mechanisms of genotoxic effects of inorganic arsenic compounds. Int. J. Occup. Med. Environ. Health, 14: 317–328

Gratzek John B., 1993. Fish anatomy, physiology, and nutrition. Tetra Press, Pets: 95 pp.

Gupta, A.K. and Rajbanshi V.K., 1986. Cytotoxicity of copper ions to certain tissues of a freshwater murrel, *Channa punctatus* (Bloch). Poll. Res., 5(3 and 4): 97-101.

Gupta A.K. and Rajbanshi V.K., 1988. Acute toxicity to cadmium to *Channa punctatus* (Bloch). Acta. Hydrochim. Hydrobiol., 16(5): 525-535.

Hans De- Smet and Ronny B., 2001. Stress responses and changes in protein metabolism in carp, *Cyprinus carpio* during cadmium exposure. Ecotoxicol. Environ. Safty. 48 (3): 255-262.

Heath A.G., 1987. Water pollution and fish physiology. CRC Press, Inc., Florida, 245 pp.

Heath A.G., 1995. Water pollution and Fish physiology. Second Ed. Lewis Publishers: 359 pp.

Hidalgo M.C., Urea E. and Sanz S., 1999. Comparative study of digestive enzymes in fish with different nutrition habits, proteolytic and amylase activity. Aquaculture, 170(3-4): 267-283.

Hilmy A., Domiaty N., Daabees A. and Latife H., 1987. Some physiological and biochemical indices of zinc toxicity in two freshwater fishes, *Clarias lazera* and *Tilapia zilli*. Comp. Biochem. Physiol., 87: 297-391.

Hilmy A.M., Shabana M.B. and Daabees A.Y.,1985. Effects of cadmium toxicity upon the *in vivo* and *in vitro* activity of proteins and five enzymes in blood serum and tissue homogenates of *Mugil cephalus*. Comp. Biochem. Physiol., 81C: 145-153.

Hilmy A.M., Shabana M.B. and Said M.M. 1981. The role of serum transminases (SGOT and SGPT) and alkaline phosphatase in relation to inorganic phosphorus with respect to mercury poisoning in *Aphanius dispar* (Teleostei) of the red sea. Comp. Biochem. Physiol., 68C: 69-74.

Hinton D.E., Kendall M.W. and Silver B.B., 1973. Use of histologic and histochemical assessment in the prognosis of the effects of aquatic pollutant : Biological methods for assessment of water quality. Am. Soc. Test. Mat., 528: 194-208.

Hoar W.S. and Randall D.J. (Eds), 1978. Fish Physiology, vol. VII. Academic Press, New York, NY., 302pp.

Hodson P.V., 1988. The effect of metal metabolism on uptake, disposition and toxicity in fish. Aquat Toxicol., 11: 3-18.

Hodson P.V., Whittle D.M., Wong P.T.S., Borgmann U., Thomas R.L., Chau Y.K., Nriagu J.O. and Hallet D.J., 1984. Lead contamination of the Great Lakes and its potential effects on aquatic biota. In: Toxic contaminants in the Great Lakes. Nriagu J.O. and Simmons M.S. (Eds.), John Wiley and Sons, Indianapolis.

Hontella A., Daniel C. and Ricard A., 1996. Effects of acute and sub acute exposures of cadmium on the internal and thyroid functions in rainbow trout, *O. mykiss*. Aquat. Toxicol., 35 (3-4): 171-182.

Hota AK, Pradhan AK., 1994. Mercuric chloride toxicity on a freshwater fish, *Channa punctatus*: effect on liver phosphatases activity. Pro. Acad. Environ. Bio., 3 (1): 87-91

Hu Y., Su L. and Snow I.T., 1998. Arsenic toxicity is enzyme specific and arsenic inhibition of DNA repair is not caused by direct inhibition of repair enzymes. Mutat. Res., 408: 203-218.

Huggett R.J., Kimerle R.A., Mehrle P.M. and Bergman H.L., 1992. In; Biomarkers: Biochemical, physiological and histological markers of anthripogenic stress. Lewis Publ., Boca Raton, FL. 347pp.

Humtsoe N., Davoodi R., Kulkarni B.G. and Chavan B., 2007. Effect of arsenic on the enzymes of the Rohu carp, *Labeo rohita* (Hamilton, 1822). Raff. Bull. Zool., 14: 17-19.

Hussein A. Kaoud, Khaled M.A. Mahran, Ahmed Rezk and Mahmoud A. Khalf., 2012. Bioremediation the toxic effect of mercury on liver histopathology, some hematological parameters and enzymatic activity in Nile tilapia, *Oreochromis niloticus*. Researcher, 4(1): 60-69.

Jackim E., Hamilin J.M. and Sonis A. 1970. Effects of metal poisoning on five liver enzymes in the killfish (*Fundulus heteroclitus*). J. Fish Res Bd. Can. 27: 383.

Jacobs J.M., Carmicheal N. and Cavanagh J.B., 1977. Ultra structural changes in the nervous system of rabbits poisoned with methyl mercury. Toxicol. Appl. Pharmacol., 39: 249-261.

James R.V., Jancy Pattu, Devakiamma G. and Sampath K., 1991. Impact of sublethal levels of mercury on glycogen and selected respiratory enzymes in *Heteropneustes fossilis* and role of water hydracinth in reduction of Hg toxicity. Ind. J. Fish, 30: 249-252.

James R., Sampath K. and Alagurathinam S., 1996. Effects of lead on the respiratory enzyme activity, glycogen and blood sugar levels of the teleost *Oreochromis mossambicus* (Peters) during accumulation depuration. Asian Fish Sci., 9: 87-100.

James R., Sampath K. and Ponmani K., 1992a. Effect of metal mixtures on activity of two respiratory enzymes and their recovery in *Oreochromis mossambicus* (Peters). Ind. J. Exp. Biol., 30: 496-499.

Janska H., Kubicz A. and Bem M., 1986. Catfish liver acid phosphatase enzyme molecules with altered kinetic properties. Comp. Biochem. Physiol., 85B: 751-758.

Jat D. and Kothari S., 2006. Distribution of zinc in the gills of experimentally exposed catfish: Histological and Histochemical study. Asian J. Exp. Sci., 20: 165-170.

Jeelani S. and Shaffi S.A., 1989. Biochemical compartmentation of fish tissues chronic toxicity of mercuric nitrate on visceral phosphomonoesterases in *Channa punctatus* (Bloch). Acta Physiol. Hung., 73: 477-482.

Jiminez B.D. and Stegeman J.J. 1990. Detoxication enzymes as indicators of environment stress on fish. Am. Fish Soc. Symp., 8: 67-79.

Karan V., Vitorovic S., Tutundzic V. and Poleksic V., 1998. Functional enzymes activity and gill histology of carp after copper sulfate exposure and recovery. Exotoxicol. Environ. Saf., 40(1-2): 49-55.

Kew M.C., 2000. Serum aminotranferase concentration as evidence of hepatocellular damages. Lancet, 355: 591-592.

Khare P., 1993. Endosulphan, rogar, mercuric chloride and lead nitrate induced histopathological and histochemical changes with stomach and intestine of *Mystus cavasius*. Ph.D. Thesis, Barktullah University, Bhopal.

Khillar V.K. and Wagh S.B., 1988. Acute toxicity of pesticides in the freshwater fish *Barbus stigma*. Histopathology of the stomach. Uttar Pradesh J. Zool., 8: 176-179.

Kim S.G., Kang J.C., 2004. Effect of dietary copper exposure on accumulation, growth and hematological parameters of the juvenile rockfish, *Sebastes schlegeli*. Mar. Environ. Res., 58: 65-82.

Kohler A., 1991. Lysosomal perturbations in fish liver as indicators for toxic effects of environmental pollution. Comp. Biochem. Physiol., 100C: 123.

Kothari S., Bhalerao S. and Jat Deepali, 2003. Protection by herbal compound against toxic effects of mercuric chloride in fish intestine. J. Ecophysiol. Occop. Hlth., 3: 51-58.

Kothari S. and Soni R., 2004a. Role of polyunsaturated fatty acids against zinc induced disturbances in *Heteropneustes fossilis* intestine. Ind. J. Environ Toxicol., 2: 40-42.

Kothari S. and Soni R., 2004b. Role of polyunsaturated fatty acids against zinc sulphate toxicity in the gills of catfish, *Heteropneustes fossilis* (Bloch). J. Tissue Res., 4: 253-256.

Kotoraman M., Laszla K., Nemcsok J. and Simon L., 2000. Effects of Cd^{2+}, Cu^{2+}, Pb^{2+} and Zn^{2+} on activities of some digestive enzymes in carp (*Cyprinus carpio*). J. Environ. Sci. Health., A35 (9): 1517-1526.

Kristoffersson R., Broberg S., Oikari A. and Pekkarinen M., 1974. Effect of a sub lethal concentration of phenol on some blood plasma enzume activity in the pike (*Esox lusius*) in brackish water. Ann. Zool. Fennici., 11: 220.

Kumari S., 1984. Studies on toxicity of heavy metals to *Channa punctatus* with special reference to nickel and chromium. Ph.D. Thesis, Merrut Univesity, Meerut.

Kuzumina V.V., 1996. Influence of age on digestive enzyme activity in some freshwater teleosts. Aquaculture, 148: 25-37.

Kuzumina V.V., Golovanova I.L., Shishin M.M. and Smirnova E.S., 2004. Influence of heavy metals on feeding behaviour and digestive process. The sixth Syrian, Egyptian conference on the chemical and petroleum engineering. Al Baath University Homes, Syria.

Kwan C.Y., Belbeck L. and Daniel E.E., 1979. Abnormal biochemistry of vascular smooth muscle membrane as an important factor in the initiation and maintenance of hypertension in rats. Blood vessels 56; 259-268.

Kwan C.Y. Ito H., 1987. Comparative studies of acid and alkaline phosphatase activities in nonvascular smooth muscle membranes. Comp. Biochem. Physiol., 86B: 482-488.

Landis W.G. and Yu M., 1995. Introduction to environmental Toxicology: impacts of chemicals upon ecological systems. Lewis Publishers, Boca Raton. 328 pp.

Larsson A., Haux C. and Sjobeck M., 1985. Fish physiology and metal pollution: results and experiences from laboratory and field studies. Ecotoxicol. Environ. Safety., 9: 250.

Leake D.S., Heald B. and Peters T. J., 1982. Properties and subcellulr localization of acid phosphosphatase activity in cultured arterial smooth muscle cells. Eur. J. Biochem., 128: 557-563.

Li-Fern H and Rajasoorya C., 1999. The elevated serum alkaline phosphatase–the chase that led to two endocrinopathies and one possible unifying diagnosis. Eur. J. Endocrinol., 140 (2): 143–147.

Lindhardt K. and Walter K., 1965. Phosphatases (Phosphomonoesterases) determination serum with p-nitrophenylphosphate. In: Metheds of enzyme Analysis. Bergemeyer (Ed.) H. U. Verlag Chemie. Acad. Press. London: 783-785.

Liu S.X., Athar M., Lippai I., Waldren C. and Hei T.K., 2001. Induction of oxyradicals by arsenic: Implication for mechanism of genotoxicity. Proc. Natl. Acad. Sci. USA, 98: 1643-1648.

Mathan R., 2006. Studies on the impact of heavy metal cadmium on certain enzymes in a freshwater teleost fish, *Cyprinus carpio*. Toxicol. Lett., 164 :157.

Mathis J.C., 1958. Preparation and properties of highly purified alkaline phosphatase from swine kidney. J. Biol. Chem., 233: 1121-1127.

Martin Deva, Prasath, P. and Arivoli S., 2008. Biochemical study of freshwater fish, *Catla catla* with reference to mercury chloride, Iran. J. Environ. Health. Sci. Eng., 5(2):109-116.

Mayes P.A., 1985. Metabolism of carbohydrates. In: Harper's review of biochemistry, 20th Edn. Lange Medical Publications, Maruzen Co. Ltd. California.

McComb R.B., Bowers G.N. and Posen S., 1979. Alkaline phosphatases. New York: Plenum.

McKim J.M., Christensen G.M. and Hunt E.P., 1970. Changes in the blood of brook trout (*Salvelinus fontinalis*) after short-term and long-term exposure to copper. J. Fish. Res. Bd. Canada, 27: 1883-1889.

Mehrotra A., 1996. Studies on the toxic effects of pesticides and heavy metals on the intestine of *Nandus nandus* (Ham.). Ph.D. Thesis, Baraktullah University, Bhopal.

Mizrahi L. and Achituv Y., 1989. Effect of heavy metal ions on the enzyme activity in the Mediterranean mussel, *Donax trunculus*. Bull. Environ. Contam. Toxicol., 42: 854-859.

Moneserrate A.J., Hamilton F., Ghoshal A.K., Porta E.A. and Hartroft W.S., 1972. Lysosomes in the pathogenesis of the renal necrosis of choline deficient rats. Am. J. Pathol., 68: 113-146.

Moore M.N., 1985. Cellular eponses to pollutants. Mar. Pollut. Bull., 16: 134.

Mukhopadhyay P.K. and Hajra A., 1986. Intestinal protease activity and liver protein synthesis in *Clarias batrachus* fed isonitrogenous diets with variable energy level. Sci. and Cult., 52: 230-233.

Naidu K.A., Abhinender K., Ramamurthi R., 1984. Acute effect of mercury toxicity on some enzymes in liver of teleost, *Sarotherodon mossambicus*. Ecotoxicol. Environ. Safety., 8: 215 – 218.

Nateson R., 1985. Effect of sevin on protein level of different tissues of *Sarotherodon nossambicus* (Peters). M.Sc. Thesis, Bharathiar University, Coimbatore, India.

Neha-Bhatkar, Vankhede, G.N. and Dhande, R.R., 2004. Heavy metal induced biochemical alterations in freshwater fish, *Labeo rohita*. J. Ecotoxicol. Environ. Monit., 14(1): 1-7.

Neskovic N.K., Poleksic V., Elezovic I., Karan V. and Budimir M., 1996. Biochemical and histopathological effects of glyphosate on carp, *Cyprinus carpio* L. Bull. Environ. Contam. Toxicol., 56; 295-302.

Nevalennyi A.N.; Bednyakov D.A., 2004. Influence of Cadmium Ions on the activities of enzymes responsible for membrane digestion in carp. Russian J. Ecol., 35(2): 128-130.

Norman A.W., Mircheff A.K., Adams T.H. and Pielvogel A.S., 1975. Effect of cadmium *in vivo* and *in vitro*, on intestinal brush border AlPase and ATPase. Bull. Environ. Contam. Toxicol., 14: 653-656.

Novikoff A.B., 1961. In: The Cell (J. Brachet and A.E. Mirsky, Eds.). 423 p. Academic Press, New York.

NRC, 1999. National Research Council, Arsenic in drinking water. Washington, DC, National Acad. Press.

NRC, 2001. National Research Council, Arsenic in drinking water 2001 update. Washington, DC, National Acad. Press.

Ozretic B. and Krajnovic-Ozretic M., 1993. Plasma sorbitol dehydrogenase, glutamate dehydrogenase and alkaline phosphate as potential indicators of liver intoxication in gray mullet (*Mugil auratus*). Bull. Environ. Contam. Toxicol., 50: 586-592.

Parvathi Kumar, Palanivel Sivakumar, Mathan Ramesh and Sarasu. 2011. Sublethal effects of chromium on some biochemical profiles of the freshwater teleost, *Cyprinus carpio*. J. Appl. Biol. and Pharm. Tech. 2: 294-300.

Passow H., Rothstein A. and Clarkson T.W., 1961. The general pharmacology of heavy metals. Pharmacol. Rev., 13: 185-224.

Pearse A.G.E., 1961. Histochemistry, theoratical and applied, Churchill, London.

Radhakrishnaiah K., Suresh A. and Sivaramakrishna B., 1993. Effect of sublethal concentrations of mercury and zinc on the energetics of a freshwater fish, *Cyprinus carpio*. Acta Biological Hungarica, 44, 375-385.

Ramalingam K, Indra D., 2002. Copper sulphate ($CuSO_4$) toxicity on tissue phosphatases activity and cabohydrates turnover in *Achatina fulica*. J Environ Bio., 23(2): 181-188

Ramesh M., Saravanan M. and Kavitha C., 2009. Hormonal responses of the fish, *Cyprio carpio* to environmental lead exposure. African J. Biotechnol., 8(17): 4154-4158.

Rani A Shobha, Sudharsan R., Reddy T.N., Reddy P.U. and Raju T.N. 2001. Effect of arsenite on certain aspects of protein metabolism in freshwater teleost, *Tilapia mossambica* (Peters). J Environ Biol., 22 (2):101-4.

Rani U.A. and Ramamurthi R., 1987. Effect of sublethal concentration of cadmium on oxidative metabolism in the freshwater teleost, *Tilapia mossambia*. Ind. J. Comp. Anim. Physiol., 5: 71-74.

Roberts K.S., Cryer A., Kay J., Soble J.F., Wharfe J.R. and Simpson W.R., 1979. The effects of exposure to sublethal concentration of cadmium on enzyme activities and accumulation of the metal in tissue and organs of rainbow and brown trout (*Salmo gairdneri and Salmo katta*). Comp. Biochem. Physiol., 62C: 135-147.

Reddy A.S., Reddy M.V. and Radhakrishnaiah K., 2008. Impact of copper on the oxidative metabolism of the fry of common carp, *Cyprinus carpio* (l.) at different pH. J. Environ. Biol., 29(5): 721-724.

Regnault M. and Batrel Y. 1987. Glutamate dehydrogenase of the shrimp Cragon cragon; effects of shrimp weight and season upon its axctivity in the oxidative and reductive functions. Comp. biochem. Physiol. 86B: 525-530.

Rema L.P. and Babu P. 1999. The effect of heavy metals on some metabolpcally important enzymes in *Oreochromic mossambicus*. Fishery Technology. 36 (1) : 8-12.

Roy P. and Saha A., 2002. Metabolism and toxicity of arsenic: A human carcinogen. Curr. Sci., 82: 38- 45.

Santhana V. and Azariah J., 2003. Effect of copper chloride on the enzyme activities of the crab, *Sesarma quadratum*. Turk. J. Zoo., 27: 253-256.

Sarasus C. and Andal S., 2005. Effects of distillery effluent on tissue phosphatase activity of freshwater teleost, *Hypophthalmichthys molitrix* (Val.) and *Catla catla* (Ham.). In: Fish Biology (Ed. A. Kumar) A.P.H. Publ. Corp., pp.145-149

Sarita, 2005. Effects of heavy metals on enzyme activity in some freshwater fishes. PhD thesis, Deptt. Zoology., CCS HAU., Hisar, Haryana: 195 + xxii.

Sastry K.V. and Gupta P.K., 1980. Changes in the activities of some digestive enzymes of *Channa punctatus*, exposed chronically to mercuric chloride. J. Environ. Sci. and Helth, Part B: Pesticides, Food Contaminants, and Agricultural Wastes, 15 (1): 109-119.

Sastry K.V. and Gupta P.K. 1978a. Effect of mercuric chloride on the digestive system of *Channa punctatus*. A histopathological study. Environ. Res., 16: 270-278.

Sastry K.V. and Gupta P.K. 1978b. Alterations in the activity of some digestive enzyme of *Channa punctatus* exposed to lead nitrate. Bull. Environ. Contam. Toxicol., 19: 549-556.

Sastry K.V. and Gupta P.K. 1978c. Effect of mercuric chloride on the digestive system of *Channa punctatus*. Bull. Environ. Contam. Toxicol., 20: 353-360.

Sastry K.V. and Gupta P.K. 1979b. The effect of cadmium on the digestive system of the toteost fish, *Heteropneustes fossilis*. Environ. Res., 19: 221-230.

Sastry K.V. and Rao D.R., 1981. Enzymological and biochemical changes produced by mercuric chloride in a teleost fish, *Channa punctatus*. Toxicol. Lett., 9 (4): 321-328.

Sastry K.V. and Rao D.R., 1984. Effects of mercuric chloride on some biochemical and physiological parameters of the freshwater murrel, *Channa punctatus*. Environ. Res., 34: 343 – 350.

Sastry K.V. and Sharma K., 1980. Effects of mercuric chloride on the activities of brain enzymes in a freshwater teleost, *Ophiocephalus* (*Channa*) *punctatus*. Arch. Environ. Contam. and Toxicol., 9 (4): 425-430.

Sastry K.V. and Shukla, V. 1990a. Effects of cadmium on the intestinal absorption of some nutrients in the teleost fish, *Channa punctatus*. Pollut. Res. 9:141-148.

Sastry K.V. and Shukla V., 1990b. Toxic effects of cadmium on some biochemical and physiological parameters in the teleost fish, *Channa punctatus*. Biojournal, 2: 325-332.

Sastry K.V. and Shukla V., 1993. Acute and chronic effects of cadmium on the rate of oxygen uptake and metabolism of fish. J. Ecobiol., 5; 295-298.

Sastry K.V. and Shukla V., 1994. Acute and chronic effects of cadmium on some haematological, biochemical and enzymological parameters in the in the freshwater teleost fish, *Channa punctatus*. Acta hydrichim hydrbiol. 22: 171-176.

Sastry K.V. and Subhadra K., 1982. Effect of cadmium on some aspects of carbohydrate metabolism in a freshwater catfish, *Heteropneustes fossilis*. Toxicol. Lett., 14(1-2): 45-55.

Sastry K.V. and Subhadra K.M. 1985. *In vivo* effects of cadmium on some enzyme activities in tissues of the freshwater catfish, *Heteropneustes fossilis*. Environ. Res., 36: 32-45.

Sastry K.V. and Sunita, 1983. Alterations in the intestinal absorption of xylose induced by heavy metals in a freshwater teleost fish, *Channa punctatus*. Poll. Res., 2: 48.

Scott G.R., Sloman K.A., Rouleau C. and Wood C.M., 2003. Cadmium distrupts behaviural and physiological responses to alarm substances in juvenile rainbow trout, *Onhorhynchus mykiss*. J. Exp. Biol., 206: 1779-1790.

Shaffi S.A., 1993. Comparison of the sublethal effect of mercury and lead on viscerol dehydrogenase system in three Indian teleost. Physiol. Res., 47: 7.

Shakoori A.R., Alam J. and others, 1990. Biochemical effects of bifenthrin (talsar) administered orally for one month on the blood and liver of rabbit. Proc. Pak. Congr. Zool., 10: 61-81.

Singh H.S. and Reddy T.V., 1990. Effect of copper sulphate on haematology, blood chemistry and hepatosomatic index of an Indian catfish, *Heteropneustes fossilis* (Bloch) and its recovery. Ecotoxicol. Environ. Saf. 20: 30-35.

Srivastava D. and Pandey K., 1982. Effect of copper on tissue acid and alkaline phosphatases in the green snakehead, *Ophiocephalus punctatus* (Bloch). Toxicol. Lett., II: 237.

Stigbrand T. and Fishman W.H., 1984. Human alkaline phosphates. Prog. Clin. Res., 166: 1-361.

Styblo M., Del Razo L.M., Vega L., Germolec D.R. and others, 2004. Comparative toxicity of trivalent and pentavalent inorganic and methylated arsenicals in rat and human cells. Arch. Toxicol., 74: 289-299.

Styblo M., Drobna Z., Jaspers I., Lin S. and Thomas D.J., 2002. The role of biomethylation in toxicity and carcinogenicity of arsenic: a research update. Environ. Health Persp., 110(suppl. 5): 767-771.

Sugawara C. and Sugawara N., 1974. Cadmium toxicity for rat intestine, especially on the absorption of calcium and phosphorus. Jap. J. Hyg., 28: 511.

Tafanelli R. and Summerfelt R.C., 1975. Cadmium induced histopathological changes in goldfish. In Pathology of fishes (Eds., W.E. Rebelin and G. Migaki). Univ Wisconsin Press: 613-645.

Tietz N.W., 1987. Fundamentals of clinical chemistry. Saunders WB Co.

Tilak K.S., Veeraiah K. and Koteswara Rao D. 2005. Biochemical changes induced by chloropyrifos, an organophosphate compound in sublethal concentrations to the freshwater fish, *Catla catla* (Ham.), *Labeo rohita* (*Hamilton*) and *Cirrhinus mrigala* (Ham). J. Environ. Biol., 26, 341-347.

Tseng C.H., 2004. The potential biological mechanisms of arsenic-induced diabetes mellitus. Toxicol. Appl. Pharma., 197: 67-83.

Valarmathi S. and Azariah J., 2003. Effect of copper chloride on the enzyme activities of the crab, *Sesarma quadratum* (F.). Turk. J. Zool., 27: 253-256.

Verma H. and Srivastava N., 2010. Hanges in certain enzymes of the ovary and liver in *Channa punctatus*. E. J. Ichthyology, 6: 1-8.

Verma H. and Srivastava N., 2008a. Zinc induced architectural alterations and accumulation in the ovary of freshwater teleost, *Channa punctatus* (Bloch). Electronic J. Ichthyol., 2:85-91.

Verma H. and Srivastava N., 2008b. Effects of sublethal concentrations of zinc on bioaccumulation and architectural alterations in the liver of fish *Channa punctatus*. J. Environ. Sociobiol., 5: 135-140.

Verma S.R., Rani S. and Dalela R.C., 1980. Effect of phenol and dinitrophenol on acid and alkaline phosphates in tissues of a fish, *Notopterous notopterus*. Arch. Environ. Contam. Toxicol., 9: 451-459.

Versteeg D.J. and Giesy J.P. 1985. Lysosomal enzyme release in the bluegill sunfish (*Lepomis macrochirus*) exposed to cadmium. Arch. Environ. Contam. Toxicol., 14: 631.

Vijayendra Babu, K.V.K. and Vasudev, T. 1984. Effect of dimecron, rogor and cuman L on AchE, and phosphatase in freshwater mussel, *Lamellidens marginalis* (Lamarck). Curr. Sci., 53 (7): 935-936.

Visvikio, I. and Rachlin, J.W. 1994. Acute and chronic exposure of *Dunaliella salina* and *Chlamydomonas bullosa* to copper and cadmium: effects of ultrastructure. Arch. Environ. Contam. Toxicol. 26 (2): 154.

Wallimann T. and Hemmer W., 1994. Creatine kinase in non-muscle tissues and cells. Mol. and Cell. Biochem., 133–134: 193–220.

Wallimann T., Wyss M., Brdiczka D., Nicolay K. and Eppenberger H.M., 1992. Intracellular compartmentation, structure and function of creatine kinase isoenzymes in tissues with high and fluctuating energy demands: the 'phosphocreatine circuit' for cellular energy homeostasis. The Biochemical J., 281 (1): 21–40.

Watanabe M., Kihara T. and Shimon R., 1981. Electron microscopy of myothelial cellsof the mouse Harderian gland. Bull. Osaka Med. School, 27: 1-7.

Weiss P., Bogden J. and Enslee E., 1986. Mercury and copper induced hepatocellular changes in the mummichog, *Fundulus heteroclitus*. Environ. Health Prespect., 65: 167.

White D.J., 1977. Histochemical and histological effects of lead on the liver and kidney of the dog. Br. J. Exp. Pathol., 58 : 101-112.

Wieser W. and Hinterleitner S., 1980. Serum enzymes in ranbow trout as tools in the diagnosis of water quality. Bull. Environ. Contam. Toxicol., 25: 188.

Willett C.S. and Burton R.S., 2003. Characterization of the glutamate dehydrogenase gene and its regulation in a euryhaline copepod. Comp. Biochem. Physiol., B, 135: 639-649.

Wu Su Mei, Chen Chun Che, Lee Yi Chun, Leu Hsien Tai and Lin Nia Sung, 2006. Cortisol and copper induced metallothionein expression in three tissues of Tilapia, *Oreochromis mossambicus* in organ culture. Zool. Studies, 45: 363-370.

Yamauchi H. Fowler B.,A. 1994. Toxicity and metabolism of inorganic and methylated arsenicals. In: Nriagu, J.O. (Ed). Arsenic in the Environment., Part-II. Human Health and Ecosystem effects. John Wiley, NY: 35-43.

Yang J. and Chen H., 2003. Serum metabolic enzyme activities and hepatocyte ultrastructure of common carp after gallium exposure. Zool. Studies., 42: 455-461.

Yora T. and Sakagami Y., 1986. Comparative biochemical study of alkaline phosphatase isoenzymes in fish, amphibians, reptiles, birds, and mammals. Comp. Biochem. Physiol., 85B: 649-658.

Yoshitomi T., Koyanra J., Lida A., Okamoto N. and Ibeda Y., 1998. Cadmium induced scale deformation in carp, *Cyprinus carpio*. Bull. Environ. Contam. Toxicol., 60(4): 639-644.

Zikic R.V., Stajn A.S., Pavlovic S.Z., Ognjanovic B.I. and Saicic Z.S., 2001. Activities of superoxide dismutase and catalase in erythrocytes and plasma transaminases of gold fish, *Carassius auratus gibelio* (Bloch) exposed to cadmium. Physiol. Res., 50: 105-111.

Zonek J., Olkowski Z., and Jonderko G., 1966. Cytochemical studies on the behaviour of thiamine pyrophosphate, $NADH_2$ tetrazolium reductase and acid phosphatase in the cerebellum of rabbit chronically poisoned with magnanese. Int. Arch. Gewerbepathol. Gewerbehyg., 22: 1-9.

Chapter 13

Synthesis of Stress Proteins and Metallothionein

1. Definition and Scope of Stress Proteins

Proteins occupy a unique position in every living organism because of their major synthesis inside the nuclear cell wall, under the control of DNA molecules through the RNAs. Their primary role is mediating various metabolic pathways in a cell as enzymes (Harper, 2006; Lehninger, 2008). Since, the living organisms encounter both natural and anthropogenic stress, synthesis of such protein represents an early response in a cell to poisoning, injury or any other damage in a target tissue (Anson *et al.*, 1991). Total tissue protein estimates, in general are non-specific measure of the amount of all proteins present in a tissue, and is considered as a most common method used to examine changes in overall protein profile in the blood or any other tissue as an estimate of state of health of an organism including fishes (Chapter-9). The physiological and biochemical alterations observed in an animal under any environmental stress can be further correlated with structural and functional changes in cellular proteins. Likewise, differences occur in the ratio of a specific protein to total proteins in various tissues/organs or a body fluid, in response to a physiological change or a disease in an animal. This generalized response has been considered to be adaptive and it represents the natural capacity of an animal including fish to respond to stress. For example, saliva proteins well recognized as a mirror of the human health, has been in use in diagnostics for more than 2000 years. Likewise, in mammals 'C-Reactive Protein' and 'Serum Amyloid-A' protein have been reported to increase by more than 100-fold following an injury (Maclntyre *et al.*, 1985). On the contrary, proteins like serum albumin are said to decline during inflammation and are usually referred as negative phase reactants (Jameson *et al.*, 1983). Salivary diagnostic tests further provide an avenue to allow detection of malignancies at a sufficiently early stage so that a treatment is likely to be successful. Testing for HIV positive (Kamat, 1999) is one another such example of powerful use of saliva proteins in the diagnosis of infectious diseases. Salivary mRNA is further said to serve as a chemical signature that a particular gene has been expressed and differences in patterns of mRNA expression in saliva could indicate presence of developing oral squamous cell carcinoma (Sato *et al.*, 2010).

Such stress oriented proteins are collectively known as stress proteins, also called as Hsps (Heat shock proteins, Welch, 1993), since these stress proteins were first evidenced in *Drosophila*, as a result of its exposure to a fairly rapid increase in temperature *i.e.* their induction appeared enhanced following hyperthermia. The cell signaling mechanism involved is the initiation of heat-shock gene transcription, also considered sensitive to a variety of other similar physical and chemical stressors including the toxic metals as well as body conditions like hypoxia, ischemia/ reperfusion, carcinogens, mutagens and teratogens (Shelton *et al.*, 1986a,b; Hensen *et al.*, 1988; Nover, 1991). Both protein degradation and synthesis are sensitive over a wide range of conditions and show changes to a variety of physical and chemical modulators, as well as chemical intoxication. Tertiary changes in tissues/cells as explained earlier in chapter 9, in fact induce changes in intra-nuclear mechanisms for the synthesis of new proteins to compensate the negative influences of a stressor, such as decreased levels of proteins are reported in liver cells of fish *C. punctatus* on exposure to metals (Jana and Bhandopadhyay, 1987).

2. Hsp's Characteristics

Many workers have expressed these Hsps as a group of highly conserved and abundantly expressed cellular proteins as observed in all living organisms (Morimoto *et al.*, 1994; Feder and Hofmann, 1999), including fish (Iwama *et al.*, 1998). The stress proteins thus, are better defined as a group of cellular proteins whose synthesis increases when the cell is subjected to any sort of stress. Stress proteins being highly conserved in evolution are said to play a similar role in organisms from bacteria to humans. Iwama *et al.*, 2004 mentioned that a fish that may present a physiological response to a stressor, however, may not show any change in cellular 'Hsp' profile. In mammals, it is known that Hsps are involved in the immune response (Young, 1990; Breloer *et al.*, 2001). The Hsps are basically involved in protecting organisms from damage as a result of exposure to any stress including harmful condition and are known to be involved in protein homeostasis under normal conditions taking on a protective and a repair role upon exposure to any adverse condition (Huggett *et al.*, 1992). *In vivo* and *in vitro* studies have shown that various stressors transiently increase production of Hsps, as protection against harmful insults. The stress protein response appears to be a ubiquitous response found in all cells and tissues studied to date, the specific stress proteins induced are dependent on the toxicant type, the magnitude and the duration of exposure in an organism and the tissue studied (Goering and Fisher, 1995 and Sanders *et al.*, 1996).

Induction of stress protein synthesis is briefly characterized as:

i. The 'Hsps' response is highly conserved across widely divergent organisms, ranging from bacteria to humans (Schlesinger *et al.*, 1982), making thus, inter- and intra-species comparisons more meaningful.

ii. A rapid and a readily detectable mechanism that occur within 1 to 3 hr after insult (Blake *et al.*, 1990), as it is one of the early metabolic change in cells, following a chemical or a physical stress.

iii. It is now well understood that as the cells have evolved mechanisms to repair DNA, they have also evolved an epigenetic repair and recycling system under stress to maintain their molecular integrity by repairing or synthesizing new proteins through epigenetic repair system, *i.e.* rapid synthesis of stress proteins.

The stress-protein genes share highly conserved sequence homologies among widely divergent organisms and several of the major stress proteins are members of gene families distinguished primarily based on apparent molecular mass. Both the prokaryotic and eukaryotic cells respond to various environmental stressors by enhancing the transcription of specific genes (Welch, 1990; Nover, 1991). Synthesis of stress proteins includes activation of the nuclear DNA system *i.e.* changes at the level of gene transcription (Baglia *et al.*, 1981). An animal when subjected to any sort of stress synthesize a new protein or a group of new proteins in affected cells/tissues or enhances intensity of some existing cellular proteins to counteract the chemical toxicity effects (Iwama *et al.*, 1998). Synthesis of Hsps is said to increase in response to a variety of physical and chemical stressors, including temperature, salinity stress, and any other chemical stress including metals and pesticides (Sanders, 1990, 1993; Clayton *et al.*, 2000; Varó *et al.*, 2002; van der Oost *et al.*, 2003). According to Hightower, 1991 the primary mechanism of toxicity involves protein denaturizing which results in molecular aggregation or/and misfolding. This causes binding of a protein called heat shock factor (HSF) in controlling genes (promoter) which activate transcription of the stress protein genes. Synthesis of stress proteins then occurs to facilitate repair of the denatured proteins. As the proteins are repaired stimulation of HSF declines, thus, slowing the induction process, so the feedback loop is completed. Same was emphasized earlier by other workers too (Goering and Fisher, 1995; Sanders *et al.*, 1996). Wu *et al.* (2003) reported arsenic induced influences on the expression of a variety of proteins involving signal transduction and gene transcription. Although scientists do not precisely know how heat shock or other adverse environmental conditions activate the heat-shock factor, a number of studies suggest that stressors that cause an increase in damaged or abnormal proteins activate this process. Biochemical conditions that alter protein conformation also affect expression of the stress response in a predictable manner. The most intriguing aspect of the regulation of induction of the stress proteins is that denatured proteins act both as the signal that activates transcription of the stress protein genes and the substrate for the proteins themselves. Hightower (1991) supported the notion that synthesis of stress proteins is related to protein damage. Heat-inducible genes include a conserved sequence referred to as the "heat-shock element" in their upstream regulatory region. Protein denaturation and misfolding is the result of weakening of polar bonds and exposure of hydrophobic groups which are further important targets of stress-induced damages.

Hsps are also termed as 'Molecular Chaperone' proteins that guide other proteins to fold and mature correctly. Chaperons (*e.g.* Hsp73) are the stress proteins that are constitutive or cognate proteins. These play an important role in protein biogenesis and maturation and in folding of nascent proteins (Gething and Sambrook, 1992); and these bind to a variety of intracellular receptors to keep them in a non-activated state (Howard *et al.*, 1990) and shuttle nascent polypeptides and proteins across intracellular membranes (Craig *et al.*, 1994). Chaperones prevent the reactive hydrophobic parts of the damaged proteins from forming non-specific complexes with normal cellular proteins (Weigant *et al.*, 1997). Such a response is believed to serve as an adaptive or a survival function involving a rapid but transient reprogramming of cellular metabolic activity to protect critical cellular macromolecules against a stressor and to promote resumption of normal physiological functions during recovery or repair phase (Burdon, 1986; Welch, 1990; Goering *et al.*, 2000). For example, cells administered a sub lethal heat insult sufficient to induce stress protein synthesis, and exhibit tolerance to a lethal heat stress (Welch, 1987). Ananthan *et al.* (1986) have evidenced that of the three pairs of native and denatured proteins injected into *Xenopus laevis* oocytes, only the denatured derivatives induced expression of a reporter gene from a heat shock promoter. Amino acid analogues that create abnormal proteins too induce synthesis of a stress protein.

3. Environmental Stress and the Hsps

Until 1990's our knowledge of how animals adapt to stress at the molecular and cellular levels has been limited. There are differences in the stress responses among species (Vijayan and Moon, 1994) and differences among stocks of the same species in their tolerance to applied stressors (Iwama *et al.*, 1999). Our understanding of the cellular stress response in fish has also increased substantially (Basu *et al.*, 2002). Currently, we have considerable knowledge about the different aspects of the physiological stress response in fish, and as stated by Barton (2002), the magnitude of this response can be influenced by the stressor, as well as genetic, developmental and environmental factors. A recent review explored functions of Hsps in various aspects of fish physiology, including development and aging, environment physiology, acclimation endocrinology, immunology, and stress tolerance (Basu *et al.*, 2002). The 'Hsp' response in fact is said to vary with the type of a stressor (Airaksinen *et al.*, 2003), the season (Fader *et al.*, 1994), the fish tissue (Smith *et al.*, 1999), as well as the fish species (Basu *et al.*, 2001), and the stage of fish development (Martin *et al.*, 2001). The susceptibility of fish to different stressors also has genetic components. Stress proteins in fishes are considered as an excellent diagnostic tool to assess the extent of tissue/organ damage in chemically polluted waters. Since, the *de novo* synthesis of stress proteins can be detected early after exposure to ant chemical or environmental agent, analysis of toxicant-induced changes in gene expression, *i.e.* alterations in patterns of protein synthesis, may be useful to develop as biomarkers of exposure and toxicity, as the biomarkers are considered important tools to enable toxicologists to reliably predict and detect exposures to xenobiotics and resultant cell injury, ultimately improving risk assessments. Since, stress proteins are part of the cell's protective strategy, this accumulation should be closely coupled with the organism's physiological state, providing an early warning of impairment at the organism level. Synthesis of stress proteins is a broad response, induced regardless of the nature of a stressor; the accumulation of these proteins has the potential to provide information on the overall impact of multiple stressors on the physiological state of a fish. The induction of stress proteins could further provide an added benefit of evaluating the extent to which an organism is stressed. Under normal conditions, several of the major stress proteins are present at low levels and function as 'molecular chaperones' to facilitate folding, assembly and distribution of newly synthesized proteins (Gething and Sambrook, 1992). Studies with various aquatic organisms and soil invertebrates suggest that the accumulation of stress protein in wild populations from contaminated environments may be a useful tool for quantifying protein damage from the environmental factors. There are a few studies that relate physiological and cellular stress responses *in vivo*. Due the interplay among the many physiological processes involved, the concentrations of different stress proteins among tissues could differ significantly with the stressor. Such differences in stress protein accumulation might be useful in identifying target tissues and evaluating the extent of damage.

4. Applications of Stress Proteins as Diagnostics

Organisms encounter many stressors in their natural environment but they only have a finite capacity to adapt to these stressors. Using protein profile as a biochemical indicator of various conditions or stressors in a natural system now holds a great promise for understanding mechanisms by which an organism respond to a rapid environmental change. Under the conditions of environmental stress, stress proteins are involved in protecting and repairing vulnerable protein targets (Sanders, 1994), thus, the cellular stress response entails the orchestrated induction of key proteins that form the basis for a cell's protein repair and recycling systems. Now a day, many such new proteins are being identified as a quick analytical diagnostic feature for various diseases or the frequently occurring cellular/physiological stress disorders or the syndromes; such as the glucose regulated proteins (Grps), another such important group of stress proteins. Grps are induced by agents or conditions that interfere with glucose and oxygen utilization, or that perturb intracellular calcium homeostasis. Due to interplay of various physiological processes involved, the concentration of stress proteins among tissues could also differ significantly as per the quantity and the quality of a stressor involved. Such differences in stress protein accumulation might be further useful in identifying a target tissue and evaluating the extent of damage. Extensive studies on model species have revealed several Hsp families (Morimoto *et al.*, 1990; Forreiter and Nover, 1998), which are named as per their molecular mass (kDa). Of these, three major families of Hsps: Hsp90 (85–90·kDa), Hsp70 (68 73·kDa), some low molecular- mass Hsps (16–24·kDa) and ubiquitin are identified in diverse phyla. Ubiquitin is a small protein that exists in all eukaryotic cells. It performs its myriad functions through conjugation to a large range of target proteins and a variety of different modifications can occur. In an unstressed cell, there is a constitutive production of these proteins, required in various aspects of protein metabolism to maintain cellular homeostasis (Fink and Goto, 1998). A dominant aspect of the change in protein profile in an animal as a part of the cellular stress response, is the change in concentration of such classes of 'Hsps' (Iwama *et al.*, 1998). The stress inducible stress protein genes, *e.g.* Hsp73 and Grp94 were said to be rapidly up-regulated as a result of adverse conditions, may be a heat shock or a toxin insult (Welch, 1993). These stress proteins bind to the critical proteins in the cytoplasm to prevent irreversible denaturation and aggregation to facilitate disaggregating of denatured proteins, and to aid in the refolding of proteins during recovery after cessation of the toxic insult.

The induction of various Hsp families in fish has been reported in cell lines, primary cultures of cells, as well as in various tissues from whole animals (Iwama *et al.*, 1999). While the majority of these studies have focused on the various effects of heat shock, there is increasing interest in the physiological and protective role of Hsps following exposure of fish to various environmental stressors. For example, increased levels of various Hsps have been measured in tissues of fish exposed to industrial effluents (Vijayan *et al.*, 1998), polycyclic aromatic hydrocarbons (Vijayan *et al.*, 1998), several metals such as copper, zinc and mercury (Sanders *et al.*, 1995; Williams *et al.*, 1996: Duffy *et al.*, 1999), pesticides (Hassanein *et al.*, 1999) and arsenite (Grosvik and Goksoyr, 1996). The Hsp response, however, is said to vary according to tissue (Smith *et al.*, 1999; Rabergh *et al.*, 2000), distinct Hsp families (Smith *et al.*, 1999) and stressor (Airaksinen *et al.*, 2003), and the sensitivity of Hsp expression can also vary with the species (Basu *et al.*, 2001; Nakano and Iwama, 2002), developmental stage (Santacruz *et al.*, 1997; Lele *et al.*, 1997; Martin *et al.*, 2001) and season (Fader *et al.*, 1994). Goering *et al.* (1993) studied hepato-toxicity of Cd in rat liver and observed that the metal induced changes in protein synthesis preceded the detection of functional and clinical injury. They evaluated the target tissue-specificity of this response utilizing cadmium chloride a well known hepato-toxicant. It was found to produce changes in the liver, but not in the kidneys of the exposed rats. Whereas, rats treated with mercuric chloride, a potent nephro-toxicant was found to show changes in kidney, but not in liver. After Hg treatment, liver protein synthesis was unaffected and no hepatotoxicity was evident. Nephrosis is a minimal or a mild (cytoplasm condensation, tubular epithelial degeneration, single cell necrosis) effect at lower exposures, and progressed to moderate or severe (nuclear pyknosis, necrotic foci, sloughing of the epithelial casts into tubular lumens) at high exposure. Goering *et al.* (2000), made a remarkable study on this aspect. They have recently evaluated the differential expression of four Hsps in the renal cortex and medulla during experimental nephrotoxic injury, using $HgCl_2$ in male Sprague-Dawley rats. Rats exposed to mercuric chloride, demonstrated expression of specific stress proteins exhibiting regional heterogeneity in response to Hg (II) exposure and a positive correlation existed between accumulation of some stress proteins and acute renal injury. In whole kidney, Hg(II) induced a time and a dose related accumulation of a heat shock protein (Hsp72) and a glucose regulated protein (grp94). Accumulation of Hsp72 was predominantly localized in the cortex and not in the medulla, while grp94 accumulated in the medulla primarily. Protein synthesis occurs in the kidney before overt renal injury can be detected. They concluded that these stress proteins may be a part of cellular defense response to nephrotoxicants. Since, Cd complexs with molecules such as cysteine (cys) or metallothionein, rats treated with Cd-cys compared to $CdCl_2$, recorded Cd concentrations increased 5-fold after a.v. injection of Cd-cys in kidney compared to $CdCl_2$, mimicking Cd distribution following chronic exposure. Interestingly, *de novo* synthesis of 70, 90 and 110 kDa proteins was enhanced in liver, but not in kidney, 4 h after injection of 2 mg Cd/kg as $CdCl_2$. It also evidenced mild single cell necrosis of hepatocytes after 8 hr of the treatment of Cd, which progressed to multifocal necrotic foci at 16 hr. Synthesis of these proteins decreased by 8-10 hr post injection. No lesions were observed at lower doses of Cd. Increases in plasma sorbitol dehydrogenase activity, a clinical indicator of hepatic injury, was also not apparent until 8 hr after exposure. There were no changes in the kidney *de novo* stress protein synthesis and also no evidence of any renal injury, suggesting Cd induced changes in liver stress proteins are target organ specific. In contrast, dose-related increases in synthesis of these proteins were observed in kidney 4 h after injection of 1 and 2 mg Cd/kg as Cd-cys, but not at lower dosages. In addition, synthesis of a 68 kDa kidney protein was inhibited at 2 mg Cd/kg as Cd-cys. The threshold for Cd-induced stress protein synthesis was shown to be between 4 and 8 μg Cd/g tissue. These alterations in protein synthesis in kidney occurred at tissue Cd concentrations lower than those which resulted in renal injury, as assessed by histopathology and PAH uptake into renal slices. It was thus, suggested that these altered patterns of protein synthesis may serve as markers of cell injury or indicators of cellular stress in target organs. Same way, Fukuda *et al.* (1996) reported that iron overload produced by injection of nephrotoxic iron complexes, was reported to result in accumulation of Hsp90 in renal proximal and distal tubules. Same was emphasized earlier by other workers too (Goering and Fisher, 1995; Sanders *et al.*, 1996). Goering *et al.*, 1992 have shown that Hg (II) induces *de novo* synthesis of stress proteins and reduces the synthesis of constitutive proteins in rat kidney prior to detection of nephro-toxicity by classical clinical and functional assays. Wu *et al.* (2003) reported arsenic induced influences on the expression of a variety of proteins involving signal transduction and gene transcription. Shelton *et al.* (1986b) also reported that arsenic and gallium induce similar patterns of multiple stress protein in rat kidney epithelial cells, whereas lead was shown to induce a 32 KDa stress protein.

5. Stress70 and cpn60 Family

The low-molecular-mass Hsps, as of stress70 have diverse functions and it has been proposed that they function as Molecular Chaperones, preventing irreversible protein aggregation (Derham and Harding, 1999). Hsp70 family is one of the most extensively studied stress protein families in animals including fish. The remarkable way in which stress proteins confer tolerance by maintaining the integrity of proteins and protein complexes and act as catalyst of protein folding and repair was best explained by Sanders 1994, with this major heat-inducible protein family. Sanders (1994) has characterized stress70 family as:

i. A large multi gene family with members residing in a number of sub-cellular compartments including cytoplasm, mitochondria, and endoplasmic reticulum.

ii. The intensity and the relative concentration of stress70 is said to be greatest in tissues most vulnerable to damage caused by the stressor.

iii. Its synthesis increases and it protects tissues from stress induced damage by binding to vulnerable proteins and preventing protein denaturation and formation of insoluble aggregates.

Hsp70 like other stress proteins is known to assist folding of the nascent polypeptide chains, and it acts as a molecular chaperone to mediate the repair and degradation of altered or denatured proteins. It maintains the proteins that need to be distributed to other sub-cellular compartments, in an unfolded state and are escorted to their destination for translocation. Once inside an organelle, the target protein interacts with another member of the stress70 family which performs a similar folding function. Studies have further demonstrated increased levels of Hsp70 in various tissues in fish exposed to pathogens (Forsyth *et al.*, 1997; Ackerman and Iwama, 2001). Rainbow trout infected with a bacterial pathogen (*Vibrio anguilarum*) showed increased levels of Hsp70 in hepatic and head kidney tissues prior to clinical signs of the disease. This study also showed that the peak in hepatic Hsp70 levels corresponded to that of plasma cortisol levels, which occurred 5·days after the challenge. However, head kidney Hsp70 levels increased significantly on the fourth day after challenge, when plasma cortisol levels were similar to the control group. It has also been suggested that stress7O plays a role in this relationship and acts as a "cellular thermometer. Given the current understanding of the regulation and function of stress response, Normally stress70 prevents incorrect folding of the newly synthesized proteins, by binding to the growing peptide chain and maintaining it in a loosely folded state until synthesis is complete. It is through that stress proteins are to confer tolerance by maintaining the integrity of proteins and protein complexes associated with critical physiological processes. According to Sanders (1994) cell lines selected for survival at high temperatures constitutively synthesize stress protein Hsp70 at high levels, whereas temperature sensitive mutants do not. Hightower *et al.* (1999) on the desert pupfish (*Cyprinodon macularius*), observed that the levels of constitutive Hsp70 and the scope for increase in Hsp correlate with the ability of the *Tidepool sculpins* to cope with these environmental changes (Nakano and Iwama, 2002). Stress70 exist in low levels in the cell, within 2 hrs after heat shock it increases in the cytoplasm and associates with the nucleus, and disperses throughout in the nucleus 2-hrs later. Within 24 hrs, levels and distribution of stress70 returns to normal. Stress 70 can also break up existing aggregates and repair proteins to complete biological activity and guide damaged proteins to lysosome for degradation. Hsp90 is active in supporting various components of the cytoskeleton, enzymes and steroid hormone receptors. Another stress protein Hsp40 is suspected to interact with stree70 and participate in the process. Stress70 disassociation is an ATP-dependent process that occurs as the protein proceeds down its folding pathway to reach its correct 3-dimention shape. In addition to the differences that we have observed in the cellular stress response of fish species with overlapping habitats, differences in the cellular stress response of a particular fish species inhabiting different geographical locations have been reported by Norris *et al.* (1995). Furthermore, seasonal variations in the Hsps response within a species and between species of fish have been shown by Dietz and Somero (1992) and Fader *et al.* (1994).

The remarkable way in which stress proteins act as catalysts of protein folding and repair is further best understood by examining two major heat-inducible protein families, stress7O and cpn6O. The role of stress 70 and cpn60 in protein repair is further substantiated by their cellular localization and distribution in response to stress. Sanders (1994) have shown exposure-response relationships between accumulation of stress7O and cpn6O and exposure concentration. This relationship occurs at contaminant levels that are lower than levels at which other measures of physiological impairment can be observed. Sanders (1994) has further specified that the intensity and relative concentrations of stress70 and cpn60 identified as the stress proteins should be the greatest in tissues that are most liable to damage caused by a particular environment stressor. For chemical stressors these factors would include the distribution of the chemical among tissues, the ability of each tissue to detoxify the contaminant and minimize cellular damage and the chemicals molecular mechanism of toxicity. Additional folding functions and assembly are carried out by the chaperones (cpn6O) family. Cpn6O is found in eubacteria, mitochondria, and plastids. Environmental stress also stimulates cpn6O synthesis and induces it to prevent aggregation of mis-folded proteins and facilitate their re-naturation and assembly into complexes. Hsp90 is active in supporting various components of the cytoskeleton, enzymes and steroid hormone receptors. Environmental chemical stress stimulates synthesis of cpn60 and induces it to prevent aggregation of miscoded proteins and facilitate denaturizing and assembly into complexes. Sanders and co-workers further demonstrated that cpn6O also localizes in the nucleus in response to heat shock. The intensity and relative concentrations of stress7O and cpn6O should be greatest in tissues that are most vulnerable to damage caused by a particular environmental stressor. Earlier to heat stress, cpn60 localizes in the mitochondria also in low levels, its abundance increases from 2-8 hrs after heat shock, and within 12-24 hrs it can be seen in the nucleus associating with the nucleolus and a discrete foci, like the coiled bodies involved probably in RNA processing. The fact that these two stress proteins interact differently with complexes in the nucleus suggests that they have distinct roles in facilitating repair of different nuclear structures. Further, although both proteins interact with the nucleolus, the nucleolar association of cpn6O occurs well after stress7O has migrated from the nucleolus. The role of stress7O and cpn6O in repair is further substantiated by their cellular localization and distribution in response to stress.

Heart and brain cells in transgenic mice, that over-express the inducible form of Hsp 70, are resistant to ischemia (Plumier *et al.*, 1997). Kristensen *et al.*, 2003 in *D. melanogaster* on their exposure to heat stress in the young age found the synthesis of Hsp70 later in life, indicating a positive effect of stress resistant later in the animals' life. It is attributed to the activities of defense/cleaning systems (Hsps, antioxidases, DNA repair *etc.*) or the changed rate of metabolism which improve the functional ability of the animals. This phenomenon is termed as 'Hormosis', and it is also well known in humans (Foneger *et al.*, 2000). Studies have demonstrated that the regulation of Hsp70 gene expression occurs mainly at the transcriptional level (Fink and Goto, 1998). Susan Lindquist (Former director, Whitehead Institute of biomedical Research, Cambridge) was a pioneer in the study of protein folding. She has shown that changes in protein folding can have profound and unexpected influences in fields as wide-ranging as human disease, evolution and nanotechnology. According to Hightower (1991) and Sanders (1994), "Lindquist's group" was the first to show that cytoplasmic stress7O moves into the nucleus in response to heat shock, where it interacts with structures such as the nucleolus, the site of ribosomal assembly, and then it returns to the cytoplasm during recovery.

Lindquist group (Jarosz and Lindquist, 2010) has also established that heat shock protein 90 (Hsp90) can reveal hidden genetic variation in fruit flies and in cress plants (Arabidopsis) under certain environmental conditions. Most of these variations are likely to be harmful, but a few unusual combinations may produce valuable new traits, spurring the pace of evolution. Hattori *et al.* (1993) showed that Hsp4O responds to heat shock in a similar manner. Elevated levels of Hsp72, a widely investigated stress protein have been associated with cyto-protection against a variety of lethal or toxic insults (Tytell *et al.*, 1994, Welch and Mizzen 1988), including protection of kidneys from ischemic injury (Chatson *et al.*, 1990). Chaperonins assemble into large "double donut"-shaped complexes that direct the higher-level folding and the assembly of subunits into complexes. A functional homolog Tcp-1 is present in the cytoplasm, where it facilitates folding of actin and tubulin, and perhaps other proteins (Kubota *et al.*, 1994). This chaperonin family appears to be weakly related to the cpn6O group, but it is not considered a stress protein because its synthesis is not induced by stress. Under environmental stress, dramatic changes occur in the stress7O and cpn6O families, which suggest their dual role in protein protection and repair.

While studying stressors and the stress proteins in aquatic organisms it is very important to establish whether experimental procedures such as handling, sampling and other physical stressors are affecting the Hsp response. While handling and sampling procedures can affect common indicators of the physiological stress response in fish, such as plasma cortisol levels, it has been demonstrated in rainbow trout that handling stress does not alter levels of hepatic Hsp70 (Vijayan *et al.*, 1997), and levels of muscle, gill, heart and hepatic Hsp60 (Washburn *et al.*, 2002). Recently, Zarate and Bradley (2003) showed that common forms of hatchery-related stressors (exposure to anesthesia, formalin, hypoxia, hyperoxia, capture stress, crowding, feed deprivation and cold stress) also did not alter levels of gill Hsp30, Hsp70 and Hsp90 in Atlantic salmon (*Salmo salar*).

6. Stress Hormones and the Hsps

Fish in response to a stressor, elicits a generalized physiological stress response involving activation of the hypothalamic-pituitary axis, leading to increased levels of stress hormones (catecholamines and cortisol) and subsequent changes that help in maintaining animal's normal or homeostatic state, such as an increase in plasma glucose levels and simultaneous increase in branchial blood flow as well as an increase in muscular activity (Barton, 2002). Stress levels of plasma cortisol are said to attenuate the heat-stress induced increase in gill Hsp 30 in cut-throat trout, *O. clarki* (Ackerman *et al.*, 2000); liver and gill Hsp 70 in rainbow trout, gill Hsp70 in tilapia (Basu *et al.*, 2001); Hsp 90 mRNA in the primary culture of rainbow trout hepatocytes and Hsp70 in primary cultures of rainbow trout hepatocytes (Sathiyaa *et al.*, 2001; Boone and Vijayan, 2002). Some studies have demonstrated that steroid hormones, including cortisol, may have a direct influence on the cellular stress response. Ackerman and Iwama (2001) also revealed that peak in hepatic Hsp70 levels corresponds to that of plasma cortisol levels, which occurred 5 days after the challenge. Recently Basu *et al.* (2003) showed a functional and structural link between Hsp70 and the glucocorticoid receptor in rainbow trout. They also showed that the glucocorticoid receptor heterocomplex contains Hsp70 and that the association between Hsp70 and the glucocorticoid receptor can be altered according to the stressor. Iwama *et al.*, 2004 suggested that if Hsps are to be used as indicators of the stress response, it will probably have to be done in a stressor- and species-specific manner, and also that how a fish respond to chronic exposures? There is also a need to know how fish respond when exposed simultaneously to multiple stressors or to sequential stressors (Schreck, 2000).

7. Activation of Heat Shock Transcription Factor and Hsp Induction

The epigenetic repair system is also termed as the cellular stress response or the heat shock response. Changes in the expression of gene products occur even at a chemical concentration lower than the toxicity threshold level (Goering *et al.*, 1992). Amino acids constitute in fact the basic units of life. Any alteration in DNA composition under environmental stress shall be ameliorating amino acids and thus, attenuations in the basic proteins or sometimes formation of a non-viable protein (Sharma and Bhatia, 1982) as the stress protein, affecting ultimately the metabolic pathway and the body composition. This stimulate the process of physiological deterioration or the early ageing, which envisages progressive loss of fitness and vitality due to several cellular and sub cellular alterations in tissues, ultimately resulting in senescence and death. In case of from acute renal injury, cellular repair and recovery is said to involve in the expression of inducible stress genes, such as those of epidermal growth factor, prostaglandin synthesis and Hsps associated with cell proliferation and regeneration (Bardella and Comolli, 1994; Cowley and Gudapaty. 1995). Hsps' translocate from cytoplasm to nucleus and nucleolus, where they bind peri-ribosomes and other nuclear protein complexes in order to protect them from damage (Arrigo *et al.*, 1988; Brown *et al.*, 1993; Ellis, 1989).

Analysis of Hsp genes and a comparison of heat shock regulatory elements from a variety of organisms led to the identification of a palindrome (A segment of double-stranded DNA in which the nucleotide sequence of one strand reads in reverse order to that of the complementary strand) heat shock element (hse, Bienz and Pelham, 1987). It has been demonstrated that Hsp induction results primarily from the binding of an activated heat shock transcription factor (hsf) to a Hse upstream of Hsp genes (Morimoto *et al.*, 1992). Recently, it has been shown in zebrafish (*Danio rerio*) that the transcriptional regulation of Hsp genes, in response to heat shock, is also mediated by an Hsf (Rabergh *et al.*, 2000). Since most of the Hsp genes do not contain introns, the mRNA is rapidly translated into nascent proteins within minutes of exposure to a stressor. Genomic sequences for Hsp70 are being elucidated in the fish, rainbow trout (O. mykiss; Kothary *et al.*, 1984), medaka (O. latipes; Arai *et al.*, 1995), zebrafish (Lele *et al.*, 1997), pufferfish (*Fugu rubripes*; Lim and Brenner, 1999) and tilapia (*O. mossambicus*; Molina *et al.*, 2000).

8. Metal Effects in Fishes and the Hsps

Toxicity induced quantitative or qualitative changes in a tissue protein profile recorded in fishes after the metal treatments are attributed largely to either proteolysis or the impaired protein synthesis. According to Mustafa and Chandra (1972) the decrease in total protein content may be due to blocking of the metabolism of amino acids by the heavy metals and such cells become incapable of synthesizing proteins. The change in protein band profile and total protein content in a fish species might actually reflect damage in DNA or protein synthesizing system by a toxicant. Heavy metals were reported to cause chromosomal damage (Bartoli *et al.*, 1991), increased DNAase activity (Joshi and Desai, 1988), and decrease in DNA and RNA levels (Chaudhary, 2004). Various workers have revealed presence of different proteins in different fish tissues; a few have been identified as stimulus specific stress proteins. Various authors (Boone and Vijayan, 2002; Tabche, 2002; Ali *et al.*, 2003; Chaudhary, 2004) have reported induction of stress proteins due to various heavy metal treatments. They demonstrated a definite qualitative alteration in protein fractions and the changes appeared influenced by the type of a chemical administered and the organism's tissue studied. In *C. carpio*, Suresh *et al.* (1991) have reported impaired protein synthetic machinery on account of common environmental pollutants like mercury, and fish exposures to such xenobiotics were attributed largely to the proteo-toxic effects of the stressors, thus, affecting cellular metabolism, and inducing changes in their protein profile. Mercury at the sub lethal concentration of 0.1mg/l in *C carpio* demonstrated excessive loss of proteins, due to proteolysis and impaired cellular metabolism, thereby leading to damaged protein integrity of the fish. These changes were more pronounced in gills during early exposures, but appeared distinctly high later in kidneys, as the period of exposure was extended from 15 to 30 days (Suresh *et al*, 1991). The 70-kDa family of heat shock proteins plays an important role as molecular chaperones in unstressed and stressed cells. The constitutive member of the 70 family (Hsc70) is crucial for the chaperoning function of unstressed cells, whereas the inducible form (Hsp70) is important for allowing cells to cope with acute stressor insult, especially those affecting the protein machinery. Using this antibody, Boone and Vijayan (2002) detected Hsc70 content in the liver, heart, gill and skeletal muscle of unstressed rainbow trout. Primary cultures of trout hepatocytes subjected to a heat shock (+15° C for 1 h) or exposed to either $CuSO_4$ (200 microM for 24 h), $CdCl_2$ (10 microM for 24 h) or $NaAsO_2$ (50 microM for 1 h) resulted in higher Hsp70 accumulation over a 24-h period. However, Hsc70 content showed no change with either heat shock or heavy metal exposure suggesting that Hsc70 is not modulated by sublethal acute stressors in trout hepatocytes. De Boeck *et al.* (2003) in a study with common carp exposed *in vivo* to copper, concluded tissue specific Hsp70 levels and cortisol-related

Hsp response. These authors also suggested species and life history-dependent Hsp responses. Feng *et al.* (2003) also reported Hsp70 expression in rainbow trout hepatocytes during 24 and 48 h exposure to 25, 50, 100 and 200 μM copper sulphate. A member of multigene family, encoding 70 kDa stress proteins, was identified by Ali *et al.* (2003) in common carp (*C. carpio*). It was induced by elevated temperature and it responded to cadmium metal treatment in a tissue; its expression was recorded as time dependent. The brown trout (*Salmo trutta*) acclimated in mining-affected habitats to different levels of Cd/Zn and Cu, showed an increase mucous secretion as well as in the transcription of Hsp70 (Hansen *et al.*, 2007). The data indicate that acclimation to chronic metal exposure involves different strategies to cope with different metals and that these strategies involve both physiological mechanisms (mucus production) as well as metal-related stress gene transcription.

Guha and Khuda-Bukhsh (2004) reported 25, 25, 19, 24, 20, 20, 24, 16 and 19 protein fractions in the dorsal muscles, ventral muscles, heart, eyes, brain, gills liver, spleen and kidney, respectively in controlled specimen of *A. testudineus*. Serum samples of *O. mossambicus* showed a total of 35 proteins (Zacharia *et al.*, 2003) and 17 protein fractions in the ovary samples of *C. gariepinus* (Chaudhary, 2004). *C. gariepinus* exposed to sub lethal concentration of cadmium, showed (Chaudhary, 2004), after 14 days of exposure, however, a polypeptide of molecular weight of 48.6 kDa disappeared and an additional band of molecular weight 29 kDa appeared after 30 days of exposure. Likewise in *Penaeus indicus*, 13 fractions were separated in male hemolymph and 14 fractions in female hemolymph (Laxmilatha and Laminarayana, 2003). Sharma, 2006 also evidenced species specific differences in no. of protein fraction in fishes; such as in *C. carpio* a total of 7, 14, 15 and 17 protein fractions were identified in normal fish liver, muscles, kidney and gill tissues respectively; whereas in *L. rohita*, the number of such protein fractions was found to be 9, 16, 19, and 9 in these tissues. Likewise, in *Penaeus indicus*, 13 fractions were separated in male haemolymph, in female haemolymp 14 fractions were observed (Laxmilatha and Laxminarayana, 2003). Hermesz *et al.* (2001) observed the Hsp90 beta gene largely constitutively expressed at a fairly high level in all the examined tissues of brain, liver and kidney of a common carp fishes and it is slightly inducible by an elevated temperature. They isolated two Hsp 90 cDNA isoforms (Hsp90 alpha and Hsp 90 beta) from this common carp (*C. carpio*). Hsp90 alpha mRNA was reported as present in the brain, but was hardly detectable in the kidney and liver of unstressed animals. In the brain, this gene was reported as greatly regulated following thermal stress; whereas, in the liver and kidney heat shock has only minor effects on its expression. Hsp90 alpha, but not Hsp90 beta, was found to respond to elevated levels of Cd and in a dose-, time- and tissue dependent manner.

Different fish species behave differently in terms of their response to the same stressor or the same heavy metal treatments. *C. carpio* on exposure to heavy metals like As, Hg, Cr and Ni showed maximum alterations in their protein profile (Sarita, 2005,) in liver, followed by kidney, blood and muscles; whereas, it showed deletion of two proteins in Cr and As + Cr treatments. *C. mrigala* liver tissue had shown maximum alterations, followed by in muscles, brain and blood; unlike *C. carpio*, its brain tissue had shown multiple alterations on exposure to heavy metals alone and in combinations. The number of proteins added or deleted, were also different in the same tissue of both the species. Cr had induced three stress proteins of 134.8, 112.2 and 28.1 kDa in muscles of *C. carpio;* whereas Hg and Ni had been shown to induce a single stress protein of 134.8 and 131.8 kDa, respectively. It was concluded that induction of stress protein synthesis was not only species specific but also highly tissue specific. For *e.g.* arsenic treatments to *C. carpio* revealed synthesis of multiple protein fractions (158.4, 154.8, 147.7, 117.4, 109.6, 93.3 and 69.1 kDa) in liver and kidney, and only two fractions in blood (81.2 and 58.8 kDa) and muscles (134.8 and 112.2 kDa). This specificity in response makes it difficult to generalize the pattern of induction of specific stress proteins by metals. These differences in protein induction might also reflect differences in the mechanisms of action or their metabolism by which specific metals elicit toxicity effects in an organism. In a recent study as reported by Sexena *et al.*, 2009, in two groups of six common carp fish, exposed to a combination of various compounds of eight heavy metals in water for two months, showed decreased concentration of plasma proteins, and it also adversely affected the fish immunity on exposure to water contaminated with heavy metals. The mean value of total plasma proteins was lower (though nonsignificantly) in case of pollutant exposed fish. Serum protein profiles on SDS – PAGE, showed that the proteins p63 and p25 disappeared, while p56 and p22 appeared in the sera of fish exposed to heavy metal polluted water. Zacharia *et al.* (2003) also revealed significant changes in protein profile through SDS-PAGE in *Oreochromis mossambicus* as induced by sublethal dose of aflatoxin B1. Aflatoxin, the only naturally occurring dietary carcinogen, is an important and frequently encountered problem seriously affecting fish production in tropical countries. Aflatoxins are mostly produced by the fungal strains of *Aspergillus flavus* and have been recently recognized as carcinogens by the International Agency for Cancer Research.

Applying proteomics to monitor marine pollution (Cristoba, 2007) is a new approach to evaluate the effects of environmental pollutants on the biota. Aquatic organisms living in coastal and estuarine areas are particularly prone to exposures to a variety of pollutants, some of which can act as peroxisome proliferators. Peroxisomes are organelles from the microbody family and are present in almost all eukaryotic cells. They participate in the metabolism of fatty acids and many other metabolites. Peroxisomes harbor enzymes that rid the cell of toxic peroxides. Peroxisome responses in particular and biomarker responses in general can be influenced by several biotic and abiotic factors. Utilizing proteomics-based techniques that permit the evaluation of hundreds to thousands of proteins in a single experiment can circumvent those drawbacks.

9. Metallothionin (MTs)

MT's are present in all eukaryotes, and some MT genes are expressed in all stages of development and they are coordinately regulated by metals, glucocorticoids and inflammatory signals. MTs could serve as reservoirs of essential metals while preventing metal toxicity, and donate the metals to apo-metalloproteins, as they are synthesized. MTs may also serve as chaperons for the synthesis of metalloproteins. A low molecular weight cytosolic protein metallothionin (MT) possess an inherent capacity to bind differentially with metals like Pb, Cd, Cr, Ni and As in various fishes. MTs are specifically induced by metals such as Cd, Hg, Ag and Cu, and are used in both vertebrates and invertebrates as a biomarker for metal exposure. Generally, liver, kidney, gills and the intestine are the main sites for MT formation in fishes which is either stored or excreted. MT is a protective protein responsible for elimination of toxic metals. An exposure of the animals to the specific metals and the ions induces the synthesis of MT and they preferentially bind these toxic metals making them harmless and thus, yielding metal scavenging effect (Rema and Philip, 1996). The reduction is up to the extent to which the essential metabolic activities would otherwise be inhibited. It explains the development of tolerance by an animal to MT transition. When MT concentration increases and overcomes the capacity of this natural detoxifying system, various adverse effects which are proportional to the metal dose level are observed.

MTs, are the low molecular weight cystein rich polypeptides, encoded by a family of genes. MTs are ubiquitous both in plants and animals and are present in the contaminated as well as uncontaminated organisms and are increasingly demonstrated to play a central role in metal metabolism. MT is involved in many cellular functions such as essential metal transport and storage, heavy metal detoxification, protection against oxidative stress, *etc.* (Kagi and Schaffer, 1988; Vasak, 2005). MTs were originally isolated and characterized by Margoshes and Vallee, 1975. These are reported to be present in a variety of organisms like prokaryotes, plants, and animals, and are said to protect biological systems by encountering the influx of metal ions. According to Roesijadi (1992) at least 34 species of elasmobranches and teleosts have been reported to have MTs present in their gills, intestine, kidney, and liver. Pelgrom *et al.*, 1994 also reported liver, kidney, gills, and intestine as the main sites for MT formation in fish. MTs possess many sulfhydryl groups due to large amounts of cystein in these molecules (Kagi and Nordberg 1979). MTs being typically metal binding proteins act as metal scavengers by forming complexes due to –SH groups, and these MTs possesses an inherent capacity to bind differentially with metals like Pb, Cd, Cr, Ni and As in various fishes. This protein has a high affinity for Cd and Hg, and its presence has been variously suggested as indicating involvement in uptake storage, transport and elimination of toxic metals and in their routine metabolism (Roesijadi, 1980). Dutton *et al.*, 1993 have described an accurate, rapid, sensitive, and simple method using mercury saturation for quantifying metallothionein (MT). Analysis of hepatic MT with high Cu content from rainbow trout demonstrated virtually complete displacement of Cu, Cd, and Zn by Hg. Analyzing brown trout (*Salmo trutta*) from two native populations from Central Norway, acclimated in mining-affected habitats to different levels of Cd/ Zn and Cu, and then transferred to a nearby lake with higher levels of Cu, Cd, and Zn than those in their respective native rivers, together with trout from a nearby unaffected river, for estimating the MT-A levels, indicated that the MT-A levels were highest in the Cd/Zn-acclimated trout both before and after transfer (Hansen *et al.*, 2007). Rose *et al.*, 2014 determined the Cd levels and MT induction in liver, kidney, brain and gill of *Clarias gariepinus* during acute Cd exposure. The results indicated that Cd exposure clearly resulted in MT induction and hence MT levels can be considered as a biomarker for acute waterborne Cd pollution. The liver and kidney Cd accumulation and MT induction levels was found to be higher than gill and brain. They recorded the Cd levels increased significantly in all tissues in the following order: liver > kidney > gill > brain, in a distinct time- and dose-dependent manner. In gills, Cd accumulation levels gradually increased rapidly to reach a peak during a period of 48 hrs and then declined gradually. The MT induction levels were found in the following order in the tissues: liver > kidney >gill> brain.

Different metals induce pattern of protein expression in the tissue, unique to a specific metal. For example, Price-Haughey and Gedamu, 1987, investigated regulation of heavy metal induced gene expression in two fish cell lines: the rainbow trout hepatoma (RTH) and *Chinook salmon* embryo (CHSE) cells. They concluded that since MT synthesis in these embryonically originated cells occurred after their treatment with 5-azacytidine, MT gene expression possibly is developmentally regulated in fish. MTs exist generally in a metal-saturated form (Hodson, 1988), thus, there is always synthesis of additional MT to detoxify any additional amount of metal within a tissue or displacement of one metal by another. It is well established that a variety of metals induce the synthesis of additional MTs (Hammer, 1986), and fish acclimation to low levels of metals, such as Zn (Hobson and Birge, 1986), or Cu (Dixon and Sprague, 1981; Buckley *et al.,* 1982), or Cd (Klavercamp and Duncan, 1987) and it can result in resistant to that metal, as evident from their raised Lc_{50} values. McCarter and Roch, 1984 recorded increases MTs in response to copper intake into the fish liver. Brown and Parsons (1978) proposed a hypothesis 'Spil over hypothesis'; it suggests that if the capacity of MTs to sequester a toxic metal is exceeded, then the metal will spill over and bind to other cellular proteins such as enzymes, thereby resulting in metal toxicity (Brown *et al.,* 1990). The value of MT in a tissue can thus, be considered as an indicator of heavy metal contamination of natural freshwater ecosystems. Some metals like lead, however, do not induce synthesis of MTs, when fed orally for a week (Juedes and Thomas, 1984), although there was synthesis of reduced glutathione, which is known as another metal binding protein in fish liver. Stress responses and changes in protein metabolism were studied by De-Smet and Blust (2001) in common carp, *Cyprinus carpio* exposed to 0, 0.8, 4 and 20 mM cadmium over a 29 day period. Cd accumulated in the tissues in the following order: kidney>liver>gills. The concentrations of Cd and Zn binding metallothionein were in the order: liver>kidney>gills. The activities of proteases were increased at day 4 in gills, liver and kidney of carp. There has been some evidence of metal excretion in fish also when returned to normal water, as indicated by the decreased order of Cd accumulation in tissues of the 8 weeks old common carp C. carpio and the order was found to be gut > kidney > liver = gill > muscle (Kraal *et al.,*1995). Whereas, in the contaminated waters reported their corresponding values in the decreasing order of: gut > gills > kidney > liver > muscle. Since, the concentrations of a metal found in tissues of these fishes reflect its external medium concentrations, these determinations become ideal tool for monitoring metal levels in water.

Chapman (1985) ranked different metals as per the ability of a fish to acclimate these in the decreasing order of: Zn> Cu> Cd> Cr; this is the reverse of their binding affinities to MTs, *i.e.* Hg> Cd> Cu> Zn (Eaton, 1985). The Cd has been reported to stimulate MT formation even at very low dose levels but its chronic exposure reduces its binding capacity and as the fishes were exposed metal free water, the recovery occurred but not completely. Cadmium used in industries and a byproduct of the metallurgy of zinc, may interfere with the metallothionein's ability to regulate zinc and copper concentrations in the body. When cadmium induces metallothionein activity, it binds to copper and zinc, disrupting the homeostasis levels (Kennish, 1992). Del Ramo *et al.*, 1987 studied toxicity and accumulation of crayfish highly resistant to Cd and a higher accumulation rate of this metal in its many tissues including mid gut gland, and they expected that this crayfish showed resistant to Cd possibly due to the presence of metal binding protein in its mid-gut gland. Mid gut gland has been termed as a major site for metal accumulation and has been suggested this as the site of metal detoxification system. Cd is known to be a ubiquitous non-essential element which possesses high toxicity to aquatic organisms. There have also been some contradictions to the role of excess synthesis of MTs in fish acclimation to metals (Roch and McCarter, 1984). They found elevated levels of MTs in field exposed fish that did not translate into higher metal tolerance, may be on account of presence of some other metal binding proteins like MTs in fish tissues (Hodson, 1988). Zhao (2010) found Cd exposure of rat lung fibroblasts (RFL6) enhanced levels of metal scavenging thiols, *e.g.*, metallothionein (MT) and glutathione (GSH), MT has also a role in buffering changes in free metal ion levels in cells by binding essential metals such as Cu and Zn (Olson, 1996; Marr *et al.*, 1996). Cu exposures also affect fecal excretion of Cd. Every molecule of MT binds 7 moles of Cd or Zn (Klaassen *et al.*, 1999). Montaser *et al.*, 2010 also reported significant increase in MT synthesis simultaneously with the metal toxicity.

Many workers (Thomson *et al.*, 1980; Mukhopadhyay and Konar, 1985) have reported metals like Zn, Cu and Cr in combination were more toxic to the fish. Compared to individual metal toxicity, Cu, Zn and Cr in combination (1:1:1) were more toxic to *O. mossambicus* and the effect increased by 9, 6 and 4 folds for Cu, Zn and Cr metals, respectively; which may be due to the synergistic or the additive effects of the ions. There are further reports of additive effects of some metals in combination with Cd, as with Cu and Hg; Cd in combination with Zn, however, reduces Cd toxicity at sub lethal concentrations. Scheuhammer and Cherian (1986) opines that during Cu exposure, Cd binding capacity of metal decreases with subsequent elimination of Cd. Essential metals like Cu and Zn which are toxic at higher concentrations are generally sequestered or involved to biochemical reactions as metabolic needs dictate, while non-essential metals (Pb and Cd) are only sequestered into a non-available intracellular compartment. In both the cases undesirable intracellular interactions would be restricted through binding to metallothionein (MT). Pb also does not induce MT synthesis in fish. Cd has an inhibitory effect on the activity of Zn-containing enzymes. Toxicity of metals does not solely depend on its concentration in the tissues. Mechitsuka and Coworkers (1987) suggested that toxicity of Cd depends on the accumulated Zn as it displaces Cd from the metallothionein molecules. Indeed Cd replace Zn in metallothionein molecule and low protein values in current investigation may result in increased Cd uptake (Gulfaraz and Ahmed, 2001; Funk *et al.*, 1987). The lower concentration of Zn in fishes may be related with higher concentration of heavy metal Cd which may be attributed with replacement of Zn with Cd due to chemical similarity. Low levels of Cd in fish organs indicate their poor storage due to poorly induced MT synthesis, at chronic levels Cd binding capacity of MT also decreases. Cosson (1994) also reported that Zn ions of MT were replaced by those of Cd when both metals were combined in the organism. This metal also showed affinity to protein SH group. This may be related with interesting pattern of interaction between metal and biochemical constitutes of these species like protein, amino acids glycogen and total lipids content of these biomarkers. According to Rafia *et al.* (2008), there are many similarities between Cd and Zn, both being from same group 11B, these metals have similar tendency to form complexes.

Mercury has an especially strong affinity for MTs, so much so that it displaces copper and Zn from the molecule (Brown and Parsons, 1978), and this has been attributed to its effect on decreasing the normal tissue concentrations of these two essential metals in salmons when exposed to $HgCl_2$ in water. Rema and Philip, 1996 showed in *O mossambicus*, a notable induction of MT in the liver of Hg treated fish, as evidenced by increased amounts of total metal bound to the protein. It showed that in case of increased influx of Zn in fish, MT predominantly binds to Zn and it redistributes essential metals to other protein requiring Zn as co factor. Price-Haughey and Gedamu, 1987 investigated regulation of heavy metal induced gene expression in two fish cell lines; the rainbow trout hepatoma (RTH) and Chinook salmon embryo (CHSE) cells. They found induction of MT synthesis to a lesser extent in RTH cells, exposed to Zn, after their exposure to Cd. The time course of MT synthesis were also different for the different metal inducers, suggesting that MT may be differently regulated in these cells. CHSE cells however, did not synthesize MT in response to these metal treatments. They further concluded that, since MT synthesis in these embryonically originated cells occurred after their treatment with 5-azacytidine, MT gene expression possibly is developmentally regulated in fish. Influx of Hg is encountered by the displacement of Zn bound to the protein and also by do novo synthesis of metal binding proteins and subsequent binding of incoming Hg. Weis (1984) mentioned that in contrast to inorganic Hg, methylHg apparently does not bind to MTs. The accumulation of metals in various intracellular compartments is thus, a function of relative rates of the metal binding and release as determined by their relative metal binding affinity and competetion among different metals for binding to MT. Metals bound to MT are either stored or excreted out of the body. A few metals like Pb do not induce MT formation, whereas Cd even at very low quantity stimulate the MT formation, but its chronic exposure reduces its binding capacity with MT.

Inter-specific variations in metal accumulation in the different fish organs (Maiti and Banerjee, 1999) could be the result of differences in their capacity for the synthesis of protein, metallothionin (MT) and their binding. Martinez-Tabche *et al.*, 2002 in a pond with rainbow trout (*Oncorhynchus mykiss*) recorded a linear relation between Ni concentration and MT synthesis, o-demethylase activity and protein concentration, and recommended these parameters as the biomarkers in order to evaluate Ni toxicity. They found these effects probably due to the physiochemical characteristics of the sediments, which may give a high capacity to accumulated metals. They suggested either not to use rustic ponds in the fish culture, or use concrete ponds to avoid the accumulation of toxic compounds, or make periodic sediments removal.

The accumulation of several anthropogenic compounds in the tissues of the clam, *Ruditapes decussatus* suggests that they possess mechanisms that allow them to cope with the toxic effects of these contaminants (Geret *et al.,* 2002). Another invertebrate, a bivalve, *Donax trunculus* found along sandy beaches of the Mediterranean Sea is also an important bioindicator of metal pollution. Haifa Bay on the northern part of the coast of Israel, an area heavily polluted with effluents from oil refineries, chlor-alkali, fertilizer and other indusrial plants is known to support a dense population of *D. trunculus*. Roth and Hornung (1977) showed that in Haifa Bay the concentration of heavy metals in sediments as well as in these bivalves had been much higher than in unpolluted areas. EL-Rayis 1986 also pointed that in relation to other marine organisms clam *R. decussatus* is a good accumulator of Cd, as cadmium (Cd) induces metallothionein (MT) synthesis only after 7 days of exposure. Before MT synthesis is induced, the other mechanisms capable of handling the excess of Cd are unknown. The Cd has been reported to stimulate MT formation even at very low dose levels but its chronic exposure reduces its binding capacity and as the fishes were exposed metal free water, the recovery occurred but not completely as recorded by Shelly (2006). She observed maximum recovery in the *C. carpio* of Cd (62.72 per cent) in gills and 48.54 per cent in muscles, and it being minimum of Ni; Cd in the gills of *C. mrigala* and of Zn in its muscles.

Determination of metal levels (Cu, Zn, Cd, Ag, Hg) in soluble and insoluble fractions of gill homogenates after 7d exposure of carp (*Cyprinus carpio*) to moderate concentrations of Cd, Ag, and Hg in water showed differences between the ability of metals to bind cytosolic ligands and HSCs, and their respective potency for MT induction in gill (Cosson, 1994). Regardless of pre-treatment, mercury gave the highest increase of gill MT, and after the decontamination MT level remained high compared to control. Cd and Ag gave similar increases, but a significant difference with control appeared only after the decontamination step with Ag, whereas, one week of contamination was enough for Cd. The experimental conditions determined the following order of potency for MT induction in gill: Hg > Cd > Ag > Zn. Zn is important for the enzyme activity and has assorted functions in protein and carbohydrate metabolism (Somasundaram *et al.,* 1984). It is vital because it forms the active sites in various metallo-enzymes (Martinez *et al.,* 1999), including DNA and RNA polymerases (Dallas and Day, 1993; DWAF, 1996). Thus, it was observed that Zn alone had least toxic effects on the crude protein level in muscles of the fishes while when present in combination with Cd, it further enhanced its toxic effects. The binding of Cd, Cu, and Zn to metallothionein in carp was studied by Van Campenhout *et al.* (2004). Very strong differences in the tissue compartmentalization and cytosolic speciation of the metals were observed. For example, over 30 per cent of cytosolic zinc was bound to MT in liver while this was only 2 per cent in the kidneys although total cytosolic levels were considerably higher. Induction of metallothionein during cadmium exposure was also tissue specific, displaying different response patterns in gills, liver, and kidney. Cadmium accumulated much stronger in liver and kidney compared to the gills and the latter also showed much lower MT levels. The renal MT-induction was more sensitive to Cd exposure than the hepatic MT induction since a significant increase of Cd-MT and total MT levels occurred at lower tissue Cd concentrations in the kidney in comparison to the liver, except for the highest Cd exposure level where a drastic 10-fold increase in hepatic Cd-MT was observed. At this Cd exposure level also an apparent spill over of zinc to the high molecular weight fraction was observed in the kidneys. Van Campenhout *et al.* (2004) further studied the effect of metal exposure on the accumulation and cytosolic speciation of metals in livers of wild populations of European eel with special emphasis on metallothioneins (MT). The distribution of the metals Cd, Cu, Ni, Pb and Zn among cytosolic fractions displayed strong differences. The cytosolic concentration of Cd, Ni and Pb increased proportionally with the total liver levels. However, the cytosolic concentrations of Cu and Zn only increased above a certain liver tissue threshold level. Cd, Cu and Zn, but not Pb and Ni, were largely associated with the MT pool in correspondence with the environmental exposure and liver tissue concentrations. Most of the Pb and Ni and a considerable fraction of Cu and Zn, but not Cd, were associated to High Molecular Weight (HMW) fractions.

Martins *et al.* (2005) made an analysis of salinity effects on the synthesis of MT concentration in the gills and the hepato-pancrease of the blue crab *Callinectes sapidus*. The results indicated that hepato-pancrease has a higher ability to synthesize MT than gills and that hepato-pancrease MT is more suitable for metal monitoring in estuarine and coastal areas than the gill MT, since its synthesis was not influenced by the salinity. It was further evidenced that *in vivo* increase in gill MT in low salinity could be an adaptive response to the higher metal bioavailability in lower salinities. The increased levels of MT like proteins (MTLP) as observed by Martins and Bianchini (2009) in anterior and posterior gills of the blue crab after *in vivo* or *in vitro* hypo-osmotic shock could also be a result of an increased production of ROS in low salinities. MTLP concentration was recorded higher in hepato-pancrease than in gills, indicating former being not affected by the hypo-osmotic shock.

REFERENCES

Ackerman P. A., Forsyth R.B., Mazur C.F. and Iwama G.K., 2000. Stress hormones and the cellular stress response in salmonids. Fish Physiol. Biochem., 23: 327-336.

Ackerman P.A. and Iwama G.K., 2001. Physiological and cellular stress responses of juvenile rainbow trout to vibriosis. J. Aquat. Anim. Health, 13: 173-180.

Airaksinen S., Rabergh C.M. I., Lahti A., and 3 others, 2003. Stressor dependent regulation of the heat shock response in zebrafish, *Danio rerio*. Comp. Biochem. Physiol., A 134: 839-846.

Ali S., Al-Ogaily N., Al-Asgah A. and Gropp J., 2003. Effect of sublethal concentrations of copper on the growth performance of *Oreochromis niloticus*. J. Applied Icthyol., 19: 183-188.

Ananthan J., Goldberg A.L., and Voellmy R., 1986. Abnormal proteins serve as eukaryotic stress signals and trigger the activation of heat shock genes. Science, 232, 522-525.

Anson J.F., Laborde J.B., Pipkin J.L., Hinson W.G., Hensen D.K., Sheehan D.M. and Young J.E., 1991. Target tissue specificity of retinoic acid - induced stress protein and malformations in mice. Teratology, 44: 19-28.

Arai A., Naruse K., Mitani H. and Shima A., 1995. Cloning and characterization of cDNAs for 70-kDa heat-shock proteins (Hsp70) from two fish species of the genus *Oryzias*. Jpn. J. Genet., 70: 423-433.

Arrigo A.P., Suhan J.P. and Welch W.J., 1988. Dynamic changes in the structure and intracellular locale of the mammalian low-molecular weight heat-shock protein. Mol. Cell. Biol., 8: 5059-5071.

Baglia F.A., Kwan S.W. and Fuller G.M. 1981. Haptoglobin biosynthesis in rats. Immunonological identification polysomes synthesizing haptoglobin and quantitation of haptoglobin in the cytoplasm of liver cells. Biochem. Biophy. Acta., 696: 107-113.

Bardella L., and Comolli R., 1994. Differential expression of c-jun, c-fox and Hsp 70 mRNAs after folic and ischemia-reperfusion injury: Effect of antioxidant treatment. Exp. Nephrol., 2: 158-165.

Bartoli S., Bonara B., Colacci A., Niero A. and Grill S., 1991. DNA damaging activity of methyl parathion. Res. Comm. Chem. Phatho. Pharmacol., 71: 209-218.

Barton B.A., 2002. Stress in fishes: a diversity of responses with particular reference to changes in circulating corticosteroids. Integer. Comp. Biol., 42: 517-525.

Basu, N., Kennedy, C. J. and Iwama, G. K. 2003. The effect of stress on the association between Hsp70 and the glucocorticoids receptor in rainbow trout. Comp. Biochem. Physiol. A 134: 655-663.

Basu, N., Nakano, T., Grau, E.G. and Iwama, G. K. 2001. The effects of cortisol on Hsp 70 levels in two fish species. Gen. Comp. Endocrinol. 124: 97-105.

Basu N., Todgham A.E., Ackerman P.A., Bibeau M.R., Nkano K., and 3 others. 2002. Heat shock protein genes and theor functional significance in fish. Gene, 295: 173-183.

Bienz M. and Pelham H.R., 1987. Mechanisms of heat shock gene activation in higher eukaryotes. Adv. Genet., 24: 31-72.

Blake, M.J., Gershon, D., Fargnoli, J. and Holbrook, N.J. 1990. Discordant expression of heat shock protein mRNAs in tissue of heat stressed rats. J. Biol. Chem. 265 : 15275-15279.

Boone A.N. and Vijayan M.M., 2002. Constitutive heat shock protein 70 (HSC 70) expression in rainbow trout hepatocytes: effect of heat shock and heavy metal exposure. Comp. Biochem. Physiol. C. Toxicol. Pharmacol., 132 (2 : 223-233.

Breloer, M., Dorner, B., More, S. H., Roderian, T., Fleischer, B., and von-Bonin, A. 2001. Heat shock proteins ias danger signals: eukaryotic Hsp60 enhances and acclerates antigen-specific IFN-gamma production in T cells. Eur. J. Immunol. 31: 2051-2059.

Brown C.R., Martin R.L., Hensen W.J., Beckmann R.P. and Welch W.J., 1993. The constitutive and stress-inducible forms of Hsp70 exhibit functional similarties and interact with one another in an ATP-dependent fashion. J. Cell. Biol., 120: 1101-1112.

Brown D.A., Bay S.M. and Harshelman G.P., 1990. Expsure of scorpion fish (*Scorpoena guttata*) to cadmium: effects of acute and chronic exposures on the cytosolic distribution of cadmium, copper and zinc. Aquat. Toxicol., 16: 295.

Brown D.A. and Parsons T.R., 1978. Relationship between cytoplasmic distribution of mercury and toxic effects to zooplankton and chum salmon (*Oncorhynchus keta*) exposed to mercury in a controlled ecosystem. J. Fish Res. Bd. Can., 35: 880.

Buckley J.T., Roch M., McCarter J.A., Rendel C.A. and Mathieson A.T., 1982. Chronic exposure of coho salmon to sublethal concentration of copper. I. Effect on growth, accumulation and distribution of copper and on copper tolerance. Comp. Biochem. Physiol., 72C: 15-19.

Burdon R.H., 1986. Heat shock and the heat shock proteins. Biochem. J., 240: 313-324.

Chatson G., Perdrizet G., Anderso C., Pleau M., Berman M. and Schweizer R., 1990. Heat shock protects kidneys against warm ischemic injury. Curr. Surg. (Nov-Dec.): 420-423.

Chowdhury M.J., Pane E.F., Wood C.M., 2004. Physiological effects of dietary cadmium acclimation and waterborne cadmium challenge in rainbow trout: respiratory, ionoregulatory, and stress parameters. Comp. Biochem. Physiol. C; Toxicol Pharmacol., 139(1-3):163-73.

Clayton M.E., Steinmann R. and Fent K., 2000. Different expression patterns of Hsps60 and 70 in Zebra mussels (*Dressena polymorpha*) exposed to copper and tributylin. Aquat. Tox., 47: 213-226.

Cowley Jr. B D. and Gudapaty S., 1995. Temporal alterations in reginal gene expression after nephrotoxic renal injury. J. Lab. Clin. Med., 125: 187-199.

Craig E A., Baxter B K., Becker J., Halliday J. and Ziegelhoffer T., 1994. Cytosolic Hsp70 of *Saccharomyces cerevisae*: roles in protein synthesis, protein translocation, proteolysis and regulation. In: The biology of Heat shock Proteins and Molecular Chaperons (R.I. Morimoto, A. Tissieres and C. Geogopoulos (Eds.), Cold Spring Harbor Lab. Cold Spring Harbor, NY.: 31-53.

Cristoba Susana 2007. Proteomics-Based Method for Risk Assessment of Peroxisome. Proliferating pollutants in the marine environment. EnvironmentalGenomics: Methods in Molecular Biology, 410: 123-135.

Dallas H.F. and Day J.A., 1993. The effect of water quality variables on riverine ecosystems: A review. Water Research Commission Project No. 351.Pretoria, South Africa. pp 240.

De Boeck G., De Wachter B., Vlaeminck A. and Blust R., 2003. Effect of cortisol treatment and/or sublethal copper exposure on copper uptake and heat shock protein levels in common carp, *Cyprinus carpio*. Environ. Toxicol. Chem., 22: 1122-1126.

Del Ramo J., Pastor A., Torreblanca A., Medina J., Diaz-Mayans J., 1989. Cd-binding protein in midgut gland of fresh water Crayfish (*Procambarus clarkii*). Biol Trace Elem Res., 21:75-80.

Derham B.K. and Harding J.J., 1999. -Crystallin as a molecular chaperone. Prog. Retinal Eye Res., 18: 463-509.

De-Smet-Hans and Blust-Ronny, 2001. Stress responses and changes in protein metabolism in carp, Cyprinus *carpio* during cadmium exposure. Ecotoxicol. Environ. Saf., 48(3): 255-262.

Dietz T.J. and Somero G.N., 1992. The threshold induction temperature of the 90 kDa heat shock protein is subject to acclimatization in eurythermal goby fishes (Genus *Gillichthys*). Proc. Natl. Acad. Sci. USA, 89: 3389-3393.

Dixon D.G. and Sprague J.B., 1981. Copper bioaccumulation and hepato-protein synthesis during acclimation to copper by juvenile rainbow trout. Aquatic. Toxicol., 1: 69.

Duffy L.K., Scofield E., Rodgers T., Patton M. and Bowyer R.T., 1999. Comparative baseline levels of mercury, Hsp70 and Hsp60. In: G. K. Iwama and others 18; Subsistence fish from the Yukon-Kuskokwim delta region of Alaska. Comp. Biochem. Physiol. Toxicol. Pharmacol., 124: 181-186.

DWAF., 1996. Department of water affairs and forestry. South African Water Quality Guidelines – Second Edition. Volume 7: Aqua. Ecosys., 159 pp.

Eaton D.I., 1985. Effect of various trace metals on the binding of cadmium to rat hepatic metallothionein determined by the Cd/hemoglobin affinity assay. Toxicol. Appl. Pharmacol., 78: 158.

Ellis R.J., 1989. The molecular chaperone concept. (Review). Semin. Cell. Biol., 1: 1-9.

EL-Rayis O.A.M., 1986. Bioaccumulation of Cd in plankton, bivalves, crustacean and in six different fish species from Mex Bay, west of Alexandria. FAO Fisheries Rep., 334 (suppl.): 50-56.

Fader S.C., Yu Z. and Spotila J.R. 1994. Seasonal variation in heat shock proteins (Hsp70) in stream fish under natural conditions. J. Therm. Biol., 19: 335-341.

Feder M.E. and Hofmann G.E., 1999. Heat-shock proteins, molecular chaperones, and the stress response: evolutionary and ecological physiology. Annu. Rev. Physiol., 61: 243-282.

Feng Q., Boone A.N., Vijayan M.M., 2003. Copper impact on heat shock protein 70 expression and apoptosis in rainbow trout hepatocytes. Comp. Biochem. Physiol., C 135: 345-355.

Fink A.L. and Goto Y., 1998. Molecular Chaperones in the life cycle of proteins: Structure, function, and mode of action. New York: Marcel Dekker.

Foneger J., Beedholm R., Clark B.F.C. and Rattan S.I.S., 2000. Mild stress induced stimulation of heat shock protein synthesis and improved functional ability of human fibroblasts undergoing aging *in vitro*. Exp. Geronto., 37: 1223-1228.

Forreiter C. and Nover L., 1998. Heat induced stress proteins and the concept of molecular chaperones. J. Biosci., 23: 287-302.

Forsyth R.B., Candido E.P.M., Babich S.L. and Iwama G.K., 1997. Stress protein expression in coho salmon with bacterial kidney disease. J. Aquat. Anim. Health, 9: 18-25.

Fowler B.A. and Silbergeld E.K., 1989. Occupational diseases – New work forces, new work places. Ann. N.Y. Acad. Sci., 572: 46-54.

Fukuda A., Osawa T., Oda H., Tanaka T., Toyokuni S., and Uchida K., 1996. Oxidative stress response in iron-induced acute nephrotoxicity: Enhanced expression of heat-shock protein 90. Biochem. Biophys. Res. Commun., 219: 76–81.

Funk A.E., Day F.A. and Brady F.O., 1987. Displacement ***of*** zinc and copper from copper-induced metallothionein by Cd and by Hg: In vivo and ex vivo studies. Comp. Biochem. Physiol. C. Pharmacol. Toxicol., 86: 1-6.

Geret F., Serafim A., Barreira L., Bebianno M.J., 2002. Effect of cadmium on antioxidant enzyme activities and lipid peroxidation in the gills of the clam, *Ruditapes decussatus*. Biomarkers,7(3):242-56.

Gething M J., and Sambrook J., 1992. Protein folding in the cell. Nature, 355: 33-45.

Goering P.L. and Fisher B.R., 1995. Metals and stress proteins. In: Handbook of Experimental Pharmacology. Vol. 115. Toxicology of Metals - Biochemical Aspects (R.A. Goyer and M.G. Cherian (Eds.). Spring Verlag. New York.: 229-266.

Goering P.L., Fisher B.R., Chaudhary P.P. and Dick C.A., 1992. Relationship between stress protein induction in rat kidney by mercuric chloride and nephrotoxicity. Toxicol. Apppl. Pharmacol., 113: 184-191.

Goering P.L., Fisher B.R., Noren B.T., Papaconstantinou A., Rojko J.L. and Marler R.J., 2000. Mercury induces regional and cell - specific stress protein expression in rat kidney. Toxicol. Sci., 52 (2): 447-457.

Goering P.L., Kish C.L. and Fisher B.R., 1993. Stress protein synthesis induced in rat liver by cadmium precedes hepatotoxicity. Toxicol. Appl. Pharmacol., 122: 139-148.

Goering Peter L., Kish Christina L. and Fisher Benjamin R., 1993. Stress protein synthesis induced by cadmium-cysteine in rat kidney. Toxicol., 85: 25-39.

Grosvik B.E. and Goksoyr A., 1996. Biomarker protein expression in primary cultures of salmon (*Salmo salar* L.) hepatocytes exposed to environmental pollutants. Biomarkers, 1: 45-53.

Gulfaraz M. and Ahmad T., 2001. Concentration level of heavy and trace metals in the fish. Displacement of Zn and Cu induced metallothionine and relevant water from Rawal and Mangal Lakes. Pak. J. Biol. Sci., 5: 414-416.

Guha B. and Khuda-Bukhsh A.R., 2004. Effect of vitamin supplementation on the recovery of ethyl methane sulphonate induced proteotoxicity in *Anabas testudineus*. Ind. J. Fish, 51 (4): 441-449.

Hansen B.H., Rømma S., Garmo Ø.A., Pedersen S.A., Olsvik P.A., Andersen R.A., 2007. Induction and activity of oxidative stress-related proteins during waterborne Cd/Zn-exposure in brown trout (*Salmo trutta*). Chemosphere, 67(11):2241-9.

Harper, 2006. Harper's review of biochemistry, 20th Edn. Lange Medical Publications, Maruzen Co. Ltd., California.

Hassanein H.M.A., Banhawy M.A., Soliman F.M., Abdel-Rehim S.A., Muller W.E.G. and Schroder H.C., 1999. Induction of Hsp70 by the herbicide oxyfluoren (goal) in the Egyptian Nile fish, *Oreochromis niloticus*. Arch. Env. Contain. Toxicol., 37: 78-84.

Hattori H., Kaneda T., Lokeshwar b., Laszlo A. and Ohtsuka K., 1993. A stress-inducible 40 kDa protein (Hsp70): purification by modified two-dimentional gel electrophoresis and co-localization with Hsc70 in heat shocked HELA cells. J. Cell Sci., 104: 629-638.

Hensen D.K., Anson J.F., Hinson W.G. and Pipkin Jr. J.L., 1988. Phenytoin-induced stress protein synthesis in mouse embryonic tissue. Proc. Soc. Exp. Biol. Med., 189: 136-140.

Hermesz E., Abraham M. and Nemcsok J., 2001. Identification of two Hsp90 genes in carp. Comp. Biochem. Physiol. C Toxicol. Pharmacol., 129: 397-407.

Hightower L.E., 1991. Heat shock, stress proteins, chaperones and proteotoxicity. Cell., 66: 191-197.

Hightower L.E., Norris C.E., DiIorio P.J. and Fielding E., 1999. Heat shock responses of closely related species of tropical and desert fish. Am. Zool., 39: 877-888.

Hobson J.F. and Birge W.J., 1986. Acclimation induced changes in toxicity and induction of metallothionein like proteins in the fathead minnow following sublethal exposure to zinc. Environ. Toxicol. Chem., 8: 157.

Hodson P.V., 1988. The effect of metal metabolism on uptake, disposition and toxicity in fish. Aquat Toxicol.,11: 3-18.

Howard K.J., Holley S.J., Yamamoto K.R., and Dustelhorst C.W., 1990. Mapping the Hsp90 binding region of the glucocorticoids receptor. J. Biol. Chem., 265: 11928-11935.

Huggett R.J., Kimerle R.A., Mehrle P.M. and Bergman H.L., 1992. In; Biomarkers: Biochemical, physiological and histological markers of anthropogenic stress. Lewis Publ., Boca Raton, FL. 347pp.

Iwama G.K., Afonso L.O.B., Todgham A., Ackerman P. and Nakano K., 2004. Are Hsps suitable for indicating stressed state in fish? J. Exp. Biol., 207: 15-19.

Iwama G.K., Thomas P.T., Forsyth R.B. and Vijayan M.M., 1998. Heat shock protein and physiological stress in fish. Am. Zool., 39: 901-909.

Iwama G.K., Thomas P.T., Forsyth R.B. and Vijayan M.M., 1998. Heat shock protein expression in fish. Rev. Fish Biol. Fish, 8: 35-56.

Iwama G.K., Vijayan M.M., Forsyth R.B. and Ackerman P.A., 1999. Heat shock proteins and physiological stress in fish. Am. Zool., 39: 901-909.

Jameson J.C., Kaplan H.A., Woloski B.M.R.N.I., Hellman M. and Ham K., 1983. Glycoprotein biosynthesis during the acute phase response to inflammation. Can. J, Biochem. Cell Biol., 61: 1041-1048.

Jana S.R. and Bandhopadhyay N., 1987. Effects of heavy metals on some biochemical parameters in freshwater fish, *Channa punctatus*. Environ. Ecol., 5: 488-493.

Jarosz D.F., Lindquist S., 2010. Hsp90 and environmental stress transform the adaptive value of natural genetic variation. Science, 330: 1820-4.

Joshi U.M. and Desai A.K., 1988. Biochemical changes in liver of fish *Telapia mossambicus* (Peters) during continuous exposure to monocrotophos. Ecotoxicol. Envi. Saf., 15: 272-276.

Juedes M.J. and Thomas P., 1984. Effects of lead on the acid-soluble thiol content of croaker, *Micropogonias undulates*. Am. Zool., 24: 137A.

Kagi J.H.R. and Schaffer A., 1988. Biochemistry of metallotheionein. Biochem., 27; 8509-8515.

Kamat H.A., Adhia M., Koppikar G.V. and Parekh B.K., 1999. Detection of antibodies to HIV in saliva. Natl Med J India, 12:159-61.

Kennish M.J., 1992. Ecology of Estuaries: Anthropogenic Effects. Boca Raton, USA: CRC Press: 494 pp.

Klassen C.D., Liu J. and Choudhary S., 1999. Metallothionein: An intraellular protein to protect against cadmium toxicity. Ann. Rev. Pharmacol. Toxicol., 39; 267-294.

Klavercamp J.F. and Duncan D.A., 1987. Acclimation to cadmium toxicity by white suckers: cadmium binding capacity and metal distribution in gill and liver cytosol. Environ. Toxicol. Chem., 6: 275.

Kothary R.K., Burgess E.A. and Candido E.P.M., 1984. The heat shock phenomenon in cultured cells of rainbow trout: Hsp70 mRNA synthesis and turnover. Biochim. Biophys. Acta., 783: 137-143.

Kraal M.H., Kraak M.H.S., Groot C.J., Davids C. and De–Groot C.J., 1995. Uptake and tissue distribution of dietery and aqueous Cd by Carp (*Cyprinus carpio*). Ecotoxicol. Environ. Safety., 31(2), 179-183.

Kristensen T.N. Sorensen J.G. and Loeschcke V., 2003. Mild heat stress at a young age in *Drosophila melanogaster* leads to increased Hsp 70 synthesis after stress exposure later in life. J. Gen., 3: 89-90.

Kubota H., Hynes G., Carne A., Ashworth A. and Willson K., 1994. Identification of six Tcp-I related genes encoding divergent sub units of the Tcp-1 containing chaperonin. Curr. Biol., 4: 89-99.

Laxmilatha P. and Laxminarayana A., 2003. Haemolymph proteins of the Indian white shrimp, *Penaeus indicus* (H. Milne Edwards). Indian J. Fish., 50 (3): 279-290.

Lehninger A.L,. 2008. Principles of Biochemistry 5th Edn. Michael M.Cox and David L. Nelson. New York.: 570-572.

Lele Z., Engel S. and Krone P.H., 1997. Hsp47 and Hsp70 gene expression is differentially regulated in a stress- and tissue-specific manner in zebrafish embryos. Dev. Genet., 21: 123-133.

Lim E.H. and Brenner S., 1999. Short-range linkage relationships, genomic organisation and sequence comparisons of a cluster of five HSP70 genes in *Fugu rubripes*. Cell. Mol. Life Sci., 55: 668-678.

Maclntyre S.S., Kushner I. and Samols D. 1985. Secretion of C-protein becomes more efficient during the acute phase response. J. Biol. Chem. 260: 4169-4173.

Maiti P. and Banerjee S., 1999. Accumulation of heavy metals in different tissues of fish *Oreochromis niloticus* exposed to waste water. Environ. Ecol., 17(4): 895-898.

Marr J.C.A., Lipton J., Cacela D., Hanse J.A., Bergman H.L., Meyer J.S. and Hogstrand C., 1996. Relationship between copper exposure duration, and rainbow trout growth. Aquat. Toxicol., 36: 17-30.

Martin C.C., Tang P., Barnrado G. and Krone P.H., 2001. Expression of the chaperonin 10 gene during zebrafish development. Cell Stress Chaperones, 6: 38-43.

Martinez M., Del Ramo J., Torreblanca A. and Diaz Mayans J., 1999. Effect of cadmium exposures on zinc level in the brine shrimp, *Artemia parthogenetica*. Aquaculture,172: 315-325.

McCarter J.A. and Roch M., 1984. Hepatic metallothionein and resistance to copper in juvenile coho salmon. Comp. Biochem. Physiol., 74: 133.

Michael D. Dutton, Malcolm Stephenson and Jack F. Klaverkamp, 1993. A Mercury saturation assay for measuring metallothionein in fish. Environ. Toxicol. & Chem., 12 (7): 1193–1202

Molina A., Biemar F., Muller F., Iyengar A., Prunet P., Maclean N., Martial J.A. and Muller M., 2000. Cloning and expression analysis of an inducible HSP70 gene from tilapia fish. FEBS Lett., 474: 5-10.

Montaser M., Mahfouz M.E., El-Shazly S.A.M., Abdel-Rahman G.H. and Bakry S., 2010. Toxicity of heavy metals on fish at Jeddah Coast KSA: Metallothionein expression as a biomarker and histopathological study on liver and gills. World J Fish Marine Sci., 2(3): 174- 185.

Morimoto R.I., Sarge K.D. and Abravaya K., 1992. Transcriptional regulation of heat shock genes. J. Biol. Chem., 267: 21987-21990.

Morimoto R.I., Tissieres A. and Georgopoulos C., 1990. The stress response, function of proteins, and prespectives. In: Stress Proeins in Biology and Medicine. Cold Spring Harbor Laboratory Press, Cold Spring Harbor, N.Y.: 1-36.

Morimoto R.I., Jurivich D.A., Kroega P.E. Mathur S.K., and 5 others, 1994. Regulation of heat shock transcription factor by a family of heat shock factors. In: The Biology of Heat Shock Proteins and Molecular Chaperons (R.1. Morimoto, A. Tissieres and C. Georgopoulos (Eds.), Cold Spring Harbour Lab. Press, NY.: 417-455.

Mustafa S. J. and Chandra S., 1972. Adenosine deaminase and protein pattern in serum and cerobro-spinal fluid in experimental manganese encephalopathy. Arch. Toxicol., 28: 279-285.

Mukhopadhyay M.K. and Konar S.K., 1985. Toxic effects of metals mixture on fish and aquatic ecosystem. Environ and Ecol., 3: 22-29.

Nakano K. and Iwama G.K., 2002. The 70-kDa heat shock protein response in two intertidal sculpins, *Oligocottus maculosus* and *O. snyderi*: relationship of Hsp70 and thermal tolerance. Comp. Biochem. Physiol., *A* 133: 79-94.

Norris C.E., Dilorio P.J., Schultz R.J. and Hightower L.E., 1995. Variation in heat shock proteins within tropical and desert species of poeciliid fishes. Mol. Biol. Evol.,12: 1048-1062.

Nover l., 1991. Heat shock Response. CRC Press, Boca Raton, FL.

Olson P.E., 1996. Metallothionein in fish; Induction and use in environmental monitoring. In: toxicology of aquatic pollution: Physiological, molecular and cellular approaches (E.W.Taylor, Ed.). Cambridge Univ. Press, Cambridge, UK: pp 187-203.

Pipkin J.L., Anson J.F., Hinson W.G., Burns E.R. and Casciano D.A., 1988. Comparative immunochemical and structural similarities of five stress proteins from various tissue types. Comp. Biochem. Physiol., B89: 43-50.

Plumier J.C. Ross B.M., Currie R.W. and others, 1997. Transgenic mice expressing the human heat shock protein 70 have improved post ischemic myocardial recovery, J. Clin. Inves., 95: 1854-1860.

Price-Haughey J. and Gedamu L., 1987. Heavy metal induced protein synthesis in fish cell lines. Experientia Suppl., 52: 465-469.

Rabergh C.M.I., Airaksinen S., Soitamo A., Bjorklund H.V., Johansson T., Nikinmaa M. and Sistonen J., 2000. Tissue-specific expression of zebrafish (*Danio rerio*) heat shock factor 1 mRNAS in response to heat stress. J. Exp. Biol., 203: 1817-1824.

Rafia Azmat, Farha Aziz and Madiha Yousfi, 2008. Monitoring the effect of water pollution on four bioindicators of aquatic resources of Sindh Pakistan. Res. J. Environ. Sci.,2 (6): 465-473.

Rema L.P. and Philip B., 1996. Metallothiolein on metallothionein like proteins and heavy metals toxicity in *Oreochromis mossambicus* (Peters). Ind. J. Exp. Biol., 34: 527-530.

Roch M. and McCarter A., 1984. Hepatic metallothionein production and resistance to heavy metals by rainbow trout, *Salmo gairdneri*. II Held in a series of contaminated lakes. Comp. Biochem. Physiol., 77C.: 77.

Roesijadi G., 1992. Metallothioneins in metal regulation and toxicity in aquatic animals. Aquat. Toxicol., 22: 81.

Rose Sumit, Vincent S., Meena B., Suresh A. and Mani R., 2014. Metallothionein induction in fresh water catfish, *Clarias gariepinus* on exposure to cadmium. Intl. Int J. Pharm. Sci., 6 (1): 377-383

Roth I., Hornung H., 1977. Heavy metal concentrations in water sediments and fish from Mediterranean coastal area, Israel. Environ Sci. & Tech., 11: 265-326.

Sanders B.M., 1990. Stres proteins: Potential as multi-tired biomarkers. In: Biomarkers of Environmental Contamination, J. F. McCathy. and L.R. Shugart (Eds.). Lewis Pub., Boca Raton.:165-191.

Sanders, B.M. 1993. Stress proteins in aquatic organisms : an environmental perspective. Crit. Rev. Toxicol., 23: 49.

Sanders B.M. 1994. Stress proteins as molecular chaprones : Implication for toxicology. Env. Helth. Pres. 102 (Number 6-7): 538-540.

Sanders B.M., Georing P.L. and Jenkins K., 1996. The role of general and metal specific cellular responses in protection and repair of metal induced demage-stress proteins and metallothioneins: Toxicology of metals (L.W. Chang, Ed.). CRC/Lewis Publishers, New York: 165-187.

Sanders B.M., Nguyen J., Martin L.S., Howe S.R. and Coventry S., 1995. Induction and subcellular localization of two major stress proteins in response to copper in the fathead minnow *Pimephales promelas*. Comp. Biochem. Physiol., C 112: 335-343.

Santacruz H., Vriz S. and Angelier N., 1997. Molecular characterization of a heat shock cognate cDNA of zebrafish, Hsc70, and developmental expression of the corresponding transcripts. Dev. Genet., 21: 223-233.

Sarita, 2005. Effects of heavy metals on enzyme activity in some freshwater fishes. PhD thesis, Deptt. Zoology., CCS HAU., Hisar, Haryana: 195 + xxii.

Sathiyaa R., Campbell T. and Vijayan M.M., 2001. Cortisol modulates HSP90 mRNA expression in primary cultures of trout hepatocytes. Comp. Biochem. Physiol., *B* 129: 679-685.

Sato J., Goto J., Murata T., Kitamori S., Yamazaki Y., Satoh A., 2010. Changes in saliva interleukin-6 levels in patients with oral squamous cell carcinoma. Oral Surg., Oral Med., Oral Pathol., Oral Radiol., Endod.110: 330-6.

Scheuhammer A.M. and Cherian M.G., 1986. Quantification of metallothioneins by a silver saturation method. Toxicol. Appl. Pharmacol., 82: 417-425.

Schlesinger M.J., Ashburner M. and Tissieres A., 1982. Heat shock proteins : From Bacteria to Man. Cold Spring Harbor Laboratory Press, Cold Spring Harbor, N.Y.

Schreck C.B., 2000. Accumulation and long term effects of stress in fish. In: The Biology of Animal Stress: Basic Principles and implications for Animal Welfare. G.P. Moberg and J.A. Mench (ed.). Walingford, CABI Publ., UK.: 147-158.

Sharma Mukta, 2006. Chronic effects of pesticides on tissue proteins and metabolic enzyme activity in *Labeo rohita* and *Cyprinus carpio*. PhD thesis, Deptt. Zoology., CCS HAU., Hisar, Haryana: 115+xxxvii.

Sharma S.P. and Bhatia P.1982. Amino acid changes in ageing bruchids. Curr. Sci., 51: 664-665.

Shelly Virmani, 2005. Effects of heavy metals on histopathological and biochemical changes in *Cyprinus carpio* (L) and *Cirrhinus mrigal* (Am). PhD thesis, Deptt. Zoology., CCS HAU., Hisar, Haryana: 195+xxxiii

Shelton K.R., Egle P.M. and Todd J.M. 1986a. Evidence that glutathione participates in the induction of a stress protein. Biochem. Biophys. Commun., 134: 492-498.

Shelton K.R., Todd J.M., Egle P.M., 1986b. The induction of stress related proteins by lead. J. Biol. Chem., 261: 1935-1940.

Smith T.R., Tremblay G.C. and Bradley T.M., 1999. Characterization of Hsp response of Atlantic salmon (*Salmo salar*). Fish Physiol. Biochem., 20: 279-292.

Somasunadaram B., King P. and Shackley S., 1984. The effect of Zinc on the ultrastructure of the trunk muscle of the larvae of *Clupea harengus* (L.). Comp Biochem. Physiol. C., 79: 311-315.

Suresh A., Sivaramakrishna B., Victoriamma P.C. and Radhakrishnaiah K., 1991. Shifts in protein metabolism in some organs of freshwater fish, *Cyprinus carpio* under mercury stress. Biochem. Int., 24(2): 379-389.

Tabche M.L., Gomes O.L., Galar M.M. and Lopez L.E. 2002. Stress proteins produced by contaminated sediments with nickel in a pond with rainbow trout *Oncorhynchus mykiss* (Pisces : Salmonide). Rev. Biol. Trop., 50(3-4): 1159-1168.

Thomson K.W., Hendricks A.C. and Crains J. Jr., 1980. Acute toxicity of Zn and Cu singly and in combination to blue gill (*Lepomis macrochrus*). Bull. Environ. Contam. Toxicol., 25: 122-129.

Tytell M., Barbe M.F., and Brown I.R., 1994. Induction of heat shock (stress) protein 70 and its mRNA in the normal and light damaged rat retina after whole body hyperthermia. J. Neurosci. Res., 38:19-31.

Van Campenhout K., Infante H. G., Adams F. andBlust R., 2004. Induction and binding of Cd, Cu, and Zn to metallothionein in carp (*Cyprinus carpio*) using HPLC-ICP-TOFMS. Toxicol Sci., 80(2):276-87.

van der Oost R, Beyer J, Vermeulen N.P.E., 2003. Fish bioaccumulation and biomarkers in environmental risk assessment: a review. Environ. Toxicol. Pharmacol.,13: 57–149.

Varó I., Serrano R., Pitarch E., Amat F., López F.J., Navarro J.C., 2002. Bioaccumulation of chlorpyrifos through an experimental food chain: study of protein HSP70 as biomarker of sublethal stress in fish. Arch. Environ. Con. Tox., 42: 229-235.

Vasak M., 2005. Advances in metallothionein structure and functions. J. Trace Elem. Med. Biol.,19:13–17.

Vijayan M.M. and Moon T.W., 1994. The stress response and the plasma disappearance of corticosteroid and glucose in a marine teleost, the sea raven. Can. J. Zool., 72: 379-386.

Vijayan M.M., Pereira C., Kruzynski G. and Iwama G.K. 1997a, 1998. Sublethal concentrations of contaminant induce the expression of Hsp 70 in two salmonids. Aquat. Toxicol., 40: 101-108.

Vijayan M.M., Pereira C., Forsyth R.B., Kennedy C.J. and Iwama G.K., 1997b. Handling stress does not affect the expression of hepatic heat shock protein 70 and conjugation heat-shock-cognate Hsc71 gene from rainbow trout. Eur. J. Biochem., 204: 893-900.

Washburn B.S., Moreland J.J., Slaughter A.M., Werner I., Hinton D.E. and Sanders B.M., 2002. Effects of handling on heat shock protein expression in rainbow trout (*Oncorhynchus mykiss*). Environ. Toxicol. Chem., 21: 557-560.

Weigant F.A.C., Rijn J.V. and Wijk R.V., 1997. Enhancement of stress response by minute amounts of cadmium in sensitized Renbar H35 hepatoma cells. Toxicol., 116: 27-37.

Weiss P., 1984. Metallothionein and mercury tolerance in the killifish, *Fundulus heteroclitus.* Ar. Environ. Res., 14: 153.

Welch W.J., 1987. The mammalian heat shock (or stress) response. A cellular defense mechanism. Adv. Exp. Med. Biol., 225: 287-304.

Welch W.J., 1990. The mammalian stress response: cell physiology and biochemistry of stress proteins. In: Stress proteins in biology and medicine (R.I. morimoto, A. Tissieres, and C. Georgopoulos (Eds). Cold Spring Harbour Laboratory Press. Cold Spring Harbor, NY: 223-278.

Welch W.J., 1993. How cells respond to stress. Sci. Am., 268: 56-64.

Welch W.J., Mizzen LA., 1988. Characterization of the thermo-tolerent cell: II. Effects on the intercellular distribution of heat-shock protein 70, intermediate filaments and small nuclear ribo-nuclear protein complexes. J. Cell Biol., 106: 1117-1130.

Will J.H., Farag A.M., Stansbury M.A., Young, P. A. Berrgman, H.L. and Petersen N.S. 1996. Accumulation of Hsp 70 in juvenile and adult rainbow trout gill exposed to metal- contaminated water and/or diet. Environ. Toxicol. Chem., 15: 1324-1328.

Williams T.P., Bubb J.M. and Lester J.N., 1994. Metal accumulation with in salt marsh environments: a review. Marine pollution bulletin, 28: 277-290.

Wu M.M., Chiou H.Y., Ho I.C., Chen C.J. and Lee T.C., 2003. Gene expression of inflammatory molecules in circulating lymphocytes from arsenic-exposed human subjects. Environ. Helth. Perspect., 11: 1429-1438.

Young R.A., 1990. Stress proteins and immunology. Annu. Rev. Immunol., 8: 401-420.

Zacharia S., Thomas P.C. and Paulton M.P., 2003. Changes in serum proteins and immune response in *Oreochromis mossambicus* induced by aflatoxin B_1. Ind. J. Fish, 50 (4): 421-426.

Zarate J. and Bradley T.M., 2003. Heat shock proteins are not sensitive indicators of hatchery stress in salmon. Aquaculture, 223: 175-187.

ZhaoYinzhi, Chen Lijun,Gao Song, Toselli Paul, Stone Phillip, and Li. Wande, 2010. The critical role of the cellular thiol homeostasis in cadmium perturbation of the lung extracellular matrix. Toxicology, 267(1-3): 60.

Chapter 14

Oxidative Stress and Arsenic Toxicity

1. Oxidative Stress

Oxygen is indispensable to metabolism and energy production in almost all forms of life. Oxidative stress (OS) is caused by an imbalance between pro-oxidants and antioxidants (Al-Gubory *et al.*, 2010). This ratio can be altered by increased levels of reactive oxygen species (ROS) and/or reactive nitrogen species (RNS), or a decrease in antioxidant defense mechanisms (Ruder *et al.*, 2009). Reactive oxygen species and reactive nitrogen species serve as signaling messengers for the evolution and perpetuation of the inflammatory process that is often associated with the condition of oxidative stress, which involves genetic regulation. Reactive oxygen species (ROS) such as hydrogen peroxide, superoxide anion, oxygen and hydroxyl radicals, can directly or indirectly damage cellular DNA and protein. Reactive oxygen species (ROS), however, generated from some life processes can cause damage to cells, tissues, and organs. Oxidative damage on account of the oxidative stress is said to be a cause of pathogenesis of a variety of chronic degenerative and neurodegenerative diseases *i.e.* damage to the tissue and the cellular components such as membranes, DNA and proteins, thus, causing pathological conditions like inflammation, sclerosis, cancer and aging. Certain amount of ROS is needed for the progression of normal cell functions, provided that upon oxidation, every molecule returns to its reduced state (Chandra *et al.*, 2009). Excessive ROS production, however, may overpower the body's natural antioxidant defense system, creating an environment unsuitable for a normal physiological reaction (Al-Gubory *et al.*, 2010). Changes in the pattern of gene expression through reactive oxygen species/ reactive nitrogen species-sensitive regulatory transcription factors are crucial components of the machinery that determines cellular responses to oxidative/ redox conditions. A progressive rise of oxidative stress due to altered reduction–

oxidation (redox) homeostasis appears to be one of the hallmarks of the processes that regulate gene transcription in physiology and patho-physiology. Cancer in fact is considered a disease of genetic progression that is often associated with specific molecular, genetic and histological changes (Bayani and Squire, 2007). Based on recently developed fish technologies and tissue microarray methods the ability to develop biomarkers that can detect the critical components of these hallmarks of cancer together are now said to provide a powerful basis for diagnosing, monitoring and predicting outcome and response to treatment. The adverse effects of OS have been well documented by various workers in male and the female individuals leading to a number of reproductive diseases including sperm quality in males and endometriosis, polycystic ovary syndrome (PCOS), and unexplained infertility in females (Al-Gubory *et al.*, 2010). Physiological levels of reactive oxygen species (ROS) play an important regulatory role through various signaling transduction pathways in folliculo-genesis, oocyte maturation, endometrial cycle, luteolysis, implantation, embryogenesis, and pregnancy (Agarwal *et al.*, 2008).

The human body is composed of many important signaling pathways. Continuous production of ROS in excess can induce negative outcome of many signaling processes thereby disrupting the physiologic processes required for cell growth and proliferation. Damage induced by ROS can occur through the modulation of cytokine expression and pro-inflammatory substrates via activation of redox-sensitive transcription. Reactive oxygen species do not always directly target the pathway; instead they may produce abnormal outcome by acting as second messenger in some intermediary reactions (Irani, 2008). Likewise, oxygen-derived free radicals such as hydroxyl and hydroperoxyl species have also been shown to oxidize phospholipids and other membrane lipid components leading to lipid peroxidation. Deleterious attacks from excess ROS may ultimately end in cell death and necrosis. These harmful attacks are mediated by the following specialized mechanisms (Burton and Jauniaux, 2010):

i. ***Opening of ion channels***: The concentration of Ca^{2+} must be tightly regulated as it plays an important role in many physiological processes. The presence of excessive amounts of ROS can increase Ca^{2+} levels. Excessive release of Ca^{2+} from the endoplasmic reticulum (ER), resulting in mitochondrial permeability. Consequently, the mitochondrial membrane potential becomes unstable and ATP production ceases.

ii. ***Lipid peroxidation***: Lipids are essential components of cell membranes that maintain structures and control the cell functions. Lipid peroxidation is the reaction of lipids with the ROS, and is implicated in several human diseases and exposures, such as cancers. This occurs in areas where polyunsaturated fatty acid side chains are prevalent. These chains react with O_2, creating the peroxyl radical, which can obtain H^+ from another fatty acid, creating a continuous reaction.

iii. ***Protein modifications***: Amino acids are targets for oxidative damage. Direct oxidation of side chains can lead to the formation of carbonyl groups.

iv. ***DNA oxidation***: Mitochondrial DNA is particularly prone to ROS attack due to the presence of O_2^- in the electron transfer chain (ETC), lack of histone protection, and absence of repair mechanisms.

2. Lipid Peroxidation

Lipid peroxidation is a process generated naturally in small amounts in the body, mainly by the effect of several reactive oxygen species (hydroxyl radical, hydrogen peroxide *etc.* It can also be generated by the action of several phagocytes. These reactive oxygen species readily attack the polyunsaturated fatty acids of the fatty acid membrane, initiating a self-propagating chain reaction. The destruction of membrane lipids and the end-products of such lipid peroxidation reactions are especially dangerous for the viability of cells, even tissues. The existing natural enzymatic (catalase, superoxide dismutase) and non-enzymatic (vitamins A and E) antioxidant defense mechanisms may e overcome, causing lipid peroxidation to take place. Since lipid peroxidation is a self-propagating chain-reaction, the initial oxidation of only a few lipid molecules can result in significant tissue damage. The length of the propagation, however, depends on the chain breaking antioxidant. In recent years it has become apparent that the oxidation of lipids, or lipid peroxidation, is a crucial step in the pathogenesis of several diseases (Mylonas and Kouretas, (1999); as well as for assessing the metal toxicity effects in animals including the fish. All the cellular membranes in animals in general, possess 50 per cent of their mass comprised of lipids, especially the phospholipids (PLFA).

Lipid peroxidation reveals free ion composition in tissues after metal toxicity in fish. Cd is a toxic heavy metal for humans, animals and plants (Wagner, 1993). Dimitrova and Tishinova (1989) in one year old common carp, *C. carpio*, exposed to cadmium sulphate for 24 hours recorded increased lipid peroxidation in the liver tissues with the increasing Cd concentration in water, whereas it appeared reduced in the control fish. It was concluded that Cd participates in the oxidation reactions in the liver cells to produce lipid peroxides. Stohs and Bagchi (1995) have stated that metals, including iron, copper, chromium, and vanadium undergo redox cycling, while cadmium, mercury, and nickel, as well as lead, deplete glutathione and protein-bound sulfhydryl groups, resulting in the production of reactive oxygen species as superoxide ion, hydrogen peroxide, and hydroxyl radical. According to Stohs and Bagchi (1995), Cd induced changes of the antioxidant status occur either by enhancing superoxide radical production and lipid peroxidation, or by reducing the enzymatic and non-enzymatic antioxidants. The mechanisms involved in Cd toxicity have been demonstrated with reference to oxidative burst in many animal cells (Lopez *et al.*, 2006; Pathak and Khandelwal, 2006).

3. Antioxidants

The human body has an intrinsic ability to fight against such oxidative stress. A wide array of antioxidants constitutes a fundamental cellular defense system against oxidative and electro-philic insults. Preservation of physiological cellular functions since depends on the homeostatic balance between oxidants and antioxidants, antioxidants act as scavengers that detoxify excess ROS, which helps in maintaining the body's delicate oxidant/antioxidant balance. There are enzymatic and non-enzymatic antioxidants available in the tissues, which play a protective role against the OS damages. Enzymatic antioxidants possess a metallic center, which gives them the ability to take on different valences as they transfer electrons to balance molecules for the detoxification process. They neutralize excess ROS and prevent damage to cell structures. Endogenous enzymatic antioxidants include SOD, catalase, glutathion peroxidase (GPx), and glutathione oxidase. The non-enzymatic antioxidants consist of dietary supplements and synthetic antioxidants, such as vitamin C, taurine, hypotaurine, vitamin E, Zn, selenium (Se), beta-carotene, and carotene *etc.*

a. Superoxide Dismutase (SODs)

SOD is one of the key enzymes that provide the first line defense against the pro-oxidants and defends against free radical species by disproportionating O_2^- into hydrogen peroxide and oxygen molecules (Bowler *et al.*, 1992) *i.e.* it catalyses the transformation of superoxide radicals to H_2O_2 and O_2 (Lauterburg *et al.*, 1983). SOD is a metalloenzyme and is an important antioxidant defense in nearly all the cells exposed to free oxygen. A specific metal cofactor is required for this superoxide-scavenging activity, SODs are thus, classified in terms of the metal to which they bind, *i.e.* as copper/zinc SOD (Cu-Zn SOD), iron SOD (Fe SOD), manganese SOD (Mn SOD), or nickel SOD. Most eukaryotic cells contain two types of SODs, with Cu-Zn SOD largely localized in the cytoplasm and Mn-SOD localized in mitochondria. The cytosols of virtually all eukaryotic cells contain an SOD enzyme with copper and zinc. SOD is fundamental to anti-oxidative reactions and it causes dis-mutation of the SO anion to H_2O_2.

b. Glutathione (GSH)

GSH is a sulfhydryl antioxidant, antitoxin, and enzyme co-factor. It is involved in many processes including protein and DNA synthesis, xenobiotics conjugation, and antioxidant protection. Glutathione (GSH) family of enzymes: GPx, GST, and GSH reductase, play multiple roles including maintenance of cells in a reduced state and formation of conjugates with some hazardous endogenous and xenobiotic compounds. The biochemical function of glutathione peroxidase is to reduce lipid hydroperoxides to their corresponding alcohols and to reduce free hydrogen peroxide to water. Depletion of GSH results in DNA damage and increased H_2O_2 concentrations; as such GSH is an essential antioxidant. During the reduction of H_2O_2 to H_2O and O_2, GSH is oxidized and glutathione reductase participates in the reverse reaction, thus, recycling the GSH (Perkins, 2006). A tripeptide glutathione in its reduced form (GSH) and glutathione S-transferases (GST) are useful biomarkers in fish, but their utility is less sustainable than in the case of CYP 1A (Van der Oost *et al.*, 2003). GSH is a known non-enzymatic molecule of all seasons as it well represents in counteracting chemical toxicity in animals (Gallagher *et al.*, 1992); and the intracellular GSH is said to present the main molecule involved in the defense of cells against oxidative stress (Otto and Moon, 1996). A cell utilizes glutathione along with ascorbic acid and vitamin in the process of scavenging free radicals by superoxide dismutase (SOD) and catalase. A number of enzymes are reported to be involved both in the synthesis of glutathione in liver as well as its conjugation with xenobiotics in liver. Several investigators have reported elevated concentrations of glutathione in liver or gill from exposures to a variety of organic chemicals like metals or the pesticides (Chatterjee and Bhatacharya, 1984; Thomas and Wofford, 1984; Gallagher *et al.*, 1992). Arabi, 2004 studied the extent of cellular damage in the gills of *C carpio*, due to metal contamination. They reported that Cu and Hg had a deleterious effect on gill membrane integrity and GSH contents in a relatively dose dependent manner.

c. Xanthine Oxidase (XO)

XO is a most free radical producing enzyme as it generates O^{2+} during reduction of O_2 to H_2O_2. An increase in XO means more ROS due to effective detoxification of NH_3 by XO. Elevation in the levels of transaminase enzymes (GPT and GOT) as well as simultaneous decrease in protein levels, clearly suggest shift to protein metabolism as supplementary source of energy and an increased XO level, later indicating retardation of purine metabolism. Increased GPT is an indication of formation of ammonia in experimental fish.

d. Catalase (CAT)

CAT is another important indicator of oxidation stress in animals, and it belongs to the cellular antioxidant system that counteracts the toxicity of reactive oxygen species (ROS). CAT is a common enzyme found in nearly all living organisms exposed to oxygen and it is known to catalyze the decomposition of hydrogen peroxide to water and oxygen. They are the heme-containing enzymes that facilitate the removal of H_2O_2, which is metabolized to O_2 and water. It is a very important enzyme in reproductive functions, and has one of the highest turnover numbers of all enzymes; one catalase molecule can convert millions of molecules of hydrogen peroxide to water and oxygen each second.

e. Glutathion S-transferase (GST)

Glutathione S-transferases (GST) are a super family of primarily soluble enzymes with functions in cellular transport, synthesis of leukotrienes and prostaglandins and defense against oxidative damage and peroxidative products of DNA and lipids (George, 1994). Glutathione S-transferases (GSTs), previously known as ligandins, comprise a family of eukaryotic and prokaryotic phase II metabolic isozymes and are best known for their ability to catalyze the conjugation of the reduced form of glutathione (GSH) to xenobiotic substrates for the purpose of detoxification. (van der Oost *et al.*, 2003). These conjugate electrophilic xenobiotics and endogenous substances with the nucleophilic tripeptide glutathione from the intracellular pool. This conjugation is likely to render the molecules more water soluble, therefore, favoring their elimination from the organism. Spectrum of potential GST inducers is variable, increased overall GST activity however, has been demonstrated in the liver of various fish species after the exposure to PCBs (Schmidt *et al.*, 2004), PAHs (Henson *et al.*, 2001), and some pesticides (Frasco and Guilhermino,2002). In a study conducted by Parvez and Raisuddin 2007, on the exposure of freshwater fish, *Channa punctatus* to a quaternary herbicide paraquat, (widely used for broadleaf weed control and is said to be highly toxic to fish), observed that the levels of reduced glutathion were significantly reduced in the liver and gill of the exposed fish and the levels of ascorbic acid were increased in liver. This study demonstrate the possibility of using such non enzymatic antioxidants as biomarkers of environmental exposures for monitoring ecotoxilogical risk assessment in fishes including the oxidative stress inducing potential of the herbicide, Paraquat in fish. Sanchez *et al.* (2005) working with three-spined stickleback, also showed that copper induced oxidative stress in fish liver before significant metal accumulation could be detected and suggested the existence of homeostatic regulation mechanisms. On the contrary, Dautremepuits *et al.* (2004) found no effect in GST activity in the head kidney and liver of carps exposed to 0.25 mg L^{-1} of copper regardless the length of exposure (4 or 8 days).

GST is involved in protecting cells against peroxidative damage, and is ubiquitous in the cytosol and microsomes of eukaryotes; the GST family thus, is classified into three superfamilies: as the cytosolic, mitochondrial, and microsomal. GST is responsible for the cellular metabolism and detoxification of a variety of cytotoxic and carcinogenic compounds by catalysis and their conjugation with glutathione-II. GSTs previously known as ligandins, comprise a family of eukaryotic and prokaryotic phase-II metabolic isozymes, best known for their ability to catalyze conjugation of the reduced form of glutathione (GSH) to xenobiotic substrates for the purpose of detoxification. GST-mediated conjugation might be an important mechanism for detoxifying pre-oxidised lipid breakdown products, which have a number of adverse biological effects when present in higher amounts. Induced GST activity indicated the role of this enzyme in protection against the toxicity of xenobiotics (Leaver and George, 1998).

f. Non-Enzymatic Antioxidants

Vitamin C (ascorbic acid) is a known redox catalyst that can reduce and neutralize ROS. Vitamin C scavenges the aqueous reactive oxygen species (ROS) by very rapid electron transfer that thus, inhibits lipid peroxidation. Its reduced form is maintained through reactions with GSH and can be catalyzed by protein disulfide isomerase and glutaredoxins. Vitamin E (γ-tocopherol) is a lipid soluble vitamin with antioxidant activity. The lipid soluble, non-enzymatic antioxidant, γ-tocopherol checks lipid peroxidation through limiting the propagation of chain reaction by reacting with lipid radicals produced during lipid peroxidation (Behrman *et al.*, 2001). The protective mechanism of vitamin E could be attributed to its antioxidant property or its location in the cell membrane and its ability to stabilize membrane by interacting with unsaturated fatty acid chain (Chang *et al.*, 1978). Glutathione, a peptide found in most forms of aerobic life as it is made in the cytosol from cysteine, glutamate, and glycine (Behrman *et al.*, 2001), is also the major non-enzymatic antioxidant found in oocytes and embryos. Its antioxidant properties stem from the thiol group of its cysteine component, which is a reducing agent that allows it to be reversibly oxidized and reduced to its stable form.

The hormone melatonin is also an antioxidant that unlike vitamins C and E and GSH, is produced by the human body. Melatonin, N-acetyl-5-methoxy tryptamine, is a hormonal product of the pineal gland that plays many roles within the body including control of reproductive functions, modulation of immune system activity, and limitation of tumor oogenesis and effective inhibition of oxidative stress (Hardeland *et al.,* 1993). One major function of melatonin is to scavenge radicals formed in oxygen metabolism, thereby potentially protecting against free radical induced damage to DNA, proteins and membranes. In contrast to other antioxidants, melatonin cannot undergo redox cycling, once melatonin is oxidized it is unable to return to its reduced state, because it forms stable end-products after the reaction occurs. Transferrin and ferritin, both iron-binding proteins, also play an important role in antioxidant defense by preventing the catalyzation of free radicals through chelation (Shkolnik *et al.,* 2011). Nutrients such as Se, Cu, and Zn are required for the activity of some antioxidant enzymes, although they have no antioxidant action themselves.

4. Oxidative Stress and Fish Protection

Hydrogen peroxide (H_2O_2) and synthetic peroxides can exert pharmacological and toxicological effects on tissues. H_2O_2 is a major component of the ROS produced intra-cellularly during physiological and pathological processes, in addition to being the cause of oxidative damage. Hydrogen peroxide has important role as a signaling molecule in the regulation of a wide variety of biological processes. The compound is a major factor implicated in the free-radical theory of aging, based on how readily hydrogen peroxide can decompose into a hydroxyl radical and how superoxide radical byproducts of cellular metabolism can react with ambient water to form hydrogen peroxide. These hydroxyl radicals in turn readily react with and damage vital cellular components, especially those of the mitochondria. At least one study has also tried to link hydrogen peroxide production to cancer.[The enhanced XO activity indicates increased oxy-radical production and this result in the peroxidation of polyunsaturated fatty acids, means damaging cell membranes. And in consonance with the increased XO activity, CAT and GST activities are also increased due to their involvement in the decomposition of superoxide radicals and H_2O_2, thereby decreasing chemical toxicity. OS also cause lipid peroxidation, protein and enzyme oxidation and GSH depletion. DNA-strand breaks result in chromosomal aberrations. All the proposed mechanisms are likely to be non-linear at low doses and there have been evidences of increase in cancer cases only at higher arsenic concentrations.

Heavy metals are competent of inducing toxicity in living organisms because of their capability of interacting with the nuclear proteins and nucleic acids causing oxidative deterioration of biomolecules (Leonard *et al.,* 2004). There have been some reports confirming enzymatic changes in fish organs in response to oxidative stress on their exposures to heavy metals. Hansen *et al.,* 2006a carried out studies in three populations of brown trout (*Salmo trutta*) exposed to different metal levels in their natural environments with respect to antioxidants metallothionein (MT), superoxide dismutase (SOD) and catalase (CAT) as well as for corresponding mRNA levels. It was concluded that the observed metal effects relies on acclimation rather than on genetic adaptation in the metal exposed populations. The data indicated that chronic exposures of brown trout to Cd, Zn and/or Cu did not involve maintenance of high activities of SOD and CAT enzymes in gills, but in livers, mRNA levels of SOD,

CAT and GPx were higher in the metal-exposed trout. However, the metal-exposed groups had higher activities of SOD enzyme in liver compared to the unexposed reference trout, and CAT activity was found to be higher in kidneys of Cu-exposed trout. The Cu-exposed trout did not seem to rely on MT production to avoid Cu toxicity in gills, but rather by keeping the Cu uptake at a low level.

Catalase, a hydrogen peroxide scavenger, catalyzes the breakdown of hydrogen peroxide to water and molecular oxygen to protect cells against the toxic effects of hydrogen peroxide. Peroxidases are enzymes that reduce a variety of peroxides to their corresponding alcohols. Glutathione peroxide is considered to play an important role in protecting membranes from damage due to lipid peroxides (LP). This observation led to the view that the major detoxification function of GPx is the termination of radical chain propagation by quick reduction to yield further radicals (Kono and Fridovich, 1982). GPxs are regarded as important components of the cellular system of defense against oxidative stress resulting from the metabolism of xenobiotics (Thomas *et al.*, 1990).

Toxic stress is known to alter the activity of SOD in the vital tissues of fish. Pharmaceutically, SOD reduces free radical damage and it also has a powerful anti-inflammatory activity. Zikic *et al.* (2001) exposed gold fish to cadmium in a concentration of 20 mg Cd/l for 1, 4, 7 and 15 days. Activity of superoxide dismutase (SOD) in the red blood cells (RBC) significantly decreased after the first day of exposure. However, the SOD activity increased after 7 and 15 days of treatment. Elevated activity of catalase (CAT) was also recorded after 15 days. Almeida *et al.*, 2002 in Nile tilapia, *O. niloticus* also concluded that oxygen free radicals are produced as a mediator of Cd toxicity, since there were changes in antioxidant enzymes glutathione peroxidase (GSH-Px) and superoxide dismutase (SOD) in the liver, red and white muscles following Cd toxicity without any change in lipoperoxide. They further indicated development of resistant, related with increased activities of antioxidant enzymes, which were important in the protection of fish against Cd damage, inhibiting lipoperoxide formation. Earlier Palace *et al.* (1993) also observed the tendency of lyses in the RBCs of rainbow trout after their exposure to Cd for 181 days. This increased cell fragility was attributed to lipid peroxidation of the erythrocyte membrane which affects membrane fluidity. They also presented evidence of stimulated anti-oxidant defenses which are mobilized to counteract increased peroxidation. The findings of Siraj and Rani (2003) have indicated that in case of Cd toxicity, the glutathion- dependent enzymes as well as other such antioxidant enzymes protect tissues against cadmium toxicity and that these antioxidants provide first line of defense against cadmium, before the induction of synthesis of any metallothionein in tissues. Study of low concentrations of cadmium-induced oxidative stress response and DNA damage in the livers of *Cyprinus carpio* var. *color*, and measuring SOD, malondialdehyde (MDA), and GSH in liver (Jia *et al.*, 2011), showed that MDA and GSH levels in all treatment groups increased significantly relative to the controls. DNA damage percentage, TL and TM also significantly increased when the Cd level was >0.41 mg/L. Positive correlations were also found between DNA damage levels and MDA levels, as well as between

GSH and MDA levels; suggesting strongly that Cd-induced DNA damage in the livers of *Cyprinus carpio* var. *color* was due to lipid peroxidation and oxidative stress.

In hepatic tissue, significant increases in glutathione peroxidase and catalase activity were observed above 4 mg Hg/kg BW (Kim *et al.*, 2012). Recently, in *Labeo rohita*, fish exposed to the LC_{50} dose of mercury (57.074 mg/L) for 96 h was also reported to cause a considerable reduction in the sulf-hydryl group of glutathione peroxidase in erythrocytes (Chitra and Jayaprakas, 2013). Inorganic mercury also increased glutathione S-transferase and glutathione reductase activity at 8 mg Hg/kg BW in hepatic tissue. Remarkably low levels of calcium and chloride, and reduced osmolality, were also observed at 8 mg Hg/kg BW. Study carried out by Kumari *et al.*, 2014 has further clearly indicated that heavy metal chromium causes oxidative stress in fishes. In the study, catalase activity was found to increase in liver, muscle, gills, and brain of the fish exposed to sublethal concentrations of hexavalent chromium. Thus, the results clearly infer chromium-induced oxidative stress. The increase in catalase activity during experimental periods is probably a response to chromium induced toxic stress and serves to neutralize the impact of increased ROS generation (John *et al.*, 2001). Similar observation was made by Li *et al.*, 2011 in the brain of rainbow trout (*Oncorhynchus mykiss*) after chronic carbamazepine treatment and in the liver, gills, and kidney of *Labeo rohita* after lethal and sublethal concentrations of malathion (Slaninova *et al.*, 2014). Increased CAT, SOD, and GR activities in all organs might suggest the crucial role of these enzymes in cell protection against the deleterious effects of chromium and development of adaptive response to chromium toxicity. The current results also contribute to improving our knowledge about possible development of oxidative stress induced by exposure to chromium metal in aquatic organisms and indicate a possible role for antioxidant systems in the prevention of induced damage.

Copper sulfate is widely used in aquaculture. Vutukuru *et al.* (2005) studied effects of copper on the metabolic rate and the gill morphology of freshwater fish, *Esomus danricus*. They recorded inhibition of super oxide dismutase (60.83 per cent) and catalase (71.57 per cent) in test fish from control along with a significant decrease in the metabolic rate, total glycogen and total protein. Fýrat and Kargýn (2010) recorded GSH level and CAT activity increased in fish exposed to 1.0 mg/l Cu for both exposure periods, while no change was detected at the lower Cu concentration in response to sublethal copper (Cu) exposure in *C. carpio*. Fish were exposed to 0.1 and 1.0 mg/l Cu for 10 and 20 days. Sampaio *et al.*, 2010 after fish exposure to this compound recorded oxidative metabolism alterations and gill tissue damage in the treated fish. Treating fish with copper doses also showed the superoxide dismutase (SOD) responsive to the increase in the aquatic copper in both the liver and the red muscle (Sampaio *et al.*, 2010). Exposure to 0.4Cu and 0.4CupH also resulted in a reduction in the Na/K- ATPase activity and an increase in metallothionein (MT) in the gills. These responses suggest that the fish triggered these mechanisms to revert to the state of blood acidosis, save energy and increase the oxygen uptake. MT is an effective biomarker, responding to copper in different pH levels and dissolved oxygen.

Sreejai and Jaya, 2010 attempted to evaluate the effect of H_2S exposure at two different concentrations on the freshwater fish, *Oreochromis mossambicus*. Hydrogen sulfide (H_2S) is a potentially lethal gas produced by anaerobic decomposition of protein and other sulfur-containing organic matter, decomposition of organic effluents from municipal sewage and many industries. It is also formed in the coir retting process usually done in estuaries using coconut husk. It may occur naturally at levels which can be inimical to fish production and survival (Colby and Smith, 1967). The toxic effects of H_2S are based on its property as a chemical asphyxiate. It binds to the mitochondrial enzyme cytochrome oxidase, blocking oxidative phosphorylation and adenosine triphosphate (ATP) production. This leads to anaerobic metabolism and development of lactic acidosis. The study also aims to find out the oxidative stress in fishes exposed to H_2S by measuring the concentration of lipid peroxidation (LP) products and the changes in the levels of antioxidants in liver, gill, kidney, and brain tissues. H_2S is reported as a a potent inhibitor of aerobic respiration. H_2S exposure causes oxidative stress in fishes and results in LP. Therefore, this causes disturbances in cell integrity and might lead to cell damage/ death (Sreejai and Jaya, 2010). In the present study, the sublethal and medium lethal concentrations of H_2S exposure to fishes show an initial elevation in SOD activity up to 24 hours, followed by reduction towards the end of the experimental period. The initial increase in SOD activity indicated the generation of superoxide radical anion, and the inhibition at the end might be due to the higher amount of oxyradical formation than that could be neutralized by the enzyme. It has also been reported in some cases that the superoxide radical by itself or after its transformation to H_2O_2 caused a strong oxidation of the cysteine in the enzyme and decrease in the SOD activity (Kappus, 1985). Although there was a significantly high (***P***<0.05) initial elevation in GSH level, the activity of this enzyme decreased significantly in liver, gill, kidney, and brain (***P***<0.001). Increased GSH level could be an adaptive mechanism to slight oxidative stress, but decreased GSH level could be due to loss of adaptive mechanisms and the oxidation of GSH to GSSG (oxidized GSH). When fish tissues are in contact with the toxicant, these were removed by conjugation with GSH directly or by means of GSTs, which decreased GSH levels. In addition, the oxidative damage caused by metabolites of the toxicant could be mediated by uncoupling of mitochondrial oxidative phosphorylation (Kalra *et al.*, 1994). The reduction of GPx activity in various tissues in the H_2S-exposed fish might be attributed to the longer influence of various organic and inorganic redox active contaminants (Santos *et al.*, 2004). The decreased level of GPx in the H_2S intoxicated fish might weaken the antioxidant defense system of the fish which would eventually affect their survival. A significantly (***P***<0.001) sharp duration-dependant decrease in GPx activity level was recorded in highest duration of exposure. The low activity of GPx in different tissues of exposed fish demonstrated the incapability of these organs in neutralizing the impact of peroxides (Fatima *et al.*, 2000). On the contrary, significant (P<0.05) duration-dependent elevation in GST activity was noted in the tissues of *Oreochromis mossambicus* intoxicated with the low and intermediate sublethal concentrations of H_2S, and at the highest sublethal concentration, all the four tissues of the test fishes exhibited inhibition in GST activity. These results are in relation with the studies reported in the Egyptian catfish–*Clarias lazera* subjected to dimethoate exposure, and the study showed strong inhibition of GST in the exposed fish (Hamed *et al.*, 1999). The reduction in GST activity noted in fish tissues at the highest exposure time of H_2S indicated the impaired detoxification mechanism of the fish under long-term exposure. CAT activity was also reported significantly ($P<0.05$) increased at the initial phase. A pro-oxidant condition elicited by the presence of toxicant could be triggering an increase in the activity of this antioxidant enzyme at the initial stages of exposure as an adaptive response (Dimitrova *et al.*, 1994). A significant decrease ($\boldsymbol{P}<0.05$) in activity of CAT was observed in 96-hours experiment. The low levels of CAT could be attributed to high production of superoxide anion radical (Alves *et al.*, 2002).

Various vitamins like Vitamins C and E have been found to reduce the toxic manifestation of heavy metals (Flora, 2002). Bassam, 2010 on exposing female catfish, *Clarias gariepinus*, collected from the Nile river at Assiut region, with mercuric chloride (MC) for 3 weeks following by normal water supplemented with copper nicotinate (CN) for 1 week, observed harmful effects in fish organs, probably due to its enhancing effect on reactive oxygen species (ROS) production in fish organs especially the respiratory and osmoregulatory organs *i.e.* gills. Also, the observed oxidative stress in ovary tissue of MC-treated fish may affect fish fertility. Supplementing fish experimental diet with Vit. E (γ-tocopherol) (100 mg/kg wet diet) for 1 week, and the addition of CN in fish diets could protect the fish *C. gariepinus* against MC-induced oxidative damage showing recovery of fish organs. It could suggest that the detoxifying mechanism of action of CN is mainly due to its scavenging activity of free radicals rather than tissue healing. The activities of antioxidants enzymes superoxide dismutase, catalase and glutathione-S-transferase were also concomitantly restored to near normal level by Spirulina supplementation to mercuric chloride intoxicated mice.

The fact that pollutants influence thryoxine release, as mediated via the peroxidase system of iodination, inhibition of head kidney peroxidase enzyme as demonstrated in teleosts (Chattarjee and Bhattacharya, 1984) further amounts to affect liver metabolism. Thomas and Lewis (1987) had also mentioned about thyroid metabolism disruption in fish at the closest site to the main pollution source. As a consequence, enhanced lipid peroxidation. DNA damage, and altered calcium and sulfhydryl homeostasis occur. Fenton-like reactions may be commonly associated with most membranous fractions including mitochondria, microsomes, and peroxisomes. Phagocytic cells may be another important source of reactive oxygen species in response to metal ions. Furthermore, various studies have suggested that the ability to generate reactive oxygen species by redox cycling quinones and related compounds may require metal ions. Simultaneously OS may also affect AChE activity due to excessive secretions of cortisol, later affecting lymphocytes leading to apoptosis, on account of the death of auto active immunity/lymphocyte cells (Green *et al.*, 1992). Apoptosis means programmed cell death which is known to play critical role in a wide variety of physiological processes during fetal development and in adult tissues like the occurrence of cancer cells. Recent studies have suggested that metal ions may enhance the production of tumor necrosis factor alpha (TNF alpha) and activate protein kinase C, as well as induce the production of stress proteins.

Antioxidant defenses in fish, however, are further reported to be very dependent on a combination of biotic and abiotic factors (Martínez-Álvarez *et al.*, 2005), which make very difficult direct inter-study comparisons. Lack of GST activity in some cases may include seasonal (Geracitano *et al.*, 2004), or any other species-specific ecophysiological differences (Livingstone, 2001). Varo *et al.*, 2007 studied effects of Cu in gilthead sea bream fish, *Sparus aurata,* known as a most important Mediterranean aquacultured fish species. These fish on exposing to copper sulphate at 0, 0.25, 0.5 and 1.5 mg L^{-1}, though showed increased lipid peroxidation, but absence of response in GST activity points towards a limited extent of the putative damage caused by the Cu treatments at the highest doses The results suggest that copper sulphate baths at the conditions tested were safe treatments for *S. aurata* fingerlings.

5. Reduction-Oxidation (Redox) Sensitive Transcriptions and Gene Expression

Reduction–oxidation/oxygen (redox)-sensitive transcription factors have gained an overwhelming backlog of interest momentum over the years, ever since the onset of the burgeoning field of free radical research and oxidative stress. Dynamic variation in pO2 and redox equilibrium regulate gene expression, apoptosis signaling and the inflammatory process, thereby bearing potential consequences for screening emerging targets for therapeutic intervention. The reason for this is that redox-sensitive transcription factors are often associated with the development and progression of many human diseases. The molecular response to oxidative stress is regulated by redox-sensitive transcription factors (Leikauf *et al.*, 2002). Reactive oxygen species and reactive nitrogen species serve as signaling messengers for the evolution and perpetuation of the inflammatory process that is often associated with the condition of oxidative stress, which involves genetic regulation. Regulation of the signaling responses is governed at the genetic level by transcription factors that bind to control regions of target genes and alter their expression (Alder *et al.*, 1999). Transcription is a process in which one DNA strand is used as a template to synthesize a complementary RNA. Transcription factors are endogenous substances usually proteins, that are effective in the initiation, stimulation or termination of the genetic transcription process. While in the cytoplasm, the transcription factor is incapable of promoting transcription. Altering gene expression thus, is the most fundamental and effective way for a cell to respond to extracellular signals and/or changes in its environment, in both short and the long term. Changes in the pattern of gene expression through reactive oxygen species/reactive nitrogen species-sensitive regulatory transcription factors are the crucial components of the machinery that determines cellular responses to oxidative/redox conditions (Alder *et al.*, 1999). A signaling event occurs, such as a change of the state of phosphorylation, which results in protein subunit translocation into the nucleus. Signal transduction therefore involves complex interactions of multiple cellular pathways.

Amongst the most important signaling pathways in the body are the mitogen-activated protein kinases (MAPK). MAPK pathways are major regulators of gene transcription in response to OS. Their signaling cascades are controlled by phosphorylation and dephosphorylation of serine and/or threonine residues. This process promotes the actions of receptor tyrosine kinases, protein tyrosine kinases, receptors of cytokines, and growth factors (Brown and Sacks, 2009). The addition of H_2O_2 (a reactive oxygen species) to this cascade can disrupt the complex and promote phosphorylation (Kamata *et al.*, 2005). Likewise, in case of 'Acute lung injury,' as a result of a cascade of cellular events initiated by either infectious or noninfectious inflammatory stimuli, an elevated level of pro-inflammatory mediators combined with a decreased expression of anti-inflammatory molecules is a critical component of lung inflammation. Expression of pro-inflammatory genes is regulated by transcriptional mechanisms. NF-κB is one critical transcription factor required for the expression of many cytokines involved in the pathogenesis of acute lung injury. Modification of transcription, and particularly of NF-κB, is likely to be a logical therapeutic target for the manipulation and treatment of acute lung injury (MacNee and Rahman, 2000). Likewise, in case of 'Sensing hypoxic conditions' a crucial transcription factor that is a master regulatory element in sensing hypoxic conditions and in integrating an adapted response via gene expression of oxygen-sensitive and redox-sensitive enzymes and cofactors is hypoxia-inducible factor-1 (HIF-1) (Semenza, 2001). The signal transduction components that link the availability of oxygen to the activation of these transcription factors are poorly defined, but are broadly believed to hinge on the free abundance of oxidants. A new horizon in chemoprevention research is the recent discovery of molecular links between inflammation and cancer. Components of the cell signaling pathways, especially those that converge on redox-sensitive transcription factors, have been implicated in carcinogenesis (Surh *et al.*, 2005). A wide variety of chemo-preventive and chemo-protective agents can alter or correct undesired cellular functions caused by abnormal pro-inflammatory signal transmission mediated by inappropriately activated factors. The modulation of cellular signaling by an anti-inflammatory phyto-chemical hence provides a rational and pragmatic strategy for molecular target-based chemoprevention.

Hansen *et al.* (2006b) while studying induction of gene transcription for proteins and enzymes involved in metal-mediated oxidative stress in brown trout transferred to a Cu-contaminated river in the Røros region in Central Norway) observed significantly increased MT-A, SOD and GR transcription along with uptake of Cu in gills, while only transcription of MT-A was found to respond in liver and kidney during the exposure, and the SOD and CAT enzyme levels affected in all tissues during the exposure. Hansen *et al.*, 2007 while investigating that how transcript levels of metallothionein (MT), Cu/Zn-superoxide dismutase (SOD), catalase (CAT), glutathione peroxidase (GPX) and glutathione reductase (GR) as well as functional protein levels of MT, SOD and CAT in brown trout tissues in trout fish during a 15-days waterborne exposure to Cd and Zn, found significant uptake of both Cd and Zn in gills during the 15 days exposure, and Cd levels were found to correlate significantly with transcript levels of MT-A, SOD, GPx and GR. Whereas, gill concentrations of Zn did not correlate significantly with the transcript levels of the stress genes studied, SOD and CAT activities increased in gills after transfer, but MT protein levels decreased. In liver, SOD activity and MT protein levels increased, while in kidney only MT protein concentrations were elevated after transfer.

6. Arsenic Toxicity and Oxygen Stress

Throughout the world, millions of people are at risk of cancer, heart disease and diabetes because of chronic exposure to heavy metals (NRC, 1999 and 2000). Exposure to arsenic (As) represents a most important problem in many parts of the World. Arsenic contamination of groundwater in the West Bengal basin in India is unfolding as one of the worst natural geo-environmental disaster to date. Recently, large population in West Bengal in India and Bangladesh has been reported to be affected with arsenic (Smith *et al.*, 2000; Guha Mazumder *et al.*, 1998). Indeed, it is estimated that over 100 million individuals are exposed to arsenic, mainly through a contamination of groundwater (Faita *et al.*, 2013). In USA, Mandal *et al.* (2004) analyzed drinking water and human urine samples for arsenic species in the patients with the signs of melanosis and keratosis. They recorded presence of 66.5, 2.24 and 1.01 mg/l in total in persons regularly exposed to arsenic contamination in drinking water containing arsenic contamination @ 29.0 mg/l. The patients with the signs of melanosis and keratosis showed presence of 7.48 mg/l of arsenic in their blood plasma and 18.7 mg/l in RBC, thus, a total of 26.3 mg/l of the arsenic in the blood.

As per the available estimates about 6 million people are exposed to high levels of arsenic alone every year through drinking water, and only 15-20 per cent of the exposed individuals show arsenic-induced skin lesions (Banerjee *et al.*, 2007). Chronic exposure of humans to high concentration of arsenic in drinking water is also associated with peripheral vascular disease, hypertension, Blackfoot disease and high risks of cancer. On the contrary, studies in the US human populations exposed to average concentrations in drinking water even upto 190 μg/l, were not found to provide any evidence of increased cancer cases (Hu *et al.*, 1998; NRC, 1999 and 2001; Wu *et al.*, 2003; Asmuss *et al.*, 2000; Schoen *et al.*, 2004). Arsenic though non-mutagenic, but is known to interact synergistically with genotoxic agents in the production of mutations and also induces chromosomal aberrations and cell proliferation. Based on NRC (2001) recommendations however, EPA revised the MCL to 10 μg/l As, which according to EPA's risk assessment is associated with mean population cancer risk of $0.63\text{-}2.99 \times 10^{-4}$. Other known cases of arsenically induced human cancers are: bronchial epithelial carcinoma, human acute promyelocytic leukemia, human monoblastoid, human osteosarcoma, human neuroblastoma. It is assumed that genetic variation might play an important role in arsenic toxicity and carcinogenicity, and ERCC2 (a nucleotide excision repair pathway gene) codon 751 Lys/Lys genotype is reported to be significantly associated with arsenic-induced pre-malignant hyperkeratosis (Banerjee *et al.*, 2007).

Arsenic is a naturally occurring metalloid, ubiquitously present in the environment in both organic and inorganic forms. Inorganic arsenic (As^{i}) is the most common form of arsenic in the environment. Arsenic predominantly exists in two oxidation states- a trivalent form: As^{III} and a pentavalent form: As^{V}. In reducing environments arsenic occurs primarily as arsenite [As^{III}]. Absorption and toxicity of arsenic depends on the form in which it is ingested. In terms of arsenic toxicity, arsenate ($AsSO_4$) and arsenite ($AsSO_3$) are two major inorganic arsenic salts important to an animal body. After ingestion, the soluble form of arsenic is easily absorbed into the blood stream through intestine, and it appears distributed in liver and other organs. Its concentration in blood comes down in a few hrs, as it is metabolized within 3-4 days and is mostly excreted in urine (Buchet *et al.*, 1981). Arsenic has no known nutritional requirement to any biological unit; yet process of its metabolism has evolved in most of the animal species, including humans. There are no specific biochemical parameters that reflect arsenic toxicity, but evaluation of clinical effects must be interpreted with knowledge of exposure history. The diagnosis of chronic arsenic poisoning must rely on the characteristic; clinical features of the typical skin lesions, debility, weight loss and neuropathy, and hair arsenic levels are only supportive of the diagnosis (Flora *et al.*, 2007).

Inorganic arsenic occur in six different biochemical entities in an animal body, and because of the species differences in their toxicological effects (Aposhian, 1997), it further make arsenic a very complex metal for the study of its fate and metabolism in living animals including fish. The organo-arsenic species, mostly of methyl derivatives such as arseno-sugars, arseno-betaine, arseno-choline and arseno-lipids (Knowles and Benson, 1983) are also wide spread in aquatic organisms. Consumption of such arsenic contaminated animals may not prove toxic to consumers, but ingestion of some of the sea-foods is stated to increase the arsenic levels in their urine (Buchet, 1994).

It is likely that metabolism of arsenic, like other toxic metals is associated with the conversion of this element to a less toxic form, followed by its accumulation or excretion from the animal cells (Roy and Saha, 2002). Interestingly, like other toxic metals (Cd and Hg), arsenic is not biomagnified in animal tissues (Goering *et al.*, 1999). A certain level of arsenic, however, remains accumulated in hairs, nails, skin, scales, faeces and sweat. About 60-90 per cent of soluble arsenic compounds are absorbed from the gastro intestinal tract following ingestion. Such chronic exposure to As is associated with adverse effects on human health, resulting in occurrence of cancers, cardiovascular diseases, neurological diseases and the rate of morbidity and mortality in populations exposed is alarming. Pi *et al.*, 2002 showed that exposure to inorganic arsenic in a contaminated area of 'Inner Mongolia' increased lipid peroxide serum levels and decreased non-protein sulfhydryl serum levels, these act as indicators of oxidative stress. Similarly, residents of 'Northeastern Taiwan' on exposure to arsenic contaminated water sources (Wu *et al.*, 2001) were reported to experience elevated levels of reactive oxidants, as assessed by the increase in blood super oxide (O_2^-), with a concomitant decrease in plasma antioxidants.

Arsenic-induced oxidative stress has been the major cause of the arsenic-nutrition interactions in an animal body. It requires nutrient-dependent defense systems, and arsenic metabolism (methylation) to counteract the arsenic induced toxicity effects. Cellular insult in response to methylated metabolites has been suggested to involve geno-toxicity with strong evidence of oxidative stress as a causative factor. The underlying mechanism of toxicity includes arsenic interaction with the sulphydryl groups and the generation of reactive oxygen species leading to oxidative stress. The toxicity of inorganic As also appears to be mediated through its ability to substitute phosphate groups, affecting enzymes that depend on this group for their activity (*e.g.*, interfering in the synthesis of ATP and DNA). Trivalent arsenic toxicity could be carried out either directly by attacking –SH groups, or indirectly through generation of reactive oxygen species (ROS) (Chen *et al.*, 1998). Arsenic generates ROS and free radicals like hydrogen peroxide (H_2O_2) (Wang *et al.*, 1996; Chen *et al.*, 1998), hydroxyl radicals species (HO-), nitric oxide (NO-) or superoxide anion (O_2-) (Lynn *et al.*, 2000), dimethyl arsinic peroxyl radical [$(CH_3)_2$ AsOO-] and dimethyl arsenic radical [$(CH_3)_2$ As-] (Yamanaka *et al.*, 1997, 2001). There is the proposal that all of these reactive species are responsible for the stress response elicited by arsenicals. Development of hyperkeratosis due to arsenic exposures often leads to the precursor of arsenic-induced skin cancer (Banerjee *et al.*, 2007), and the underlying mechanism of As toxicity includes arsenic interaction with the sulphydryl groups and the generation of reactive oxygen species leading to oxidative stress.

Oxidative stress theory for arsenic carcinogenicity can be partially explained by its ability to cause cancer at high rates in the lung, bladder and skin. The first oxidative theory of arsenic carcinogenesis that includes a detailed metabolic pathway was presented by Yamanaka *et al.* (1990). Di-methylarsine (a trivalent arsenic form) is a minor *in vivo* metabolite of DMA (a pentavalent arsenic form) produced by a process of reduction *in vivo* (Yamanake and Okada, 1994). Dimethylarsine can react with molecular oxygen to form $(CH_3)_2As^-$ radicals and superoxide anions. This $(CH_3)_2As^-$ can add another molecule of molecular oxygen and form the $(CH_3)_2$AsO- radical. Hydroxyl radical may be produced via cellular iron and other transition metals. Exposure to these free radicals can lead to DNA damage (single strand breaks), (Kitchin, 2001). A dose dependent relationship between cumulative arsenic exposure and prevalence of diabetes mellitus was also documented among the residents in case of endemic arseniasis in Taiwan (Lai *et al.*, 1994; Nakagawa *et al.*, 2002; Kang *et al.*, 2003) and among copper-smelter workers (Rahman *et al.*, 1998) and art-glass workers (Rahman *et al.*, 1996). Tseng (2004) reported that arsenite has high affinity for sulfhydryl groups and thus, can form covalent bond with disulfide bridges in the molecules of insulin, insulin receptors, glucose transporter and enzymes involved in glucose metabolism *e.g.*, pyruate dehydrogenase and α-ketoglutarate dehydrogenase. Tseng (2004) also reported the potential biological mechanisms of arsenic induced diabetes mellitus and reported that arsenate can substitute phosphate in the formation of adenosine triphosphate (ATP) and other phosphate intermediates involved in glucose metabolism, which could theoretically slow down the normal metabolism of glucose, interrupting the mechanism of production of energy, and interfering with the ATP dependent insulin secretion. It has also been reported that arsenite can increase nicotinamide adenine di-nucleotide oxidase activity and induces superoxide, which than cause oxidative deoxyribonucleic and (DNA) damage in human vascular smooth muscle cells (Lynn *et al.*, 2000).

Mukherjee *et al.*, 2004 revealed increased amylase activity and production of nitrite and malon-dialdehyde along with significant reduction in the number of cells in the islets of animals treated with arsenic-trioxide to controls; and an estimated enhanced activity of oxidative stress producing enzymes and/or by their products in pancreatic tissue of mammals due to their exposure to arsenic trioxide. Recently, a mechanism had been proposed (Andrew *et al.*, 2003) which suggest that generation of arsine gas from methylated metabolites, particularly DMA^{III} and trimethylarsenic oxide (TMAO), may also contribute to arsenic-induced genotoxicity. ROS induces PolyADP-riboxylation, thus, inducing DNA strand-breaks, and NAD depletion. Wu *et al.*, 2003, evidenced that arsenic can influence the expression of a variety of proteins, involving signal transduction and gene transcription. Inorganic-As can directly interact with proteins. There is evidence that arsenic interactions with critical enzymes may inhibit DNA repair system, which along with chromosomal aberrations may have implications for carcinogenesis. Chromosomal damage from arsenic exposure is a well established feature of arsenic toxicity. Formation of micronuclei and other forms of chromosomal damage, such as chromosome deletion and aneuploidy, is evident in both *in vitro* studies as well as in animal models (Basu *et al.*, 2002). *In vitro* studies have suggested arsenic induced inhibition, indirectly or directly through gene modulation, in the activity of DNA ligases (Hu *et al.*, 1998). Schoen *et al.*, 2004 had also stated about several such modes of action, including generation of oxidative stress, perturbation of DNA methylation patterns, inhibition of DNA repair, and modulation of signal transduction pathways proposed to characterize arsenic toxicity. It has been also shown that arsenite can compete with zinc in metal binding proteins displaying vicinal dithiols contained in zinc fingers of DNA binding and repair proteins and transcription factors (Asmuss *et al.*, 2000).

Earlier studies focused majorly on how arsenic inhibit nucleotide excision repair (NER) (Hartwig *et al.*, 1997; Andrew *et al.*, 2003, 2006), the major pathway for repairing bulky distortions in DNA double helix. Decreased excision repair capacity is also associated with arsenic exposure as arsenite is known to inhibit more than 200 enzymes, including DNA ligase (Hu *et al.*, 1998). Interaction of arsenic with SH-groups induces structural modification in proteins, leading to the inactivation of many enzymes (Akter *et al.*, 2005). Arsenic mediated oxidative damage in enzymes is also reported to interfere with the DNA repair mechanisms, such as nucleotide excision repair (Andrew *et al.*, 2006) and Base Excision Repair (Sykora and Snow 2008), by either inhibiting ligation or down regulating the gene expression of DNA repair enzymes, particularly the DNA polymerase b (Ding *et al.*, 2009). Sykora and Snow, 2008 has revealed that 'Base excision repair' (BER) is crucial for development and for the repair of endogenous DNA damage. Arsenic, a well-established human carcinogen, is known to produce oxidative DNA damage, which is repaired primarily by BER, whilst high doses of arsenic can also inhibit DNA repair. Exposures of some tissues (lung fibroblasts and keratinocytes) to sodium arsenite, As^{III}, exhibited significant dose-dependent down regulation at doses of As(III) above 1 microM. They provided evidence that changes in BER due to low doses of arsenic could contribute to a non-linear, threshold dose response for arsenic carcinogenesis.

7. Methylation and Metabolism of Arsenic

Several *in vitro* studies have provided evidence of arsenic toxicity through an oxidative stress mode of action. The highly reactive species of inorganic arsenic are potentially more toxic to the living world. Trivalent species, *i.e.*, methylarsonous acid and arsenite, are highly inhibitory, with 50 per cent inhibitory concentrations of 9.1 and 15.0 μM, respectively; whereas pentavalent species are generally nontoxic

(Sierra-Alvarez *et al.*, 2004). Methylation of arsenic has long been regarded as a detoxification process because the pentavalent methylated arsenic metabolites: mono-methylarsonic acid ($MMAs^V$) and di-methylarsinic acid ($DMAs^V$) are much less toxic and excreted more readily than As (III). Methylated organoarsenicals, chiefly monomethylarsonic acid [MMA^V] and dimethylarsinic acid [DMA^V], are also introduced into the environment as defoliants and herbicides in agriculture or as biotransformation products of inorganic arsenicals (Bentley and Chasteen, 2002; Hughes, 2002).

In the process of arsenic metabolism, bio-methylation is considered as the primary detoxification mechanism. A large number of yeast, fungi, algae, plants and animals were found to transfer inorganic arsenic compounds to the methyl derivatives (Aposhian, 1997; Styblo and Thomas, 1997; Thomas *et al.*, 2001). The inorganic metabolism further dependent on the folate supply (Spiegelstein *et al.*, 2003), both folate deficiency and arsenic exposure have been reported to increase congenital malformations in animals including humans. Following intake of arsenate (As^v+) from soil-water/food system in an animal body, inorganic arsenic is readily absorbed through the gastrointestinal tract, and bio-transformed in the liver and other tissues, by alternating reduction and oxidative methylation. In humans absorbed inorganic pentavalent arsenic (As^V) is bio–transformed to trivalent arsenic (As^{III}) through sequential addition of methyl groups, acquired from S-adenosyl methionine (SAM). Trivalent (As^{III}) form of arsenic undergoes methylation to form less toxic compounds that are excreted in urine, but some inorganic arsenic is excreted in the urine unchanged (Hall, 2002; Hopenhayn Rich *et al.*, 1993).

Inorganic arsenic is metabolized by a following sequential process, as described by Vahter (1994):

i. *Arsenate ($AsSO_4$) metabolism*: Arsenic enters a cell as arsenate via its phosphate carrier system; once inside the cell, it competes with phosphates *i.e.* it can bind to ADP forming ADP-arsenate; and is rapidly hydrolyzed thereafter. Some of it can be reduced enzymatically to arsenite ($AsSO_3$) by arsenate reductase (Buchet *et al.*, 1981; Radabaugh *et al.*, 2000), Aposhian *et al.*, 2004 revealed that this reduction process is catalyzed by the arsenate-reductase (mono-methyl-arsenic acid (MMA^V) reductase)/PNP protein.

ii. *Arsenite ($AsSO_3$) metabolism*: Arsenite is chemically more reactive than arsenate and its metabolism is more complex. Its diffusion in a eukaryotic cell is not a simple process, as it was earlier believed. Recently, it has been suggested that a transport system brings it into the cell (Liu *et al.*, 2002). Once it is in the cell, there are many pathways it can follow: it can be oxidized to less reactive arsenate and/or it can bind to thiols such as glutathione (GSH) and thiol containing proteins. It binds more strongly to di-thiols, such as lipoic acid and thiols that are in very close proximity to each other as a part of the 3-dimentional protein structure.

iii. *Methylation Process*: Inorganic arsenic is methylated to monomethylarsonic acid ($MMAs^V$) and finally to dimethylarsenic acid ($DMAs^V$), by the donar S-adenosyl-methionine (SAM), as explained by Carter *et al.*, 2003. This reaction is catalyzed by the methyltransferase in the presence of glutathion (Styblo and Thomas, 1997), followed by excretion through urine.

Aposhian *et al.* (2004) studied the enzymology involved in bio-transformation (biomethylation) of inorganic arsenic to dimethylarsinous acid [DMA^{III}] and established that mono-methylarsonons acid [MAA^{III}] and DMA^{III} appears in urine of people chronically exposed to arsenic. They can inhibit various enzymatic functions including glycolysis and TCA (Tricarboxylic acid) cycle by binding to sulfhydryl groups of enzymes. Pentavalent arsenic compound (As^{v}) can uncouple mitochondrial oxidative phosphorylation. Recent studies have suggested that epigenetic mechanisms may also mediate toxicity and carcinogenicity resulting from arsenic exposure (Wang *et al.*, 2007), since arsenic exposures has been shown to alter methylation levels of both global DNA and gene promoters; histone acylation, methylation, and phosphorylation; and miRNA expression, in studies analyzing mainly a limited number of epigenetic end points. Styblo *et al.* (2002) also concluded that trivalent forms of arsenic are more toxic than the pentavalency forms, with MMA^{III} being the most toxic arsenic metabolite. Cytotoxiity assays revealed the following order of toxicity of arsenicals: $MMA^{III} > iAs^{III} > MMA^{V} + DMA^{V}$ (Petrick *et al.*, 2000). Martinez *et al.*, 2011 confirmed that metabolism of arsenic generates a variety of genotoxic and cytotoxic species, damaging DNA directly and indirectly, through the generation of these reactive oxidative species and induction of DNA adducts, DNA strand breaks and cross links, and there is inhibition of the DNA repair process itself. Since SAM is the methyl group donor used by DNA methyl-transferases to maintain normal epigenetic patterns in all human cells, arsenic is also postulated to affect maintenance of normal DNA methylation pattern, chromatin structure, and genomic stability. Faita *et al.*, 2013 has recently reviewed the genotoxic effects of As in the cells, and has discussed the importance of signaling and repair of arsenic-induced DNA damage.

The traditional belief that bio-methylation was a detoxification pathway of inorganic As, has been questioned (Styblo *et al.*, 2002). Maintenance of proper DNA methylation is critical to the regulation of gene transcription. Continuous arsenic exposures in fishes/animals may cause DNA's hypo-methylation (arsenic modified DNA) due to continuous methyl depletion, facilitating aberrant gene expression that result in carcinogenesis. Such alterations in DNA methylation may alter gene transcription and allow for unregulated cell growth. Although there are no such direct evidence of arsenic interference in alterations in DNA methylation by the arsenic, indirect evidence has instead led researchers to hypothesize that because arsenic metabolism and DNA methylation share a common cofactor, *i.e.* S-adenosylmethionine (SAM), arsenic exposure may cause excessive demand on SAM, thus, resulting in perturbation of normal DNA methylation pattern (Zhao *et al.*, 1997). The hyper-methylation of DNA occurred concomitantly with reduced SAM levels, which also resulted in cell transformations (Zhao *et al.*, 1997). *In vitro* studies show that arsenic can disrupt normal DNA methylation through the hypo- and hyper-methylation of key genes, resulting in cell transformation and unregulated growth in cells. Information regarding arsenic's mode of action supports the presence of protective cellular mechanisms in living organisms that can mitigate arsenic's toxicity. Arsenic is also not mutagenic as it does not interact with DNA directly (Gradecka *et al.*, 2001). Nesnow *et al.*, 2002 suggested that ROS (reactive oxygen radicals) are formed concomitantly with the oxidation of DMA^{III} to DMA^{V}, and exposure to ROS inhibitors could reverse the genotoxicity of both MMA ^{III} and DMA ^{III} forms in a DNA nicking assay. Noda *et al.*, 2002 stated that despite the presence of oxidative damage, the observed lesions do not result in mutations, presumably due to effective DNA repair. DNA adducts formation induced by oxidative stress and its subsequent repair is consistent with a nonlinear approach to arsenic toxicity.

8. Protection against Arsenic Toxicity

Under-nourished persons or feeding on low beta-carotene diet were said to be more susceptible to arsenic toxicity (NRC, 1999). Diets low in methyl groups (Choline or Methionine) can further enhance arsenic induced alterations in DNA methylation (Okoji *et al.*, 2002). Likewise, selenium deficient diet enhances arsenic retention in body (Thomas *et al.*, 2004). Several *in vitro* studies demonstrated that cells can acquire tolerance to both inorganic arsenic and DMA following pre-exposure to inorganic arsenic (Liu *et al.*, 2001). Arsenic induced tolerance is consistent with the 'Hormesis' theory. Hormesis is a recently advanced new theory, appreciated by the toxicologists. It holds that low doses of most chemicals, through the subtle activation of protective cellular mechanisms, are associated with a therapeutic effect (Calabrese and Baldwin, 2003a).

Selenium is an essential element trace element for mammals and through seleno-proteins it participates in various biological processes such as antioxidant defence (Zwolak and Zaporowska, 2012). Studies have revealed that selenium may reduce arsenic accumulation in the organism and protect against arsenic related skin lesions. A selenium deficient diet increased arsenic induced changes in female reproductive system. Cheng *et al.*, 2007 also demonstrated application of lipoic acid (LA) as a useful protective agent against arsenic-induced glial cell toxicity and reversing arsenic-induced damage in human brain. α-lipoic acid (LA) is a naturally occurring antioxidant and it functions as a cofactor in several multienzyme complexes (Reed, 1974). It is a thiol-compound naturally occurring in plants and animals, which is thought to be a strong antioxidant and possess neuroprotective effects, and it was found to completely inhibited U-118 cells autophagic cell death induced by AS_2O_3. Recently, Lai *et al.* (2011) have observed an important role of DNA polymerase b in repairing arsenic induced DNA damage since its deficiency exacerbates arsenic induced genotoxic effects and its over expression effectively protects the cell from arsenic induced oxidative stresses.

Chelation therapy with chelating agents like British Anti Lewisite (BAL), sodium 2,3- dimercaptopropane 1-sulfonate (DMPS), meso 2,3 dimercaptosuccinic acid (DMSA) *etc.*, is considered to be the best known treatment against arsenic poisoning (Flora *et al.*, 2007). The treatment with these chelating agents however, is compromised with certain serious drawbacks/side effects. The studies show that supplementation of antioxidants along with a chelating agent prove to be a better treatment regimen. Likewise, Sharma *et al.* (2007) have shown that

Spirulina treatment augments the antioxidants defense mechanism in mercuric chloride induced toxicity and provides evidence that it may have a therapeutic role in free radical mediated diseases. Oxidative stress induced by mercuric chloride (5 mg/kg body weight i.p.) in mice substantially increases the lipid peroxidation (LPO) level along with corresponding decrease in the reduced glutathione and various antioxidant enzymes in liver and increase in serum transaminases activity. Supplementation of Spirulina (800 mg/kg body weight orally, in olive oil, along with mercuric chloride) for 40 days resulted in decreased LPO level, serum glutamate oxaloacetate and serum glutamate pyruvate transaminase activity along with increase in liver GSH level. GSH plays a critical role in both the enzymatic and non-enzymatic reduction of pentavalent arsenicals to trivalent and in the complexation of arsenicals to form arsenicothiols during methylation process (Scott *et al.*, 1993). The interaction of arsenic with glutathione and its related enzymes by changing their redox status and this may lead to the alterations of their biological function. Inactivation of GSH related enzymes could have deleterious effects on the detoxification processes and other critical cellular processes involving GSH mediated redox regulation.

REFERENCES

Agarwal A., Gupta S., Sekhon L. and Shah R., 2008. Redox considerations in female reproductive function and assisted reproduction: from molecular mechanisms to health implications. Antioxid. Redox Signal, 10:1375–1403.

Alder V., Yin Z., Tew K.D., Ronai Z., 1999. Role of redox potential and reactive oxygen species in stress signaling. Oncogene, 18: 6104-6111.

Al-Gubory K.H., Fowler P.A. and Garrel C., 2010. The roles of cellular reactive oxygen species, oxidative stress and antioxidants in pregnancy outcomes. Int. J. Biochem. Cell Biol., 42:1634–1650.

Almeida J.A., Diniz Y.S., Marques S.F., Faine L.A., Ribas B.O., Burneiko R.C. and Novelli E.L., 2002. The use of the oxidative stress responses as biomarkers in Nile tilapia (*Oreochromis niloticus*) exposed to *in vivo* cadmium contamination. Environ Int., 27(8): 673-9.

Alves SR, Severino PC, Ibbotson DP. Effect of furadan in the brown mussel, *Perna perna* and in the mangrove oyster, *Crassostrea rhizophorae*. Mar Environ Res. 2002; 54:5. [PubMed]

Andrew A.S., Burgess J.L., Meza M.M. and others, 2006. Arsenic exposure is associated with decreased DNA repair *in vitro* and in individuals exposed to drinking water arsenic. Environ. Hlth. Perspec., 114 (8): 1193–1198.

Andrew A.S., Karagas M.R. and Hamilton J.W., 2003. Decreased DNA repair gene expression among individuals exposed to arsenic in United States drinking water. Int. J. Cancer, 104(3): 263-8.

Aposhian H.V., 1997. Enzymatic methylation of arsenic species and other new approaches to arsenic toxicity. Ann. Rev. Pharmcol. Toxicol., 37: 397-419.

Aposhian H.V., Gurzau E.S., Le X.C. and others, 2000. Occurrence of monomethylarsonous acid in urine of humans exposed to inorganic arsenic. Chem. Res.Toxicol., 13 (8): 693–697.

Aposhian H.V., Zakharyan R.A., Avram M.D., Reyes A.S. and Wollenberg M.L., 2004. A review of the enzymology of arsenic metabolism and a new potential role of hydrogen peroxide in detoxification of the trivalent arsenic species. Toxicol. Appl. Pharmacol., 198: 327-335.

Asmuss M., Mullenders L.H., Eker A. and Hartwig A., 2000. Differential effects of toxic metal compounds on the activities of Fpg and XPA, two zinc finger proteins involved in DNA repair. Carcinogenesis, 21: 2097-2104.

Banerjee M., Sarkar J., Das J.K., Mukherjee A., Sarkar A.K., Mondal L. and Giri A.K., 2007. Polymorphism in the ERCC2 codon 751 is associated with arsenic-induced premalignant hyperkeratosis and significant chromosome aberrations. Carcinogenesis, 28(3): 672-6.

Bassam Al-Salahy M., 2010. Physiological studies on the effect of copper nicotinate (Cu–N complex) on the fish, *Clarias gariepinus* exposed to mercuric chloride. Fish Physiol. Biochem., 37(3): 373-385.

Basu N., Todgham A.E., Ackerman P.A., Bibeau M.R., Nkano K., and 3 others. 2002. Heat shock protein genes and the functional significance in fish. Gene, 295: 173-183.

Bayani Jane and Squire Jeremy A., 2007. Application and interpretation of FISH in biomarker studies. Cancer Lett., 249: 97–109.

Behrman H.R., Kodaman P.H., Preston S.L. and Gao S., 2001. Oxidative stress and the ovary. J. Soc. Gynecol Investigations, 8: S40–S42.

Bentley R. and Chasteen T.G., 2002. Microbial methylation of metalloids: arsenic, antimony, and bismuth. Microbiol. Mol. Biol. Rev., 66: 250-271.

Brown M/D. and Sacks D.B., 2009. Protein scaffolds in MAP kinase signaling. Cell Signal, 21: 462–469.

Buchet J.P., 1994. Assessment of exposure to inorganic arsenic following ingestion of marine organisms by volunteears. Environ. Res. 66: 44-51.

Buchet J.P., Lauwerys R. and Roels H., 1981. Comparison of the urinary excretion of arsenic metabolites after a single oral dose of sodium arsanite, monomethylarsenate, or dimethylarsenate in man. Intl. Arch. Occup. Environ. Hlth., 48: 301-306.

Burton G.J. and Jauniaux E., 2010. Oxidative Stress. Best Pract. Res. Clin. Obstet. Gynaecol., 25: 287–299.

Carter D.E., Aposhian H.V. and Gandolfi A.J., 2003. The metabolism of inorganic arsenic oxides. Pallium arsenide, and arsine: a toxicological review. Toxicol. Appl. Pharmacol., 193(3): 309-34.

Chandra A., Surti N., Kesavan S. and Agarwal A. 2009 Significance of oxidative stress in human reproduction. Arch. Med., 5: 528–542.

Chang L.W., Gilbert M. and Sprecher J., 1978. Modification of methyl mercury neurotoxicity by vitamin E. Environ. Res., 17: 356-366.

Chen Y.C., Lin-Shiau S.Y. and Lin J.K., 1998. Involvement of reactive oxygen species and caspase 3 activation in arsenite induced apoptosis. J. Cell Physiol., 177: 324-33.

Chen Wang, Gao Z., Breuil T., Y. C., Hiratsuka Y., 1995. Biologicaly degradation of resin acids in wood chips by wood-inhabiting fungi. Appl. Environ. Microbiol., 61: 222-225.

Cheng Tain-Junn, Wang Ying-Jan, Kao Wei-Wan, Chen Rong-Jane and Ho Yuan-Soon, 2007. Protection against arsenic trioxide-induced autophagic cell death in U118 human glioma cells by use of lipoic acid. Food and Chem. Toxicol.., 45 (6): 1027–1038.

Chitra S. and Jayaprakash K., 2013. Effect of Mercury on blood components of freshwater edible fish, *Labeo rohita*. J. Acad. Indus. Res. Vol. 1(12): 774-777.

Colby P.J. and Smith L.L., Jr. 1967. Survival of wall eye egg and fry on paper fiber sludge deposits in Rainy Rivers, Minnesota. Trans. Amer. Fish Doc. 96: 296.

Dimitrova M.S., Tishinova V., Velcheva V., 1994. Combined effect of zinc and lead on the hepatic superoxide dismutase-catalase system in carp (*Cyprinus carpio*) Comp. Biochem. Physio., 108: 46.

Faita Francesca, Cori Liliana, Bianchi Fabrizio, and Andreassi Maria Grazia, 2013. Arsenic-Induced Genotoxicity and Genetic Susceptibility to Arsenic-Related Pathologies. Int. J. Environ. Res. Pub. Hlth., 10(4): 1527–1546.

Fatima M., Ahmad I.I., Sayeed I.I., Athar M. and Raisuddin S., 2000. Pollutant induced over activation of phagocytes is concomitantly associated with peroxidative damage in fish tissues. Aquat. Toxicol. 49: 243–50.

Fýrat Ö. and Kargýn F., 2010. Individual and combined effects of heavy metals on serum biochemistry of Nile Tilapia, *Oreochromis niloticus*. Arch. Environ. Contam. Toxicol., 58: 151–157.

Flora S.J.S., 2002. Nutritional components modify metal absorption, toxic response and chelation therapy. J. Nutri. Environ. Med., 12: 53-67.

Flora S.J.S., Bhadauria Smrati, Kannan G.M. and Singh Nutan, 2007. Arsenic induced oxidative stress and the role of antioxidant supplementation during chelation: A review. J. Environ. Biol., 28(2): 333-347.

Goering P.L., Aposhian H.V., Mass M.J., Cebrian M., Beck B.D. and Walkes M.P., 1999. The enigma of arsenic carcinogenesis: role of metabolism. Toxicol. Sci., 49: 5–14.

Gradecka D., Palus J., and Wasowicz W., 2001. Selected mechanisms of genotoxic effects of inorganic arsenic compounds. Int. J. Occup. Med. Environ. Hlth., 14: 317–328.

Guha Mazumder D.N., Ghoshal U.C., Saha J., Santra A., De B.K., Chatterjee A., Dutta S., Angle C.R. and Centeno J.A., 1998. Randomized placebo controlled trial of 2,3-dimercapto succinic acid in therapy of chronic arsenicosis due to drinking arsenic contaminated subsoil water. Clin. Toxicol., 36: 683-690.

Hall A.H., 2002. Chronic arsenic poisoning. Toxicol. Lett., 128: 69-72.

Hamed R.R., Elawa S.E. and Farid N.M., 1999. Evaluation of detoxification of enzyme levels in Egyptian cat fish, *Clarias lazera*, exposed to dimethoate. Bull. Environ. Contam. Toxicol., 63: 796.

Hardeland R., Reiter R.J., Poeggeler B. and Tan D.X., 1993. The significance of the metabolism of the neurohormone melatonin: Antioxidative protection and formation of bioactive substances. Neurosci. Biobehav. Rev., 17: 347-357.

Hartwig A., Groblinghoff U.D., Beyersmann D., Natarajan A.T., Filon R. and Mullenders L.H., 1997. Interaction of arsenic(III) with nucleotide excision repair in UV-irradiated human fibroblasts. ,18(2): 399-405.

Hopenhayn-Rich C., Smith A.H. and Gordan H.N., 1003. Human studies do not support the methylation threshold hypothesis for the toxicity of inorganic arsenic. Environ. Res., 60: 161-177.

Hu Y., Su L. and Snow E.T., 1998. Arsenic toxicity is enzyme specific and its effects on ligation are not caused by the direct inhibition of DNA repair enzymes. Mutation Res., 408 (3): 203–18.

Hughes M.F., 2002. Arsenic toxicity and potential mechanism of action. Toxicol. Lett., 133: 1-16.

Irani K., 2000. Oxidant signaling in vascular cell growth, death, and survival: a review of the roles of reactive oxygen species in smooth muscle and endothelial cell mitogenic and apoptotic signaling. Circ. Res., 87: 179–183.

Jia Xiuying, Zhang Hangjun and Liu Xiaoxu, 2011. Low levels of cadmium exposure induce DNA damage and oxidative stress in the liver of Oujiang colored common carp, *Cyprinus carpio* var. *color*, Fish Physiol. Biochem., 37 (1): 97-103.

John S., Kale M., Rathore N. and Bhatnagar D., 2001. Protective effect of vitamin E in dimethoate and malathion induced oxidative stress in rat erythrocytes. J. Nutri. Biochem., 12(9): 500–504.

Kalra J., Mantha S.V. and Prasad K., 1994. Oxygen free radicals: Key factors in clinical diseases. Lab Med. Int., 11:13–9.

Kamata H., Honda S., Maeda S., Chang L., Hirata H. and Karin M., 2005. Reactive oxygen species promote TNFalpha-induced death and sustained JNK activation by inhibiting MAP kinase phosphatases. Cell, 120: 649–661.

Kappus H., 1985. Lipid peroxidation: Mechanisms, analysis, enzymology and biological relevance. In: Sies H. (Ed.). Oxidative stress, London: Academic Press, pp 310.

Kang J., Zhang Y., Chen J., Chen H., Lin C., Wang Q. and Ou Y., 2003. Nickel-induced histone hypoacetylation: the role of reactive oxygen species. Toxicol. Sci., 74: 279-286.

Kim Jun-Han, Lee Jung-Sick and Kang Ju-Chan, 2012. Effect of inorganic mercury on hematological and antioxidant parameters on olive flounder, *Paralichthys olivaceus*. Fish Aquat. Sci., 15(3): 215-220.

Kitchin K.T., 2001. Recent advances in arsenic carcinogenesis: mode of action, animal model system and methylated arsenic metabolites. Toxicol. Appl. Pharmacol., 172: 249-261.

Knowles F.C. and Benson A.A., 1983. The biochemistry of arsenic. Trends Biochem. Sci., 8: 178-180.

Kono Y. and Fridovich I., 1982. Superoxide radical inhibits catalase. J. Biol. Chem., 57: 51.

Kumari Kanchan, Khare Ankur, and Dange Swati, 2014. The applicability of oxidative stress biomarkers in assessing chromium induced toxicity in the fish *Labeo rohita*. Biomed Res Int., 2014: 782493.

Lai Y., Zhao W., Chen C., Wu M., Zhang Z., 2011. Role of DNA polymerase beta in the genotoxicity of arsenic. Environ. Mol. Mutagen., 52: 460–468

Lai M.S., Hsueh YM, Chen CJ, and others, 1994. Ingested inorganic arsenic and prevalence of diabetes mellitus. Am. J. Epidemiol, 139: 484-492.

Lauterburg B.H., Smith C.V., Hughes H. and Mitchell J.R., 1983. Determinants of hepatic glutathione turnover: toxicological significance. In: Lamble J.W. (Ed.). Drug metabolism and distribution, Amsterdam: Elsevier Biomedical Press; pp180.

Leaver M.J., George S.G., 1998. A piscidine glutathione S-transferase which efficiently conjugates the end products of lipid peroxidation. Mar. Environ. Res., 47: 46–74.

Leikauf G.D., McDowell S.A., Wesselkamper S.C., Hardie W.D., Leikauf J.E., Korfhagen T.R. and Prows D.R. 2002. Acute lung injury: functional genomics and genetic susceptibility. Chest., 121: 70S-75S.

Leonard S.S., Harris G.K., Shi X., 2004. Metal-induced oxidative stress and signal transduction. Free Radical Biol. and Med., 37(12):1921–1942.

Li Z.H., Zlabek V., Velisek J., Grabic R., Machova J. and Randak T., 2010. Modulation of antioxidant defence system in brain of rainbow trout (*Oncorhynchus mykiss*) after chronic carbamazepine treatment. Comp. Biochem. Physiol., C-Toxicology and Pharmacology, 151(1):137–141.

Liu S.X., Athar M., Lippai I., Waldren C. and Hei T.K., 2001. Induction of oxyradicals by arsenic: Implication for mechanism of genotoxicity. Proc. Natl. Acad. Sci. USA., 98: 1643-1648.

Liu Z., Shen J., Carbrey J.M., MUkhopadhay R., Agre P. and Resen B.P., 2002. Arsenic transport by mammalian aqua-glyceroporines AQP7 and AQP9. Proc. Natl. Acad. Sci. USA., 99: 6053-6058.

Lushchak OV, Kubrak OI, Nykorak MZ, Storey KB, Lushchak VI, 2008. The effect of potassium dichromate on free radical processes in goldfish: possible protective role of glutathione. Aquatic Toxicology. 87(2):108–114.

Lynn S., Gurr J.R., Lai H.T. and Jan K.Y., 2000. NADH oxidase activation is involved in arsenite-induced oxidative DNA damage in human vascular smooth muscle cells. Circ. Res., 86: 514-519.

MacNee W. and Rahman I., 2001. Is oxidative stress central to the pathogenesis of chronic obstructive pulmonary disease? Trends Mol. Med., 7: 55-62.

Mandal B.K., Ogra Y., Anzai K. and Suzuki K.T., 2004. Speclation of arsenic in biological samples. Toxicol. Appl. Pharmacol., 198: 307-318.

Marlatt V., Hewitt M. and Van Der Kraak G., 2006. Utility of *in vitro* test methods to assess the activity of xenoestrogens in fish. Environ. Toxicol. Chem., 12: 3204-3212.

Martinez Victor D., Vucic Emily A., Adonis Marta, Gil Lionel and Lam Wan L., 2011. Arsenic biotransformation as a cancer promoting factor by inducing DNA damage and disruption of repair mechanisms. Mol. Biol. Int., Vol. 2011: 11 pp.

Mukherjee S., Das D., Darbar S. Mukherjee M. Das A.S. and Mitra C., 2004. Arsenic trioxide generate oxidative stress and islet cell toxicity in rabbit. Curr. Sci., 86: 854-857.

Nakagawa Y., Akao Y., Morikawa H., Hirata I., Katsu K., Naoe T., Ohishi N. and Yagi K., 2002. Arsenic trioxide-induced apoptosis through oxidative stress in cells of colon cancer cell lines. Life Sci., 70: 2253-2269.

NRC., 1999. National Research Council, Arsenic in drinking water. Washington DC., National Acad. Press.

NRC., 2001. National Research Council, Arsenic in drinking water 2001 update. Washington DC., National Acad. Press.

Nesnow Stephen, Roop Barbara C., Lambert Guy, Kadiiska Maria, Mason Ronald P., Cullen William R. and Mass Marc J., 2002. DNA Damage Induced by Methylated Trivalent Arsenicals Is Mediated by Reactive Oxygen Species. Chem. Res. Toxicol., 15(12):1627-34.

Okoji R.S., Yu R.C., Maronpot R.R. and Froines J.R., 2002. Sodium arsenite administration via drinking water increases genome wide and Ha-ras DNA hypomethylation in Methyl deficient C57BL/6J mice. Carcinogenesis, 23(5): 777- 785.

Palace V.P., Majewski H.S. and Klaverkamp J.F., 1993. Interactions among antitoxicant defences in liver of rainbow trout (*Oncorrhyncus mykiss*) exposed to cadmium. Canad. J. Fish Aq. Sci., 50: 156.

Patil V.K. and David M., 2013. Oxidative stress in freshwater fish, Labeo rohita as a biomarker of malathion exposure. Environ. Monit. Assess.,185: 10191–10199.

Perkins A.V., 2006. Endogenous anti-oxidants in pregnancy and preeclampsia. Aust. N. Z. J. Obstet. Gynaecol., 46: 77–83.

Petrick J.S., Ayala-Fierro F., Culen W.R., Carter D.E. and Aposhian H.V., 2000. Monomethylarsenous acid (MMA^{III}) is more toxic than arsenite in Chang human hepatocytes. Toxicol. Appl. Pharmacol., 163: 203-207.

Pi J., Yamauchi H., Kumagai Y. and others, 2002. "Evidence for induction of oxidative stress caused by chronic exposure of Chinese residents to arsenic contained in drinking water". Environmental Health Perspectives, 110 (4): 331–6.

Radabaugh T.R., Sampayo-Reyes A., Zakharyan R.A. and Aposhian H.V., 2000. Arsenate reductase: II. Purine nucleotide phosphorylase in the presence of dyhydrolipoic acid is a route for reduction of arsenate to arsenite in mammalian system. Chem. Res. Toxicol., 13: 26-30.

Rahman M., Tondel M., Chaudhary I.A. and Axelson O., 1998. Relations between exposures to arsenic, skin lesions and glucosuria. Occop. Environ. Med., 56: 277-281.

Rahman M., Wingren G. and Axelson O., 1996. Diabetes mellitus among Swedish art glass workers- An effect of arsenic exposure. Scand. J. Work, Environ. Health, 22: 146-149.

Reed L.J., 1974. Multienzyme complexes. Acc. Chem. Res., 7: 40-46.

Roy P., and Saha A., 2002. Metabolism and toxicity of arsenic: A human carcinogen. Curr. Sci., 82: 38- 45.

Ruder E.H., Hartman T.J. and Goldman M.B. 2009. Impact of oxidative stress on female fertility. Curr. Opin. Obstet. Gynecol., 21: 219–222.

Santos M.A., Pacheco M., Ahmad I., 2004. Antioxidant responses to in situ bleached craft pulpmill effluent outlet exposure to Anguilla anguilla. Environ Int., 30: 301.

Sampaio Fernanda Garcia, Boijink Cheila, Santos Laila R. B. dos, Yoshioka Eliane Tie, Kalinin Ana, Rantin Francisco Tadeu, 2010. The combined effect of copper and low pH on antioxidant defenses and biochemical parameters in neotropical fish pacu, Piaractus mesopotamicus (Holmberg, 1887). Ecotoxicol., 19(5): 963-76.

Schoen A., Beck B., Sharma R. and Dube E., 2004. Arseni toxicity at low doses: Epidemiological and mod of action considerations. Toxicol. Appl. Pharmacol., 198; 253-267.

Scott N., Hatlelid K.M., Mackinzie N.E. and Carter D.E, Reaction of arsenic (III) and arsenic (V) species with glutathione. Chem. Res. Toxicol., 6, 102-106 (1993)

Semenza G.L., 2001. HIF-1 and mechanisms of hypoxia sensing. Curr. Opin. Cell Biol., 13:167-171.

Sharmahttp://www.sciencedirect.com/science/article/pii/S0278691507002062 - aff1 M.K., Sharmahttp://www.sciencedirect.com/science/article/pii/S0278691507002062 - aff2 Ambika, Kumar Ashok and Kumar Madhu, 2007. Spirulina fusiformis provides protection against mercuric chloride induced oxidative stress in Swiss albino mice. Food and Chemical Toxicology, 45 (12): 2412–2419.

Shkolnik K, Tadmor A, Ben-Dor S, Nevo N, Galiani D, Dekel N., 2011. Reactive oxygen species are indispensable in ovulation. Proc Natl. Acad. Sci. USA., 108:1462–1467.

Siraj B.P. and Rani U.A., 2003. Cadmium induced antioxidant defence mechanism in a freshwater teleost, Oreochromis mossambicus. Ecotoxico. environ. Safety., 56: 218-221.

Slaninova A., Helena M., Hostovsky M. and others, 2014. Effects of subchronic exposure to N,N-Diethyl-m- toluamide on selected biomarkers in common carp (*Cyprinus carpio* L.) BioMed. Res. Intl., 2014: 8.

Smith A.H., Arroyo A.P., Guha Mazumder D.N., Kosnett M.J., Hernadez A.L., Beeris M., Smith M.M. and Moore L.E., 2000. Arsenic induced skin lesions among Atacameno people in Northern Chile despite good nutrition and centuries of exposure. Environ. Hlth. Perspect., 108, 617-620.

Spiegelstein O., Lu X., Le X.C., Troen A., and 4 others., 2003. Effects of dietary intake of folate intake and folate binding protein-! (Follop1) on urinary speciation of sodium arsenate in mice. Toxicol Lett., 145(2); 167-74.

Sreejai R. and Jaya D.S., 2010. Studies on the changes in lipid peroxidation and antioxidants in fishes exposed to hydrogen sulfide. Toxicol. Int., 17(2): 71–77.

Styblo M., Drobna Z., Jaspers I., Lin S. and Thomas D.J., 2002. The role of biomethylation in toxicity and carcinogenicity of arsenic: a research update. Environ. Health Persp., 110(suppl. 5): 767-771.

Styblo M. and Thomas D.J., 1997. In: Arsenic: Exposure and Health Effects (eds. Abernathy C.O. *et al.*) Chapman and Hall UK,: 283-295.

Surh Y.J., Kundu J.K., Na H.K. and Lee J.S., 2005. Redox-sensitive transcription factors as prime targets for chemoprevention with anti-inflammatory and antioxidative phytochemicals. J Nutr., 135(12) Suppl: 2993S-3001S.

Sykora Peter, Snow Elizabeth T., 2008. Modulation of DNA polymerase β-dependent base excision repair in cultured human cells after low dose exposure to arsenite. Toxicol. and Appl. Pharmacol., 228(3): 385-94.

Thomas J.P., Maiorino M., Ursini F. and Girotti A.W., 1990 Protective action of phospholipid hydroperoxide glutathione peroxidase against membrane damaging lipid peroxidation. J. Biol. Chem., 265: 454–61.

Vahter Marie E., 2007. Interactions between arsenic-induced toxicity and nutrition in early life. J. Nutr., 137 (12): 2798-2804.

Vutukuru S. S., Suma Ch., Madhavi K. Radha, Juveria, Pauleena J. Smitha, Rao J. Venkateswara, and Anjaneyulu Y., 2005. Studies on the development of potential biomarkers for rapid assessment of copper toxicity to freshwater fish using *Esomus danricus* as model. Int. J. Environ. Res. Public Hlth., 2(1): 63–73.

Wang T.S., Kuo C.F., Jan K.Y. and Huang H., 1996. Arsenite induces apoptosis in Chinese hamster ovary cells by generation of reactive oxygen species. J. Cell. Physiol., 169: 256-268.

Wang Y.H., Wu M.M., Hong C.T., Lien L.M., Hsieh Y.C., Tseng H.P., Chang S.F., Su C.L., Chiou H.Y. and Chen C.J., 2007. Effects of arsenic exposure and genetic polymorphisms of p53, glutathione S-transferase M1, T1, and P1 on the risk of carotid atherosclerosis in Taiwan. Atherosclerosis,192: 305–312.

Wu M.M., Chiou H.Y., Ho I.C., Chen C.J. and Lee T.C., 2003. Gene expression of inflammatory molecules in circulating lymphocytes from arsenic-exposed human subjects. Environ. Helth. Perspect., 11: 1429-1438.

Wu M.M., Chiou H.Y., Wang T.W. and others, 2001. Association of blood arsenic levels with increased reactive oxidants and decreased antioxidant capacity in a human population of northeastern Taiwan. Environ. Helth. Perspect., 109 (10): 1011–1017.

Yamanaka K., Hoshino M., Okamoto M., Sawamura R., Hasegawa A. and Okada S., 1990. Induction of DNA damage by dimethylarsine, a metabolite of inorganic arsenics, is for the major part likely due to its peroxyl radical. Biochem. Biophys. Res. Commun., 168: 58-64.

Yamanaka K., Hayashi H., Tachikawa M., Kato K., Hasegawa A., Oku N. and Okada S., 1997. Metabolic methylation is a possible genotoxicity-enhancing process of inorganic arsenics. Mutat. Res., 394: 95-101.

Zhao C.Q., Young M.R., Diwan B.A., Coogan T.P. and Waalkes M.P., 1997. Association of arsenic induced malignant transformation with DNA hypomethylation and aberrant gene expression. Proc. Natl. Acad. Sci. USA, 94(20): 10907-10912.

Zikiæ R.V., Stajn A.S., Pavloviæ S.Z., Ognjanoviæ B.I., Saièiæ Zorica S., 2001. Activities of superoxide dismutase and catalase in erythrocytes and plasma transaminases of goldfish (*Carassius auratus* gibelio Bloch.) exposed to cadmium. Physiol. Res., 50(1):105-11.

Zwolak I. and Zaporowska H., 2012. Selenium interactions and toxicity: a review. Cell biology and toxicology, 32:185–92.

Index

D

E

F

N

O

P

R

S

T

U

V

W

X

Z

www.ingramcontent.com/pod-product-compliance
Ingram Content Group UK Ltd.
Pitfield, Milton Keynes, MK11 3LW, UK
UKHW021442280726
14060UKWH00001BA/205